Coastal and Estuarine Studies

Managing Editors:
Malcolm J. Bowman   Richard T. Barber
Christopher N.K. Mooers   John A. Raven

Coastal and Estuarine Studies

39

Mohammed I. El-Sabh
Norman Silverberg (Eds.)

Oceanography of a Large-Scale Estuarine System

The St. Lawrence

Springer-Verlag

Berlin Heidelberg New York London
Paris Tokyo Hong Kong Barcelona

Managing Editors

Malcolm J. Bowman
Marine Sciences Research Center, State University of New York
Stony Brook, N.Y. 11794, USA

Richard T. Barber
Duke Marine Laboratory
Beaufort, N.C. 28516, USA

Christopher N. K. Mooers
Ocean Process Analysis Laboratory
Institute for the Study of the Earth, Oceans and Space
University of New Hampshire
Durham, NH 03824-3525, USA

John A. Raven
Dept. of Biological Sciences, Dundee University
Dundee, DD1 4HN, Scotland

Contributing Editors

Ain Aitsam (Tallinn, USSR) · Larry Atkinson (Savannah, USA)
Robert C. Beardsley (Woods Hole, USA) · Tseng Cheng-Ken (Qingdao, PRC)
Keith R. Dyer (Merseyside, UK) · Jon B. Hinwood (Melbourne, AUS)
Jorg Imberger (Western Australia, AUS) · Hideo Kawai (Kyoto, Japan)
Paul H. Le Blond (Vancouver, Canada) · L. Mysak (Montreal, Canada)
Akira Okubo (Stony Brook, USA) · William S. Reebourgh (Fairbanks, USA)
David A. Ross (Woods Hole, USA) · John H. Simpson (Gwynedd, UK)
Absornsuda Siripong (Bangkok, Thailand) · Robert L. Smith (Covallis, USA)
Mathias Tomczak (Sydney, AUS) · Paul Tyler (Swansea, UK)

Editors

Mohammed I. El-Sabh
Université du Québec à Rimouski
Rimouski, Québec, G5L 3A1, Canada

Norman Silverberg
Institut Maurice-Lamontagne Ste-Flavre
Québec, G5H 3Z4, Canada

ISBN 0-387-97406-7 Springer-Verlag New York Berlin Heidelberg
ISBN 3-540-97406-7 Springer-Verlag Berlin Heidelberg New York

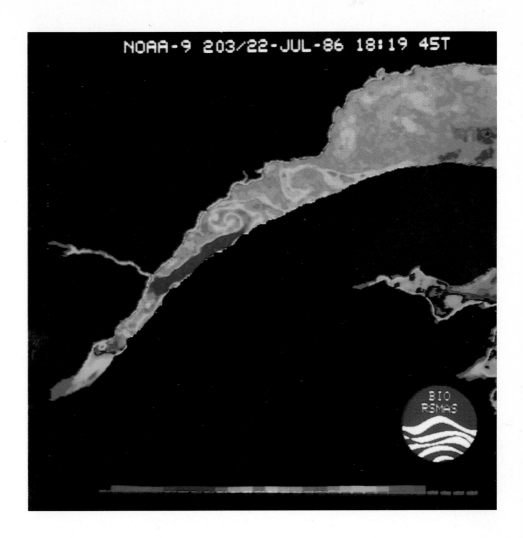

NOAA-9 Satellite infrared thermal image of the St.Lawrence Estuary obtained on July 22, 1986, showing frontal zones and mesoscale features discussed in this volume. The temperature ranges from 5°C (dark blue) to 17°C (dark green) to 25°C (dark red).

TABLE OF CONTENTS

Chapter 1

The St. Lawrence Estuary: Introduction

Mohammed I. El-Sabh[1] and Norman Silverberg[2]

[1]Département d'Océanographie, Université du Québec à Rimouski, Rimouski (Québec) Canada G5L 3A1

[2]Institut Maurice-Lamontagne, Department Fisheries and Oceans, Box 1000, Mont-Joli (Quebec) G5H 3Z4

1. Why this book ?

Coastal regions, and particularly the transition zone between fresh and salt water called estuaries, continue to attract the interest of scientists and governments. In an age of growing awareness of interactive processes affecting the entire planet, those occuring at the frontier between the extensively manipulated continental home of the human race and the once-pristine world ocean merit such continued attention. Estuaries, however, are particularly complicated environments, with cycles of motion that may actually never be repetitive, and each estuary is somewhat different from its neighbours.

Since the appearance of "Estuaries" (Lauff, 1967), other, mainly symposium- or workshop-inspired, volumes have been published (e.g. Cronin, 1975; Kennedy, 1980, 1982, 1984; Kjerfve, 1978, 1988; Ketchum, 1983; van de Kreeke, 1986; Wiley, 1976, 1978; Wolfe, 1986). It is difficult, however, to find accounts of the physical, biological, chemical and geological characteristics of individual estuaries all in one place, and large estuaries, such as Chesapeake Bay, Long Island Sound, the Strait of Georgia and the Skagerrak, seem not to have been treated as coherent entities. Nor has the St. Lawrence.

The many shallow (<100 m), small- to moderate-sized (km to 10's of kms) estuaries so common in the world typically exhibit a residual two-layered circulation pattern, with landward flow in the lower layer. Where estuarine channels are large enough to be influenced by the earth's rotation (Coriolis effect), transverse flow may be superimposed on the longitudinal. In such instances, a two-dimensional description may be inadequate to describe the circulation and mixing patterns and their influence on sedimentation, contaminant dispersal and the distributions of phyto-, zoo- and ichthyoplankton communities.

The St. Lawrence is among the world's most intensively studied estuaries but it is rarely referred to in the estuarine literature, and was left unmentioned in Officer's (1976) survey

Coastal and Estuarine Studies, Vol. 39
M. I. El-Sabh, N. Silverberg (Eds.)
Oceanography of a Large-Scale Estuarine System
The St. Lawrence
© Springer-Verlag New York, Inc., 1990

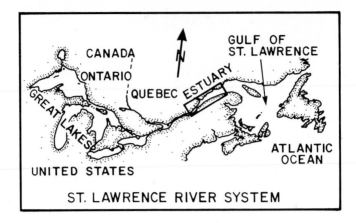

Figure 1. General map showing the extent of the St.Lawrence waterway.

of more than 60 estuaries around the world. The St. Lawrence has been investigated with steadily increasing effort in all disciplines for over half a century. This has resulted in several hundred primary journal publications, and the proceedings of two symposia (El-Sabh et al., 1979; Lacroix et al., 1985). We felt that the accumulated body of knowledge was important enough to merit greater accessibility in the form of a single volume of summary and synthesis.

The book attempts to discuss all of the aspects of the St. Lawrence and to serve as an example of a large-scale estuary. Our purpose is to relate what we know and do not know about the different processes in this system. It is our hope that this volume will represent a timely and useful reference for those who teach and conduct research on estuaries, and serve as a useful document for legislators, planners and other persons charged with the management of our fisheries and other marine resources.

2. Historical notes

The Gulf and Estuary of the St.Lawrence, the St. Lawrence River and the Great Lakes (Figure 1) provide a navigable route to the heart of the North American continent which encouraged the settlement and development of much of Canada and Central U.S.A. The estuary was navigated in 1535 by the French explorer Jacques Cartier, and the first European settlements were established by Samuel de Champlain in 1608. The first hydrographic chart was produced by the young James Cook in 1759, and a number of scientific publications about the St. Lawrence appeared in the 19th century (e.g. Kelly, 1832; Lyell, 1846; Laflamme, 1885). Serious oceanographic research really started with the work done between 1931 and 1938 at Université Laval's Station Biologique du Saint-Laurent at Trois-Pistoles (Figure 2a). Armed with a 15.6 m research vessel, the

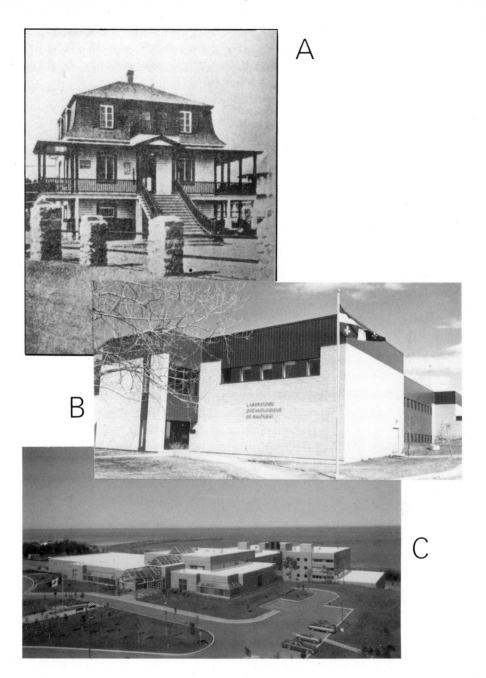

Figure 2. Research laboratories which have contributed towards the advancement of our knowledge of the St.Lawrence Estuary: (A) Station biologique du Saint-Laurent at Trois-Pistoles (1931-1938); (B) Laboratoire Océanologique de Rimouski (1977-) and (C) Institut Maurice-Lamontagne at Pointe-aux-Cenelles (1987-).

"Laval ", the Station's pioneering studies produced the first description of water masses, circulation, temperatures and salinities, and the distribution of chemical parameters, plankton and benthos. It was a long time before anything approaching the same scale of work was again undertaken. There were a few local investigations in various parts of the Estuary, but research did not begin to regain momentum until the 1960's. The Marine Sciences Centre of McGill University and the Bedford Institute of Oceanography were largely responsable for this renewal. Intensive multidsciplinary research programs in the SLE have only been carried out during the last two decades, with universities playing an increasing role. An important event was the establishment in 1970 of GIROQ (Groupe Interuniversitaire de Recherches Océanographiques du Québec) which included researchers from Laval, McGill and Université de Montréal. Multidisciplinary teaching and research programs at the Université du Québec à Rimouski and INRS-Océanologie soon followed, beginning in 1972. Both organizations are housed at the Laboratoire Océanologique de Rimouski (Figure 2b). Most recently, in 1987, the opening of the Institut Maurice-Lamontagne by Fisheries and Oceans Canada at Pointe-aux-Cenelles (Figure 2c), 40 km east of Rimouski, has provided a major new facility for coastal and estuarine research in eastern Canada.

3. A large estuary

The St.Lawrence River (Figure 1) is among the largest on the North American continent, with a drainage basin of approximately 1 320 000 km^2, and an average fresh water discharge of 11 900 m^3 s^{-1} at Québec City, second only to that of the Mississippi and represents more than 1% of all the fresh water supply of the world. The St. Lawrence Estuary (SLE), because of a combination of long-term erosion along the frontier between the Appalachian and Canadian Shield geological provinces and more recent Quaternary glacial and post-glacial events, occupies part of a depression which extends almost 1500 km inland from the Atlantic Ocean. The SLE's great length is matched by widths as large as 60 km and depths of over 350 m (Figure 3).

Although tidal influences are felt as far upstream as Montreal, the SLE, with an area of 10 800 km^2, is defined here as the area stretching from the landward limit of salt intrusion near Ile d'Orléans, to Pointe-des-Monts, where its channel opens into the Gulf of St.Lawrence (Figure 3). Because of its grand size in all three dimensions, the St.Lawrence displays very important variations in both space and time, some of which are more commonly found in continental shelf and oceanic systems. These include upwelling zones, major tidal and density fronts, and 10-50 km diameter eddies, . The system, in fact, includes several types of estuary. From Ile d'Orléans to the head of the Laurentian Trough and the confluence of the Saguenay, the Upper Estuary, with its abundant shoals, turbidity maximum, and strong salinity gradients, resembles many well-mixed and partially stratified estuaries. The Lower Estuary is fairly unique; because of its great dimensions and unimpeded connection with

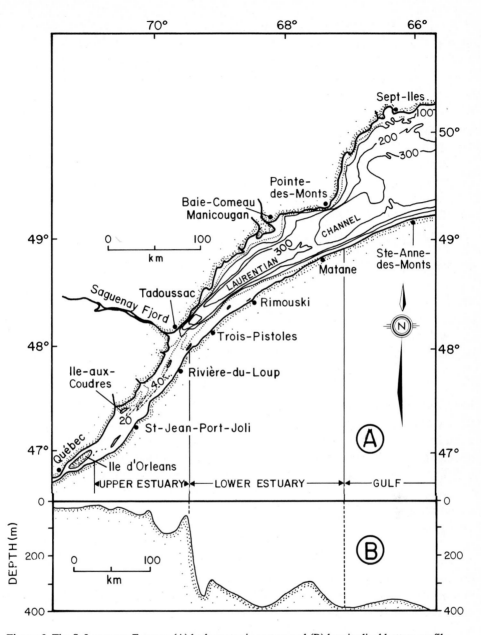

Figure 3. The St.Lawrence Estuary: (A) bathymetry in meters and (B) longitudinal bottom profile.

the mixed Labrador and Slope waters from the Atlantic Ocean, its character is more oceani
than most estuaries. The Saguenay, with its steep walls, deep basins and very marke
pycnocline is similar to many other fjords, save that despite the 20 m sill at the entrance, th
deep waters remain very well oxygenated.

The major forcing functions of winds, tides and fresh water run-off contribute about equally to the physical regime of the SLE and allow the low frequency current field to manifest extremely complex behavior. The renewal time for the surface waters is several months, while that for the bottom waters is several years. This has a strong influence on the ecology of the estuary and is particularly useful in geochemical investigations.

Other unique aspects of the sub-arctic St.Lawrence are the long winters with ice cover and the extensive modification of the fresh water run-off regime. The province of Quebec is among the largest producers of hydroelectric power in the world, and the influence of the unseasonal releases of fresh water from behind the power dams due to high winter demands for electricity may be an important aspect of the overall oceanography of the system.

4. Contents of the volume

We attempted to be as comprehensive as possible in this volume, and bring together reports dealing with the many interrelated aspects of estuarine research in the St.Lawrence. The chapters address a very wide range of topics. Some are summaries of published and unpublished work, others synthesise the knowledge of a given topic or report on the results of promising ongoing research.

Hydrodynamic phenomena, some of which may be unique to large-scale estuaries, set the framework for the transport, distribution and reaction processes that biological, chemical, geological, fisheries and engineering scientists are interested in. The first four chapters are directed toward the physics of estuarine circulation, mixing and their variability in space and time. In going from the Upper to the Lower Estuary, the emphasis shifts from high frequency to low frequency motions, from estuarine to oceanic characteristics. A series of numerical models, increasing in complexity from one-dimensional to three-dimensional baroclinic, are described in Chapter 2 and are used to synthesize the available knowledge of tides, tidal currents and associated phenomena in the SLE. Meteorologically and buoyancy induced subtidal salinity and velocity variations are discussed in Chapter 3. Phenomena such as coastal jets, baroclinic shelf waves, internal Kelvin waves, baroclinic eddies created by adjustment to transient forcing, fronts, wind-induced upwelling and unstable shear waves become possible in a large-scale estuaries (Frontispiece). These topographic waves and mesoscale features are dealt with in Chapters 4 and 5.

Chapters 6 - 11 deal with the geological and chemical aspects of the SLE. Chapter 6 describes the sediment and suspended particulate matter distributions and illustrates how peculiarities in drainage basin configuration, climatic and hydrographic factors, and recent geological history may explain the remarkably low sediment contribution to the Gulf of St.Lawrence and Atlantic Ocean. A discussion on the principal agents of nearshore sediment dynamics for the SLE, such as tides, waves and the formation of ice in winter, follows in Chapter 7.

In many of the world's estuaries, shallow depths, short distances, rapid mixing, short residence times, and, in some cases, repeated resuspension of bottom sediments, make it difficult to isolate and observe chemical phenomena. The spatial and temporal separations inherent in the size of the SLE provide a valuable natural laboratory for studying processes influencing the reactivity and transport of trace chemicals, nutrients, suspended particulate matter (Chapter 8) and organic matter (Chapter 9). Chapter 10 demonstrates how progress in our understanding of early diagenesis and sediment-water interactions in the marine environment has been aided by studies in the Laurentian Channel, where many phenomena are as readily discernible as in the deep-sea, and where logistics are simple in comparison.

Ecological stress and pollution of the coastal environment are often the consequence of increased concentration of human population and industries along our waterways. The volume of the SLE compartment and of the magnitude of the tidal exchange have protected this estuary to some extent, but it has not been immune, as attested by the number of beaches closed to clam gathering and the sad condition of the beluga whale population. Chapter 11 summarizes what is known about the level of chemical contamination in the SLE system.

Chapters 12 - 16 deal with the SLE from a biological view point. The imprint and dominant role of the physical oceanographic regime on the ecology of phytoplankton (Chapter 12), the composition and distribution of zooplankton (Chapter 13) and the species composition and dynamics of ichthyoplankton in the estuary (Chapter 14) are reviewed and discussed. Considerable work has also been done on both the littoral and subtidal bottom fauna in the SLE. Chapter 15 provides a short overview of these benthic studies, with enough references to permit further pursuit of this essential aspect of the St. Lawrence. As illustrated in Chapter 16, the effect of hydrodynamic processes operating within the estuary appear to exert strong influences on the abundance of commercial and other fish species beyond its limits downstream from the Gulf of St.Lawrence to possibly the Gulf of Maine.

The 170 km long, 280 m deep Saguenay Fjord, a fair-sized estuary in its own right, is a major tributary to the St.Lawrence. Chapter 17 reviews the oceanographic and geological characteristics gleaned by researchers over the past several decades and focuses on some of the unique sedimentological, biological and geochemical features of this fjord.

The final chapter takes stab at highlighting the unique or particularly revealing features of this large estuary, pontificates on the state of our knowledge, and otherwise exploits the advantages of the editors who have been the first to profit from reading this wonderful collections of review papers. As a complement to this book, reference should be made to the recent volumes by Strain (1988) and Therriault (1990) devoted to the Gulf of St.Lawrence.

Acknowledgments

The publication of this volume was achieved through the cooperative efforts of many individuals. We sincerely thank each author for his contribution and are especially grateful for the congenial way in which they responded to suggestions and editorial comments. We also wish to thank the reviewers who took time from their already busy schedules to contribute to the scientific quality of this volume. The motivation for producing this book developed over 17 years of contact with colleagues and students at the Université du Québec at Rimouski and with other scientists enamoured with the St.Lawrence, who have watched with us as new scientific information overwhelmed our filing systems and our individual perspectives. We have received important technical support and encouragement from our institutions, the Université du Québec at Rimouski and the Maurice-Lamontagne Institute. In particular, our heartful appreciation go to the following people: Bjorn Sundby, Lionel Corriveau, Richard Fournier, Jocelyne Gagnon and Bruno Langlois. Yves Gratton kindly provided the satellite image used as the Frontispiece. We dedicate this book to all the individuals, scientists, students, technicians, and ships personnel, and to the various funding organizations who have contributed towards the advancement of our knowledge of the St.Lawrence.

REFERENCES

Cronin, L.E. (Editor). 1975. Estuarine Research, Vols. 1-2, Academic Press, New York, 738 & 587 pp.

El-Sabh, M.I., E. Bourget, M.J. Bewers and J.C. Dionne. (Editors). 1979. Oceanography of the St.Lawrence Estuary. Naturaliste canadien, vol. 106(1), 276 pp.

Kelly, W. 1832. On the temperature, fogs and mirages of the River St.Lawrence. Trans. Lit. Hist. Soc. Québec, Vol. 3(1): 1-45.

Kennedy, V.S. (Editor). 1980. Estuarine Perspectives. Academic Press, New York, 533 pp.

Kennedy, V.S. (Editor). 1982. Estuarine Comparisons. Academic Press, New York, 709 pp.

Kennedy, V.S. (Editor). 1984. The Estuary as a filter. Academic Press, New York, 511 pp.

Ketchum, B.H. (Editor). 1983. Estuaries and Enclosed Seas. Elsevier Scientific Publ. Co., Amsterdam, 500 pp.

Kjerfve, B. (Editor). 1978. Estuarine Transport Processes. University of South Carolina Press, Columbia, 331 pp.

Kjerfve, B. (Editor). 1988. Hydrodynamics of Estuaries, Vols. 1 and 2. CRC Press Inc., Boca Raton, Florida, 163 and 125 pp.

Lacroix, G., E, Bourget, J.-C. Therriault, J. Lebel, P.-H. LeBlond and W.C. Leggett. 1985. St.Lawrence Estuary: Oceanographic and Ecological Processes. Naturaliste canadien, Vol. 112(1), 161 pp.

Laflamme, J.C.K. 1885. Le Saguenay: essai de géographie physique. Soc, Géogr. Québec, Bull. 1(4): 47-65 p.

Lauff, G.H. (Editor). 1967. Estuaries. Am. Assoc. Adv. Sci., Publ., 83, 723 pp.

Lyell, C. 1846. On the packing of the ice in the River St.Lawrence. Quater. J. Geol. Soc. London. Vol. 2: 422-427.

Officer, C.B. 1976. Physical Oceanography of Estuaries and Associated Coastal Waters. John Wiley & Sons, New York, 465 pp.

Strain, P.M. (Editor). 1988. Chemical Oceanography in the Gulf of St.Lawrence. Canadian Bull. Fish. Aquatic Sci., Vol. 220, 190 pp.

Therriault, J.C. (Editor). 1990. Workshop/Symposium on the Gulf of St.Lawrence: small ocean or big estuary? Canadian Bull. Fish. Aquatic Sci. (in press).

Van de Kreeke, J. (Editor). 1980. Physics of Shallow Estuaries and Bays. Springer-Verlag, Berlin, 280 pp.

Wiley, M. (Editor). 1976. Estuarine Processes, Vol. 1 and 2. Academic Press, New York, 541 and 428 pp.

Wiley, M. (Editor). 1978. Estuarine Interactions. Academic Press, New York, 603 pp.

Chapter 2

Mathematical Modelling of Tides in the St. Lawrence Estuary

Mohammed I. El-Sabh[1] and Tad. S. Murty[2]

[1]Département d'Océanographie, Université du Québec à Rimouski, 300 Allée des Ursulines, Rimouski, Québec, G5L 3A1, Canada

[2]Institute of Ocean Sciences, P.O. Box 6000, Sidney, British Columbia V8L 4B2

ABSTRACT

This paper is an attempt to synthesize current knowledge about tides and tidal modelling efforts in the St. Lawrence Estuary. A sequence of numerical models, increasing in complexity from simple one-dimensional to baroclinic, is discussed. The performance of these models and their limitations are pointed out. In the last section, some recommendations are given for future work to improve our present knowledge of the tidal properties over the estuary. These include complementary *in situ* observations of offshore tidal elevations and principal tidal currents, in addition to new numerical modelling experiments combining all the positive attributes of the previous models.

Key Words: barotropic, baroclinic, tides, currents, model, fronts, mixing, eddies, residual.

1. Introduction

The drainage basin of the St. Lawrence system encompasses an area of 1.5×10^6 km^2 and a population of about 30 million people. It contains the Great Lakes, the St. Lawrence River which extends from the outflow of Lake Ontario to Quebec City, the St. Lawrence Estuary which begins at the landward limit of the salt water intrusion near Ile d'Orléans and extends 400 km seaward to Pointe-des-Monts, and the Gulf of St. Lawrence (Figure 1). These waters are prolific in marine life and supporting more than 40% of all Canadian sea-fish landings, the importance of which is presently being enhanced by extensive developments in aquaculture. Increases in the discharge of domestic and industrial effluents and in the density of marine traffic, both commercial and recreational, emphasize the need for a better understanding of the basic phenomena that occur within the system and for predictive capability with respect to major contingencies likely to occur in the future. Strongly tidal, variously stratified, and topographically complex, the system presents an immense variety of fluid dynamical problems varying

Coastal and Estuarine Studies, Vol. 39
M. I. El-Sabh, N. Silverberg (Eds.)
Oceanography of a Large-Scale Estuarine System
The St. Lawrence
© Springer-Verlag New York, Inc., 1990

widely with respect to spatial and temporal scales. This paper is primarily concerned with the numerical modelling studies undertaken to achieve an understanding of tidal propagation in the St. Lawrence Estuary (SLE). A general account of the physical oceanography of the SLE may be found in El-Sabh (1988). Other aspects of the oceanography of the region are dealt with in other parts of this volume.

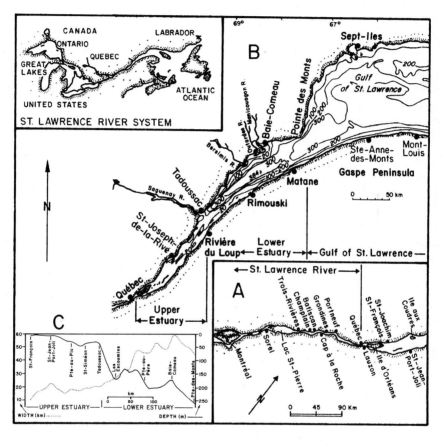

Figure 1. General map of the St. Lawrence system, showing sea level stations.

The area between Trois-Rivières and Quebec City (Figure 1) represents the main tidal portion of the St. Lawrence River. The Upper SLE landward from the mouth of the Saguenay River, varies in width from 2 km near Ile d'Orléans to 24 km near Tadoussac (Figure 1). It has an uneven topography with several disconnected channels and troughs separated by ridges and islands. The southern side is shallower and characterized by depths of less than 10m. The north side has a nearly continuous channel with depths increasing from about 10m in the shoals near Ile d'Orléans to 120m in the deep basin at the north-eastern extremity of the Upper Estuary. In the Lower St. Lawrence Estuary (LSLE) between Tadoussac and Pointe-des-Monts the beginning of the Laurentian Channel forms a deep central trough with depths more than 300m. It will be shown that the complex topography of the SLE presents a major difficulty in tidal modelling.

The barotropic, or surface, tides of the St. Lawrence system have been the subject of an intensive amount of modelling effort over the last 20 years. The work has seldom been restricted to the SLE, but has usually been concerned with the whole tidal portion from Cabot Strait at the entrance to the Gulf of St. Lawrence to Montreal, which is the maximum upstream penetration of a measurable tide, using either physical, analytical or numerical models. Funke and Crookshank (1979) developed a hybrid model for the St. Lawrence, in which a hydraulic (physical) model and a numerical model run simultaneously and interactively by exchanging information at their mutual interface. The original St. Lawrence hydraulic model, which was partially dismantled (due to space requirements for other projects), represents the reach from Pointe-au-Père to Montreal. The dismantled portion is from Neuville to Montreal. A one-dimensional numerical model was then developed for this dismantled portion and was dynamically coupled to the remaining part of the physical model. Although the simulated results from this hybrid model were quite satisfactory, physical models applied to large coastal systems suffer, in general, from two major disadvantages. One of these concerns the effects of the earth's rotation and consequent requirement that the model be constructed on a rotating and virtually vibration-free, rigid platform. The other concerns the requirement for a practicable depth of working fluid and the need to greatly exaggerate the vertical scale with respect to the horizontal scale of the model.

Previous analytical tidal models include those of Vincent (1965), Partenscky and Louchard (1967), Partenscky and Warmoes (1970) and Partenscky and Marche (1974). Although such models adequately predict the observed distortion of the tidal wave as it progresses upstream, they could not include the complex topography in a refined manner, as would be possible in a numerical model. In the next section, observed tidal characteristics in the St. Lawrence Estuary are briefly described. In the remainder of the paper, a sequence of numerical models increasing in complexity, is discussed.

2. Tidal characteristics of the St. Lawrence Estuary

The St. Lawrence Gulf and estuary are not large enough to generate independent tides (i.e. due to tidal potential). Hence the tides in this system are co-oscillating tides with the Atlantic Ocean. Several excellent general descriptions of the propagation of the barotropic (surface) tides in the whole St. Lawrence system can be found in Farquharson (1970) and Godin (1979, 1980). In this section we shall summarize the major tidal characteristics in the SLE.

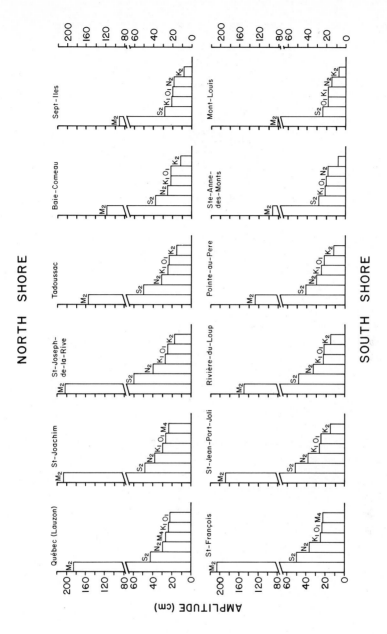

Figure 2. Observation points of maximum amplitude for the main tidal constituents in the St. Lawrence estuary.

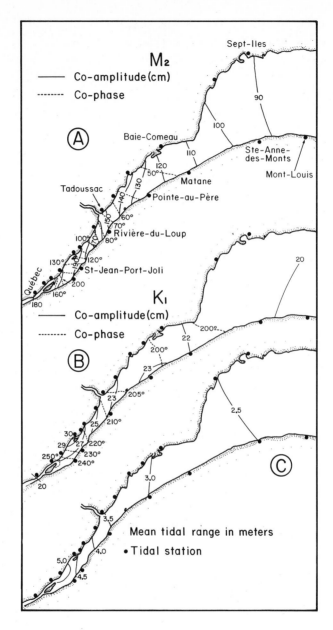

Figure 3. Observed co-phase and co-amplitude (cm) lines for (A) M_2 and (B) K_1 tidal constituents in the St. Lawrence Estuary. Phase values are relative to time zone (GMT + 5hrs). (C) Mean tidal range in meters.

Examination of the maximum amplitude for the six important tidal constituents observed at all stations along both shores of the estuary (Figure 2) reveals the strong semi-diurnal pattern of tides in this region. The main tidal constituent is M_2, which has a period of half a lunar day. The next constituent in importance is S_2, which has a period of half a solar day.

The ratio of the amplitudes between these two constituents (M_2/S_2) has a constant value of 3.3 throughout the estuary. The third constituent is N_2, which is linked to the variation of the moon's distance from the earth, and which goes through a complete cycle each month. The maximum amplitude of the M_2 constituent is found between Isle-aux-Coudres and the eastern tip of Ile-d'Orléans (Figure 3A). By the time the wave has reached Ile d'Orléans, the rotational effects have disappeared and the wave crest is virtually horizontal. There the channel splits and the depth decreases. Friction takes over and the tide acquires an increasingly shallow-water character. It takes eleven hours to progress from Quebec to the western extremity of Lac St. Pierre, beyond which the semi-diurnal tide is no longer detectable.

The co-phase lines indicate that the M_2 tide enters the estuary first along the north shore. At the estuary mouth, near Les Méchins, the M_2 tide has a phase lag of $48°$, which increases to $54°$ at Pointe-au-Père, to $80°$ at Rivière-du-Loup and to $134°$ at St. Jean-Port-Joli. For a discussion of time lags associated with the occurrence of low water in the St. Lawrence system see LeBlond (1970).

Because the diurnal tide has a small amplitude in the St. Lawrence Gulf and Estuary, it is of secondary importance (Figure 2). It takes a progressive character in the SLE (Figure 3B), increasing in amplitude as far as the Ile d'Orléans. Thereafter it is increasingly delayed and distorted by friction and it is no longer detectable in the middle of Lac St. Pierre. Furthermore, progress of tides in the area between Ile d'Orléans and Lac St. Pierre is slowed down by the freshwater discharge. High and low water move more and more slowly upstream and then rapidly decrease in amplitude past Quebec City where the tide rises fast, but falls slowly, giving the impression of a saw-tooth signal.

Semi-monthly tides are created mainly by the nonlinear interaction of M_2 and S_2 and reflect the downstream occurrence of spring and neap tides by rises and falls in the local mean level. The effect becomes apparent at Cap-à-la-Roche (Figure 4) and reaches a maximum at Trois-Rivières where the amplitude exceeds that of the regular daily tide. It is faint but still perceptible in Montreal where effects of the river discharge are also apparent.

The interval between the time of the new moon or full moon (when the equilibrium semi-diurnal tide is a maximum) and the time of the local spring tide is referred to as the age of the tide. A localized increase in the age of the tide is a good indication of resonances at that location. It is with this view that Murty and El-Sabh (1985) studied the geographical distribution of the age of the semi-diurnal and diurnal tides in the St. Lawrence system and other estuarine waterbodies using the amplitudes and phases of tidal constituents M_2, S_2, N_2, K_1 and 0_1. The age of the semi-diurnal tide in the St. Lawrence Estuary was found to increase gradually from 40 hours at the mouth to 49 hours near Quebec City. Interestingly, the age of the diurnal tide is negative everywhere, with values between minus twenty (-20) hours at the mouth and minus thirty (-30) hours at St. Jean-Port-Joli. The negative values for the age may arise from an unusual combination of the normal modes.

For more discussion on the existence of the negative ages and its causes, the reader is referred to the review by Murty and El-Sabh (1985). The tendency for the age of the tide to increase towards the head of the SLE, and of other estuaries within the St. Lawrence system, supports the conjecture of Webb (1973) that shallow-water effects may be causing this increas in age near estuary heads.

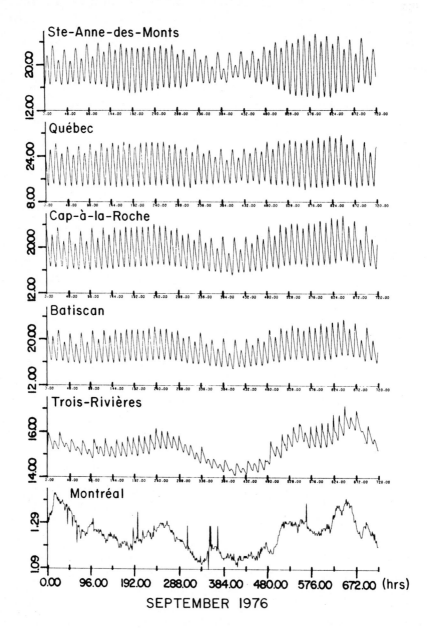

Figure 4. Creation of a semimonthly tide in the upper reaches of the St. Lawrence River caused by the succession of spring and neap tides downstream (From Godin, 1979).

3. Barotropic models

In recent years considerable success has been achieved in the use of hydrodynamical numerical methods to simulate barotropic tides in coastal seas and estuaries. Although the basic procedures are relatively simple, difficulties are encountered in the application of such techniques to problems concerning strong tidal flows in topographically complex systems such as the St. Lawrence. It is usually presumed that variations in phase speed and partial reflections of a long wave propagating through the system can be adequately represented by a sufficiently detailed finite-difference approximation to the actual geometry. Energy losses from the wave are then empirically accounted for by adjusting the frictional coefficients in the momentum equations to optimize the agreement between the observed and computed tidal elevations around the interior of the waterbody. It is important to ensure, as far as possible, that such adjustments do not compensate for inadequacies in the theoretical assumptions, schematic representation of topography, or in prescribing boundary conditions. A prerequisite is thus a sufficiency of observations, of both elevations and velocities, to establish the standards of performance which could in principle be achieved by the model. If necessary, the sensitivity of the computed fields to the various approximations is then systematically investigated and improvements made until these standards are attained or the need for a radically different type of model has been demonstrated. In this section, we will present some examples of the numerical studies that employed a one- and two-dimensional model to simulate barotropic tidal motions in the SLE. The various limitations encountered and subsequent modifications are also discussed.

3.1. The one-dimensional tidal models:

The shallow-water equations of continuity and momentum (Dronkers, 1964) applied at a cross-section of a one-dimensional channel are, respectively,

$$\frac{\partial Q}{\partial x} + B\frac{\partial h}{\partial t} - Q_T = 0 \tag{1}$$

$$\frac{\partial U}{\partial t} + U\frac{\partial U}{\partial x} + g\frac{\partial h}{\partial x} + F + Q_T \cdot \frac{U}{A} = 0 \tag{2}$$

where

    x = distance along the median line of the channel,
    U = velocity,
    h = water elevation above I.G.L.D. (International Great Lakes Datum),
    D = average depth of a cross-section below the I.G.L.D.
    B = average storage width of the cross-section,

Q = horizontal transport = U.A,

A = area of cross-section (= B.D),

Q_T = tributary freshwater discharge per unit length,

F = bottom friction term = $(gU|U|)/C^2D$, C is the Chezy friction factor,

g = acceleration due to gravity.

Both Godin (1971) and Prandle and Crookshank (1972, 1974) independently developed a one-dimensional model for tidal propagation in the St. Lawrence River and Estuary. Neglecting the variation in freshwater discharge, and the bifurcation of the channel at Ile d'Orléans, the contribution of the tributaries downstream of Montreal such as the Richelieu, the St. Maurice, the Saguenay and the Manicuoagan Rivers, a modified version of Equations 1 and 2 was written by Godin (1971) in terms of a finite-difference scheme. The area between Montreal and Pointe-des-Monts was divided into 53 subchannels each of which was schematized as a channel of rectangular section of width B and depth D. At Pointe-des-Monts an elevation boundary condition was specified at all times, which embodies approximations for observed values of M_2, S_2, N_2, K_1 and O_1 at section 1. At Montreal the discharge was taken to be constant at all times, with a value of 8.5×10^3 m^3/s, corresponding to the average discharge of the St. Lawrence River

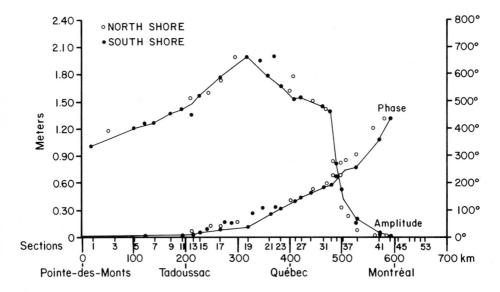

Figure 5. Observed and computed M_2 amplitude and phase changes along the St.Lawrence Estuary (modified from Godin, 1971).

at Montreal. Figure 5 shows as a continuous curve the calculated amplitude and phase change of M_2 using a Chezy coefficient of C = 58 m$^{1/2}$ /s. The analyzed values based on observations are shown by dots; those which are squared were derived from stations located on the north shore. Although the calculated amplitudes and phases of M_2 fall consistently below those observed upstream of Tadoussac, Godin (1971) considered the overall agreement to be adequate and sufficient for a study of the fundamental parameters influencing the propagation of the tide in the St. Lawrence. His results, obtained from this one-dimensional model, indicate that increases in the freshwater discharge do not significantly alter the tide between Pointe-des-Monts and Ile d'Orléans; further upstream, the tidal amplitude is slightly reduced and the progress of the tide is retarded by up to 3 hours.

The one-dimensional model developed by Prandle and Crookshank (1972, 1974) for the area between Pointe-au-Père and Montreal paid special attention to both the discharge of freshwater from the different tributaries and the bifurcation of the channel at Ile d'Orléans. First, the study area was divided lengthwise into 121 sections; the length of each section lay in the range of 2-12 miles (3.2-19.2 km) and was determined by using the stability criterion:

$$\frac{\Delta x}{\Delta t} \geq U + c \tag{3}$$

where c, the wave celerity, is equal to $(gD)^{1/2}$. A staggered grid with central finite-differences in the explicit step is used in this numerical model. Since the contribution from the non-linear term $U\frac{\partial U}{\partial x}$ in Equation 2 may not be very significant, Prandle and Crookshank (1972) simply ignored this term immediately adjacent to junctions and boundaries, rather than complicating the equations. Initially, the water was assumed to be at rest. The seaward boundary condition was the specification of tidal elevation at Pointe-au-Père. The landward boundary condition at Montreal involved the specification of the river discharge as a function of time. The model was then calibrated by tuning the friction coefficients for each section in order to reproduce, as closely as possible, a set of reliable hourly water-levels recorded at 14 stations along the channel. No attempt was made to force the model to exactly reproduce any particular gauge where it was suspected that the recorded data were in error. Satisfactory agreement can be seen in Figure 6 when comparing the model predictions of sea-level elevations with actual recordings at three stations in the SLE. Furthermore, a most important measure of the accuracy of the simulation is obtained by comparing the model results with velocity measurements available at St. François for an eight-day period.

In assessing the accuracy of the two above-mentioned one-dimensional models, it should be remembered that there is the problem of advection of saltwater into the freshwater regime. This forces a spatial gradient of density and resultant velocities. In this case Equations 1 and 2 are not adequate. Furthermore, a variation in water level across a particular cross-section can also occur due to centrifugal forces. As the river widens, Coriolis force also becomes important and there are definite gradients in the elevations and currents across the width of the channel. Hence, some difference must be expected between the mean cross-sectional value of the level considered in these one-dimensional models and a single-point reading of a gauge which is generally located on the shore. Finally, the model predictions are limited by the accuracy of the boundary conditions. Of particular concern is the accuracy of the freshwater discharges used in these models.

3.2 The two-dimensional barotropic tidal models

For a simulation of the overall tidal propagation in the St. Lawrence Estuary, the one-dimensional models described above are extremely useful. However, when a more detailed study of a localized area is required these models are ineffective. Recourse must be made to a two-dimensional model, for which the Coriolis effect can be taken into account and the flow variations about the transverse axis can be simulated.

During the last 25 years, in connection with the increase of computer facilities, two-dimensional numerical modelling of tides has been intensively developed and successfully applied to investigate tidal motions in many coastal areas. The number of such models presented in the literature is enormous, and even when restricted to a particular area a complete list of such models is difficult to prepare. In this paper devoted to the St. Lawrence Estuary, we shall refer particularly to the following three numerical models: Prandle and Crookshank (1972, 1974); Levesque et al. (1979) and Pingree & Griffiths (1980), respectively referred to hereafter as PC, LME and PG. All these models employ a rectangular grid system. We then present the preliminary results obtained by the authors (El-Sabh and Murty, in preparation) using an irregular triangular grid technique.

The basic equations for all of these approaches are the Navier-Stokes equations and the equation of continuity, written either in Cartesian or polar coordinates. In practice, most of the tidal models used are restricted to a vertically-averaged version of these equations, resulting from the shallow-water approximation,. This reduces the problem to a two-dimensional system involving u and v, the eastward and northward mean-depth velocity components, and the sea surface elevation. The fully non-linear form of the depth-averaged equations of momentum and continuity applicable to barotropic tidal motions in a rotating sea may be written in spherical polar coordinates as follows:

$$\frac{\partial u}{\partial t} + \frac{u}{R \cos\phi} \frac{\partial u}{\partial \lambda} + \frac{v}{r} \frac{\partial u}{\partial \phi} - \frac{uv}{R} \tan\phi - 2\Omega \sin\phi \, v$$

$$+ \frac{\tau_\lambda}{\rho(h+D)} + \frac{g}{R \cos\phi} \frac{\partial \zeta}{\partial \lambda} = \frac{g}{R \cos\phi} \frac{\partial P}{\partial \lambda}$$

(4)

$$\frac{\partial v}{\partial t} + \frac{u}{R\cos\phi}\frac{\partial v}{\partial \lambda} + \frac{v}{R}\frac{\partial v}{\partial \phi} - \frac{u^2}{R}\tan\phi - 2\Omega\sin\phi\, u$$

$$+ \frac{\tau_\phi}{\rho(h+D)} + \frac{g}{R}\frac{\partial \zeta}{\partial \phi} = \frac{g}{R}\frac{\partial P}{\partial \phi} \tag{5}$$

$$\frac{\partial \zeta}{\partial t} + \frac{1}{R\cos\phi}\left[\frac{\partial}{\partial \lambda}(h+\zeta)u + \frac{\partial}{\partial \lambda}(h+\zeta)\,v\cos\phi\right] = 0 \tag{6}$$

where,

λ,ϕ = east longitude and latitude, respectively;

t = time;

ζ = elevation of the sea surface above the undisturbed level;

h = the undisturbed water depth;

R = the radius of the earth;

Ω = the angular speed of the earth's rotation;

g = the acceleration due to gravity;

u,v, = eastward and northward mean depth components of current;

P = the generating tidal potential;

τ_λ and τ_ϕ are the eastward and northward components of the bottom stress τ_B

In the derivation of these equations it is assumed that the velocities are independent of depth.

In order to solve Equations 4-6. appropriate conditions have to be specified at the boundaries. On any land boundary, the normal velocity component must be zero; along the open boundaries, the most appropriate boundary condition is to prescribe the flux of water, but such a condition is difficult to estimate. It is easier to specify the sea level elevation deduced from observations at the coast. Some simplifications are applied to Equations 4-6 for practical applications. El-Sabh et al. (1979) calculated the independent tide in the SLE and found it negligibly small. Thus, with the exception of the PG model, the other models did not include this term in their calculations. Furthermore, the dimensions of the estuary are such that a Cartesian coordinate system would suffice to express Equations 4-6. PC adaopted a Cartesian system,model, and PG retained a spherical polar system. In order to save computational efforts, Equations 4 and 5 are sometimes linearized by ignoring the nonlinear acceleration terms. Such a simplification is largely justified when the modelled

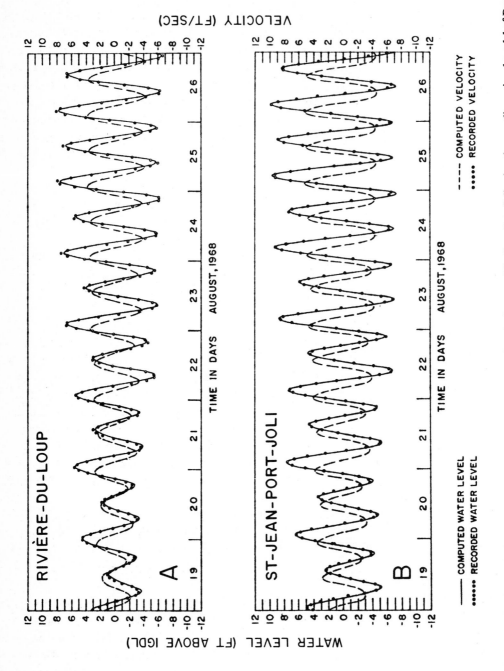

Figure 6. Computed and recorded values of water level and velocity at Rivière-du-Loup and St. Jean-Port-Joli using the one-dimensional model of Prandle and Crookshank (1972).

area is not too shallow and does not present important velocity gradients. Charnock and Crease (1957) suggested through dimensional analysis that the nonlinear advective terms only become important when the free surface height is of the same order of magnitude as the water depth. Both PC and LME used this approximation. However, analysis of the current measurements taken in the SLE (e.g. Godin, 1971; Gagnon and El-Sabh, 1980; Muir, 1982) has pointed out the existence of harmonic and interaction constituents in the spectrum, particularly in the Upper Estuary, which cannot be reproduced in the models if these nonlinear terms are excluded.

The bottom friction parameterisation is a very important factor for the success of these models. Several laws are classically used, but they must all postulate at least a quadratic velocity dependence in the form

$$\tau_B = \rho \, C_D \, u \, |u|, \tag{7}$$

where C_D is the drag coefficient, and u is the tidal current. The friction terms in PC were represented by the Chezy formula, with the water depth replacing the hydraulic radius:

$$\tau_B = \frac{gu \, |u|}{C^2_D \, (h+D)} \tag{8}$$

A constant friction coefficient was used throughout this model. PG, on the other hand, found that bottom friction based on (7) may be seriously underestimated in regions of low M_2 tidal current. As a practical expedient the bottom stress was then linearized for speeds less than $s \sim 20$ cm/s, such that

$$\tau_B = \phi \, C_D \, (\, |u| + s) \, u \tag{9}$$

The system 4-6 of differential equations is transformed into a set of finite difference equations for numerical integration. The precision of the numerical solutions depends on the properties of the finite difference scheme adopted. The numerical scheme in all three models (PC, LME and PG) used the staggered mesh and leap-frog method with linear numerical analogues centred in space and time. For the exact forms of the finite difference forms used in these models see Prandle and Crookshank (1972), Freeman and Murty (1976) and Pingree and Griffiths (1978, 1979).

With the finite difference techniques, rectangular grids must have constant spacing. As the size of the mesh greatly influences the precision of the results, it is often difficult to conciliate the fineness of the mesh size with the computational time requirements. Regular grid-sizes of 3.7 km and 0.9 km respectively were used by PC in their large-scale model and the more detailed small-scale portion near Ile d'Orléans. The grid size used by the LME model was 3.7 km in the longitudinal direction and 2.8 km in the latitudinal direction, while that used by PG has the dimension of 5 minutes in latitude (about 5 nautical miles) and 10 minutes in longitude.

3.2.1 The PC Results

A large-scale two-dimensional model for the SLE extending between Pointe-au-Père and the eastern tip of Ile d'Orleans was developed by PC. At the upstream boundary, this model was combined wth the one-dimensional model described earlier for the St. Lawrence River, thus extending the simulation as far as Montreal. At the junction with Saguenay River, the model was also combined with another one-dimensional representation of the tidal reach of the Saguenay River (Figure 7). Although the tidal discharge of the Saguenay River is small compared to that of the St. Lawrence River at their junction, its influence on the surface layer velocity pattern is quite significant in this area. The one-dimensional portions were represented by section lengths of one to four miles (1.6 to 6.4 km). Boundary conditions were specified as water elevation at Pointe-au-Père and recorded freshwater discharge at the other openings, namely at Montreal and at the head of the Saguenay River. Starting from rest, PC have carried on their simulations until cyclic convergence was achieved after only two semi-diurnal tidal cycles.

The model was run for the spring tide of 13 August, 1968 and the results were presented as velocity fields corresponding to mean-depth currents at hourly stages over one tidal cycle. We present on Figure 8 the computed velocity vector for each grid point, along with the water level contours with reference to high and low water stages at Pointe-au-Père. Superimposed on the mathematical results are solid black arrows which represent the measured surface tidal currents that would be found in the summer season for an average tide, as reported in the Tidal Current Atlas (Anonymous, 1939) . The computed velocity vectors are, in general, smaller in magnitude than the observations. PC attributed this to the fact that density variations in the estuary were not considered and that the velocity vectors obtained from the model represent the vertically averaged velocity components. A further complication arises in the St. Lawrence at the mouth of the Saguenay River. Here the discharge of the Saguenay is largely received by the upper layer. This results in cross currents directed towards the south shore being initiated during the ebb period of the tide. They are most noticeable in field data obtained one hour after low water at Pointe-au-Père. The mathematical model, which does not consider the vertical dimension, shows only a small local effect caused by the tidal discharge of the Saguenay. The discrepancy between the calculated and observed velocities may also be partly attributable to the innacuracies we have since found to exist in the 1939 tidal atlas .

25

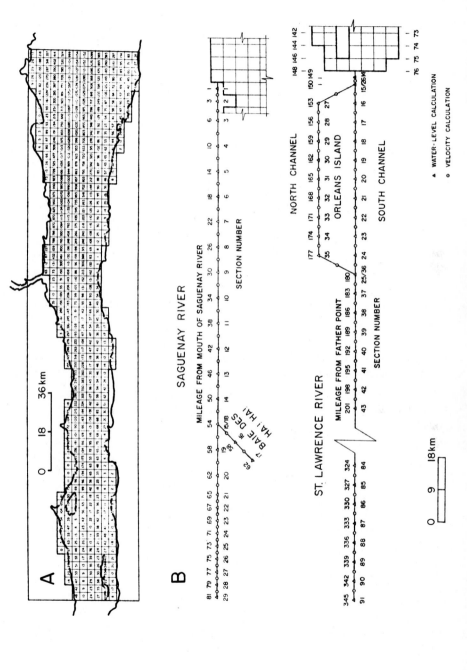

Figure 7. (A) Rectangular grid model employed by Prandle and Crookshank (1972). (B) Scheme for coupling the one- and two-dimensional models used by Prandle and Crookshank (1972).

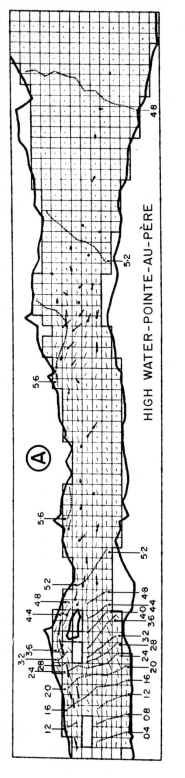

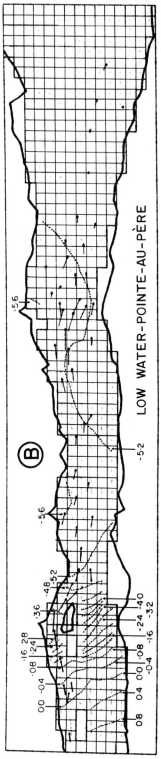

SCALE OF VELOCITY VECTOR IN FT/SEC

5 0 5 10

→ RECORDED SURFACE VELOCITY

······· COMPUTED WATER-LEVEL CONTOUR

COMPUTED VELOCITY VECTOR

Figure 8. Computed water level contour and velocity vector in the St. Lawrence estuary with reference to H.W. (A) and L.W. at Pointe-au-Père (modified from Prandle and Crookshank (1972).

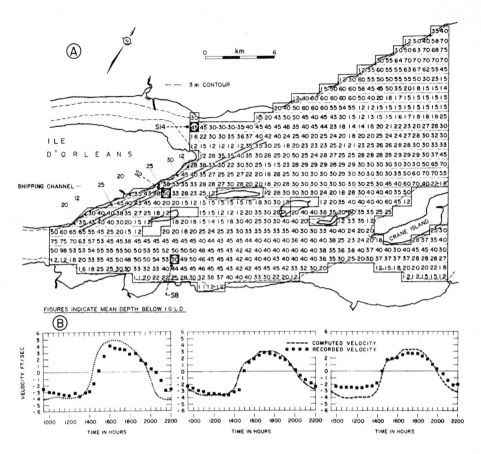

Figure 9. (A) Fine-grid model for the region seaward of Ile d'Orléans, (B) Calculated and recorded current velocity at three stations near the eastern tip of Ile d'Orléans (modified from Prandle and Crookshank (1972).

A useful feature of the numerical models is the capability of determining the sensitivity of the output to a particular parameter. This capability was exploited by PC in comparing the model output with, and without, the Coriolis force included in the equation of motion. The importance of this term in the SLE was clearly demonstrated. Furthermore, a more refined version of the PC two-dimensional model, with a grid size of 1.6 km, has been used by Ouellet and Cerceau (1976) to study the influence of tide on the salinity distribution in the Upper Estuary between the eastern tip of Ile d'Orléans and Rivière-du-Loup. Comparison between the calculated and observed salinity data led the authors to the conclusion that only a model which would take the third (vertical) axis into consideration, would lead to a better representation of the salinity distribution in the SLE.

In order to investigate the flow distribution and the effect of major engineering works in the proximity of the seaward end of Ile d'Orléans, PC also developed a more detailed small-scale two-dimensional model with grid size of 0.8 km for the area of interest. The model was formulated using an explicit finite-difference scheme, assuming vertical homogeneity in terms of density and velocity and did not include the convective terms. The Coriolis term was included in the model and was found to have no discernable effect and, hence, was omitted in subsequent runs. The 2-D model was coupled with the one-dimensional model described earlier for the St. Lawrence River. A time step of 35 seconds was used. The initial conditions for the 1-D section were taken from outputs from previous runs of the model. The water level across the 2-D section was taken as the value given at the downstream boundary of the 1-D program. The velocities in the 2-D section were taken as zero. These approximations enabled cyclic conditions to be obtained in the model after only two semi-diurnal tidal periods. The model was run for the spring tide of 10 August, 1968. A comparison of computed and recorded velocities at three stations (Figure 9) shows reasonable agreement within the accuracy of the recordings and the analogy between a single point recording and a cross-sectional mean value of velocity.

3.2.2. The LME results

This two-dimensional model used the linearized form of the momentum and continuity Equations 4-6 in a spherical polar coordinate system to study the tidal propagation in the SLE. The model extends from Ile-aux-Coudres on the upstream side to Pointe-des-Monts, and at both ends it is open. Although freshwater discharge from rivers is not included, this study is the first tidal model for the SLE that makes use of real data for the five important tidal constituents, namely M_2, S_2, N_2, K_1 and O_1. A simple scale analysis showed that the acceleration, Coriolis and surface gradient terms were of the order of 10^{-4} m s^{-2} whereas the bottom friction term were about a tenth of this value. However, in very shallow water with depths less than 10m, bottom friction becomes more important. With known boundary conditions, and taking into account the topography, depths variations, Coriolis force and linear bottom friction, the original version of the model developed by Levesque et al. (1979) generated the time variations of water level oscillations and horizontal movement for the rest of the estuary. The quadratic form of the bottom friction, atmospheric pressure and wind stress were added in a subsequent version of the model (El-Sabh et al., 1979). Starting from rest, the model was run for 196 hours of real time, considering that the dynamics had reached equilibrium after a spin up phase of 72 hours. LME retained the last three days of simulation to estimate the amplitudes and phases of the simulated constituents all over the domain. Cotidal maps for each of the five main harmonic constituents M_2, S_2, N_2, K_1 and O_1 were produced and compared with shore-based gauge observations.

One of the difficulties encountered by LME in their efforts to model the SLE was the choice of the boundary limits, particularly at the upstream end. Two experiments were carried out; one by Levesque et al. (1979), who selected the limit at a line just upstream of Ile-aux-Coudres, and the other by El-Sabh et al. (1979), who extended their model upstream to the eastern tip of Ile d'Orléans (St. François). Comparison of the results obtained from these two models with observations for the main semi-diurnal and diurnal tidal

TABLE 1

Observed and Calculated Amplitudes and Phases for M2 and K1 Tidal Constituents at Selected Stations in the St. Lawrence Estuary Using Two Different Upstream Boundary Limits.   Phases Values are Presented Relative to Time Zone (GMT + 5) hours.

Station	Amplitude (cm)			Phase (°)		
	Observation	Model 1 *	Model 2 **	Observation	Model 1 *	Model 2 **
M_2 Semi-diurnal Tidal Constituent						
Upper Estuary:						
Cap-aux-Oies	193	230	227	103	100	108
St. Jean-Port-Joli	187	175	190	134	136	135
St. Siméon	162	182	185	89	84	85
Rivière-du-Loup	151	155	150	80	78	80
Lower Estuary:						
Tadoussac	155	155	160	68	67	74
Forestville	141	140	143	52	57	56
Bail-Comeau	119	123	127	47	49	52
Pointe-au-Père	127	130	138	54	59	55
Matane	114	112	116	49	53	52
	150	156	160	75	76	77
K_1 Diurnal Tidal Constituent						
Upper Estuary:						
Cap-aux-Oies	27	29	29	219	224	220
St. Jean-Port-Joli	26	25	27	244	244	241
St. Siméon	25	26	26	216	216	213
Rivière-du-Loup	23	24	24	208	217	211
Lower Estuary:						
Tadoussac	24	24	24	205	211	207
Forestville	24	23	23	201	208	204
Baie-Comeau	22	22	23	199	205	201
Pointe-au-Père	24	22	23	204	209	206
Matane	22	21	22	191	206	204
	24	24	25	210	216	212

* Upstream limit at Ile-aux-Coudres (Levesque et al., 1979)
** Upstream limit at St. François (El-Sabh et al., 1979)

constituents M_2 and K_1 (Table 1) indicates that although both models overestimate the M_2 amplitudes, particularly in the Upper Estuary (Cap-au-Oies), the selection of Ile-aux-Coudres by LME as the upstream limit gave better agreement. The discrepancy between the calculated and observed values can be attributed to several factors including - (1)

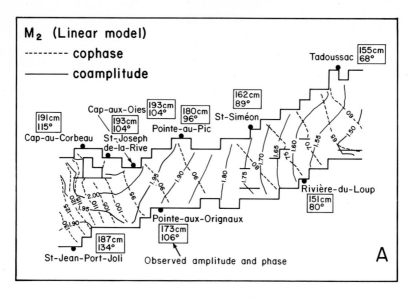

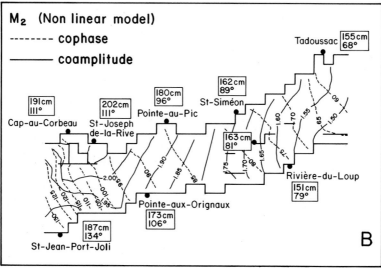

Figure 10. Computed M_2 co-phase and co-amplitude lines in the Upper St. Lawrence Estuary using linear (A) and nonlinear (B) numerical models. Values in squares represent observations at coastal stations.

the constrictions and shallowness in the area between Ile-aux-Coudres and Quebec City by a misrepresentation by the rectangular grid size of 2.8 x 3.7 km, - (2) the nonlinear damping interactions between the different tidal constituents which were not included in the equations of motion and (3) the large tidal range in the SLE which reaches its maximum value of about 6m in this area. In order to investigate the effect of the nonlinear advective terms, we carried out two experiments, with and without these terms, using a Cartesian coordinate system for Equations 4-6, following the principles discussed in Henry (1982), quadratic friction and a finer rectangular grid net for the area between Pointe-des-Monts and Ile-aux-Coudres. As can be seen in Figure 10, which shows the model outputs for the M$_2$ tide in the Upper Estuary, the two solutions gave almost similar results, except that co-amplitude lines in the nonlinear case are shifted in the upstream direction and towards the north shore.

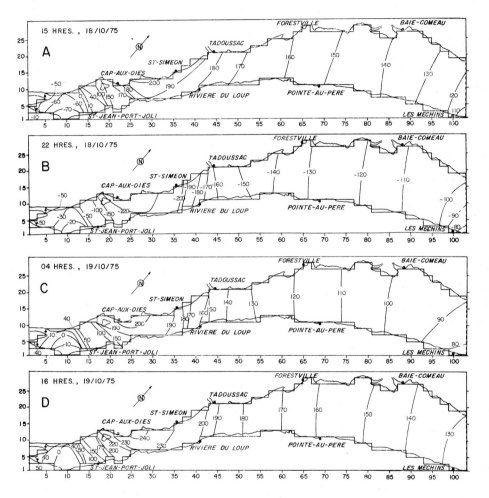

Figure 11. Water level contours in the St. Lawrence Estuary on 18-19 October, 1975 (from Murty and El-Sabh, 1980).

The water level contours at the flood and ebb tides with reference to Pointe-au-Père were computed by LME for October 18-19, 1975 (Figure 11). These contours, which illustrate the propagation of the wave fronts across the estuary, were prepared using the water level at each grid point as generated by the linear model and taking into account the five tidal harmonic constituents. The tide propagates progressively along the north shore up to Tadoussac during flood tide. In this part of the estuary, differences in water level between the north and south shores are less than 5 cm. The highest level is found near Cap-aux-Oies. The levels then decrease rapidly southward in the direction of St-Jean-Port-Joli. Upstream of this station ebb tides start to occur. The large difference in water level between both shores at the upstream boundary of the model can be attributed, in part, to the changes in the axial direction of the estuary and in part to the abrupt rise in bottom topography just beyond Tadoussac where the estuary becomes narrow. As expected, this picture is reversed during ebb conditions; the surface water attains transverse slopes of exceptional steepness in the upstream part, with high levels along the south shore. In the lower part of the estuary, the decrease is moderate and the surface profile of the wave is regular, with small differences between both shores.

TABLE 2

Observed and calculated Mean Tidal Streams (cm/s) in the St. Lawrence Estuary (From Levesque et al., 1979)

Tidal Const.	M_2			S_2			N_2			K_1		
	Calcu-lated		Obser-ved	Calcu-lated		Obser-ved	Calcu-lated		Obser-ved	Calcu-lated		Obser-ved
Region	M*	C**	F***	M	C	F	M	C	F	M	C	F
Ile-aux-Coudres	112	116	120	41	30	29	32	20	21	28	8	7
Rivière-du-Loup	90	89	–	24	24	–	17	16	–	13	7	–
Tadoussac	95	69	–	39	4	–	27	13	–	21	6	–
Pointe-au-Père	18	14	12	5	4	3	3	3	3	3	1.4	1.3
Les Méchins	20	18	16	8	6	6	6	4	4	12	1.8	1.5

* M = Levesque et al. (1979) model results.
** C = Forrester (1972).continuity estimations
*** F = Forrester (1972).direct observations

Tidal streams in the SLE were also calculated by LME and the model results were compared with those obtained by Forrester (1972). From observed surface elevations and the continuity considerations, Forrester (1972) estimated the average tidal streams for seven harmonic constituents through several sections across the SLE. For quantitative comparison, Table 2 shows Forrester's results for selected regions together with values obtained by the LME model and those deduced from direct current measurements in three of the cross-sections. In comparing the calculated and observed data, it must be remembered that the current or tidal stream calculated by the model, or by continuity for a particular cross-section, is the average value for the entire water column, and that the current meter results are values for a single depth. Furthermore, the model results reported in Table 2 for a particular region are the mean of all grid points over the entire region. Forrester (1972) concluded that tidal streams calculated by continuity are believed to be more accurate estimates of the average flows than could be obtained by direct current measurements. Except for the K_1 tidal constituent, Table 2 shows that the amplitudes of the semi-diurnal tidal streams are generally of the same order of magnitude as those calculated by Forrester. The general discrepancy of the K_1 tidal current can be explained by the fact that the horizontal motion computed from the LME model is not accurate near the two open boundaries, especially for the diurnal tide. Note that the two stations, Ile-aux-Coudres and Les Méchins, are at the open boundaries of the model. At Pointe-au-Père the agreement was much better, while at Rivière-du-Loup and Tadoussac no observational data exist for comparison.

It is well known from the literature, that an oscillatory tidal current in coastal waters can transfer energy from short period (of the order of a day) to long-term residual motion (or secondary flow) through interaction with the topography. Nihoul and Ronday (1975) showed the importance of the tidally-generated residual circulation (TGRC) in the North Sea, and Tee (1976, 1977) examined the problem in the Minas Basin of the Bay of Fundy. Kuipers and Vreugdenhil (1973) and Ianniello (1979) identified certain terms in the vorticity equation as contributing to the phenomenon under consideration. Zimmerman (1978) studied the problem using an analytical model topographic generation of residual motion and concluded that, for the existence of residual eddies, pronounced spatial inhomogeneity of tidal stress is required. An irregular bottom topography can contribute to the required inhomogeneity of the tidal stress. A simple scale analysis of the terms identified by Kuipers and Vreugdenhil (1973) in the vorticity equation indicates that some gyres can be attributed, at least in part, to TGRC. These gyres include the following: the gyres reported by El-Sabh (1976) in the Gulf of St. Lawrence, the different eddy patterns described by El-Sabh et al. (1982) and summarized in Ingram and El-Sabh (this volume) in the St. Lawrence Estuary, the eddies in the south-eastern part of the Northwest passage reported by Murty and Fissel (1978), those reported by Crean et al. (1988) in the Strait of Georgia and the circulation pattern identified by Murty et al. (1980) in the Masset Inlet can be attributed, in part, to TGRC. Such a tidally-generated residual circulation is important in computing surface drift. This problem was examined by Murty and El-Sabh (1980) for the St. Lawrence Estuary using the following expression

$$\text{TGRC} = \sum_{\text{one tidal cycle}} \left\{ (h+\eta) \left[\frac{\partial}{\partial y} \left(\frac{U}{h+\eta} \right) - \frac{\partial}{\partial x} \left(\frac{V}{h+\eta} \right) \right] - f \right\} \left\{ \frac{\partial}{\partial x} \left(\frac{U}{h+\eta} \right) + \frac{\partial}{\partial y} \left(\frac{V}{h+\eta} \right) \right\} \qquad (10)$$

U, V and η are respectively the transport components and the deviation of the free surface from its equilibrium level. Values for these variables were obtained from the LME linear model and the residual circulation pattern was calculated (Figure 12A). The magnitude of these residual currents varies from less than 1 cm/s to a maximum of 22 cm/s. However, average values are of the order of 4-5 cm/s, which are in reasonable agreement with the observations reported by El-Sabh (1979) and Koutitonsky and El-Sabh (1985). The results obtained indicate that, whereas the oscillatory tidal currents do not show small scale eddy patterns, the residual motion associated with tides consists of eddies with a scale of 12-30 km. A tongue-like feature extends 80 km upstream from the mouth of the estuary and shows a cyclonic pattern. It is of interest to note that evidence for the existence of the predicted cyclonic eddy, near the north shore between Forestville and Baie-Comeau, has been found from the path followed over a two-day period by a parachute drogue set to drift at a depth of 10m (Figure 12B).

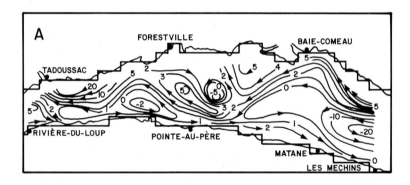

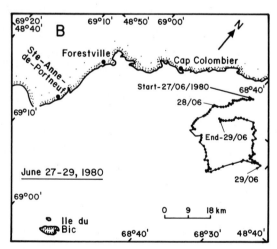

Figure 12. (A) Tidally-generated residual currents in the Lower St. Lawrence Estuary. Numerals indicate velocities in cm/s (From Murty and El-Sabh, 1980). (B) Trajectory of a near-surface drifter, showing the Rimouski cyclonic eddy observed in June 1980.

One of the practical problems that pointed out to us the necessity of including TGRC is the mathematical modelling of the movement of oil slicks. Aubin et al. (1979) modeled numerically the movement of two observed oil slicks in the Upper Estuary using a finer grid than the one used by LME. We believe that the discrepancy between the observed and computed trajectories can be attributed to TGRC, in addition to other factors, such as wave action, internal tide, stratification, etc.

3.2.3. The PG results

PG developed a two-dimensional barotropic model of the M_2 tide in the St. Lawrence Gulf and Estuary system to predict frontal regions separating areas of tidally mixed waters from areas showing pronounced summer stratification. The fully nonlinear equations of momentum and continuity (4-6) were used. Although, as mentioned earlier, in partially enclosed shelf seas the tide generating force is generally sufficiently small to be neglected in comparison to the tide driven at the open sea boundaries, PG felt that in the weak M_2 tidal environment of the Gulf of St. Lawrence (with tidal current amplitudes generally less than 20 cm/s), the tide generating force had to be included. Thus, the M_2 tidal forcing of this model was provided by two sources; by boundary values of tidal elevations at the entrance to Cabot Strait and the Strait of Belle Isle, and by the tide generating force. The output from the model was Fourier analysed to determine the M_2 elevations and phases on each grid square and the results were presented in the form of a chart of the M_2 co-phase and co-amplitude lines for the whole system. This chart, when compared to those published by Farquharson (1970) and Godin (1979) based on extrapolation of observed values, shows the main characteristics of the M_2 tides in the study area. According to PG, the model results agree with observations, with standard deviations of 8 cm for amplitude and 10° in phase. However, a careful examination of that solution shows that large discrepancies occur in the estuary region, with amplitudes smaller than observed values at coastal stations. The discrepancies in amplitude reach values of more than 45 cm upstream of Tadoussac in the Upper Estuary. The large-scale coarse mesh used by PG in their model may provide an explanation for this discrepancy.

In the numerical model, the local mean dissipation of tidal energy per unit area due to work done against bottom friction, is simply the scaler product of the bottom stress and the tidal velocity averaged over time. PG evaluated the Simpson-Hunter (1974) parameter, S defined by Equation 11, for the St. Lawrence Gulf and Estuary system based on values for the bottom friction given by Equation (9) and M_2 tidal currents obtained from their model. We show in Figure 13A contours of S as obtained for the SLE. Note that values of S in the neighbourhood of S~1.5 define transitional regions between waters that are well mixed, S < 1, and waters that are well stratified, S > 2, during the summer months. According to PG, the cool feature associated with the frontal zone stretching from the mouth of the Saguenay River towards Rimouski and observed in satellite thermal images (see El-Sabh, 1988, and Gratton et al., 1988, for more details) is likely to be of tidal origin.

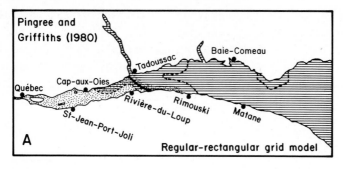

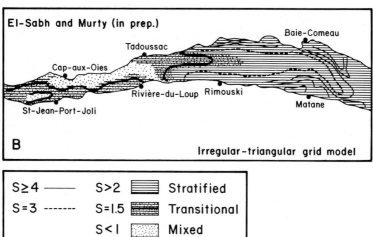

Figure 13. Contours of the stratification parameter S with bold line (S = 1.5) identifying frontal regions. (A) using the regular rectangular grid model of Pingree and Griffithes (1980) and (B) using the irregular-triangular grid model shown in Figure 14.

3.2.4. Finite-Element and Irregular-Grid Finite-Difference Models

For completeness, this section will be concerned mainly with a few unpublished attempts that have been made using the finite-element approach to tidal modelling in the SLE. We will deal later with some preliminary results obtained with irregular-grid finite-difference techniques, which combine the best features of both the finite-element and the finite-difference methods. Unlike the finite-difference scheme used in the PC, LME and PG models, for these two techniques the variables satisfying the governing equations and boundary conditions are approximated by piecewise polynomials. Their main advantage is

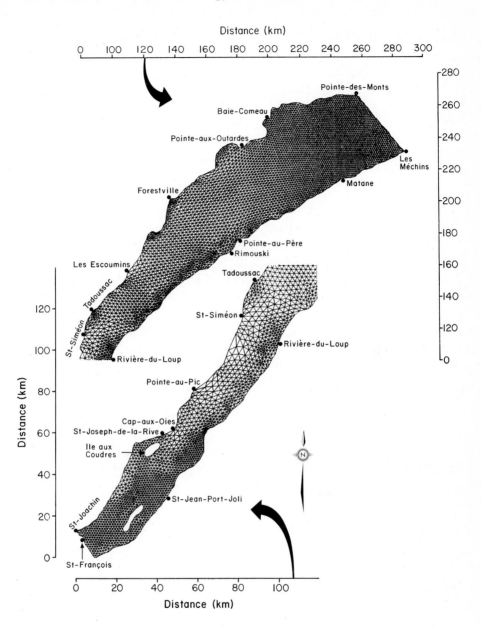

Figure 14. Irregular-triangular grid developed by the authors for modelling of tides and tsunami in the St. Lawrence Estuary.

the highly flexible grid of irregular sized and shaped triangular elements, so that real water bodies can be modelled more realistically.

Two independent attempts were made using the finite-element method to simulate the barotropic tidal propagation in the SLE; the first was by LeProvost and El-Sabh using a similar technique to that described in LePrevost and Rougier (1981), the second was a model developed by Koutitonsky et al. (1986). In both studies, the estuary was subdivided into triangular elements, giving a total of 486 elements in the first case and 380 elements in the latter. In locations where topographic details were not important, a coarser grid was been used. These models assumed a homogeneous vertical density structure and have, as a result, been vertically integrated. Verification of the two models consisted of matching calculated with observed water levels. Good agreement was obtained for the available results, which are currently limited to the production of cotidal maps for M_2 tide.

Irregular-grid finite-difference techniques have many similarities with finite-element models, the first similarity being that both techniques can make use of identical grids of irregular-sized and shaped triangular elements. Following Thacker et al. (1980), we applied a technique for automatic construction of a mesh with irregular-triangular grids for tide, storm surge and tsunami models in the SLE between Pointe-des-Monts and the eastern tip of Ile d'Orléans. The grid consisted of 5826 grid points and provides larger triangular elements in deeper water and smaller triangles in shallow areas, particularly in the Upper Estuary (Figure 14). A linear vertically-integrated model was then developed using the finite-difference scheme and the irregular grid. For modelling of tides, observed tidal constitutents were specified at the mouth of the estuary. Separate runs are made for each of the five important tidal constituents and also for the total tide. Co-amplitude, co-phase lines and tidal current ellipses are constructed for each of the five tidal harmonics. Tidally-generated residual currents and the Simpson-Hunter stratification parameter were also calculated. Comparison of the model results with observations shows excellent agreement. For example, the M_2 amplitude agrees within 3% on the average. A complete description of the model and discussion of the results obtained will be published elsewhere (El-Sabh and Murty, in preparation), but some of the results regarding the calculations of the Simpson-Hunter parameter, S, will briefly be discussed here.

Contours of S, evaluated from the irregular-triangular finite-difference model for spring tidal conditions, are shown in Figure 13B. For comparison with the PG model, we show the results as obtained by using the M_2 constituent as the only tidal forcing of the model at the mouth. Following Simpson and Hunter (1974), the parameter S was calculated using the following equation

$$S = \log_{10} \frac{(h+\zeta)}{C_D|u^3|} \tag{11}$$

where $(h+\zeta)$ is the mean water depth measured from the free surface, u is the M_2 tidal current amplitudes and C_D is the bottom drag coefficient.

It is clear that the results obtained for frontal regions, by better representing the bottom topography and coastline using the irregular triangular grid model, correspond more closely to observations as reported in the literature (see El-Sabh, 1988; Ingram and El-Sabh, 1990, for more details). Significant numerical predictions for tidally well-mixed regions (S < 1) are found in the northern half of the Upper Estuary between Cap-aux-Oies and Tadoussac, and in the area around Ile-aux-Lièvres. As discussed earlier (Table 2), this is the area where tidal currents in the SLE attain their maximum values. The numerical model also indicates that tidally well-stratified regions are found seaward of Rimouski. As expected, the model results (not shown here) show frontal movement during the spring-neap cycle of tidal mixing. It is of interest to note that meso- and large-scale distribution of phytoplankton biomass and biological productivity in the SLE are closely associated with regional differences in tidal mixing as shown in Figure 13B. (See Therriault et al., this volume, for more details).

4. Baroclinic models

The above described tidal models involved vertically integrated equations. The approach provides a useful tool for studying not only tidal dynamics and the simulation of barotropic tides and streams in a complex system, but also the associated tidal residual circulation. However, these models did not reveal the characteristic internal features of the estuarine circulation. This section addresses the different approaches employed to achieve this objective for the SLE.

4.1. The laterally-integrated models

Three models were developed to study different aspects of the vertical circulation using two-dimensional grids arranged vertically. The governing equations were averaged across the width of the channel and thus Coriolis effects were excluded. The two models developed by Forrester (1967) and by Tee and Lim (1987) use a simplified (flat or stepwise) bottom topography. In the former, an attempt was made to reproduce the M_2 constituents in all observed tidal heights and axial streams by the superposition of surface progressive and standing waves, plus an internal standing wave, all of M_2 tidal period, along the axis of the channel. This simple model provided good agreement between prediction and observation for the surface elevation and mean stream, and only qualitative agreement for the upper-layer stream relative to the mean stream. Forrester attributed this to the fact that the internal wave is very sensitive to the choice of densities and thicknesses for the two layers. The second model by Tee and Lim (1987) was mainly developed to investigate the laterally-averaged seasonal estuarine circulation and, in particular, the freshwater pulse in the SLE. Results from this model are discussed elsewhere in this book (Tee, 1990) and will not be dealt with any further.

Borne deGrandpré and El-Sabh (1980) developed and applied a real-time two-dimensional baroclinic model to the SLE which, unlike the above mentioned models, incorporates the actual bottom topography (Figure 1C) to study and simulate the vertical circulation of tides in the system. With relevant tidal elevations and salinity conditions at the mouth and landward boundaries, the model permits the calculation of the water height, vertical and longitudinal velocity and the salinity distribution throughout a tidal cycle. It was inspired by the principle developed by Hamilton (1975) and applied to the Rotterdam Waterway. Since an account of the mathematical basis and computational scheme for the model has been published (Hamilton, 1975), only the basic features will be given here.The governing equations, which express the conservation of volume, momentum and salt content may be written as:

$$\frac{\partial(bu)}{\partial x} + b\frac{\partial w}{\partial z} = 0,$$

$$(12)$$

$$\frac{\partial \zeta}{\partial t} + \frac{I}{b}\frac{\partial}{\partial x}\left(b\int_{-\zeta}^{H} u\, dz\right) = 0,$$

$$\frac{\partial u}{\partial t} + u\frac{\partial u}{\partial x} + w\frac{\partial u}{\partial z} + g\frac{\partial \zeta}{\partial x} + ag(z+\zeta)\frac{\partial \overline{S}}{\partial x} - \frac{\partial}{\partial z}\left[N_z \frac{\partial u}{\partial z}\right] = 0 \qquad (13)$$

$$\frac{\partial S}{\partial t} + u\frac{\partial S}{\partial x} + w\frac{\partial S}{\partial z} - \frac{I}{b}\frac{\partial}{\partial x}\left[bK_x \frac{\partial S}{\partial x}\right] - \frac{\partial}{\partial z}\left[K_z \frac{\partial S}{\partial z}\right] = 0 \qquad (14)$$

where b(x) is the cross-section width, H(x) is the undisturbed depth and $\zeta(x,t)$ is the elevation of the free surface above the undisturbed reference level. $u(x,z,t)$ and $w(z,x,t)$ are the velocity components in the x- and z-directions, respectively, N_z the vertical eddy viscosity coefficient, K_x, K_z the horizontal and vertical eddy diffusion coefficients, g the acceleration due to gravity, $S(x,z,t)$ salinity and $\overline{S}(x,z,t)$ is given by:

$$\overline{S}(x,z,t) = \frac{I}{(z+\zeta)}\int_{-\zeta}^{z} S(x,z,t)\, dz \qquad (15)$$

The density ρ is related to salinity by the linear equation of state:

$$\rho = \rho_0(I+\alpha S)$$

where ρ_0 is the density of freshwater and α is constant taken as 0.78×10^{-3}. There is no salt flux through the surface or bottom, therefore

$$\left[K_z \frac{\partial S}{\partial z} \right]_{z = -\zeta} = 0$$

$$\left[K_z \frac{\partial S}{\partial z} \right]_{z = H} = 0 \tag{16}$$

It is also assumed that there is no wind stress on the surface

$$\left[N_z \frac{\partial u}{\partial z} \right]_{z = -\zeta} = 0 \tag{17}$$

The shearing stress at the bottom is taken to be related by a quadratic law to the velocity:

$$\left[N_z \frac{\partial u}{\partial z} \right]_{z = H} = -k u_\Delta |u_\Delta| \tag{18}$$

where k is the friction coefficient and u_Δ is the velocity at 1m above the bottom. In addition, the vertical velocities at the bottom, $w(x, z = H, t)$ are zero.

A finite difference initial-value method was used to solve the basic governing Equations (12) - (14). The finite difference equations were obtained using a finite difference grid in the vertical plane of the estuary. While the bottom was fixed, the surface water varied with time and could move through the grid points. The numerical scheme, which uses only one time step, was essentially explicit, except for the viscosity and diffusion terms which were developed in an implicit manner. Furthermore, a central finite difference scheme was used, except for the advective term for salt in Equation (14), which used forward or backward differences depending on the state of the tide (ebb or flood, respectively).

The area covered by the model, between Ile d'Orléans and Pointe-au-Père, was divided into 66 cross-sections. These sections are separated by a distance of 5.6 km, which corresponds to the horizontal grid spacing dx. In order to have better resolution, values for the vertical grid size (dz) were chosen as 4 and 20m for the Upper and Lower Estuaries, respectively. Calibration of the model was carried out by adjusting the friction, eddy viscosity and eddy diffusion coefficients. The model then was used to calculate the vertical circulation during different stages of the tide in May 1979 (Borne deGrandpré et al., 1981) and for September 1979 (Borne deGrandpré and El-Sabh, 1980). These two months respectively represent periods of high and low freshwater discharge in the estuary, and hence different stratification conditions.

Figure 15 shows the vertical circulation along the LSLE, as deduced by the model, at different hours of the tidal period on 25 May 1979. These figures indicate that, in general, water flow reflects the geometry of the bottom topography. The highest velocities were obtained in the Upper Estuary, reaching a value of 200 cm/s or more, particularly in the region upstream of Pointe-au-Pic. In the Lower Estuary, maximum current speeds are of

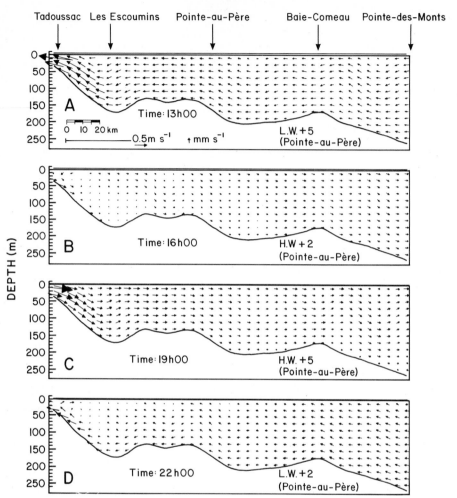

25 May, 1979

Figure 15. Vertical circulation in the St. Lawrence Estuary on 25 May 1979 (from Borne de Grandpré et al., 1981).

the order of 25 cm/s, except in the region near Tadoussac. The abrupt rise in bottom topography may explain the increase in current speeds at that location. Close examination of Figure 15 also indicates that there exist successive minima and maxima of the surface currents associated with different phases of the tidal period. These minima and maxima correspond to a vertical oscillation of the velocity vectors and reflect the presence of an internal tide, previously reported by Forrester (1974), which propagates seaward from the inland end of the Laurentian Channel near Tadoussac. Borne deGrandpré et al. (1981) used the model to estimate both barotropic (surface tidal current) and baroclinic tides

43

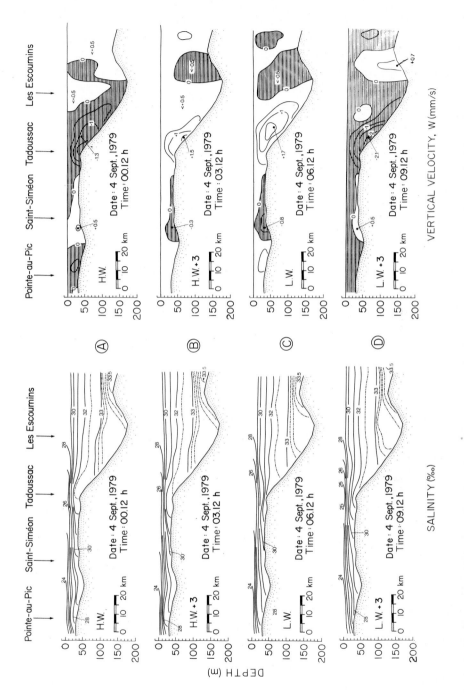

Figure 16. Salinity distribution and vertical velocity component (w) computed by Borne de Grandpré and El-Sabh (1980) in the St. Lawrence estuary.

(internal tidal currents) in that region. They confirmed the ability of the model to reproduce not only the estuarine vertical circulation but also the observed internal tides in the SLE. Furthermore, the model was used by Borne deGrandpré and El-Sabh (1980) to obtain, for the first time, estimates for the vertical velocity component, w, in the area of intense upwelling and mixing, and the associated salinity distribution near the head of the Laurentian Channel (Figure 16). The maximum vertical velocity (upward or downward) of 2-3 mm/s occured near Tadoussac during high and low water slack periods, respectively. The model results also showed that vertical oscillation of the intermediate layer during a tidal cycle may reach a value up to 50m, which is in excellent agreement with observations (Ingram, 1975; Forrester, 1974; Therriault and Lacroix, 1976). Other features of the vertical estuarine circulation which have previously been observed in the estuary, such as the two-layer residual circulation, the longitudinal distribution of both tidal range and water levels at high and low tides, and the effect of horizontal salinity gradients on the vertical velocity profiles were reproduced by this model.

Although the two-dimensional numerical model described above permits the calculation of the instantaneous vertical and longitudinal distribution of currents, salinity and tidal elevation throughout a tidal cycle, much work needs to be done to improve the model, particularly with regard to prediction of salinity, by using a more realistic formulation for the turbulence coefficients. The fact that the governing equations used here were laterally averaged provides another limitation to the application of the present model to the St. Lawrence estuary. Variations in the lateral circulation may be important in the Lower Estuary where the channel is wide enough for the earth's rotation to have influence and where Coriolis force plays an important role. Also in the Upper Estuary where the complex geometry contributes to the existence of features such as the tidal elevation differences between north and south shores, the deep penetration of isohalines along the north shore and the different paths for tidal propagation during ebb and flood. In fact, occasional transverse currents, with mean velocities varying between 7 and 30 cm s^{-1}, have been reported both in the Lower St. Lawrence Estuary (El-Sabh, 1979) and in the Upper Estuary (d'Anglejan & Ingram, 1976; Ingram & d'Anglejan, 1977). Apart from this, the present model may be considered as a first step towards the development of a three-dimensional numerical model for the St. Lawrence Estuary. Further modelling efforts by El-Sabh and Borne deGrandpré are now in progress to extend this 2-D vertical model and develop a time-dependent three-dimensional model. As a first step, we divided the estuary into three longitudinal sections (Figure 17) and solved the nonlinear governing equations (12 - 14) by a finite difference initial-value method. This model will permit the calculation in real time of the water height, vertical, lateral and longitudinal salinity and velocity distributions and hence will lead to a better understanding of tidal dynamics and related physical processes occurring in the St. Lawrence Estuary.

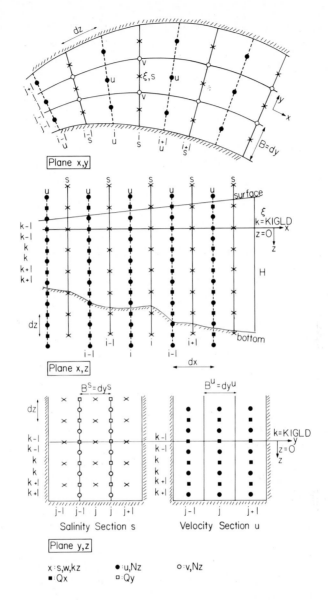

Figure 17. Mathematical notations and schematization of the three-dimensional numerical model developed by Borne de Grandpré and El-Sabh.

5. Conclusions

The one-dimensional and laterally integrated tidal models have several obvious short-comings, since they cannot represent the two-dimensional nature of the system and allow for the inclusion of the earth's rotation effects. The various versions of the two-dimensional models also have some disadvantages, namely that the nonlinear advective terms, and the shallow water effects are not properly included. The irregular triangular grid models, although they provide a better topographic resolution, cannot include stratification effects, which appear to be necessary to reproduce the tides accurately in the St. Lawrence estuary.

It is clearly evident from the above discussion that a fully three-dimensional model, combining all the positive properties of the previous models, and coupled with more *in-situ* observations of offshore tidal elevations and currents, is required to simulate barotropic and baroclinic processes within the St. Lawrence Estuary. This system is demanding of any scheme for a variety of reasons including the topographical complexity of the sea bottom, the strong tidal streams in the Upper Estuary, and the variously stratified estuarine circulation. The model should be

- completely nonlinear, including advective terms carefully treated in the finite difference scheme adopted,
- applied with the highest possible space resolution,
- driven along time in such a way that the nonlinear interactions between main tidal constituents will be correctly reproduced,
- using improved bottom friction laws and turbulence coefficients.

Work on such a model had been initiated by the authors in collaborative effort with Dr. J.A. Stronach. The three-dimensional model under development is an adaptation of an 8-level barocline model developed earlier for the system consisting of the Strait of Georgia, Juan de Fuca Strait and Puget Sound. Since the model is implicit, there are no restrictions on the time step, other than considerations of accuracy of tidal phase propagation.

The implicit solution of the surface gravity wave propagation component is done by means of a successive over-relaxation (S.O.R.) technique applied to the continuity equation. In this equation, the divergence at the present time level, averaged with the divergence at a future time level, gives rise to changes in sea surface over a single time step. The divergence at the future level is based on future water level, and this is the reason for the implicitness of the technique. The continuity equation is thereby transferred into an elliptic equation for the water level field at the advanced timestep.

The vertical viscosity scheme is also semi-implicit, using a Crank-Nicholson solution for the implicit component. The vertical eddy viscosity is dependant upon shear and the Richardson number. Horizontal advection is computed along characteristics. The vertical spacing, while uniform in the horizontal planes, is allowed to vary in the vertical. The model is flexible enough to either increase or decrease the number of levels in the adaptation to the St. Lawrence Estuary, depending upon its density stratification characteristics.

Acknowledgements

Two reviewers provided us with valuable comments on an earlier version of the manuscript for which we are grateful. We also thank D. Booth and N. Silverberg for their extremely useful remarks on the final text. We would like to thank our colleagues and graduate students for their collaboration, fruitful discussion and encouragement throughout the different stages of our efforts to model the tidal propagation in the St. Lawrence Estuary. In particular, we wish to thank R.F. Henry, J.A. Stronach, C. LeProvost, J.-F. Dumais, L. Levesque, F. Aubin, J.-M. Brian, J. Chassé and B. Tessier. We also wish to express our appreciation to Lorena Quay for typing the manuscript and Johanne Nöel for preparing the drawings. This study was financially supported by operating grant from the Natural Science and Engineering Research Council of Canada to M.I. El-Sabh.

REFERENCES

Anonymous, 1939. Tidal Current Charts, St. Lawrence Estuary: Orleans I. to Father Point, Tidal Publication No. 21, Canadian Hydrographic Service, Dept. Mines and Technical Surveys, Ottawa, 34 pp.

Aubin, F., T.S. Murty and M.I. El-Sabh, 1979. Numerical simulation of the movement and dispersion of oil slicks in the upper St. Lawrence estuary: Preliminary Results, Naturaliste Can., 106:37-44.

Borne deGrandpré, C. and M.I. El-Sabh, 1980. Etude de la circulation verticale dans l'estuaire du Saint-Laurent au moyen de la modélisation mathématique. Atmos. Ocean., 18:304-321.

Borne deGrandpré, C. and M.I. El-Sabh, and J.C. Salomon, 1981. A two-dimensional numerical model of the vertical circulation of tides in the St. Lawrence estuary. Estuarine Coastal Shelf Sci., 12:375-387.

Charnock, H., and J. Crease, 1957. Recent advances in Science. North Sea surges, Sci. Prog. 45:494-511.

Crean, P.B., T.S. Murty and J.A. Stronach, 1988. Numerical simulation of oceanographic processes in the waters between Vancouver Island and the mainland. Oceanogr. Mar. Biol. Ann. Rev., 26:11-142.

d'Anglejan, B.F.. and R.G. Ingram, 1976. Time-depth variations in tidal flux of suspended matter in the St. Lawrence estuary, Estuarine Coastal Mar. Sci., 4:401-416.

Dronkers, J.J., 1964. Tidal Computations, North-Holland Publishing Company, Amsterdam. 518 pp.

El-Sabh, M.I., 1976. Surface circulation pattern in the Gulf of St. Lawrence, J. Fish. Res. Board Can., 33:124-138.

El-Sabh, M.I., 1979. The lower St. Lawrence estuary as a physical oceanographic system. Naturaliste Can., 106:55-73.

El-Sabh, M.I., 1988. Physical oceanography of the St. Lawrence estuary. In: Hydrodynamics of Estuaries, Vol. II (B. Kjerfve, Ed.). CRC Press, 61-78.

El-Sabh, M.I., T.S. Murty and L. Levesque, 1979. "Mouvements des eaux induits par la marée et le vent dans l'estuaire du Saint-Laurent", Naturaliste Can., 106:89-104.

El-Sabh, M.I., H.-J. Lie and V.G. Koutitonsky, 1982. Variability of the near surface residual current in the lower St. Lawrence estuary, J. Geophys. Res., 87:9589-9600.

Farquharson, W.I., 1970. Tides, tidal streams and currents in the Gulf of St. Lawrence. Bedford Inst. Oceanogr. AOL Report 1970-5, 145p.

Forrester, W.D., 1967. Currents and geostrophic currents in the St. Lawrence estuary. Bedford Inst. Oceanogr. Report 1967-5, 175p.

Forrester, W.D., 1972. Tidal transports and streams in the St. Lawrence estuary. Int. Hydrographic Rev., XLIX(1):95-108.

Forrester, W.D., 1974. Internal tides in the St. Lawrence estuary. J. Mar. Res., 32:55-66.

Freeman, N.G. and T.S. Murty, 1976. "Numerical modeling of tides in Hudson Bay". J. Fish. Res. Board Can., 33:2345-2361.

Funke, E.R., and N.L. Crookshanke, 1979. A hybrid model of the St. Lawrence River estuary. p.2855-2869. Proc. 16th Coastal Eng. Conf., August 27-September 3, 1978, Hamburg, W. Germany.

Gagnon, M., and M.I. El-Sabh, 1980. Effets de la marée interne et des oscillations de basse fréquence sur la circulation otière dans l'estuaire du Saint-Laurent. Naturaliste Can., 107:159-174.

Godin, G., 1971. Hydrodynamical studies on the St. Lawrence River. Mar. Sci. Br., Dept. Energy, Mines Res., Ottawa, Report 18, 116p.

Godin, G., 1979. La Marée dans le Golfe et l'estuaire du Saint-Laurent. Naturaliste Can. 106:105-121.

Godin, G., 1980. Cotidal charts for Canada. Mar. Sci. Info. Dir., Dept. Fish. Oceans, Ottawa. Report 55, 93p.

Gratton, Y., G. Mertz and J.A. Gagné, 1988. Satellite observations of tidal upwelling and mixing in the St. Lawrence Estuary. J. Geophys. Res., 93:6947-6954.

Hamilton, P., 1975. A numerical model of vertical circulation of tidal estuaries and its application to the Rotterdam Waterway. Geophys. J.R. Astronom. Soc. 40:1-21.

Henry, R.F., 1982. Automated programming of explicit shallow-water models. Part I. Linearized models with linear or quadratic friction. Pac. Mar. Sci., Inst. Ocean Sci., Dept. Fish. Oceans, Victoria, B.C. Report 3, 70 p.

Ingram, R.G., 1975. Influence of tidal-induced vertical mixing on primary productivity in the St. Lawrence estuary. Mem. Soc. R. Sci. Liège, 6:59-74.

Ingram, R.G. and B. d'Anglejan, 1977. On the importance of cross-channel suspended matter flux in the upper St. Lawrence estuary. Proc. Symp. Modelling Transp. Mech. Oceans and Lakes, Environ. Canada MS Rep. 43:149-159.

Ingram, R.G. and M.I. El-Sabh, 1990. Fronts and mesoscale features in the St. Lawrence estuary. In: "Oceanography of a large-scale estuarine system: The St. Lawrence Coastal and Estuarine Studies, Springer-Verlag (this volume).

Ianniello, J.P., 1974. Tidally induced residual currents in estuaries. J. Phys. Oceanogr., 9:962-974.

Koutitonsky, V.G. and M. I. El-Sabh, 1985. Estaurine mean flow estimation revisited: application to the St. Lawrence estuary. J. Marine Res., 43, 1-12.

Koutitonsky, V.G., B. Côté and C. Torro, 1986. Modélisation hydrodynamique de la circulation en relation avec les débarquements de maquereaux dans le golfe du Saint-Laurent. Unpublished Rapport d'etape No. 1, Contrat OSD85-00239, Dept. Fish. Oceans, Ottawa.

Kuipers, J. and C.B. Vreugdenhil, 1973. "Calculations of two-dimensional horizontal flow", Delft Hydraulics Lab., Report on Basic Res., S163, Pt. 1, 44p.

LeBlond, P.H., 1978. On tidal propagation in shallow rivers. J. Geophys. Res., 83:4717-4721.

LeProvost, C., G. Rougier and A. Poncet, 1981. Numerical modelling of harmonic constituents of the tides, with application to the English Channel. J. Phys. Oceanography, 1123-1138.

Levesque, L., T.S. Murty and M.I. El-Sabh, 1979. Numerical modeling of tidal propagation in the St. Lawrence estuary. Int. Hydrographic Rev., 56:117-132.

Muir, L.R., 1982. Internal tides in a partially mixed estuary. Unpublished Report No. 9, Canada Center for Inland Waters, Burlington, Ontario, 177p.

Murty, T.S. and M.I. El-Sabh, 1980. Tidally-generated residual motion in the St. Lawrence Estuary, p.127-145. Proc. Thirteen Annual Simulation Symp., (V.P. Roy, R.G. Cumings, C. Hammer and W. Malamphy, Eds.), Tampa, FL.

Murty, T.S. and M.I. El-Sabh, 1985. The age of tides. Oceanogr. Mar. Biol. Ann. Rev., 23:11-103.

Murty, T.S. and D.B. Fissel, 1978. "Circulation in Lancaster Sound", presented at the Pacific Northwest Marine Sciences Workshop, Parksville, B.C., Canada, February 1978.

Murty, T.S., F.G. Barber and J.D. Taylor, 1980. "Role of advective terms in tidally-generated residual circulation", Limnology and Oceanography, 25:529-533.

Nihoul, J.C.J. nd F.C. Ronday, 1975. "The influence of the tidal stress on the residual circulation", Tellus, 27:484-489.

Ouellet, M.Y. and J. Cerceau, 1976. Simulation of the salinity distribution in the St. Lawrence estuary by a two-dimensional mathematical model. p.1249-1269. Proceedings of the 15th Coastal Engineering Conference. Am. Soc. of Civil Engineers.

Partensky, H.W. and L. Louchard, 1967. Etude sur la variation cyclique de la salinité moyenne dans l'estuaire du Saint-Laurent. Univ. de Montréal, Ecole Polytech., Div. d'Hydraulique. Rapp. Cons. Nat. Rech., Ottawa, 147p.

Partensky, H.W. and C. Marche, 1974. Etude de la déformation progressive de l'onde de marée dans l'estuaire du Saint-Laurent. Rapport de la section hydraulique, Ecole Polytechnique, Université de Montréal, 70 p.

Partensky, H.W. and J.C. Warmoes, 1970. Etude des marées dans l'estuaire du Saint-Laurent à l'aide d'un modèle mathématique linéarisé. Rapport soumis au Cons. Nat. de Rech. du Can., 677 p.

Pingree, R.D. and D.K. Griffiths, 1978. Tidal fronts on the shelf seas around the British Isles, J. Geophys. Res., Chapman Conference Special Issue, 83:4615-4622.

Pingree, R.D. and D.K. Griffiths, 1979. Sand transport paths around the British Isles resulting from M2 and M4 tidal interactions, J. Mar. Biol. Ass. U.K., 59:497-513.

Pingree, R.D. and D.K. Griffiths, 1980. A numerical model of the M2 tide in the Gulf of St. Lawrence. Oceanologica Acta, 3:221-225.

Prandle, D. and N.L. Crookshank, 1972. Numerical studies of the St. Lawrence River. Mech. Eng. Rep. MH-109, Hydraulic Lab., Nat'l. Res. Counc., Ottawa, Canada, 104 pp.

Prandle, D. and N.L. Crookshank, 1974. Numerical model of the St. Lawrence Estuary. J. Hydraulics Div., ASCE, Vol. 100 (HY4). Proc. Paper 10472:517-529.

Simpson, J.H. and J.R. Hunter, 1974. Fronts in the Irish Sea, Nature, London, 1250:404-406.

Tee, K.T., 1976. Tide-induced residual current, a two-dimensional numerical nonlinear tidal model, J. Marine Res., 34:603-628.

Tee, K.T., 1977. Tide-induced residual current-verification of a numerical model, J. Phys. Oceanogrphy, 7:396-402.

Tee, K.T., 1990. Meteorologically and buoyancy induced sub-tidal salinity and velocity variations in the St. Lawrence estuary. In: Oceanography of a large-scale estuarine system: The St. Lawrence. Coastal and Estuarine Studies, Springer-Verlag (this volume).

Tee, K.T. and T.H. Lim, 1987. The freshwater pulse - a numerical model with application to the St. Lawrence estuary. J. Mar. Res., 45:871-909.

Thacker, W.C., A. Gonzalez and G.E. Putland, 1980. A method for automating the construction of irregular computational grids for storm surge forecast models. J. Comput. Phys., 37:371-387.

Therriault, J.C. and G. Lacroix, 1976. Nutrients, chlorophyll and internal tides in the St. Lawrence estuary. J. Fish. Res. Board Can. 33:2747-2757.

Vncent, R. 1965. An investigation of the tidal characteristics of the St. Lawrence estuary by a mathematical model. M.Sc. Thesis, Université Laval, 119 p.

Webb, D.J. 1973. On the age of the semi-diurnal tide. Deep-Sea Res., 20:847-852.

Zimmerman, J.T.F., 1978. "Topographic generation of residual circulation by oscillatory (tidal) currents". Geophysical Astrophysical Fluid Dynamics, 11:35-47.

Chapter 3

Meteorologically and Buoyancy Induced Subtidal Salinity and Velocity Variations in the St. Lawrence Estuary

Kim-Tai Tee

Physical and Chemical Sciences, Department of Fisheries and Oceans, Bedford Institute of Oceanography
P.O. Box 1006, Dartmouth, Nova Scotia , Canada B2Y 4A2

ABSTRACT

Meteorologically and buoyancy induced subtidal salinity and velocity variations in the St. Lawrence Estuary are reviewed using 1979 and 1982 current meter data and results of 2-D numerical models. The meteorological forcing shows oscillations with periods of 10-15 and 40-50 days, and a long-term trend with a time scale of approximately 130 days. The salinity data indicates that the 10-15 day forcing induced an interfacial oscillation which propagated anticlockwise in the lower estuary as a free internal Kelvin wave. However, the velocity data provide a different picture of wave propagation. The 40-50 day forcing induces an oscillation that occurs initially near the transition region between the Upper and Lower estuaries, and then propagates toward upstream and downstream regions. The oscillation also propagates toward the surface in the lower estuary. The long-period salinity variation in the Upper estuary observed in 1982 is induced mainly by wind-forcing.

The contribution of ice to the subtidal salinity variation in the St. Lawrence Estuary is insignificant in most of the area, except near the head of the estuary. Freshwater runoff from the St. Lawrence River produces a seasonal pulse in the estuary. The formation, distribution and propagation of the pulse are reviewed using results of a 2-D numerical model. Near the mouth of the estuary, large eddies with a time scale of about 80 days were observed. The forcing of these eddies is unclear; it may be associated with freshwater runoff from the St. Lawrence River or induced by instability in the transverse density front at the mouth of the estuary. The runoff from the Saguenay and Manicouagan rivers, although small compared to total runoff into the St. Lawrence Estuary, is found to contribute significantly to the salinity variations in the Lower estuary.

Key Words: Subtidal oscillations, Estuarine circulation, Meteorological forcing, Buoyancy forcing

Coastal and Estuarine Studies, Vol. 39
M. I. El-Sabh, N. Silverberg (Eds.)
Oceanography of a Large-Scale Estuarine System
The St. Lawrence
© Springer-Verlag New York, Inc., 1990

1. Introduction

Observations of wind-induced subtidal oscillations have been reported by Weisberg and Sturges (1976) and Weisberg (1976) for the West Passage and Providence River of Narragansett Bay, by Elliott (1978), Elliott et al. (1978), Wang and Elliott (1978), Wang (1979a,b) and Vieira (1986) for Chesapeake Bay and the Potamac Estuary; by Wong and Garvine (1984) for the Delaware Estuary; by Smith (1977) for Corpus Christi Bay, by Schroeder and Wiseman (1985) for Mobile Bay and other estuarine systems on the northern Gulf of Mexico, and by Walters (1982) for South San Francisco Bay. In these studies, the estuaries were shallow, generally less than 15 m, and the observed velocity and sea level response to atmospheric forcing was generally independent of the vertical density variation. Some significant vertical variations of the subtidal current were induced by bottom friction and sea level variation (Wang, 1979b; Vieira, 1986; Elliott, 1978; Walters, 1982). The periods of current and sea level oscillation discussed in these studies were mostly between 2 and 30 days. Seasonal variations of sea level data were discussed by Wang (1979a) and Walters (1982).

The St. Lawrence Estuary is long and deep (Fig. 1). Its length from Quebec City, where salt intrusion ends, to Pointe-des-Monts, where the coastline suddenly diverges, is about 400 km. An abrupt rise in the floor of the Laurentian Channel near Tadoussac separates the Upper estuary, with typical depths of 20-50 m, and the Lower estuary, with typical depths of 300-350 m. The bottom topography in the Upper estuary is fairly complex with a deep channel in the northern section and a shallow channel in the southern section.

The response of the St. Lawrence Estuary to atmospheric and buoyancy forcings is very different from those observed previously in shallow estuaries. This is because the estuary includes the deep Laurentian Channel, where frictional effects are small and the stratification is strong. It is expected that the response has a strong baroclinic component associated with the vertical density gradient (sections 4, 5).

Comprehensive reviews of the physical oceanography of the St. Lawrence Estuary have been provided previously by El-Sabh (1979, 1988). Here, we review only the meteorologically and buoyancy induced subtidal salinity and velocity variations in the estuary. Using results of more recent studies by Tee (1989a,b) and Mertz et al. (1988, 1989), we can provide more detailed and coherent review of the subtidal salinity and velocity variation in the St. Lawrence Estuary than those carried out previously.

There are two sets of long-term (>100 days) current meter data for the St. Lawrence Estuary, one near the mouth of the estuary carried out in 1979 (El-Sabh et al., 1982; Mertz. et al., 1988, 1989); and another, carried out in 1982, involved simultaneous observations in the Upper and Lower estuary (Tee, 1989a,b). These two data sets provide most of the information about the observed meteorologically and buoyancy induced subtidal salinity

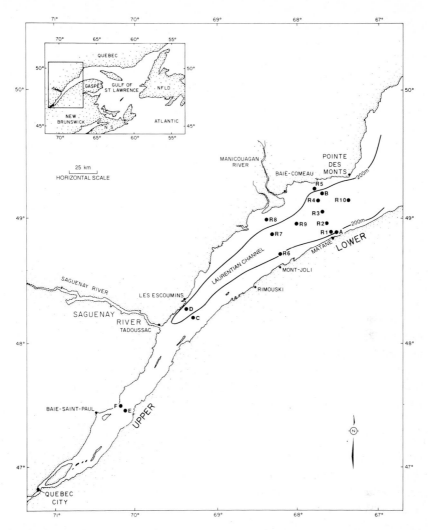

Figure 1. The St. Lawrence Estuary. Stations A - F: 1982 observations (Tee, 1989a, b); stations R1 - R10: 1979 observation (El-Sabh et al., 1982).

and velocity variations in the St. Lawrence Estuary. The following section describes briefly the field observations and data analysis of the two sets of measurements.

2. Field Observation and Data Analysis

In the 1982 program, six current meter stations were located along the St. Lawrence Estuary from April to September: stations A and B near Pointe-des-Monts and stations C and D near Les-Escoumins in the Lower estuary, and stations E and F near Baie Saint-Paul in the Upper estuary (Fig. 1). There were two Aanderaa RCM-5 current meters at all the stations except station C, where there were four. At the stations in the Lower estuary (A-D) the devices were moored at 20 and 30 meters from the surface. Another two at station C were moored at 40 and 50 m. In the Upper estuary, the current meters were moored at 14 and 19 m at station E and 16 and 21 m at station F. All the current meters functioned properly, except those at station F where the lower current meter was lost, and the upper current meter contained a short record (62 days) which is not used in the following discussion.

The data were sampled at 30 minute intervals. The currents were resolved into along-channel and cross-channel components. The along-channel axis is directed from the head to the mouth of the estuary, and the cross-channel axis from the north to south shores. To study the subtidal oscillations, the time-series of salinity, temperature and current components were low-passed with a Cartwright filter (half power at 35 h) and subsampled at 6 h intervals. To examine low-frequency oscillations in the St. Lawrence Estuary, the low-passed time-series of salinity, velocity and forcing functions (Section 3) were fitted (least squares method) to polynomials. An example of the low-passed (solid) and low-frequency (dashed) wind forcing is shown in Figure 2. To carry out spectral analysis of the data and forcing functions, the low-passed records were divided into four blocks, with each block containing 34.5 days of data.

In the 1979 observations, ten current meter stations (R1 to R10, Fig. 1) were located between Mont-Joli and Pointe-des-Monts, near the mouth of the estuary. The measurements extended from May to September, but because of some instrument failures, the actual data record lengths varied between 7 and 132 days. The recording intervals were 30 min for the meter at R7 and 20 min for all other meters. Most of the current meters were above 30 m. Some measurements were also taken at intermediate and deeper levels along the Matane line (across the estuary). Here, only the results from the upper layer, where the effects of meteorological and buoyancy forcings on the circulation were analyzed, are reviewed. To examine subtidal variations, the data were first smoothed using a symmetric low-pass filter of the Lanczos Taper type to remove high frequency fluctuations, and then the hourly smoothed series were filtered with a 39-weight Doodson's X0 filter.

3. The meteorological and buoyancy forcing

The meteorological forcing in the St. Lawrence Estuary has subtidal variations with periods of 10 to 15 and 40 to 50 days (El-Sabh et al., 1982; Mertz et al., 1988; Tee, 1989a,b). An

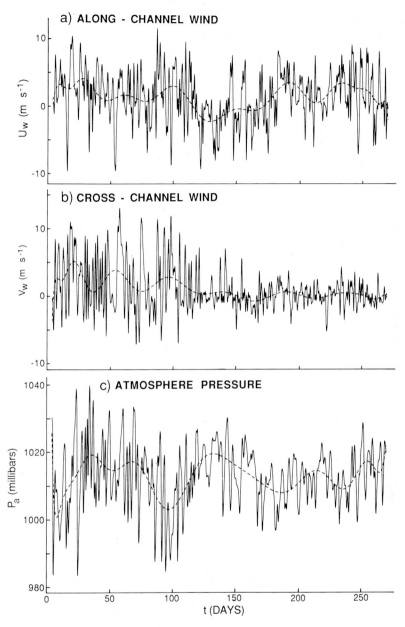

Figure 2. The (a) along-channel (u_w), (b) cross-channel (v_w) component of the wind velocity, and (c) the atmospheric pressure (P_a) at Mont-Joli. Solid: low-passed data; dashed: low frequency data obtained by fitting (least squares method) the low-passed data to a 20th order polynomial.

example of the variation is shown in Figure 2, which includes the low-passed data (solid) and the long-period variation (dashed) of the atmospheric pressure and the along- and across-channel winds at Mont Joli from January to September (days 0 to 270), 1982. Figure 2 also indicates a long-term trend which has downstream winds for 170 days < t < 248 days and upstream winds for 117 days < t < 170 days. During the period of observation (>110 days), the cross-channel component of the wind is much smaller than the along-channel component, especially at low frequency. In the following, only the along-channel wind is used.

There are two major buoyancy forcings in the St. Lawrence Estuary. The first is the freshwater runoff from the St. Lawrence, Saguenay and Manicouagan rivers. The second is ice formatioh and melting. Using a numerical model described in section 4.1., it was found (Tee, 1989b) that the contribution of the second forcing on the salinity variation in the St. Lawrence Estuary was quite small, less than 5% of the total salinity variation in most of the area, except near Quebec City where it contributed to a maximum of about 18%. In the following, the effect of the second forcing on the salinity variations in the Upper and Lower estuaries will not be discussed.

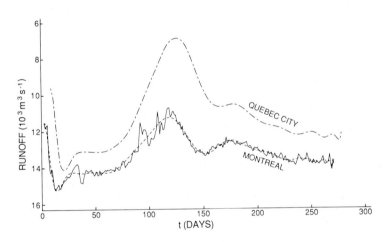

Figure 3. Freshwater runoff through Montreal (about 190 km upstream of Quebec City) and Quebec City. Solid: daily runoff through Montreal; dashed: low frequency runoff through Montreal obtained by fitting (lease square method) the daily data to a 20th order polynomial; dash dotted: low frequency runoff through Quebec City.

The freshwater runoff into the St. Lawrence Estuary is contributed mainly from the St. Lawrence River (78-80%, Tee and Lim, 1987; El-Sabh, 1988). The contributions from the Saguenay River and the Manicouagan River System are, respectively, about 10 to 13%, and 9 to 10%. The runoff from the St. Lawrence River is characterized by seasonal peaks in spring and fall, while that from the Saguenay has a series of short pulses (10-20 days) between May and July (Tee, 1989a). The runoff from the Manicouagan River System, because it is highly regulated for hydroelectric power, is more or less uniform throughout the year. Figure 3 shows an example of the 1982 freshwater runoff through Montreal (low-passed, solid curve). The dashed-curve in the figure is the low-frequency curve obtained by fitting (least squares method) the low-passed data to a 20th-order polynomial. The runoff past Quebec City (dash dotted, Fig. 3) is estimated from the mean ratio (1960-1972) of monthly discharges at Quebec City to those at Montreal. A phase lag of 7 days is applied for the runoff to arrive at Quebec City from Montreal (F. Jordan, Bedford Institute of Oceanography, Canada, Private Communication). In the numerical simulation of the salinity variation in the St. Lawrence Estuary (section 4.1.), only the runoff at Quebec City is used.

4. The meteorologically induced variations

The importance of meteorological forcing to the circulation in the St. Lawrence Estuary has been demonstrated by Murty and El-Sabh (1977), Ingram (1979), El-Sabh et al. (1982), Therriault and Levasseur (1986) and Mertz et al. (1988). From a 4-day drogue-track near Rimouski, Murty and El-Sabh observed a northward transverse current after the passage of cyclonic weather systems. They attributed the development of the current to the adjustment of pressure and velocity fields set up previously by wind forcing. From current meter moorings near the mouth of the Saguenay Fjord in the spring of 1973 and the winter of 1974, Ingram (1979) found the average winter surface current rotated counter-clockwise by 90 degrees with respect to the spring value. He attributed this change of current direction to the change in meteorological forcing. Average cross-channel winds during the spring 1973 observations were zero, whereas the equivalent value for the winter 1974 observations was 5 km/hr, directed toward the south shore. From hydrographic data taken on a grid of 29 stations in the Lower St. Lawrence Estuary at approximately monthly intervals on 16 occasions during 1979-1980, Therriault and Levasseur found that a relationship between freshwater runoff and surface salinity was absent during the fall of the surveyed years. They proposed that the increase of surface salinity in the fall is due to wind mixing between surface and lower layers. A more detailed examination of the meteorologically-forced salinity and velocity variations, using 1979 and 1982 measurements, is described in the following subsections.

4.1. The 10-15 day oscillation

Figure 4 shows the correlation between the along-channel wind stress at Baie-Comeau and the along-channel current at station R8 during the 1979 observation program (El-Sabh et al. 1982). An approximately 12-day oscillation, meteorologically forced, is indicated by two significant correlations at time lags of 2.4 and 14 days. Recent analysis of current meter data at station R10 by Mertz et al. (1988) showed a similar result (wind leads the current by 2 or 3 days).

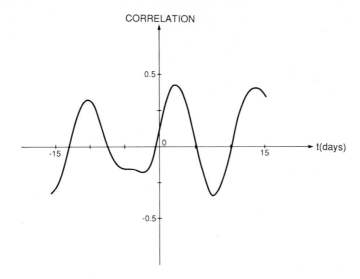

Figure 4. Cross-correlation function between the along-channel wind stress and the along-channel velocity at 12 m at station R8. A positive time lag indicates the advance of wind stress relative to current. The 95% significance level for the correlation function is $\pm\, 0.153$ (El-Sabh et al., 1982).

The 1982 current meter data revealed some characteristics of the 11-12 day oscillation. Figure 5 shows the coherence and phase of the salinity (s) and velocity (u, v) at stations A - D in the Lower estuary with the along-channel wind velocity (u_w) at Mont-Joli. The horizontal axis on this figure is the distance measured anticlockwise from station B. The locations of the stations on the axis are indicated by the names of the stations (A, B, C, D). From Figure 5a, we can see that the coherences of s, u and v with u_w are high, with most of them significant at the 95% confidence level. The only exception is the small cross-channel current at station A, which does not have significant coherence with the wind.

Figure 5b shows that the phase of s with respect to u_w increases with the distance from station B. Because the mean horizontal salinity gradient in the Lower estuary is small (Tee and Lim, 1987), the increase/decrease of the salinity indicates upward/downward displacement of the halocline. For the two-layer system characteristic of the area, this vertical displacement can be approximated by the vertical motion of the interface. The increase of the phase lag with distance from stations B to A in Figure 5b indicates anticlockwise propagation of an internal gravity wave. From the distances and time lags between the stations, the phase speed (c_p) of the wave can be computed. The result is that $c_p = 0.83$ m s^{-1} between stations B and D, in the northern section, and 0.68 m s^{-1} between stations C and A in the southern section.

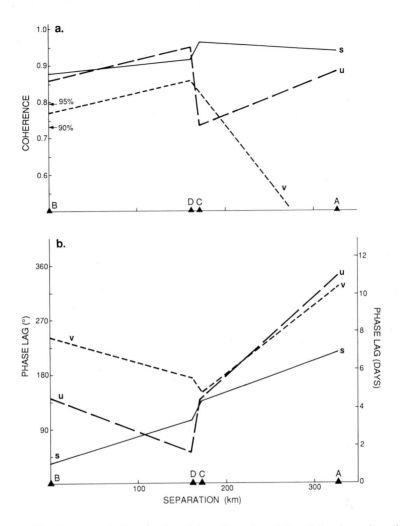

Figure 5. The (a) coherence and (b) phase lag of the salinity (s, solid), and the along-channel (u, long-dashed) and cross-channel (v, short-dashed) velocities with the along-channel wind velocity at an 11.5 day period. The 90% and 95% confident levels are indicated on the vertical axis. The horizontal axis is the distance measured anticlockwise from station B. The locations of the stations on the axis are indicated by the names of the stations (A, B, C, D). The phase lags are shown in terms of degrees and days.

The observed phase speed of the internal gravity wave is much higher than the mean currents in the Lower St. Lawrence Estuary, which have values generally between 0.1 and 0.2 m s⁻¹. It is also much higher than the phase speed of a baroclinic shelf wave which, for a wave period of 11.5 days, is about 0.1 m s⁻¹ (Lie and El-Sabh 1983). The phase speed of an internal gravity wave for a two-layer model is $c_p = (g'H_1H_2/H)^{1/2}$, where $g'=gp'/p_0$ is the reduced gravity, H_1 and H_2 are the depths of the upper and lower layers,

H is the total depth, p_0 is the average density of the two layers, and p' is the difference in density between the upper and lower layers. By taking typical values of $p'/p_0 = 2.271 \times 10^{-3}$, $H_1 = 30$ m (depth of the halocline) and $H_2 = 300$ m, we obtain $c_p = 0.78$ m s^{-1}, which agrees very well with the observed phase speed in the northern and southern sections. The observed anticlockwise propagation with the shallow water to the right is consistent with the characteristics of an internal Kelvin wave. The predominately along-channel flow in the observed velocity is also consistent with a Kelvin wave hypothesis.

However, the observed velocity data do not support the propagation of a free Kelvin wave detected from the salinity data. From Figure 5b, we can see that the phases of u and v at downstream locations (stations A and B) lag those at upstream locations (stations C and D), which indicates the downstream propagation along both shores. The inconsistency of wave propagation between salinity and velocity data may arise because the wave consists of both free and forced components. Also, there may be complex interaction between the internal Kelvin wave and strong frontal structure that existed at the head of the Laurentian Channel (near stations C and D) and near the mouth of the St. Lawrence Estuary (near stations A and B, Fig. 1). A two-layer baroclinic model, including both the along- and across-channel variation, would be useful for understanding the wind-forced 10-15 days oscillations in the St. Lawrence Estuary. By using Farquharson's (1966) short current meter records (two to three cycles of 10-15 day forcing), Mertz et al. (1988) suggested that the wind-induced flows are parallel to the wind near shore, but opposite to the wind offshore. The two-layer model would also be useful to verify this suggestion.

In the Upper estuary, the coherence between the wind stress and the velocity and salinity at station E is insignificant at the 11.5 day period. However, the wind is significantly coherent with the along-channel current (about 0.80 confidence level) and the salinity (about 0.90 confidence level) at a period of 17.3 days. This result indicates that the meteorologically induced oscillations behave quite differently between the shallow Upper estuary and the deep Lower estuary.

4.2. The long-period variations

As indicated in section 3, the atmospheric forcing has oscillations with a period of 40 to 50 days, and a long-term trend blowing downstream for 170 days < to < 247 days and upstream for 117 days < t < 170 days. Comparison of the low-frequency components of the along-channel wind veloctity, the atmospheric pressure, and the salinity for the (1982) upper current meters is shown in Figure 6. We can see that there are some apparent correspondences between the salinity variation and the atmospheric forcing. This result indicates that the long-period oscillation is likely to be induced by wind forcing. In the 1979 measurement, Mertz et al. (1988) found that the along-channel current at station R10 was significnatly coherent with the along-channel wind at the 40 to 80-day bands.

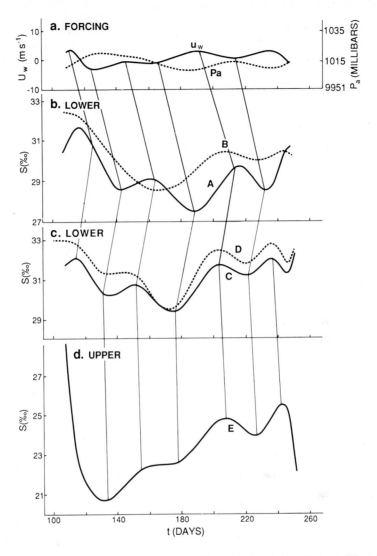

Figure 6. The low-frequency oscillations of the atmospheric forcing and the salinity at 20 m. (a) The along-channel wind velocity (u_w, solid), and the atmospheric pressure (P_a, dashed); (b) the salinity at stations A (solid) and B (dashed) in the Lower estuary; (c) the salinity at stations C (solid) and D (dashed) near the head of the Lower estuary; (d) the salinity at station E in the Upper estuary. The forcing and propagation of the 40- to 50-day oscillation is shown.

The salinity maxima or minima occurred initially at stations C and D, and then propagated upstream and downstream to stations A, B and E. From the time lags and distance between the stations, we obtain a propagation speed of about 0.17 m s^{-1} downstream, and 0.25 m s^{-1} upstream. These speeds are of the same order as the along-channel current generated by freshwater runoff and wind. The vertical variation of salinity data also indicates that the salinity maxima or minima at stations A, B, C and D occur earlier at deeper current meters, and those at stations E occur slightly later in deeper water. An example of the upward propagation of the maxima or minima at station C is shown in Figure 7.

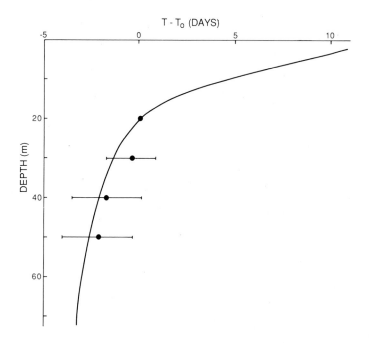

Figure 7. The relative time lags (T-T_0) of the salinity variation at station C. T_0 is the time lags at 20 m of the same station. Solid curve: computed data; solid dotted: observed data at station C. The result indicates the upward propagation of the salinity variation.

There is also some evidence that the long-term salinity variation at station E was forced by the along-channel wind. The increase of salinity for t >175 days corresponded to downstream wind. The increase of salinity during these periods cannot be accounted for by the freshwater runoff because the runoff at these periods remained basically uniform.

Because of short current meter records, we cannot provide an accurate description of the correlation between long-period atmospheric forcing and salinity data using spectral analysis. To confirm the observed correlation, a two-dimensional cross-sectional averaged numerical model, used by Tee and Lim (1987) to study the freshwater pulse in the St.

Lawrence Estuary, has been modified to include wind forcing and ice formation and melting. Because of the simplicity of a 2-D numerical model, the freshwater runoff includes only that through the head of the estuary. A steady state estuarine circulation for January 7, 1982 was first obtained by using the observed wind, ice and runoff on that date. The computation was then continued for nine months using the long-period forcings described in section 3. Detailed descriptions of the numerical model and the computed results are given in Tee and Lim (1987) and Tee (1989b).

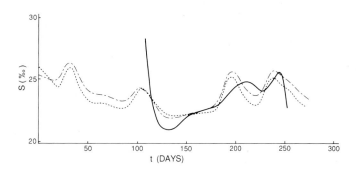

Figure 8. Low frequency variation of salinity at 14 m at station E in the Upper estuary. Solid: observed data; dash dotted: computed data with all forcings; dotted: computed data with wind-forcing only. The result indicates that wind is the dominant forcing for low-frequency variation of salinty at the station.

Figure 8 shows the comparison between observed and computed salinity at 14 m at station E in the Upper estuary. By comparing the solid (observed S) and dashed (computed S) curves, we can see that the observed long-period salinity variation at station E can be reproduced reasonably well. To examine the importance of each forcing, a computation was carried out by varying the wind-forcing, but keeping the ice and runoff at their January 7 values. The result for salinity at 14 m at station E is shown in Figure 8 as a dotted curve. By comparing the dash-dotted (all forcings) and dotted (wind forcing only) curves, we can see that the observed long-period salinity variation is induced mainly by wind forcing.

The comparison between observed and computed salinity variations at stations A, B, C and D in the Lower estuary has also been carried out. It was found (Tee, 1989b) that although the salinity variation was reproduced qualitatively, the amplitudes of the modeled oscillation were substantially smaller than that detected. This may be because the freshwater runoff from the Saguenay and Manicougan rivers are not included in the numerical model. The absence of these additional runoff inputs in springtime results in a small vertical salinity gradient near the surface in the Lower estuary, which in turn produces a small salinity variation from wind-induced vertical displacement of halocline. The importance of the

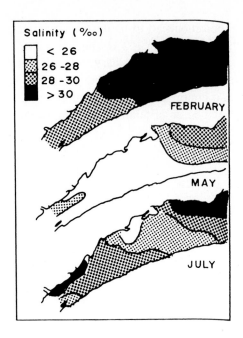

Figure 9. Spatial patterns of salinity variation in the photic zone of the Lower St. Lawrence Estuary for the months of February, May and July of 1980. The result indicates the significance of the Manicouagan River outflow to the seasonal variation of salinity in the estuary (Therriault and Levasseur, 1985).

Manicougan River to the salinity near the surface in the Lower estuary has been shown (Fig. 9) by Therriault and Levasseur (1985).

Although the amplitude of the salinity variation in the Lower estuary cannot be simulated quantitatively, the propagation of the salinity maxima or minima shown in Figure 6 is reproduced very well. Figure 10 shows the computed salinity variation at current meter stations A, B, C, D and E. The upper curve is for stations A and B, the middle curve for stations C and D, and the lower curve for station E. By comparing Figs. 6 and 10, we can see that the occurrences of the observed salinity maxima and minina (Fig. 6) which are earlier at stations C and D, and later at stations A, B and E are simulated well by the numerical model (Fig. 10). The numerical model also reproduces the upward propagation of the observed salinity maxima and minima at stations A, B, C and D in the Lower estuary. An example of the good comparison at station C is shown in Figure 7.

5. The buoyancy induced variations

Freshwater runoff in the St. Lawrence Estuary has been identified as one of the important factors controlling phytoplankton production in the Lower Estuary (Therriault and

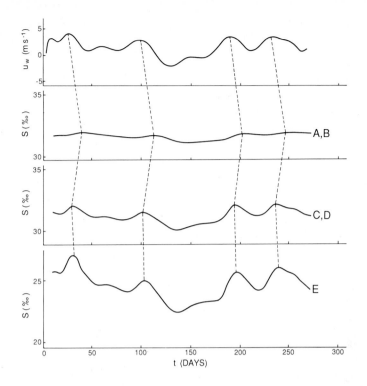

Figure 10. The computed low-frequency salinity variation at 20 m. (a) The along-channel wind velocity; (b) the salinity at stations A and B; (c) the salinity at stations C and D; and (d) the salinity at stations E and F. The forcing and propagation of the 40-to-50 day oscillation is illustrated. The result agrees with the observation (compare to Fig. 6).

Levasseur, 1985, 1986). Its inter-annual variability has an impact on fishery production in the Gulf of St. Lawrence (Sinclair et al., 1986).

The freshwater pulse, characterized by a salinity minimum, can be induced by seasonal variation of freshwater runoff. Using a 2-D cross-sectional averaged numerical model and a mean (1950-1976) seasonal runoff through the head of the estuary, Tee and Lim (1987) simulated the distribution, propagation and dissipation of the seasonal pulse. The result showed that the pulse was formed initially at two surface locations, and then propagated generally toward the ocean and deeper water. The only exception was that the arrival time for the deep water pulse (25 m or deeper) increased upstream toward the slope region (head of the Laurentian Channel). The amplitude of the pulse generally decreased in the direction of pulse's propagation. The only exeption is that, in the far upstream region, the amplitude increased from surface to deep water. Dynamics associated with the distribution, formation and propagation of the pulse were discussed. The simulation of the freshwater pulse for the 1982 runoff (Fig. 3) has also been carried out (Tee, 1989b). The results are similar to those computed previously (Tee and Lim, 1987).

As indicated in section 4.2., because of the absence of freshwater runoff from the Saguenay and Manicouagan rivers in the numerical model, the observed salinity variations in the Lower estuary, which include the signals of seasonal pulse, are significantly stronger than those computed numerically. To improve the simulation of the freshwater pulse, future modelling can be carried out by separating the St. Lawrence estuarine system into three branches: the main branch of the St. Lawrence Estuary, and the other branches of the Saguenay and Manicouagan rivers. The salinity and velocity in each branch can be inter-related through appropriate boundary conditions at the sections joining the branches.

The 2-D numerical model cannot simulate some observed cross-channel salinity and velocity variations in the Lower estuary. During the 1979 observation period, El-Sabh et al. (1982) found that the near-surface circulation system in the observational area (between Mont-Joli and Pointe-des-Monts in the Lower estuary (Fig. 1) could be classified into three types of patterns (Fig. 11): (1) an anticyclonic eddy centered between the Estuary-Gulf boundary and the Baie-Comeau cross-section (region I) accompanied by a cyclonic eddy located to the west between the two cross-sections at Baie-Comeau and Rimouski (region II), Fig, 11a, (2) a cyclonic eddy in region I with an anticyclonic eddy in region II (Fig. 11c), and (3) the transition phase between phases 1 and 2 (Fig. 11b). The intensity of these eddies was related to sea level changes in the estuary, which in turn correlated with atmospheric pressure. The occurrence of these eddies repeated in about 80 days. Using a two-layer unforced-wave model, Lie and El-Sabh (1983) suggested that the observed eddies were formed by the superposition of two fourth order baroclinic shelf waves propagating in opposite directions. Recently, Mertz et al. (1989) proposed that the change of eddy structure in the Lower estuary was caused by unstable wave activities in a transverse density front at the mouth of the estuary.

In the 1982 observations, the low-frequency variation of the along-channel velocity at stations A and B in the Lower estuary showed strong variation across the section. Figure 12 shows that the along-channel flow at station A in the southern section [u(A)] correlated inversely with that at station B in the northern section [u(B)]. These strong opposite flows on either side of the estuary are consistent with the eddy motion near the mouth of the estuary observed in 1979. The period between the maxima in u(B), or the minima in u(A) is about 73 days, consistent with approximately 80 days observed in 1979. Figure 12 shows the comparison between the runoff (R) at Montreal, the salinity at station B[S(B)], and the along-channel velocities at stations A and B. The time origin of R is increased by 50 days so that maximum R at day 114 coincides with minimum salinity at day 164. We can see from the figure that the low-frequency variations of R (short dashes) are well correlated with those of the salinity (solid), and the along-channel velocities. Maximum

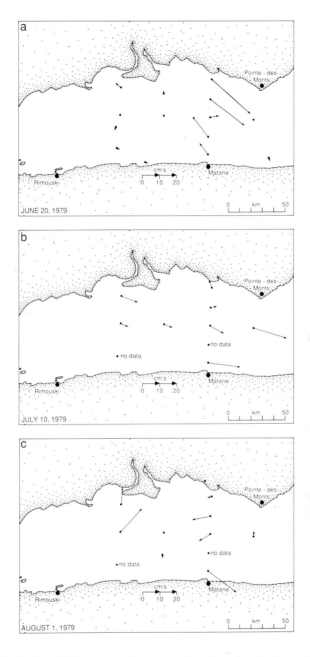

Figure 11. Spatial distribution of daily mean residual current in the Lower St. Lawrence Estuary on June 20, July 10 and August 1, 1979. The result indicates strong eddy motion in the area (El-Sabh, 1988).

runoff at days 164 and 237 correspond to minimum salinity at station B, along-channel velocity at station A, and to maximum along-channel velocity at station B. The opposite correspondence can be seen for the minimum runoff at day 210 (Fig. 3). This result

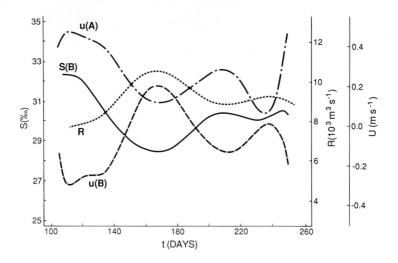

Figure 12. The correspondence between the freshwater input at Montreal (R), the along-channel currents at 20 m at stations A and B [u(A), u(B)], and the salinity at 20 m at station B[s(B)]. The time origin of R is increased by 50 days so that the maximum R at day 114 coincides with the minimum salinity at day 164.

strongly suggests that the 70-80 day oscillations in the salinity and velocity in the Lower estuary are forced by seasonal variation of the freshwater inputs into the St. Lawrence Estuary. Koutitonsky et al. (1985) has suggested that the observed 80-day oscillation in 1979 was also induced by buoyancy forcing. This buoyancy forcing of low-frequency velocity variation does not agree with previously suggested forcings by meteorological parameters (El-Sabh, 1982) and by instability at the transverse density front at the mouth of the estuary (Mertz et al., 1989). A three-dimensional model, forced by seasonal runoff and wind stress, is probably required to understand the forcing mechanism for the observed long-period (70-80 days) salinity and velocity variations in the estuary.

The observed salinity data in the St. Lawrence Estuary also show short-period pulses with time scale of 10 to 15 days (Tee, 1989a). However, these pulses were found to be induced mainly by wind-forcing. The pulses occur only during the high runoff season, which indicates that the runoff produces an indirect forcing by providing a source of low salinity water near the surface.

6. Summary and Conclusions

Meteorologically and buoyancy induced subtidal salinity and velocity variations in the St. Lawrence Estuary are reviewed using 1979 and 1982 current meter data and results of 2-D numerical models. The meteorological forcing shows oscillations with periods of 10-15 and 40-50 days, and a long-term trend with a time scale of about 130 days. From salinity data, it was found that the 10-15 day forcing induced an interfacial oscillation which propagated anticlockwise in the Lower estuary as a free internal Kelvin wave. However, the velocity data provide a different picture of wave propagation. A two-layer model including cross-channel variation would be useful to examine dynamics of this 10-15 day oscillation.

The 40-50 day forcing produces an oscillation which occurs initially near the transition region between the Upper and Lower estuaries, and then propagates toward upstream and dowstream regions. In the Lower estuary, the propagation is found to be toward the surface. These results are confirmed by a 2-D numerical model. The model also confirms that the long-period salinity variation in the Upper estuary observed in 1982 is induced mainly by wind-forcing.

From results of a 2-D numerical model, it was found that the forcing of the salinity variation in the St. Lawrence Estuary due to ice formation is small (<5%) in most of the area, except near the head of the estuary where it can contribute 10-20%. The freshwater runoff from the St. Lawrence River produces a seasonal pulse in the estuary. The formation, distribution and propagation of the pulse are reviewed using the results of a 2-D numerical model. Near the mouth of the estuary, large eddy motions with a time scale of about 80 days were observed. Although the forcing of these eddies was attributed to atmospheric events or instabilities in the transverse density front at the mouth of the estuary, there is strong evidence that the eddies may be associated with freshwater runoff from St. Lawrence River. A 3-D numerical model would be useful to clarify the forcing mechanisms of this 80-day oscillation. Although the combined runoff from the Saguenay and Manicouagan rivers is only about 20% of the total runoff into the St. Lawrence Estuary, there is some evidence from both observation and numerical models that it contributes significantly to the salinity variation in the Lower estuary.

REFERENCES

Elliott, A.J. 1978. Observation of the meteorologically induced circulation in the Potomac Estuary, Estua. Coastal Mar. Sci. 6, 285-299.

Elliott, A.J., Wang, D.P., and D.W. Pritchard. 1978. The circulation near the head of Chesapeake Bay, J. Mar. Res. 36, 643-655.

El-Sabh, M.I. 1979. The lower St. Lawrence Estuary as a physical oceanography system, Naturaliste Can., 106, 53-73.

El-Sabh, M.I. 1988. Physical oceanography of the St. Lawrence Estuary, in "Hydrodynamics of Estuaries", edited by B. Kjerfve, vol. II, CRC Press, 61-78.

El-Sabh, M.I., Lie, H.J., and V.G. Koutitonsky. 1982. Variability of the near surface residual current in the lower St. Lawrence Estuary, J. Geophys. Res, 87, 9589-9600.

Farquharson, W.I. 1966. St. Lawrence estuary current surveys. Bedford Institute of Oceanography Technical Report, BIO 66-6, 84 pp.

Ingram, R.G. 1979. Water mass modification in the St. Lawrence Estuary, Naturaliste Can., 106, 45-54.

Koutitonsky, V.G., R.E. Wilson and M.I. El-Sabh. 1985. On low frequency current variability in the lower St. Lawrence Estuary, 19th Annual Congress, Canadian Meteor. Oceanogr. Society, Montreal, 51.

Lie, H.J. and M.I. El-Sabh. 1983. Formation of eddies and transverse current in a two-layer channel of variable bottom with application to the lower St. Lawrence Estuary, J. Phys. Oceanogr. , 13, 1063-1075.

Mertz, G., M.I. El-Sabh and V.G. Koutitonsky. 1988. Wind-driven motions at the mouth of the lower St. Lawrence Estuary, Atmosphere-Ocean, 26, 509-523.

Mertz, G., M.I. El-Sabh and V.G. Koutitonsky. 1989. Low frequency variability in the lower St. Lawrence Estuary, J. Mar. Res., 47, 285-302.

Murty, T.S. and M.I. El-Sabh. 1977. Transverse currents in the St. Lawrence Estuary: A theoretical treatment, in "Transport Processes in Lakes and Oceans", edited by R.J. Gibbs, 35-62.

Schroeder, W.W. and W.J. Wiseman, Jr. 1985. Low-frequency shelf-estuary exchange processes in Mobile Bay and other estuarine systems on the northern Gulf of Mexico, Proceedings of the Eighth Biennial International Estuarine Research Conference, edited by D.A. Wolfe, Published by Academic Press Inc. 355-367.

Sinclair, M., G.L. Bugden, C.L. Tang, J.C. Therriault and P.A. Yeats. 1986. Assessment of effects of freshwater runoff variability on fisheries production in coastal waters, in "The role of freshwater outflow in coastal marine ecosystems" edited by S. Skreslet, NATO ASI Series, vol. G7, 139-160.

Smith, N.P. 1977. Meteorological and tidal exchanges between Corpus Christi Bay, Texas, and the Northwestern Gulf of Mexico, Estuar. Coastal Mar. Sci. 5, 511-520.

Tee, K.T. 1989a. Subtidal salinity and velocity variations in the St. Lawrence Estuary, J. Geophys. Res., 94, 8075-8090.

Tee, K.T. 1989b. Modeling long-period salinity and velocity variations in the St. Lawrence Estuary (In preparation).

Tee, K.T. and T.H. Lim. 1987. The freshwater pulse - a numerical model with application to the St. Lawrence Estuary, J. Mar. Res., 45, 871-909.

Therriault, J.C. and M. Levasseur. 1985. Control of phytoplankton production in the lower St. Lawrence Estuary: light and freshwater runoff, Naturaliste Can., 112, 77-96.

Therriault, J.C. and M. Levasseur. 1986. Freshwater runoff control of the spatio-temporal distribution of phytoplankton in the lower St. Lawrence Estuary (Canada), In "The role of freshwater outflow in coastal marine ecosystems" edited by S. Skreslet, NATO ASI Series, vol. G7, 251-260.

Vieira, M.E.C. 1986. The meteorologically driven circulation in mid-Chesapeake Bay. J. Mar. Res., 44, 473-493.

Walters, R.A. 1982. Low-frequency variation in sea level and currents in South San Francisco Bay, J. Phys. Oceanogr., 12, 658-668.

Wang, D.P. 1979a. Subtidal sea level variations in the Chesapeake Bay and relations to atmospheric forcing, J. Phys. Oceanogr., 9 413-421.

Wang, D.P. 1979b. Wind-driven circulation in the Chesapeake Bay, Winter 1975, J. Phys. Oceanogr., 9, 564-572.

Wang, D.P. and A.J. Elliott. 1978. Nontidal variability in the Chesapeake Bay and Potomac River: Evidence for non-local forcing, J. Phys. Oceanogr., 8, 225-232.

Weisberg, R.H. 1976. The nontidal flow in the Providence River of Narragansett Bay: A stochastic approach to estuarine circulation, J. Phys. Oceanogr., 6, 721-734.

Weisberg, R.H. and W. Sturges. 1976. Velocity observation in the West Passage of Narragansett Bay: A partially mixed estuary, J. Phys. Oceanogr. 6, 345-354.

Wong, K.C. and R.W. Garvine. 1984. Observations of wind-induced subtidal variability in the Delaware Estuary, J. Geophys. Res., 89, 589-597.

Chapter 4

Fronts and Mesoscale Features in the St. Lawrence Estuary

R. Grant Ingram[1] and Mohammed I. El-Sabh[2]

[1]Department of Meteorology, McGill University, Montreal, Québec, H3A 2K6, Canada
[2]Département d'Océanographie, Université du Québec à Rimouski, Rimouski, Québec, G5L 3Al, Canada

ABSTRACT

Mesoscale features found in the St. Lawrence estuary range from those encountered in much smaller estuaries to those occuring on continental shelves. The influence of freshwater input, wind forcing and tidal processes on the distribution of temperature, salinity and current fields are discussed. Mechanisms responsible for variability of frontal phenomena, eddy formation and coastal jet dynamics are considered as a function of seasonal changes in the forcing variables. Comparisons of features in the St. Lawrence and in similar regions elsewhere are presented

Key Words: St.Lawrence, estuary, river, tide, wind, front, eddy, jet, coastal, Gaspe Current, circulation, mixing.

l. Introduction

The St. Lawrence River is one of North America's major rivers, channeling waters from the Great Lakes drainage basin and downstream tributaries into the Gulf of St. Lawrence and finally into the northwest region of the Atlantic Ocean. Although tidal influences are felt as far upstream as Montreal, salt water intrusion is generally limited to the area downstream of Quebec City. The region of interest stretches from the upstream limit of salt intrusion to the mouth of the estuary, assumed to be in the area of Pointe des Monts (Fig. 1). This area can be subdivided on the basis of bathymetry and hydrographical features as follows:- (i) the Upper St. Lawrence, with large tides and low salinities; (ii) downstream of Ile aux Coudres to the confluence of the St. Lawrence and the Saguenay Rivers, characterized by intermediate salinities and variable topography ranging from shallow plateaus to deep channels (over 100 m deep); and (iii) the Lower Estuary (sometimes referred to as the Maritime Estuary) stretching from the Saguenay to Pointe des Monts. The major bathymetric feature of the Lower Estuary is the Laurentian Channel with depths

Coastal and Estuarine Studies, Vol. 39
M. I. El-Sabh, N. Silverberg (Eds.)
Oceanography of a Large-Scale Estuarine System
The St. Lawrence
© Springer-Verlag New York, Inc., 1990

exceeding 300 m. Advection of continental shelf and slope waters from the northwest Atlantic occurs within the channel as far inland as the mouth of the Saguenay.

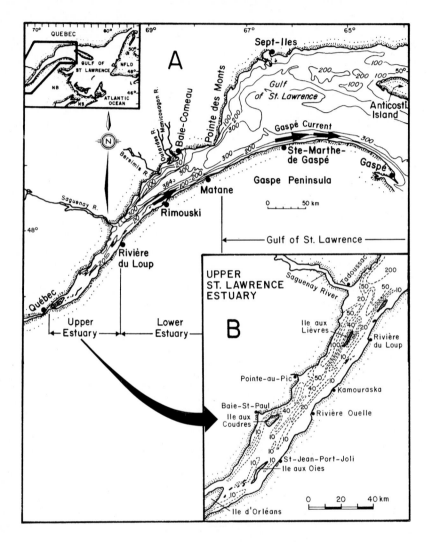

Figure 1. Place map showing area of interest, major freshwater sources and bathymetry in meters.

Salinities range from near zero at Quebec to over 34 g/kg in the deep waters of the Laurentian Channel. The surface and along channel salinity distribution for mid-summer conditions is shown in Figure 2. The manner in which fresh and sea water mix varies greatly in different estuaries. The extremes range from an offshore plume with no salt water intrusion to broad rivers where oceanic waters advance far inland (riverine). The ratio of the mean downstream freshwater velocity to the root mean square tidal velocity approximately determines the degree to which either of the above modes occur (Garvine, 1975). From the data shown above, the St. Lawrence is in the riverine mode.

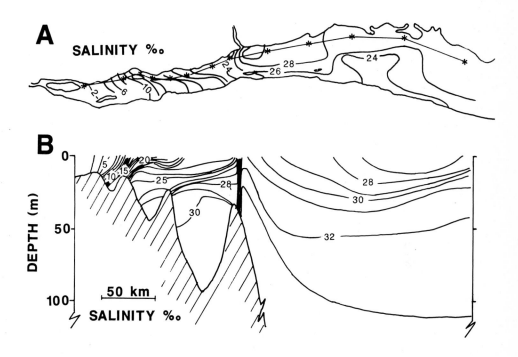

Figure 2. Surface (A) and along-channel (B) salinity distribution in late June 1975, recorded over a six day period (adapted from Kranck, 1979; Greisman and Ingram, 1977).

The transition from fresh to oceanic waters within a river occurs both gradually or rapidly depending on the region. Bottom topography plays the major role in determining the location of sharp gradients. In the St. Lawrence, strong salinity gradients are observed in the shoaling region near Ile aux Coudres and near the head of the Laurentian Channel. In support of the observational evidence, both Pingree & Griffiths (1980) and El-Sabh & Murty (1990) predicted the presence of fronts in these areas using a two-dimensional numerical model of the M_2 tides to separate well mixed and stratified areas. At the head of the Laurentian Channel, semidiurnal upwelling of subsurface water generates large spatial gradients of density (Greisman and Ingram, 1977; Forrester, 1974; Ingram, 1975; 1979 and 1983) and input of nutrients to the euphotic zone (Steven, 1974; Therriault and Lacroix, 1976).

The major sources of freshwater, in order of magnitude, are the St. Lawrence River, Saguenay Fjord and the Manicouagan Complex (regulated for hydroelectric purposes). The temperature of the river sources varies seasonally from 0 °C to over 20 °C. Discharge also changes over the annual cycle with a maximum in mid-spring. Mean annual discharge of the St. Lawrence near Quebec is of the order of 10^4 m^3/s, amounting to about 80% of the total freshwater input to the St. Lawrence estuary (El-Sabh, 1988). In regard to mesoscale features, fresh water sources can be assumed to vary slowly in time and space for periods less than fortnightly. Saline waters in the estuary originate from the Gulf of St. Lawrence, where T-S characteristics for salinities greater than 33 g/kg are relatively constant over the year. For near surface waters, salinities in the gulf range down to 26 g/kg, while temperatures vary from -1.7 °C to 20 °C (Forrester, 1964; El-Sabh, 1979). Similar to freshwater sources, we assume T-S properties of the sub-surface gulf waters change slowly (excludes seasonality) for the phenomena of interest. Because of high amplitude internal tides in the Laurentian Channel (Forrester, 1974), diurnal and semi-diurnal temporal variability of T and S are observed at a given location.

The aim of the present paper is to examine both frontal and mesoscale features in the St. Lawrence estuary. The phenomena of interest persist for periods from a few hours to several days or longer.

2. Upper Estuary

2.1. Salt intrusion region -Ile d'Orleans to Ile aux Coudres

The upstream limit of salt intrusion for weak runoff and spring tides occurs in the Quebec City region during winter. At other times of the year, the upstream limit is found downstream of Ile d'Orléans. This area is characterized by shallow depths (typically 10 m) and larger surface tides than adjacent regions. This part of the estuary is the least well known from an oceanographic point of view. Between Ile d'Orléans and Ile aux Coudres, the salinity and circulation patterns indicate the presence of a strong salinity gradient. Aided by the null bottom velocity zone associated with the wedge, turbidity values are much higher here (turbidity maximum zone) than either up or downstream (d'Anglejan, 1981; Kranck, 1979; Silverberg and Sundby, 1979). Similarly, the area serves as "trap" for larval fish and other properties, as residence times are much larger in this region (Dodson et al., 1989). Although Pingree and Griffiths (1980) and Silverberg and Sundby (1979) indicate this entire region to be well mixed, Kranck (1979) and Greisman and Ingram (1977) show significant vertical stratification in deeper areas and cross-channel density gradients.

2.2. Ile-aux-Coudres to Saguenay

In this region, two deep (> 50 and > 150 m) basins separated by a shallow sill (< 40 m) characterize the northern side of the estuary (Fig. 1b). On the southern side, a shallow channel (~20 m) and broad plateaus (< 10 m) are found. The complex bathymetry and variable water mass characteristics lead to the formation of front-like features in at least three regions: the Pointe-au-Pic sill, along the middle of the channel downstream of Ile-aux-Lievres and offshore of some shallow bays on the southern coast. We define a front as an area of rapid spatial change in density, often associated with convergent surface flow.

The shallow South Channel is an area of considerable tidal mixing, while in the deeper North Channel, surface and internal tides propagate freely up to the sill facing Pointe-au-Pic (Muir, 1980). A front is often found along the mid-channel boundary between the two channels. El-Sabh & Murty (1990) predict frontogenesis in this area. Mertz et al., (1988b) have also found evidence of a distinct boundary separating the North and South Channel using satellite data.

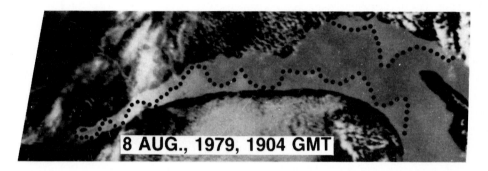

Figure 3. Infrared satellite image over the St. Lawrence estuary on August 8, 1979 showing boundary between two distinct (temperature) water masses (adapted from Mertz et al., 1988b).

Upstream of this region, a skewed frontal feature is found near Pointe-au-Pic, which is further downstream on the southern shore than to the north. Mertz et al. (1988b) show this as a continuous feature which joins the mid-channel downstream front (Fig. 3). D'Anglejan (1981) found that the front separates water masses of distinctly different turbidities. Local frontogenesis associated with differential bottom topography also occurs on the outer limit of Baie Ste. Anne and other bays along the south coast (d'Anglejan et al. 1981).

2.3. Mouth of the Saguenay Fjord

The presence of two distinct water masses (St. Lawrence and Saguenay), rapid depth changes (Fig. 1) and tidally induced upwelling lead to a multiplicity of fronts in this area (Fig. 4 and Ingram, 1976, 1985). Ingram (1976) observed periodic (semidiurnal) generation of a convergent surface front at high tide on the Ile Rouge bank. The Bank is a sandy 20-35 m deep feature downstream of a small island separating the Laurentian Channel and the 60-100 m deep South Channel. Changes of 5 σ_t units and 100 cm/s in velocity perpendicular to the frontal boundary were observed across the 2 m wide surface transition zone. Density contrast is amplified by the presence of tidally induced upwelling of sub-surface waters at the head of the Laurentian Channel at or near each high tide.

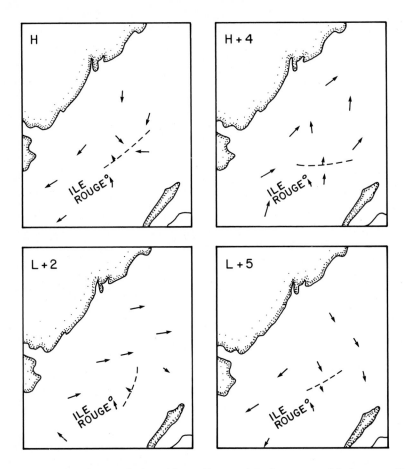

Figure 4. Frontal locations (broken line) near the confluence of the Saguenay and St. Lawrence Rivers as a function of tidal phase (hours after high (H) and low (L) tide) over the semi-diurnal period. Arrows indicate local surface flow (adapted from Ingram, 1985).

Features of the upwelling have been described by Hachey et al. (1956), Ingram (1975) and Reid (1977). Flotsam accumulation in the convergence zone and the presence of numerous sea birds made frontal passage a visually striking event. The longitudinal (parallel to the estuarine axis) front formed on the northern boundary of the Ile Rouge Bank is similar in many respects to that described by Huzzey and Brubaker (1988) in the York River. Although these fronts were usually generated once every semi-diurnal tide cycle, doublets sometimes occurred (Ingram, 1985). The timing and depth of the lighter water intrusion over this Bank suggests that the simple or double fronts may result from the shoaling of internal solitons (Ingram, 1978) generated in the South Channel. In this hypothesis, the internal density interface associated with one or two solitons breaks the surface as the waves propagate up the slope and onto the shallow bank. Wave dissipation would account for the intermittency of the second front.

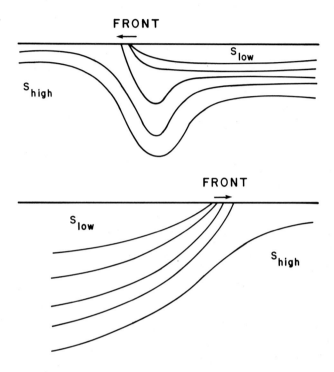

Figure 5. Proposed cross-sectional isohaline (S) structure for two frontal types observed in the Ile Rouge area (adapted from Ingram, 1985).

Ingram (1985) also found three other distinct periods within the semidiurnal tide cycle suitable for the generation of fronts in the estuarine area near the mouth of the Saguenay River. During the ebb tide, a distinct plume was formed immediately offshore of the Saguenay, with front-like features on the lateral boundary with an opposing cross-flow along the St. Lawrence estuary. Fronts similar to these were observed by Garvine and Monk (1974) at the mouth of the Connecticut River. The fronts observed in the St. Lawrence were of two types (Fig. 5): a shallow well defined intrusion of less dense water or a deeper more diffuse transition to less dense water. A general discussion of coastal and estuarine fronts can be found in Bowman and Esaias (1977). During breakup of the frontal features, strong horizontal shears lead to the generation of small scale (50-500 m) eddies of lighter (in a density sense) and more turbid water which were often advected downstream with the mean flow into the Lower Estuary. Denman and Platt (1975) have noted a high degree of patchiness in temperature and chlorophyll values in the area immediately downstream of the upwelling zone, which may have resulted from the passage of these small scale eddies.

3. Lower Estuary

Because the Lower St.Lawrence estuary (LSLE) has a width (20-50 km) considerably greater than its internal Rossby radius ($\lambda_R \cong 10$ km, Lie and El-Sabh, 1983; Mertz et al., 1988b), a more complicated circulation can be expected. According to Garvine (1986) large-scale estuaries such as the LSLE are strongly influenced by Coriolis effects. Thus, mesoscale phenomena such as coastal jets, internal Kelvin waves, baroclinic eddies and unstable waves are possible. Indeed, recent satellite thermal observations, direct current measurements and water mass analyses have demonstrated the complex character and strong spatial and temporal variability of the mesoscale features in the Lower Estuary. As seen in Figure 6, these features include the presence of a cold water mass centered near the head of the Laurentian Channel produced by intense vertical mixing and internal tides described earlier; several lateral density fronts, including one near Pointe des Monts and another in the Ile du Bic area; and a series of cold and warm cores, of about 50 km in diameter, occupying most of the central region.

Close examination of extensive satellite imagery covering several years (El-Sabh, 1985; Lavoie et al., 1985; Lacroix, 1987; Gratton et al, 1988 and Proulx, 1989) shows that the appearance of cold water, with a minimum temperature between 2° and 7 °C, at the head of the Laurentian Channel, is a common feature. These studies also showed that the cold water can occur episodically up to 100 km downstream of this area. Therriault and Lacroix (1976), Ingram (1983, 1985), Lacroix (1987) and Proulx (1989) found modulation of the cold water intensity with the neap-spring tidal cycle near the head of the channel. Furthermore, satellite imagery shows that the cold water is not always confined in this area but may occur in two different configurations: one adjacent to the south shore, the other centered over the mid-channel region and sometimes occupying virtually the entire upstream half of the maritime estuary (Lacroix, 1987; Gratton et al., 1988, Proulx, 1989). Wind mixing, internal tides and advection by the mean outflow from both the Saguenay River and upper estuary were found to be the most significant physical factors contributing to variations of the cold water zone.

Near the estuary-gulf boundary, Tang (1980) found a persistent quasi-permanent cold front during summer 1978. He suggested that this frontal system, referred to here as the Pointe-des-Mont cold front, is formed by the low salinity estuarine water encouering the more saline and denser water of the northwestern Gulf of St.Lawrence. El-Sabh (1985), Lavoie et al. (1985), Lacroix (1987) and Proulx (1989) confirmed the presence of the Pointe-des-Monts cold front (with a minimum temperature of 2° to 4°C) using infrared satellite images taken over several years (Figs. 6 and 7). The density difference of the two water masses creates a horizontal density gradient that drives an anticyclonic along-front geostrophic flow

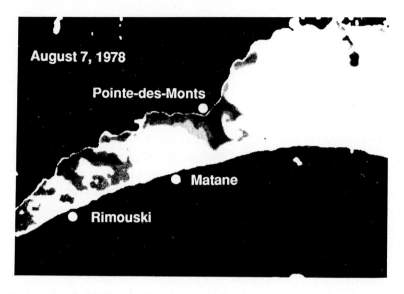

Figure 6. HCMM satellite thermal infrared imagery of the lower St. Lawrence taken on August 7, 1978. Lighter shades represent warmer temperatures (from Lavoie et al., 1985).

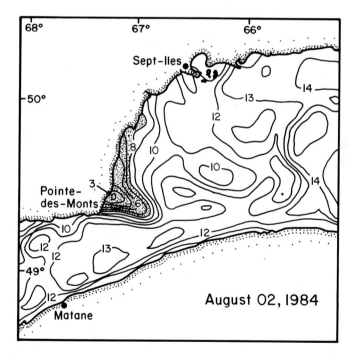

Figure 7. Sea surface temperature in °C reproduced from a NOAA-7 thermal image observed on 2 August, 1984 near the mouth of the LSLE. Shaded areas represent surface temperatures lower than 8°C. (modified from Lacroix, 1987).

around the mouth. This flow carries the estuarine water downstream and upon reaching the south shore it turns eastward and joins the strong buoyancy-driven jet moving seaward along the south shore of the Lower Estuary, to form the Gaspe Current. This current, with surface speeds of up to 1 m/s, flows along the Gaspe Peninsula and is considered the most dominant feature of the circulation in the Gulf (El-Sabh, 1976; Tang, 1980a; Dickie and Trites, 1983; El-Sabh and Benoit, 1984; Benoit et al., 1985).

Time series of satellite thermal imageries taken in 1978, 1979, 1984 and 1985 (Lacroix, 1987; Mertz et al., 1989b, 1989; Proulx, 1989) suggest that the cold Pointe-des-Monts front may advance into the Gulf of St.Lawrence, regress into the estuary at other times, or disappear altogether. Lacroix (1987) and Proulx (1989) discussed in detail the intensity and space-time variations of the cold Pointe-des-Monts front. They found the variability was highly correlated to the passage of low-pressure systems over the region, with periods ranging from 10 to 15 days (Murty and El-Sabh, 1977). These authors also suggested that advection towards the estuary mouth of (wind-driven) upwelled intermediate cold layer along the north shore of the Gulf between Pointe-des-Monts and Sept-Iles (Fig. 7), could contribute, in part, to the formation and variability of the Pointe-des-Monts front.

Mechanisms for the formation of eddy motions in the LSLE have been of interest in recent years. El-Sabh (1979) proposed the existence of two eddies; a large anticyclonic one with its center between Pointe-des-Monts and Rimouski and a smaller cyclonic eddy between Rimouski and Tadoussac. The presence of these eddies has recently been supported by current meter observations. Farquharson (1966) was the first to point out the existence of an anticyclonic eddy, 50 km in diameter, from the 5-day drift of a parachute drogue at 75 m between Pointe-des-Monts and Baie-Comeau. He found that the strength of the eddy varied with the fortnightly period. He related changes in magnitude to the southerly transverse currents observed at the mouth, presumably caused by variations of the neap-spring tidally induced residual currents. However, reanalysis of the current meter data taken by Farquharson at the Estuary-Gulf boundary (Mertz et al., 1988a) showed that wind-driven motions, and not forcing due to lunar phase changes, exerted a stronger influence on the exchange between the Gulf and estuarine waters. Meteorological forcing produces subtidal variations of the currents in the 10- to 15-day range. A similar correlation has also been found by Tee (1989).

In 1979, a mesoscale current meter survey was undertaken to resolve the presence of eddies and transverse currents previously reported in the area between Pointe-des-Monts and Rimouski. The experiment and the near surface subtidal variability were described by El-Sabh et al. (1982). They interpreted the very low frequency (70 to 80 days) circulation pattern in terms of an estuary-wide (50 km in -diameter) eddy occupying the area near the mouth (Fig. 8). Although the relationship was uncertain, the strength and rotational sense of the eddy were found to be correlated to sea-level variations. Lie and El-Sabh (1983), using a two-layer unforced-wave numerical model, suggested that the observed eddies at the seasonal time scale were formed by the superposition of two fourth-order baroclinic

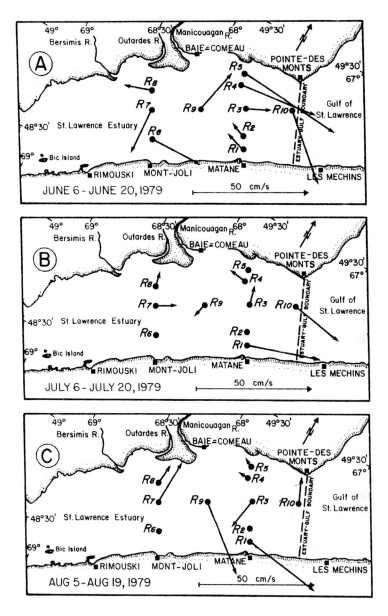

Figure 8. Modes of the very low frequency circulation pattern in the lower St. Lawrence estuary observed in 1979 (based on data from El-Sabh et al., 1982).

shelf waves propagaing in opposite directions. Although the forcing of the oscillation was suggested by El-Sabh et al. (1982) to be of atmospheric origin, more recently, it has been suggested that the observed very low frequency variability, trends and eddy motion near the mouth are more likely to be induced by freshwater runoff (Koutitonsky et al., 1990; Tee and Lim, 1987; Tee, 1989). Both Koutitonsky et al. (1990) and Tee (1989) showed that regulated freshwater, particularly from the Saguenay and Manicouagan reservoirs, is discharged into the estuary as large pulses and not at a constant yearly rate. Using objective analysis of the 1979 data set and a reduced-gravity numerical model, Koutitonsky et al. (1990) re-examined the seasonal circulation variability in the LSLE in relation to freshwater runoff. Their analysis revealed the presence of an estuary-wide anticyclonic eddy between Pointe-des-Monts and Baie-Comeau which lasted for 40 days in June as a response to buoyancy forcing by high river runoff (Fig. 8a). The mechanism proposed by Koutitonsky et al. (1990) for its formation was a geostrophic adjustment of lighter surface waters in a channel whose width was much larger than the internal radius of deformation. This structure undergoes a major transition in July (Fig. 8b), resulting in wave-like instabilities of the current field (Mertz et al., 1988a). After this shift, the mean outflow near Baie-Comeau becomes much weaker, allowing Gulf waters to enter the estuary along the north shore, with enhancement of outflow along the south shore, producing a cyclonic eddy near the mouth (Fig. 8c). Such changes in the basic circulation pattern at seasonal time scales may be masked or perturbed by other events at synoptic time scales. For example, the presence of very strong winds blowing from the Gulf towards the estuary resulted in a reversal of the circulation pattern near the mouth to cyclonic eddy motion (Mertz et al., 1990). Following storm passage, the circulation returns to anticyclonic.

Smaller scale eddies may also be present in the Lower Estuary. Murty and El-Sabh (1980) studied the tidally generated residual motion resulting from interaction of the tide with topography. They showed that while the predicted oscillatory tidal currents do not show small-scale eddy patterns, the residual motion associated with tides consists of eddies of scales 12 to 30 km. El-Sabh and Gagnon (1984) had found evidence for a 12 km wide cyclonic eddy by tracking near-surface drifters over a 3-day period in 1980 near the north shore (Fig. 9c).

There seems to be ample evidence to show that the estuarine south-shore jet, driven by discharge from the upper estuary and Saguenay river, occasionally bifurcates near Rimouski, with its separated portion flowing along the north shore. The north shore branch may remain there due to the Coanda effect. Carstens et al. (1984) described this effect in terms of lateral entrainment. A river plume will be drawn toward a coast and tend to attach there. This could lead to the formation of a north shore jet until it meets the corner at Pointe-des-Monts, detaches from the coast and turns right (due to Coriolis effects). It ultimately rejoins the south shore jet near the mouth of the estuary. Surface distributions of salinity (Figure 2a), temperature, density and chemical variables in the Ile du Bic - Rimouski region often suggest a veering of the flow towards the north shore (Greisman and Ingram, 1977; El-Sabh, 1977; Kranck, 1979; Bewers and Yeats, 1979). Forrester (1970) reported a significant northward cross-channel current off Rimouski and noted that "the outflow above 50 m in that region appears to be almost equally strong along both shores, and weaker in the middle". This is a departure from the ideal two-layer estuary. The presence of a northerly cross-channel residual current near Rimouski was observed by El-Sabh (1979) by tracking a 10 m deep parachute drogue from near Ile du Bic to the north

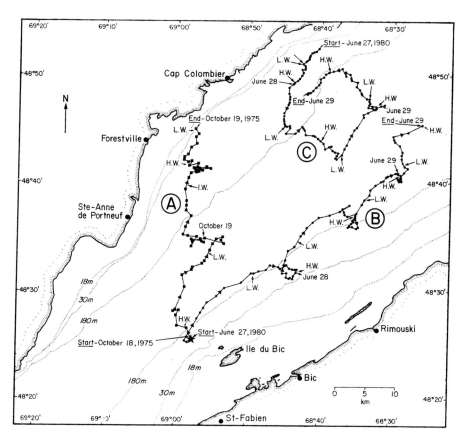

Figure 9. Trajectories of the near-surface drifters observed in 1975 and 1980 in the Ile du Bic - Matane area (adapted from El-Sabh and Gagnon, 1984).

shore (Fig. 9a). Neu (1970), Therriault and Levasseur (1985), Larouche et al. (1987) and El-Sabh (1977, 1983) have all reported the presence of relatively fresher water along the north shore of the Lower Estuary. Farquharson (1966) and El-Sabh et al. (1982) reported southward motion at the mouth of the estuary. Finally, measurements taken in the area between Mont-Joli and Pointe-des-Monts also showed the mean circulation in the upper 30 m to be characterized by two coastal currents flowing seaward parallel to both shores, with southward cross-channel flow at the mouth (Koutitonsky & El-Sabh, 1985). A curious behavior of the estuarine south-shore jet as observed during summer 1978 is demonstrated in Figure 10. Current meter records at Station S_1 showed a weak downchannel mean flow until early July; after this time, the flow was predominantly up-estuary, with three large upstream pulses. Mertz et al. (1989) discussed these records and showed that these upstream pulses along the south shore might have developed due to an unstable wave growing in the transverse southward segment at the mouth. Upstream of the Rimouski section, the south-shore estuarine jet apparently behaves in a more conventional manner.

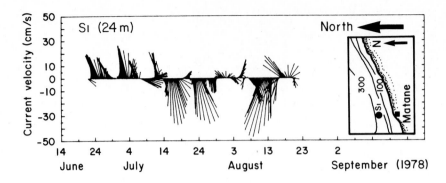

Figure 10. The subtidal current observed in 1978 along the south shore of the LSLE off Matane (adapted from Mertz et al., 1989).

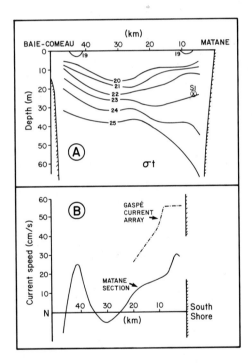

Figure 11. Observed σ_t distribution and calculated geostrophic currents at the Matane section compared with Gaspé Current data for June 1978 (adapted from Mertz et al., 1988b).

Lateral density distributions observed during June 24-26, 1978 along the Matane section were used to obtain the σ_t section shown in Figure 11a and the geostrophic current velocities at 20-m depth in Figure 11b. As can be seen, a general rise of the isopycnals toward the north shore occurs, with a zone near the south shore where the isopycnals in the depth range 20 to 30 m rise toward the south shore. Thus, based on geostrophy, one sees that the core of the south-shore jet is at about 30 m depth, with its speed decreasing from 30 m to 20 m and remaining roughly constant above 20 m. The geostrophic currents were referenced to the current observed at S_1. These results show the estuarine jet splits into shore-hugging and mid-channel components. A study of the surface dynamic topography in this region (Tang, 1980) during September 1978 showed that the mid-channel jet is part of a quasi-permanent loop or meander. The bifurcation of the jets takes place just upstream of the Matane section. In fact, the current meter placed at S_1 (Fig. 10) recorded a near zero mean (record length of 2 months) showing that during much of the summer the surface currents looped outside this location.

The formation of loops or meanders in a jet may occur due to an instability of the flow field. Since the south-shore jet is a near-surface phenomenon with strong lateral and vertical shears, one anticipates finding mesoscale features associated with instability (wave-like meanders and eddies). Indeed, unstable waves are often observed in thermal satellite images of the Lower Estuary (El-Sabh, 1985, 1988; Lavoie et al., 1985; Lacroix, 1987; Mertz et al., 1988, 1989). A typical occurrence of unstable waves, similar in appearance to backward breaking waves described by Griffiths and Linden (1981), Stern et al. (1982) and Chao and Ikao (1987), is shown in Figure 12. A mid-estuary longitudinal front separates two distinct water masses, the southern half of the estuary being colder than the northern part. The frontal demarcation line appears to have four crests, with an average spacing of 75 km. This value corresponds to the wavelength of the topographic waves observed by Lie and El-Sabh (1983). The amplitude of the wavelike perturbation increases seaward as does the estuarine width. Mertz et al. (1988b) showed that the south-shore jet was both baroclinically and barotropically unstable and was related to meteorological forcing (atmospheric pressure gradient and wind) over the estuary. The time scale for these disturbances is on the order of 10 to 15 days (El-Sabh, 1985). It is of interest to note that low-passed time series incorporating all major forcing fields (tidal, atmospheric and buoyancy) acting in the LSLE show an oscillation at this time scale (El-Sabh and Gagnon, 1984; El-Sabh et al., 1982; Mertz et al., 1988a; Tee, 1989). The relative contribution of driving forces to the space-time variations of currents, eddy motions and unstable waves, particularly at the synoptic time scales is unknown. As concluded by El-Sabh (1988), broad scale and intensive field experiments are needed, which if coupled with three-dimensional numerical models, will provide a better understanding of the mesoscale variability observed in the LSLE.

4. The Gaspe Current

As mentioned earlier, the discharge of the St. Lawrence River drives an intense buoyant jet along the south shore of the LSLE. This jet exits the estuary and flows along the Gaspe Peninsula where it is augmented by the Anticosti cyclonic gyre in the northwest Gulf of

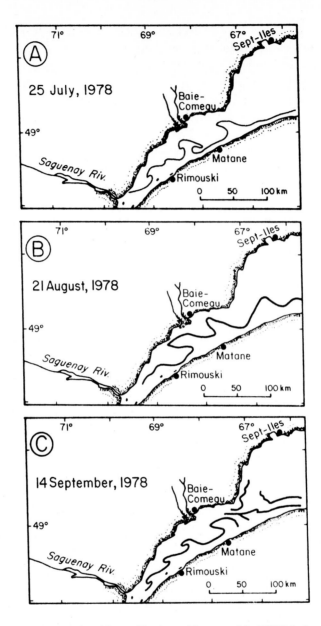

Figure 12. Mesoscale features observed in satellite thermal images of the LSLE during summer 1978. The line in the estuary demarcates the boundary between cool water along the south shore and warmer water offshore (modified from Mertz et al., 1989).

St. Lawrence (El-Sabh, 1976) to form the well known Gaspe Current. This buoyancy-driven coastal jet is particularly strong along the Gaspe Peninsula and reaches a geostrophic transport of 3×10^5 m³/s (Benoit et al., 1985). Given that the current width is only about 20 km, the Rossby number of the flow has a maximum value of 0.5, which is somewhat greater than that of the Gulf Stream.

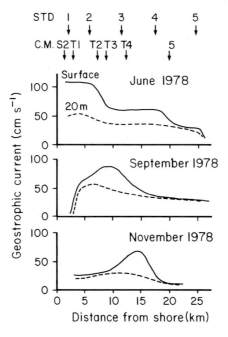

Figure 13. The seasonal evolution of the (geostrophic) near surface lateral profile of the Gaspe Current (adapted from Benoit et al., 1985).

Based on an extensive current meter array and numerous hydrographic measurements taken in 1978, a comprehensive description of the structure, seasonal characteristics and dynamics of the Gaspe Current was completed (Tang, 1980; El-Sabh and Benoit, 1984; Benoit et al., 1985). Since the current is buoyancy driven, its properties are strongly influenced by the seasonal variation of the freshwater discharge. Diminishing discharge from June to November results in a continuing increase of salinity and decreasing current strength from a maximum speed of 110 cm/s in June to 60 cm/s in November (Fig. 13). High vertical shears exist in the upper 40 m of the water column. During the same period, the width of the current decreases, and the position of the current maximum shifts from nearshore to about 14 km offshore. Furthermore, because the Gaspe Current is in geostrophic balance laterally, isopycnals in the current zone develop a strong tilt and intersect the surface to form a strong front about 10 km from shore (Tang, 1980a). This front can be frequently identified in infrared satellite images as a narrow band of cold surface water running parallel to the shore. From the results of water mass analysis, Tang (1983) suggested that the cold surface water could be caused by upwelling at the outer edge of the Gaspe Current by an Ekman suction mechanism. Recently, Mertz and El-Sabh (1989a) examined satellite thermal images and temperature data from a current meter array to show that the cold anomaly at the front is strongly influenced by wind activity in the vicinity of the Gaspe Current. They also show that thermal variability of the Gaspe Current is characterized by three classes of fluctuations: (i) wind-driven warming during strong eastward wind pulses, (ii) dramatic cooling episodes generated by frontal upwelling and (iii) large temperature jumps during periods of weak wind, followed by high frequency oscillations, which may be associated with episodes of baroclinic instability.

Current measurements (Benoit et al., 1985) and geostrophic calculations (El-Sabh, 1976) indicate that the Gaspe Current is unidirectional most of the time during summer. The core of the Gaspe Current occasionally moves offshore, where a wavy pattern in the flow develops. Tang (1980b) and Mertz et al. (1988b) discussed unstable wave development in the Gaspe Current for high runoff conditions. Simple model calculations and energy transfer considerations suggest that both baroclinic and barotropic instability mechanisms are important. The dynamically relevant scales for baroclinic instability (internal Rossby radius) and for barotropic instability (current half-width) are both about 10 km. These are the preferred scales for unstable wave growth. Therefore, one anticipates that unstable disturbances will have a wavelength of about $2\pi.10km = 60$ km, as is observed. Recently, Mertz and El-Sabh (1989b) discussed the occurrence and evolution of an autumnal current instability, when runoff is only 60% of its maximum spring value. The observed configuration differs considerably from that in summer apparently because of the offshore location of the jet core in early autumn. A stability analysis (Mertz and El-Sabh, 1989b) suggests that Gaspe Current instabilities may grow somewhat faster in autumn. Wind action has also been mentioned as a factor triggering unstable wave episodes in the Gaspe Current.

Clearly much work remains to be done in order to achieve an understanding of instabilities in the Gaspe/LSLE current system. One particular aspect mentioned by Mertz and El-Sabh (1989b) is the problem of unstable wave variability: the wavelike features are not always present even though the Gaspe Current is likely to be unstable throughout summer and autumn. These authors proposed an examination of vacillation cycles in a numerical model of the Gaspe Current. If a steady state is attained, perturbations representing wind forcing could be applied to the model. A numerical model would also allow the incorporation of features such as coastline curvature and varying bottom topography.

5. Conclusions

The mean circulation in the area upstream of the Saguenay is predominately downstream both at the surface and at greater depths on the southern shore. Sub-surface upstream flow occurs on the north side. Downstream of the Saguenay, the salinity regime initially resembles the discharge front transition to a coastal front, as described by Garvine (1987). In this analogy, the South Channel acts as a low salinity source for the wider Lower Estuary. Horizontal (lateral) undulations of the boundary separating the lighter and heavier water masses and eddy-like features become more pronounced downstream of the Ile du Bic, in an area where channel width exceeds the internal Rossby radius. Vorticity considerations promote cross-channel (cyclonic vorticity) flow as the lighter water enters the progressively deeper waters of the Laurentian Channel (see Mertz and Gratton, this volume, for discussion on topographically modified motions). At the downstream end of the estuary, the Gaspe Current dominates the flow conditions.

The large physical dimensions of the Lower Estuary, combined with the fact that variability associated with the spring-neap tidal cycle, the atmospheric pressure and buoyancy (regulated) forcing act within the same time range produces a complicated current field. This may help explain the complex distributions of phytoplankton production, biomass, and chemical and geochemical variables observed by several authors, since these are controlled mainly by physical processes. For example, Therriault and Levasseur (1985), based on a two-year study of phytoplankton production, were able to divide the Lower Estuary into four distinct regions. Each region corresponds to one of the mesoscale features described earlier and is dominated by different hydrodynamical processes. Region I, corresponding to the buoyancy-driven outflow from the Upper Estuary and the Saguenay along the south shore, is the least productive and is characterized by high turbidity and instability of the water column. Processes in Region II are mainly controlled by internal tide activity and vertical mixing at the head of the Laurentian Channel. Region III, occupying most of the central part of the estuary, is the most productive. It is under the stabilizing influence of the freshwater plume of the Manicouagan and Outardes rivers and eddy-like motion. Region IV, at the Estuary-Gulf boundary, is the most stable (vertically) of all four sub-areas, as suggested by lower observed nutrient values. Therriault and Levasseur concluded that the Lower Estuary cannot be considered as a single entity since clear regional differences in the hydrodynamics and phytoplankton yield are found.

The frontal features found in the St.Lawrence are similar to those found in other coastal and estuarine areas. A review of different types can be found in Bowman and Esaias (1977). The most common front in the shallower regions of the estuary is that formed at the boundary of well-mixed and stratified water masses. The strong tidal current regime and variable bathymetry in the upstream areas of the St.Lawrence provides suitable conditions for frontogenesis in many locations. Near the Saguenay fjord, at least two different mechanisms of frontogenesis are at work. Surface intrusion of lighter water occurs either as a shallow (compared to local depth) well defined feature, with evidence of a rotor (deep lobe) at the leading edge on some occasions, or a deep layer with a much broader transition of water type. Garvine and Monk (1974) and Garvine (1974) have discussed the physical characteristics and dynamics of similar small scale fronts formed on the lateral boundaries of the Connecticut River plume as it enters Long Island Sound. Frontal phenomena associated with the Gaspe Current has been discussed by Tang (1980a).

In summary, mesoscale features in the St.Lawrence estuary range from those found in much smaller estuaries to those characteristic of shelf waters. The large length scales (horizontal and vertical) in the Lower Estuary region provide a similar geometry to that found on open shelves in spite of two lateral boundaries. Vertical stratification in the St. Lawrence is somewhat larger. Thus, analogies between physical processes occurring the the St. Lawrence estuary and elsewhere are appropriate for both estuarine and open coastal situations.

Acknowledgements

Paul H. LeBlond and R.E. Wilson provided useful comments on the initial manuscript for which we are grateful. Thanks to Ann Cossette for typing the manuscript and Johanne Noel for preparing the drawings. This study was supported by operating grants to R.G. Ingram and M.I. El-Sabh from the Natural Sciences and Engineering Research Council of Canada (NSERC).

REFERENCES

Benoit, J., M.I. El-Sabh and C.L. Tang. 1985. Structure and seasonal characteristics of the Gaspe Current, J. Geophys. Res., 90, 3225-3235.

Bewers, J.M. and P.A. Yeats. 1979. The behavior of trace metals in estuaries of the St. Lawrence basin. Naturaliste Canadien, 106, 149-161

Bowman, M. and W. Esaias. 1978. Oceanic fronts in coastal processes. Springer-Verlag, New York, 114 p.

Carstens, T., T.A. McClimans and J.H. Nilsen. 1984. Satellite imagery of boundary currents, in: Remote Sensing of Shelf Sea Hydrodynamics. (J. Nihoul, Ed.), Elsevier, Amsterdam, 235-256.

Chao, S.-Y. and T.W. Kao. 1987. Frontal instabilities of baroclinic ocean currents with application to the Gulf Stream. J. Phys. Oceanogr., 17, 792-807.

d'Anglejan, B. 1981. On the advection of turbidity in the St. Lawrence middle estuary. Estuaries 4: 2-15.

d'Anglejan, B., R.G. Ingram and J.P. Savard. 1981. Suspended sediment exchanges between the St. Lawrence estuary and a coastal embayment. Marine Geology 40: 85-100.

Denman, K. and T. Platt. 1975. Coherences in the horizontal distributions of phytoplankton and temperature in the upper ocean. Mem. Soc. Royal Sc. Liege, 7:13-30.

Dickie, L. and L.M. Trites. 1983. The Gulf of St. Lawrence, in Estuaries and Enclosed Seas (B. H. Ketchum, Eds.), Elsevier, Amsterdam, 403-475.

Dodson, J., J.C. Dauvin, R.G. Ingram and B. d'Anglejan. 1989. Abundance of larval rainbow smelt in relation to the maximum turbidity zone and associated macroplanktonic fauna of the middle St. Lawrence estuary. Estuaries 12:66-81.

El-Sabh, M.I. 1976. Surface circulation patterns in the Gulf of St. Lawrence, J. Fish. Res. Board Can., 33, 124-138.

El-Sabh, M.I. 1977. Circulation pattern and water characteristics in the lower St. Lawrence estuary, in: Proc. Symp. Modelling of Transport Mechanisms in Oceans and Lakes, (T.S. Murty, Ed.), Marine Sciences Directorate, Dept. of Fish. and the Environment, Ottawa, MS Rept Series, 43, 243-248.

El-Sabh, M.I. 1979. The lower St. Lawrence estuary as a physical oceanographic system. Naturaliste Canadien, 106 (1), 55-73.

El-Sabh, M.I. 1985. Variability of surface currents in a wide and deep estuary: The St. Lawrence as a case study. Estuaries, 8, 28, 67A.

El-Sabh, M.I. 1988. Physical oceanography of the St. Lawrence estuary. in: Hydrodynamics of Estuaries, vol. II (B. Kjerfve, Ed.).CRC Press, Baco Raton, Florida, 61-78.

El-Sabh, M.I. and J. Benoit. 1984. Variabilité spatio-temporelle du courant de Gaspé. Sciences et Techniques de l'Eau, 17, 55-64.

El-Sabh, M.I. and M. Gagnon. 1984. Du regime de courant du secteur Bic-Pointe-au-Père. 2nd Symposium on the Oceanography of the St. Lawrence Estuary. Université Laval, Québec. May 14-17, 1984.

El-Sabh, M.I. and T.S. Murty. 1990. Mathematical modelling of tides in the St. Lawrence estuary. in: Oceanography of a Large-scale Estuarine System: the St. Lawrence Coastal and Estuarine Studies, Springer-Verlag (this volume).

El-Sabh, M.I., H.-J. Lie and V.G. Koutitonsky. 1982. Variability of the near surface residual current in the lower St. Lawrence estuary. J. Geophys. Res., 87, 9589-9600.

Farquharson, W.I. 1966. St. Lawrence estuary current surveys. Bedford Inst. Oceanogr., Rept. Ser. 66-6. Unpublished manuscript, 84 p.

Forrester, W. D. 1964. A quantitative Temperature-Salinity study of the Gulf of St. Lawrence. Bedford Institute Rept. BIO 64-11. Dartmouth, N.S. Canada, 21 p.

Forrester, W.D. 1970. Geostrophic approximation in the St. Lawrence estuary. Tellus, 11 (1), 53-65.

Forrester, W. D. 1974. Internal tides in the St. Lawrence estuary. J. Mar. Res. 32:55-66.

Garvine, R.W. 1974. Dynamics of small scale oceanic fronts. J. Phys. Oceanogr., 4: 557-569.

Garvine, R.W. 1975. The distribution of salinity and temperature in the Connecticut River estuary. J. Geophys. Res. 80: 1176-1183.

Garvine, R.W. 1986. The role of brackish plumes in open shelf waters, in: The Role of Freshwater Outflow in Coastal Marine Ecosystems (S. Skreslet, Ed.), Springer-Verlag, Berlin, 47-65.

Garvine, R.W. 1987. Estuary plumes and fronts in shelf waters: A Layer Model. J. Phys. Oceanogr. 17, 1877-1896.

Garvine, R.W. and J.D. Monk. 1974. Frontal structure of a river plume. J. Geophys. Res., 79: 2251-2259.

Gratton, Y., G. Mertz and J.A. Gagné. 1988. Satellite observations of tidal upwelling and mixing in the St. Lawrence Estuary. J. Geophys. Res., 93, 6947-6954.

Greisman, P. and R.G. Ingram. 1977. Nutrient distribution in the St. Lawrence estuary. J. Fish. Res. Board Can., 34: 2117-2123.

Griffiths, R.W. and P.F. Linden. 1981. The stability of buoyancy-driven boundary currents. Dyn. Atmos. Oceans, 5, 281-306.

Hachey, H.B., L. Lauzier and W.B. Bailey. 1956. Oceanographic features of submarine topography. Transactions Royal Society Canada, 50: 67-81.

Huzzey, L.M. and J.M. Brubaker. 1988. The formation of longitudinal fronts in a coastal plain estuary. J. Geophys. Res. 93: 1329-1334.

Ikeda, M. and J.R. Apel. 1981. Mesoscale eddies from spatially growing meanders in an eastward-flowing oceanic jet using a two-layer quasigeostrophic model. J. Phys. Oceanogr., 11, 1638-1661.

Ingram, R.G. 1975. Influence of tidal-induced vertical mixing on primary productivity in the St. Lawrence estuary. Mem. Soc. Roy. Sci. Liege, 7: 59-74.

Ingram, R.G. 1976. Characteristics of a tidal-induced estuarine front. J. Geophys. Res., 81: 1951-1959.

Ingram, R.G. 1978. Internal wave observations off Isle Verte. J. Mar. Res. 36: 715-724.

Ingram, R.G. 1979. Water mass modification in the St. Lawrence estuary. Naturaliste Canadien, 106: 45-54.

Ingram, R.G. 1983. Vertical mixing at the head of the Laurentian Channel. Estuar., Coastal and Shelf Sci., 16: 333-338.

Ingram, R.G. 1985. Frontal characteristics at the head of the Laurentian Channel. Naturaliste Canadien, 112: 32-38.

Koutitonsky, V.G. and M.I. El-Sabh. 1985. Estuarine mean flow estimation revisited: application to the St. Lawrence estuary. J. Marine Res., 43, 1-12.

Koutitonsky, V.G., R.E. Wilson and M.I. El-Sabh. 1990. On the seasonal response of the lower St. Lawrence estuary to buoyancy forcing by regulated runoff. Submitted to Estuar. Coastal and Shelf Sci.

Kranck, K. 1979. Dynamics and distribution of suspended particulate matter in the St. Lawrence estuary. Naturaliste Canadien, 106: 163-173.

Lacroix, J. 1987. Etude descriptive de la variabilité spatio-temporelle des phenomenes physiques de surface de l'estuaire maritime et de la partie ouest du golfe du Saint-Laurent à l'aide d'images thermiques du satellite NOAA-7. M. Sc. Thesis, Université du Québec à Rimouski, 186 p.

Larouche, P., V.G. Koutitonsky, J.-P. Chanut and M.I. El-Sabh. 1987. Lateral stratification and dynamic balance at the Matane transect in the lower St. Lawrence estuary. Estuar. Coastal and Shelf Sci., 24, 859-871.

Lavoie, A., F. Bonn, J.M.M. Dubois and M.I. El-Sabh. 1985. Structure thermique et variabilite du courant de surface de l'estuaire maritime du Saint-Laurent a l'aide d'images du satellite HCMM. Canadian Journal of Remote Sensing, 11, 70-84.

Lie, H.-J. and M.I. El-Sabh. 1983. Formation of eddies and transverse currents in a two-layer channel of variable bottom with application to the lower St. Lawrence estuary. J. Phys. Oceanogr., 13, 1063-1075.

Mertz, G. and M.I. El-Sabh. 1989a. Thermal variability of the Gaspé Current. Workshop/Symposium, The Gulf of St. Lawrence: Small ocean or Big Estuary ? Maurice-Lamontagne Institute, Mont-Joli, Qué., 14-17 March, 1989.

Mertz, G. and M.I. El-Sabh. 1989b. An autumn instability event in the Gaspe Current. J. Phys. Oceanogr.: 19, 148-156.

Mertz, G., M.I. El-Sabh and V.G. Koutitonsky. 1989. Low frequency variability in the lower St. Lawrence estuary. J. Marine Res., 47: 285-302.

Mertz, G., M.I. El-Sabh and V.G. Koutitonsky. 1988a. Wind-driven currents at the mouth of the St. Lawrence estuary. Atmosphere-Ocean, 26: 509-523.

Mertz, G., M.I. El-Sabh, D. Proulx and A.R. Condal. 1988b. Instability of a buoyancy-driven coastal jet: The Gaspe Current and its St. Lawrence precursor. J. Geophys. Res., 93: 6885-6893.

Mertz, G., V.G. Koutitonsky, Y. Gratton and M.I. El-Sabh. 1990. Wind-induced eddy motion in the Lower St. Lawrence estuary. Submitted to Estuar. Coastal and Shelf Sci.

Muir, L. 1980. Internal tides in a partially mixed estuary. Canada Centre for Inland Waters Manuscript Rept. no. 9, Burlington, Ont. 177 pp.

Murty, T.S. and M.I. El-Sabh. 1977. Transverse curents in the St. Lawrence estuary: A theoretical treatment. In: Transport Processes in Lakes and Oceans (R.J. Gibbs, Ed.) Plenum Publishing Corporation, New York, N.Y., pp. 35-62.

Murty, T.S. and M.I. El-Sabh, 1980. Tidally-generated residual motion in the St. Lawrence Estuary. Proc. Thirteen Annual Simulation Symp., (V.P. Roy, R.G. Cumings, C. Hammer and W. Malamphy, Eds.), p. 127-145.

Neu, H.J.A. 1970. A study on mixing and circulation in the St. Lawrence estuary up to 1964. Bedford Inst. Oceanogr. AOL Rept. 1970-9. Unpublished manuscript, 31 p.

Pingree, R.D. and D.K. Griffiths. 1980. A numerical model of the M_2 tide in the Gulf of St. Lawrence. Oceanologica Acta, 3: 221-225.

Proulx, D. 1989. Etude synoptique des variations spatio-temporelles de la structure thermique de l'estuaire maritime et du secteur nord-ouest du golfe du Saint-Laurent, pour la periode estivale de 1985, à l'aide des images satellites NOAA-9. M.Sc. thesis, Université du Québec à Rimouski, 226 p.

Reid, S.J. 1977. Circulation and mixing in the St. Lawrence estuary near Ile Rouge. Bedford Inst. Oceanogr. Data Rept. Series BI-R-77-1, Dartmouth, Nova Scotia, 36 p.

Silverberg, N. and B. Sundby. 1979. Observations in the turbidity maximum of the St. Lawrence Estuary. Can. J. Earth Sci., 16, 939-950.

Stern, M.E., J.A. Whitehead and B.-L. Hua. 1982. The intrusion of a density current along the coast of a rotating fluid. J. Fluid Mech., 123, 237-265.

Steven, D.M. 1974. Primary and secondary production in the Gulf of St. Lawrence. McGill University Marine Sciences Manuscript Rept. 26, 116 p., Montreal, Canada.

Tang, C.L. 1980a. Mixing and circulation in the northwestern Gulf of St. Lawrence. J. Geophys. Res., 85, 2787-2796.

Tang, C.L. 1980b. Observation of wavelike motion of the Gaspe Current. J. Phys. Oceanogr., 10: 853-860.

Tang, C.L. 1983. Cross-front mixing and frontal upwelling in a controlled quasi-permanent density front in the Gulf of St. Lawrence. J. Phys. Oceanogr., 13: 1468-1481.

Tee, K.T. 1989. Subtidal salinity and velocity variations in the St. Lawrence estuary. J. Geophys. Res., 94: 8075- 8090.

Tee, K.T. and T.H. Lim. 1988. The freshwater pulse - a numerical model with application to the St. Lawrence estuary. J. Marine Res., 45, 871-909.

Therriault, J.-C. and G. Lacroix. 1976. Nutrients, chlorophyll and internal tides in the St. Lawrence estuary. J. Fish. Res. Board Can., 33: 2747-2757.

Therriault, J.-C. and M. Levasseur. 1985. Control of phytoplankton production in the lower St. Lawrence estuary. Naturaliste Canadien, 112, 77-96.

Chapter 5

Topographic Waves and Topographically Induced Motions in the St. Lawrence Estuary

Gordon Mertz[1] and Yves Gratton[2]

[1]Northwest Atlantic Fisheries Centre, Fisheries and Oceans Canada, P.O. Box 5667, St. John's, Newfoundland, A1C 5X1

[2]Institut Maurice-Lamontagne, Pêches et Océans Canada, B.P. 1000, Mont-Joli, Québec, G5H 3Z4

ABSTRACT

The St. Lawrence Estuary exhibits an estuarine character between Quebec City and the Saguenay Fjord (Upper estuary) and a maritime character between the Saguenay Fjord and the Gulf of St. Lawrence (Lower estuary). Topographically modified motions are mostly supra-inertial in the former section and sub-inertial in the latter. The shallow Upper estuary is dominated by high-frequency motions such as internal waves and tides, which are generated by interaction of the tidal flow with sills, banks and especially the shoaling region at the head of the Laurentian Trough. The deeper Lower estuary exhibits a variety of strong low-frequency motions including topographic waves and unstable shear waves.

Key Words: Internal waves, internal tides, unstable waves, eddy

1. Introduction

Oceanic motions are affected by topographic features over a great frequency range. Tidal flow over sills can generate internal waves and internal tides. At low-frequencies, vortex stretching due to bottom slopes (in the presence of rotation) can give rise to topographic Rossby waves. The St. Lawrence Estuary is remarkable in that the full spectrum of oceanic variability, from internal waves to low-frequency wind-driven currents and unstable waves can be found within its confines. It is the purpose of this review paper to discuss the role that topographic features play in the dynamics of the St. Lawrence Estuary.

The St. Lawrence Estuary (Fig. 1) is divided into two segments based on the characteristics of the bottom topography. The Upper estuary, extending from Ile d'Orléans to Tadoussac, is rather narrow (2 to 24 km wide) and exhibits very uneven bottom topography. One finds deep troughs (the North and South channels), banks, islands and ridges in this area. Owing to the relatively shallow mean depth of the Upper estuary, tidal ranges and currents are very large (up to 10 m and 3 m s^{-1}, respectively). In contrast, the Lower St. Lawrence Estuary

Coastal and Estuarine Studies, Vol. 39
M. I. El-Sabh, N. Silverberg (Eds.)
Oceanography of a Large-Scale Estuarine System
The St. Lawrence
© Springer-Verlag New York, Inc., 1990

(LSLE) is much deeper on average, with a comparatively smooth bottom. The main feature of the LSLE's bottom topography is the deep (300 m) Laurentian Trough underlying the bulk of the estuary. This channel intrudes from the Atlantic Ocean, through the Gulf of St. Lawrence, and into the LSLE. On the south shore of the LSLE, in its upstream half, there is a well developed shelf characterized by a gentle bottom slope, compared to the rapid drop into the Laurentian Trough. Tidal currents are on the order of 30 cm s^{-1} on the shelf; they are weaker in the deeper portion of the LSLE.

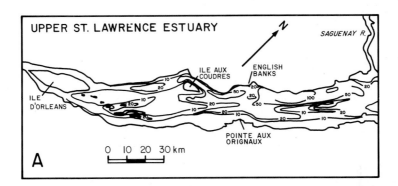

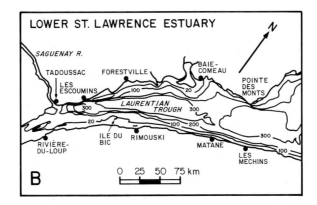

Figure 1. The St. Lawrence Estuary with bathymetry (in meters).

The LSLE has a pronounced maritime character. Specifically, its width (30 to 50 km) is equal to several internal Rossby radii, the scale characterizing baroclinic motions. This implies that the LSLE can accomodate such features as baroclinic shelf waves and unstable shear waves. The emphasis in going from the Upper to Lower estuaries thus shifts from high-frequency to low-frequency motions, from estuarine to oceanic characteristics.

Perhaps the most important topographic feature of the St. Lawrence Estuary, taken as a whole, is the termination of the Laurentian Trough near the mouth of the Saguenay. This area marks the transition from Lower to Upper estuary. In only 20 km the bottom shoals from a depth of 300 m to a depth of about 50 m. To the upstream propagating barotropic tide, this region appears step-like. This feature couples the baroclinic and barotropic motions to yield strong internal tides, propagating upstream and downstream from the shoaling region.

The plan of this paper is to discuss topographically generated and modified motions in order of decreasing frequency. We proceed from (i) internal waves to (ii) internal tides and upwelling and finally conclude with (iii) vorticity waves including topographic waves and unstable waves.

2. Internal waves

In the Upper estuary the St. Lawrence River's runoff is mixed with salt water penetrating from the Gulf of St. Lawrence. Surface values of the salinity progress from about 0 %oo at Québec City to 26 to 28 %oo near the mouth of the Saguenay (Neu, 1970). The shallow mean depth of the estuary implies that it is a zone of strong tidal dissipation (Forrester, 1969); a portion of the energy dissipated goes into mixing. Mixing events will commence when the local shear is large enough to produce small-scale instability and thus a pocket of turbulence. Local enhancement of the shear may be produced by internal tides or internal waves.

Internal waves are most easily pictured in a two layer fluid: the density difference between layers allows the propagation of interfacial waves completely analogous to surface gravity waves. In reality the density profile $\rho(z)$ exhibits continuous stratification and not sharp layers. This allows vertical as well as horizontal propagation of the internal waves and limits their frequency to the range $f < \omega < N$, where N is the Brunt Vaisala frequency ($N^2 = - g \rho^{-1} \, \partial\rho /\partial z$) and f is the local inertial frequency (the Coriolis parameter). In fact the M_2 internal tide falls into this frequency range, and it is a free internal wave. However, for the Upper St. Lawrence Estuary internal waves are observed with periods of several minutes (Deguise, 1977) and thus we distinguish them from the much longer period internal tides, which we will discuss in the next section.

Internal waves may be generated by wind action or by disruption of a flow field by an obstacle. The Upper estuary, with its rough bottom, and strong tidal flows is an ideal locale for internal wave generation. The propensity for a flow field to generate disturbances when perturbated by an obstacle may be gauged by the internal Froude number $F = (\pi U)/(ND)$ where D is the total depth of the fluid and U is the characteristic speed of the flow. The Froude number measures the ratio of the flow speed to the long internal gravity wave speed (π/ND). If F satisfies $0.5 < F < 1$, a first mode lee wave may be formed. That is, stationary spatial undulations will form on the interface on the lee side of an obstacle in the flow (see Baines, 1979). If the flow field is oscillatory, like a tidal flow, then the lee waves will form as the flow speed increases and the criterion $0.5 < F < 1$ is satisfied, and as the flow slackens the lee waves are released, and propagate in the upstream direction as internal waves. In practise, more than just one mode may be involved and the process of internal wave generation is quite complex (for an excellent discussion of these effects see Farmer and Freeland, 1983).

Deguise (1977) has described measurements of internal wave activity in the Upper estuary. A field program was undertaken specifically to measure internal wave motions near English Bank (indicated by an arrow in Fig. 1). This bank interrupts the nearly continuous channel along the north shore of the Upper estuary. The channeled flow is forced over the bank, so that it acts like an obstacle to the flow field. Tidal currents are strong (up to 1.5 m s^{-1}), and the stratification is moderate, both factors tending to promote the development of internal waves. The bank obstructs 70% of the total depth of the channel.

Deguise's observations may be summarized as follows: (i) high frequency internal oscillations are detectable in the density field. The typical periods are two to three minutes. (ii) The oscillations may attain very large amplitudes (σ_t excursions of up to three units), sometimes larger than the internal tides. (iii) The internal waves occur in pulses near the peak of flood or ebb tides and can be detected on both the upstream and downstream sides of the bank. (iv) The waves show features of nonlinearity, taking the form of a series of "soliton-like" waves.

The relevance of these observations to the dynamics of the Upper estuary lies in the fact that these oscillations are energetic and are characterized by relatively small spatial scales, making them good candidates for contributing to shear instability and mixing. Since the Upper estuary exhibits many pronounced topographic features there may be a number of zones here where internal wave generation takes place.

Ingram (1978) has observed internal waves trains in aerial photographs near Isle Verte (Fig. 1) which is close to the junction of the Upper and Lower estuaries and near the south shore. These waves appear as bands in the photographs, running perpendicular to the channel axis and represent areas of flow convergence where flotsam accumulates or turbidity variations due to the flow field. Field observations in the area show large temperature jumps (up to 4 °C) corresponding to the passage of flotsam lines. The time between pulses is on the order of several minutes, clearly illustrating the high-frequency nature of these phenomena. Ingram (1978) believes that these internal waves may develop when light waters carried downstream on the ebbing tide intrude upon dense water residing in the South channel (remnant from the previous flood tide). The resulting internal density front steepens in time and the action of dispersion and nonlinearity yields internal waves. Lee and Beardsley (1974) have discussed generation mechanisms of this type.

Work on internal waves in the St. Lawrence Estuary is still in its infancy. The complexity of the bottom topography in the Upper estuary makes the interpretation of data from this region a formidable problem. Classic studies of internal wave generation, such as those conducted in Knight Inlet (see Farmer and Freeland, 1983), focus on areas where a single sill is responsible for much of the internal wave activity. In the Upper estuary there are many possible sources of internal waves. Mixing processes are much less vigorous in the LSLE. Nevertheless, it is likely that breaking internal waves contribute to the mixing of the waters of this body. Gargett (1984) has noted that the breaking internal waves may make an important contribution to mixing in bodies as diverse as estuaries, lakes and the upper oceans.

3. Internal tides and upwelling.

In the coastal ocean, the generation of internal tides by the interaction of the barotropic tide with an abrupt shelf break is a well known process and has been recently reviewed by Baines (1982). It is essentially a scattering process by which the incident barotropic tidal wave will, through interaction with the topography, generate not only a reflected and transmitted barotropic component, but also a set of reflected and transmitted baroclinic waves. The major differences between the continental shelf and the St. Lawrence are the stratification, the width of the generating region and the amplitude of the internal tides.

The topography of the head region of the Laurentian Trough includes sills and banks as well as a general shoaling. As noted by Therriault and Lacroix (1976), these conditions are ideal for internal tide generation, especially since the pycnocline intersects the top of certain banks during part of the tidal cycle.

Huntsman (1923) first suggested that tidal energy propels the mixing processes at work in the St. Lawrence Estuary. As the Laurentian Trough shoals, tidal velocities increase dramatically, suggesting that dissipative processes such as mixing may be active in the shoaling region. Moreover, the head of the Laurentian Trough is situated very near the confluence of the Saguenay Fjord and the St. Lawrence Estuary. Thus, in this zone, mixing of Upper estuarine waters, Saguenay waters, and Gulf of St. Lawrence waters upwelled from the intermediate layer is occurring. Because of these features, this region has attracted great interest, particularly the upwelling and the internal tide activity.

Forrester (1969) and Reid (1977) have reported on the complex interplay of the tides with topographic features such as sills and banks in the head region. Of specific interest here, Forrester (1974) and Ingram (1975) have found evidence that strong internal tides can be propagated seaward from the shoaling region. Muir (1979; 1981) has collected the available hydrographic data for the Upper estuary and shown that strong internal tides and overtides are present, which were most likely to have originated in the head region. When intermediate-depth water rises into the surface layer, with the cresting internal tides, fronts may be formed in the head region (Ingram, 1976, 1985) with concomitant large shears which promote mixing. These internal tides are of great interest to biologists because they raise nutrient-enriched waters into the surface zone. Therriault and Lacroix (1976) have conducted an extensive field program here to measure the nutrient volumes raised by the internal tides. Ingram (1983) has shown that nutrification of the surface waters occurs on a semi-diurnal basis. Of some significance is the fact that much of the water comprising the Gaspe Current has its origin here. The upward nutrient flux mixed into the surface waters is comparable to the nutrient flux received by the Gulf of St. Lawrence via the Gaspe Current (Reid, 1977; Steven, 1974).

Satellite thermal images of the St. Lawrence Estuary show that the head region is frequently cold (Lavoie et al., 1985; Lacroix, 1987; Gratton et al., 1988). It is presumed that the cold spot is a product of the upwelling and mixing processes in action. These authors note that the cold water is not always confined to the head region, but may be found up to 100 km downstream of it (see Fig. 2). These observations strongly suggest that mixing, and perhaps upwelling, may occur in other areas than the head region. Since internal tides are observed in the Laurentian Trough (Forrester, 1974) and on the shelf off Rimouski (Gagnon, 1977), one suspects that they play a role in creating these cold anomalies. We will now discuss the generation mechanisms for these internal tides before returning to a discussion of satellite observations of the cold anomaly and the role of internal tides in generating it.

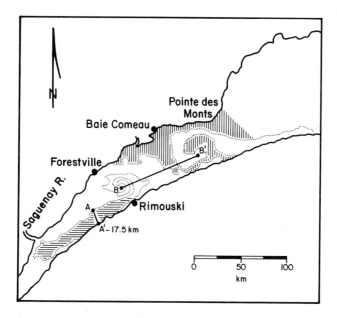

Figure 2. A sketch of the sea surface temperature reproduced from a NOAA-9 thermal image (July 22, 1986) of the LSLE. The horizontally shaded region depicts the south shore cold anomaly (5°C). The maximum width of the anomaly (A-A') is 17.5 km. Note the presence of a cyclonic eddy (B) and a backward breaking wave (B'). The temperature ranges from 5°C in the horizontally shaded region to about 15°C in the vertically shaded region.

Borne de Grandpré et al. (1981) developed a laterally averaged two-dimensional numerical tidal model to study the influence of topography and stratification (salinity only) on tidal elevations and currents. They show that the internal tides propagated roughly as in a two-layer system. Borne de Grandpré and El-Sabh (1980) had used the model to show that the largest vertical velocities were found at the head of the Laurentian Trough. This led Gratton and Saucier (1987) to examine the internal tide generation mechanism at the head of the Laurentian Trough analytically. They used a linear, inviscid, two-layer model without rotation to study the scattering of an incoming barotropic tide over a rapidly shoaling bottom. A similar model with a sill had been developed by Blackford (1978) for a combined Saguenay Fjord-Laurentian Trough cross-section.

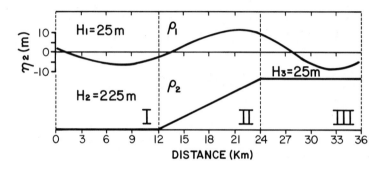

Figure 3. Internal tide generation model (Gratton and Saucier, 1987). A deep water semidiurnal tide (1 m) incident on the linear shoaling region generating internal tides (10 m) propagating upstream and downstream; the resulting interface displacement is shown.

The geometry of the model used by Gratton and Saucier (1987) is shown in Figure 3. It consists of two flat-bottom channels linked by a zone of rapid shoaling. The main results are the following: The shape of the shoaling region has little effect (less than 3%) on the amplitude and phase of surface and interface elevations. Only the fractional depth change is important and this simple model is sufficient to account for most of the tidal amplification between Pointe-au-Père and Tadoussac. The stratification has a drastic effect on the phases and amplitudes of the reflected and transmitted internal tides. By doubling the reduced gravity one finds changes of up to 50 % in the phase angles. Since the stratification is known to change with the neap-spring cycle, this explains the variety of phases found by Muir (1979) for the Upper estuary and by Forrester (1974) for the Lower estuary. The amplitude will also be modified by the neap-spring cycle. An incoming M_2 barotropic wave of 1.28 m (at Pointe-au Père; based on a harmonic analysis of nine years of sea-level data) will cause an interface displacement of 31 m at the head of the Laurentian Trough near spring tide, when the envelope due to the beating of the M_2 (24 m internal tidal amplitude) and S_2 (7 m internal tidal amplitude) tides achieves its maximum. Since the model uses an upper layer of 25 m, this will result in a surfacing of the interface. Therefore, any next-generation model should include a surfacing interface. The model results are sufficient to explain the cold zones observed at the head of the Laurentian Trough.

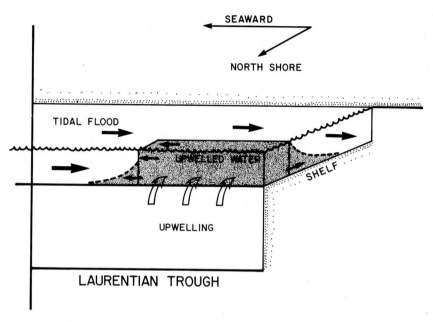

Figure 4. A conceptual depiction of how water upwelled onto the shelf might interact with the shelf's tidal circulation.

As noted earlier, cold water is not always confined to the head region. Figure 2 shows a band of cold water along the south shore, about 100 km long and 10-15 km wide. Its presence is somewhat unexpected since tidal activity is much weaker as one leaves the energetic head region. A cold anomaly was also found in hydrographic data by Gagnon (1977). He reports finding (in July 1974) a blob of cold water near the Bic Island Plateau (the bulge in bottom contours near Rimouski) resting on the shelf (30 m deep here). The surface expression of this feature is a cold region about 10 km across. Gagnon notes that the winds were not favorable to upwelling during the period of the observation. Gagnon refers to this blob as an upwelling spot and presumes that internal tides may account for its presence. We note that the shelf topography in the upstream half of the LSLE may promote upwelling activity remote from the head region. The internal tide, with a strong cross-channel baroclinic component (Forrester 1974), propagates seaward through the Laurentian Trough and raises intermediate-depth water twice daily. As this water rises with the cresting internal tide it will encounter the shelf break, where the Trough "walls" give way to the gentle slope of the shelf. The rising dense water may intrude onto the shelf and interact with the local topography or tidal circulation. Figure 4 is a conceptual depiction of how an upwelled cold mass might interact with the tides. Large shears between the spreading mass and tidal circulation may arise. Shear instabilities might fragment the large upwelled mass into blobs; mixing processes due to shear instabilities may allow the blobs to persist even as the pycnocline falls with the waning internal tide. Complete understanding of the mechanism creating cold blobs and promoting the growth of the cold anomaly along the south shore must await the acquisition of more detailed hydrographic information and the development of sophisticated models. However, we can safely say that internal tides must make a substantial contribution to the upwelling and mixing processes involved in the formation of the anomaly.

4. Vorticity waves.

Motions characterized by frequencies lower than the local inertial frequency ($\omega < f$) are governed by the conservation of angular momentum and displacements of the free surface play a less significant role in subinertial dynamics. The focus shifts to the dynamics of vorticity. The conservation of potential vorticity reads

$$\frac{d}{dt} \frac{(\xi_i + f)}{h_i} = 0 \ , \ \text{where} \quad \frac{d}{dt} = \frac{\partial}{\partial t} + u_i \frac{\partial}{\partial x} + v_i \frac{\partial}{\partial y} \quad , \quad \xi_i = \frac{\partial v_i}{\partial x} - \frac{\partial u_i}{\partial y} \ ,$$

and ξ_i is the relative vorticity of the motion in the layer i of thickness h_i. Increases in the layer thickness (or the total depth of the fluid for a homogeneous fluid) create vortex stretching, enhancing the fluid's relative vorticity. The interaction of vortex columns induced by vortex stretching can produce coherent "vorticity-wave" type motions (Longuet-Higgins, 1972).

A barotropic (homogeneous) fluid with no mean flow may support vorticity waves due to the gradient of planetary vorticity (Rossby waves) or due to the variations of the bottom topography (topographic Rossby waves). Rossby waves may have a truly planetary scale, being found in the atmosphere as standing oscillations circling the globe. Continental shelf waves, in contrast, are laterally confined to shelf scales, although they may propagate great distances along the shelf. Studies of shelf waves began when investigators (Hamon, 1964) found that sea level does not always respond barometrically to atmospheric pressure variations over continental shelves. Atmospheric disturbances can excite shelf wave modes, accounting for this discrepancy (Robinson, 1964; Mysak, 1967). Gill and Schumann (1974) showed that shelf waves are strongly excited by "weather band" variations in the alongshore wind stress. A good review of shelf wave motions is provided by Mysak (1980).

When horizontally or laterally sheared mean flows are present, a spectrum of shear modes develops. In the simplest case, a stratified fluid with vertical shear will exhibit sloping layer interfaces, due to the thermal wind relation. These interface slopes act like topographic variations and create vorticity waves. The shear modes can interact with one another or with the topographic modes to produce waves which extract energy from the mean flow and grow in time. These are unstable waves.

In the St. Lawrence, observational records are usually short and of limited spatial resolution. The largest temporal and best spatial coverages were obtained during 1978 in the Gaspé Current (Benoît et al., 1980) and during 1979 in the Rimouski-Matane region (El-Sabh et al., 1982; Koutitonsky, 1985). In general, the kinetic energy is almost evenly partitioned between tidal and low-frequency motions. The most energetic non-tidal spectral bands are 2-5 days, 8-15 days, and 70-80 days. In all bands, low-frequency motions appear to be surface-intensified (El-Sabh et al., 1982; Koutitonsky, 1985; Benoît, 1980). For the Rimouski-Matane region most of the kinetic energy is found in a broad band centered at 12 days (Koutitonsky, 1985). The horizontal coherence scale is at least on the order of 50 km for the 2-5 day and 8-20 day bands, although the coherence is sometimes weak. The coherence scales appear to be shorter at depth (Koutitonsky, 1985).

Once one allows for the shorter scales inherent to channel motions, the characteristics (large coherence scales) of the low-frequency motions in the LSLE correspond more closely to open ocean situations (MODE group, 1978) than to other estuaries or straits such as the Strait of Georgia (between Vancouver Island and Vancouver, B.C.; see the review by LeBlond, 1983). In the MODE experiment correlation scales were found to be 140 km at 500 m, 70 km at 1500 m and 55 km at 4000 m, with irregular patterns of coherence in the array. In the Strait of Georgia (Chang et al., 1976; Yao et al., 1982), the low-frequency energy is also surface intensified and concentrated in a broad peak centered at 10-25 days. The peaks are broader and correlation scales appear to be shorter than 10 km in the horizontal and 100 m in the vertical.

In the St. Lawrence all the modelling effort has been directed toward explaining the low-frequency motions in terms of topographic waves, with emphasis on the Rimouski-Matane area of the LSLE. Pelchat (1986) studied the propagation of free barotropic topographic waves in a variable breadth channel. The contribution of Pelchat's analytical model was to show that the wavenumber, phase velocity, and group velocity will be greatly modified by the geometry of the estuary along the direction of propagation. They will adjust, within a wavelength, to the changes in the estuary width.

In all other studies relevant to the LSLE, the estuary is modelled as an infinitely long and straight channel on an f-plane. Lie and El-Sabh (1983), numerically, and Gratton and LeBlond (1986), analytically, introduced stratification effects by using a two-layer system. These two-layer models (with that of Gratton, 1983) were the first attempts at modelling topographic wave propagation in a channel with non-monotonic topography; i.e. with a reversal of bottom slope. For the first time, some of the low-frequency features of the Strait of Georgia and the LSLE could be accounted for, namely: the short temporal and spatial scales (Strait of Georgia), the surface intensification of the currents, and the observed cross-channel velocities.

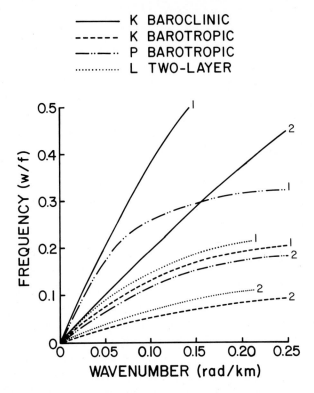

Figure 5: Comparison of dispersion relations (first two modes) for the models of Lie and El-Sabh (1983): L; Koutitonsky (1985): K; and Pelchat (1986): P.

The most thorough analysis and modelling effort in the LSLE was that of Koutitonsky (1985). He developed a continuously stratified topographic wave numerical model. He found that the first two empirical modes (Kundu and Allen, 1976) corresponded to his calculated topographic wave modes. He also found that the first empirical mode is coherent with the atmospheric forcing at 3-4 days and 8 days. Koutitonsky's results are the first to show a strong coherence between winds and low-frequency current measurements. He also showed (see Fig. 5) that two-layer models were underestimating the frequency and the phase speed of the topographic waves in the Rimouski-Matane section.

Koutitonsky's (1985) results also raised a lot of questions. The structure of his topographic waves was seemingly more like Kelvin waves than topographic waves. Most of the structure in the predicted velocity fields was within 15-20 km of the coast: that is, shore-trapped. This raises questions about the 50 km coherency scale. It also suggests that motions at larger lateral and longitudinal scales may be more barotropic than baroclinic, if topographic effects are the most important ones. On the other hand, other aspects of vorticity dynamics may take precedence at larger spatial scales, namely shear effects. All these results show that a high resolution, large spatial coverage sampling program is needed in the LSLE.

To complete the topic of vorticity waves we will briefly discuss unstable waves. A comprehensive treatment of instability in fluid flows may be found in Drazin and Reid (1981). As we noted in the introduction to this section, instability arises when the natural modes of the system interact, yielding a class of waves capable of extracting energy from the mean flow. Instabilities are generally classified into two types: (i) baroclinic instability in which the potential energy due to the slope of the isopycnals (as required by geostrophy when vertical shear is present) feeds growing waves., and (ii) barotropic instability, where the kinetic energy of the mean flow is released to disturbances. Hart (1974) has analyzed the cases in which each type of instability dominates. Roughly speaking, barotropic instability is more important for currents confined to a thin upper layer.

Lavoie et al. (1985), Lacroix (1987), and Mertz et al. (1988) have reported satellite thermal images showing "backward breaking" wave type features in the LSLE. Figure 2 here is reproduced from a satellite thermal image and captures a backward breaking wave in the downstream half of the LSLE and a cyclonic eddy in the upstream half. The term backward breaking wave was introduced by Griffiths and Linden (1981) to describe features evolving in an unstable boundary current in their laboratory simulations. A weak region of cyclonic vorticity wraps around a strong anti-cyclone to form a sharp crested streamer pointing upstream ("backwards"). Griffiths and Linden noted that the development of unstable waves is quite typical of buoyant boundary jets. This establishes the connection with the LSLE, where a buoyant jet flows along the south coast. Mertz et al. (1988) have shown that this jet is unstable by applying simple barotropic and baroclinic instability models. They concluded that both baroclinic and barotropic instability contribute to the growth of the observed wave features. Both types of instability are characterized by a length scale of about 10 km (the internal Rossby radius of deformation for baroclinic instability, and the half width of the current for barotropic instability). Thus, the expected wavelength for unstable waves is $2\pi * 10$ km or about 60 km, which agrees with observations.

The south shore jet of the LSLE joins the cyclonic circulation of the Northwest Gulf of St. Lawrence to form the Gaspé Current. Here, too, unstable wave activity has been reported (Tang, 1980; Lacroix, 1987; Mertz et al., 1988). The combined system consisting of the south shore jet and the Gaspé Current stretches over 400 km and must rank with other, better-known, unstable buoyant jets such as Australia's Leeuwin Current (Griffiths and Pearce, 1985) and the Norwegian Coastal Current (Carstens et al., 1984) in terms of interest to the oceanographic community.

5. Conclusions

The shallow, rough-bottomed Upper estuary is a true estuary where strong tidal flows over the irregular topography produce most of the mixing observed in the St. Lawrence Estuary. Internal tides and waves induced by the interaction of the flow field with the bottom topography contribute to the complex circulation found here. The LSLE has a maritime character, being much deeper and wider than upstream areas. The sub-tidal variability is as strong as the dominant M_2 tide and largely consists of mainly wind-forced topographic waves and perhaps shear modified (unstable) topographic waves. Variability of this sort is typical of oceanic domains, emphasizing the maritime nature of the Lower St. Lawrence Estuary.

Much work remains to be done on the phenomena discussed in this paper. The complexity of the Upper estuary suggests that an understanding of its dynamics will be achieved only through numerical models. In the Lower estuary, processes are more likely to be amenable to "quasi-linear" treatments than in the Upper estuary and we anticipate that great progress will be made in understanding its dynamics in the near future.

REFERENCES

Baines, P.G. 1979. Observations of stratified flow over two-dimensional obstacles in fluid of finite depth. Tellus, 31, 351-371.

Baines, P.G. 1982. On internal tide generation models. Deep-Sea Res., 29, 307-338.

Benoît, J., M.I. El-Sabh and G.L. Tang 1985. Structure and characteristics of the Gaspé Current. J. Geophys. Res., 90, 3225-3236.

Blackford, B.L. 1978. On the generation of internal waves by tidal flow over a sill - a possible nonlinear mechanism. J. Mar. Res., 36, 529-549.

Borne de Grandpré, C. and M.I. El-Sabh 1980. Etude de la circulation verticale dans l'estuaire du Saint-Laurent au moyen de la modélisation mathématique. Atmos.-Oceans, 18, 304-321.

Borne de Grandpré, C., M.I. El-Sabh and J.C. Salomon 1981. A two-dimensional numerical model of the vertical circulation of tides in the St. Lawrence Estuary. Estuar. and Coastal Shelf Sci., 12, 375-389.

Carstens, T., T.A. McClimans and J.H. Nilsen 1984. Satellite imagery of boundary currents. In Remote Sensing of Shelf Sea Hydrodynamics, J.C. Nihoul, Ed., Elsevier, Amsterdam, pp. 235-256.

Chang, P., S. Pond and S. Tabata 1976. Subsurface currents in the Strait of Georgia, west of Sturgeon Bank. J. Fish. Res. Board Can., 23, 2218-2241.

Deguise, J.C. 1977. High frequency internal waves in the St. Lawrence Estuary. M.Sc. Thesis, McGill University, Montreal, Canada, 93 pp.

Drazin, P.G. and W.H. Reid 1981. Hydrodynamic Stability. Cambridge University Press, London, 527 pp.

El-Sabh, M.I. 1988. Physical Oceanography of the St. Lawrence Estuary. In Hydrodynamics of Estuaries, B. Kjerfve ed., CRC press, Boca Raton, vol. 2, 61-78.

El-Sabh, M.I. 1979. The lower St. Lawrence Estuary as a physical oceanographic system. Natur. Can., 106, 55-73.

El-Sabh, M.I., H.-J. Lie and V.G. Koutitonsky 1982. Variability of the near-surface residual current in the lower St. Lawrence Estuary. J. Geophys. Res., 87, 9589-9600.

Farmer, D. and H. Freeland 1983. The physical oceanography of fjords. Prog. in Oceanogr., 12, 147-219.

Forrester, W.D. 1972. Tidal transports and streams in the St. Lawrence River and Estuary. Int. Hydrogr. Rev., 49, 95-108.

Forrester, W.D. 1974. Internal tides in the St. Lawrence Estuary. J. Mar. Res., 32, 55-66.

Gagnon, M. 1977. Etude d'océanographie physique dans la région de Rimouski: Estuaire du Saint-Laurent. M. Sc. Thesis, Université du Québec à Rimouski, Rimouski, Canada, 159 pp.

Gargett, A.E. 1984. Vertical eddy diffusivity in the ocean interior. J. Mar. Res., 42, 359-393.

Gill, A. and E.H. Schumann 1974. The generation of long shelf waves by the wind. J. Phys. Oceanogr., 4, 83-90.

Gratton, Y. 1983. Low-frequency vorticity waves over strong topography. Ph. D. Thesis, Unniversity of British Columbia, Vancouver, Canada, 143 pp.

Gratton, Y. and P.H. LeBlond 1986. Vorticity waves over strong topography. J. Phys. Oceanogr., 16, 151-166.

Gratton, Y., G. Mertz and J.A. Gagné 1988. Satellite observations of upwelling and mixing in the St. Lawrence Estuary. J. Geophys. Res., 93, 6947-6954.

Gratton, Y. and F. Saucier 1987. Internal tide generation at the head of the Laurentian Trough. CMOS XXI annual meeting, St-John's, Canada. Abstract only.

Griffiths, R.W. and P.F. Linden 1981. The stability of buoyancy-driven boundary currents. Dyn. Atmos. Oceans, 5, 281-306.

Griffiths, R.W. and A.F. Pearce 1985. Instability and eddy pairs on the Leeuwin Current south of Australia. Deep-Sea Res., 32, 1511-1534.

Hamon, B.V. 1962. The spectrum of mean sea level at Sydney, Coff's Harbour, and Lord Howe Island. J. Geophys. Res., 67, 5147-5155.

Hart, J.E. 1974. On the mixed stability problem for quasi-geostrophic ocean currents. J. Phys. Oceanogr., 4, 349-356.

Huntsman, A.G. 1923. The influence of tidal oscillations on vertical circulation in estuaries. Trans. R. Soc. Can., Ser. 3, 17, 11-14.

Ingram, R.G. 1975. Influence of tidal-induced vertical mixing on primary productivity in the St.Lawrence Estuary. Mém. Soc. R. Sci. Liège, 7, 59-74.

Ingram, R.G. 1976. Characteristics of a tide-induced estuarine front. J. Geophys. Res., 81, 1951-1959.

Ingram, R.G., 1978. Internal wave observation off Ile Verte. J. Mar. Res., 36, 715-724.

Ingram, R.G. 1983. Vertical mixing at the head of the Laurentian Channel. Estuar. Coastal Shelf Sci., 16, 333-338.

Ingram, R.G. 1985. Frontal characteristics at the head of the Laurentian Channel. Nat. Can., 112, 31-38.

Koutitonsky, V.G. 1985. Subinertial coastal-trapped waves in channels with variable stratification and topography. Ph. D. Thesis, State University of New York at Stony Brook, Stony Brook, N.Y., 169 pp.

Koutitonsky, V.G. and R.E. Wilson 1988. Coastal trapped waves in continuously stratified channels. Part I: Numerical description of behavioral properties. J. Phys. Oceanogr., 18, 652-661.

Kundu, P.K. and J.S. Allen 1976. Some three-dimensional characteristics of low-frequency current fluctuations near the Oregon coast. J. Phys. Oceanogr., 6, 181-199.

Lacroix, J. 1987. Etude descriptive de la variabilité spatio-temporelle des phénomènes physiques de surface de l'estuaire maritime et de la partie ouest du Golfe du Saint-Laurent à l'aide d'images thermiques du satellite NOAA-7. M. Sc. Thesis, Université du Québec à Rimouski, Rimouski, Canada, 186 pp.

Lavoie, A., F. Bonn, J.M.M. Dubois and M.I. El-Sabh 1985. Structure thermique et variabilité du courant de surface de l'estuaire maritime du Saint-Laurent à l'aide d'images du satellite HCMM. Can. J. Remote Sensing, 11, 70-84.

LeBlond, P.H. 1983. The Strait of Georgia: Functional anatomy of a coastal sea. Can. J. Fish. Aquat. Sci., 40, 1033-1063.

Lee, C.Y. and R.C. Beardsley 1974. The generation of long nonlinear internal waves in weakly stratified shear flow. J. Geophys. Res., 79, 453-462.

Lie, H.-J. and M.I. El-Sabh 1983. Formation of eddies and transverse currents in a two-layer channel of variable bottom with application to the Lower St. Lawrence Estuary. J. Phys. Oceanogr., 13, 1063-1075.

Longuet-Higgins, M.S. 1972. Topographic Rossby waves. Mém. Soc. R. Sci. Liège, 6e Ser. 2, 11-16.

Mertz, G., M.I. El-Sabh, D. Proulx and A. Condal 1988. Instability of a buoyancy-driven current: The Gaspé Current and its St. Lawrence precursor. J. Geophys. Res., 93, 6885-6893.

Mode Group 1978. The mid-ocean dynamics experiment. Deep-Sea Res., 25, 859-910.

Muir, L.R. 1979. Internal tides in the middle estuary of the St. Lawrence. Nat. Can., 106, 27-36.

Muir, L.R. 1981. Variability of temperature, salinity and tidally-averaged density in the middle estuary of the St.`Lawrence. Atmos.-Ocean, 19, 320-336.

Mysak, L.A. 1967. On the theory of continental shelf waves. J. Mar. Res., 25, 205-227.

Mysak, L.A. 1980. Recent advances in shelf wave dynamics. Rev. Geophys. Space Phys., 18, 211-241.

Neu, H.J.A. 1970. A study on mixing and circulation in the St.Lawrence Estuary up to 1964. Atlant. Oceanogr. Lab. Bedford Inst., Rep. Ser. 1970-9, Dartmouth, Canada, 31 pp.

Pelchat, B. 1986. Ondes de vorticité dans des canaux de largeur variable. M. Sc. Thesis, Université du Québec à Rimouski, Rimouski, Canada, 119 pp.

Reid, S.J. 1977. Circulation and mixing in the St. Lawrence Estuary near Ile Rouge. Bedford Inst. Oceanogr. Report Ser. BI-R-77, Dartmouth, Canada, 36 pp.

Robinson, A.R. 1964. Continental shelf waves and the response of sea level to weather systems. J. Geophys. Res., 69, 367-368.

Steven, D.M., 1974. Primarary and secondary production in the Gulf of St. Lawrence. Mar. Sci. Center, McGill University Rep. No 26, Montreal, Canada, 116 pp.

Therriault, J.-C. and G. Lacroix 1976. Nutrients, chlorophyll, and internal tides in the St. Lawrence Estuary. J. Fish. Res. Board Can., 33, 2747-2757.

Tang, C.L. 1980. Observation of wavelike motion of the Gaspé Current. J. Phys. Oceanogr., 10, 853-860.

Yao, T., S. Pond and L.A. Mysak 1982. Low-frequency subsurface current and density fluctuations in the Strait of Georgia. Atmos.-Ocean, 20, 343-356.

Chapter 6

Recent Sediments and Sediment Transport Processes in the St. Lawrence Estuary

Bruno d'Anglejan

Department of Geological Sciences, McGill University, Franck Dawson Adams, 3450 University
Montréal, Québec, Canada H3A 2A7

ABSTRACT

Glacial erosion as well as the combined effects of the post-Wisconsin submergence and uplift have shaped the physiography and determined the bathymetry of the St. Lawrence Upper and Lower Estuaries. On the channel floor, ice contact facies overlain by post-glacial transgressive (Goldthwait) muds form two distinct stratigraphic units under gravel and sand lag deposits, and contemporaneous estuarine silty clays. The upper region is mostly shallow (depths less than 30 m) and has the characteristics of a macrotidal, relatively well-mixed estuary, with a turbidity maximum controlled by the gravitational circulation, tidal and wind-induced turbulence, and a seasonal cycle of deposition and erosion on mudflats and marshes. Advective transport of fresh turbid waters takes place along the south shore, with a negligible net deposition and a mean transit time of the suspended matter estimated at around one year. The lower partially mixed region is dominated by the deep (300 m to 350 m) Laurentian Trough. It is the site of active sedimentation, with accumulation rates of between 1.5 mm/a and 4 mm/a, and up to 60 meters of Holocene fine-grained deposits. Particle distribution and transport is determined by the three-layer vertical structure of watermasses found in the summer months, by inputs from the Upper Estuary and the Saguenay, and by biological production. Of the 3.6×10^6 tonnes of suspended load introduced yearly from the river into the estuary, a significant fraction settles over the Lower Estuary. Basic particle composition is inherited from the Champlain and Goldthwait deposits eroded in the St. Lawrence Lowlands, and is influenced by geochemical transformations during estuarine mixing as well as by biogenic and diagenetic processes.

Key Words: Quaternary history; geomorphology; sediments; sedment transport; particulate suspended matter; sediment mass balance; marshes and tidal flats.

1. Introduction

Both in terms of physical dimension and water transport, the estuary of the St. Lawrence is among the largest in the world. Yet, due to peculiarities in drainage basin configuration, climatic and hydrographic factors and recent history, its present sediment contribution to the Gulf of St. Lawrence is remarkably low. Some other aspects which make it uniquely interesting among estuaries of comparable size are: 1) the deep imprint that glaciation, as

Coastal and Estuarine Studies, Vol. 39
M. I. El-Sabh, N. Silverberg (Eds.)
Oceanography of a Large-Scale Estuarine System
The St. Lawrence
© Springer-Verlag New York, Inc., 1990

well as its recovery from it, has left on its physiography and on its sedimentological characteristics; 2) its location at the lower limit of the subarctic region in eastern Canada, where annual freezing and ice decay exert a major influence on hydrographic, sedimentary and biogeochemical processes; 3) the large tidal range, as much as 5 m near the river entrance, and the large water volumes displaced by the tides relative to the freshwater input. In this respect, the St. Lawrence Estuary contrasts sharply with microtidal estuaries studied further south along the eastern seaboard of the United States (Meade, 1972), in which gravitational flow dominates and determines sediment retention.

This paper reviews information gathered over the last fifteen years on the post-glacial history of the estuary and on its Holocene record. The result of many studies on recent sedimentation and suspended sediment transport processes are also presented and synthesized.

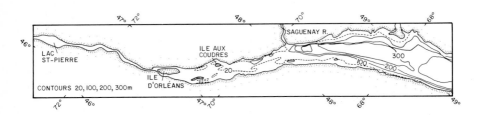

Figure 1. General bathymetry of the estuary.

2. Physiography

The estuary is subdivided on the basis of depth, circulation and hydrography, into the Upper Estuary, between Quebec City and the Saguenay, and the Lower or Maritime Estuary, between the Saguenay and the opening into the Gulf of St. Lawrence near Pointe-des-Monts. They cover areas of 3,470 km^2 and 9,350 km^2, respectively and have an overall length of 400 km. The total area is about 1% of the St. Lawrence drainage basin. The regional physiography has been examined in detail by Loring and Nota (1973). To a first approximation, the increases in width and cross-sectional area away from the river exit are exponential (Forrester, 1972). The width is 2 km near Ile d'Orleans, 24 km near the Saguenay, and reaches 50 km at Pointe-des-Monts (Figure 1).

The raised Laurentian plateau made up of the Precambrian Shield crystalline rocks terminates over a steep coastline to the north, while the south shore which borders along the Cambro-Silurian sedimentary rocks of the Appalachians is much more subdued. The contact between the two provinces is set along a major structural boundary, known as Logan's Line, separating undisturbed Ordovician rocks overlying the basement to the north from highly distorted beds of the same age to the south. This boundary approximately follows the centre of the estuary.

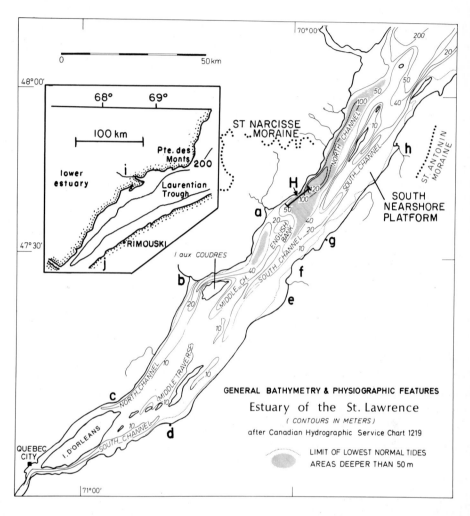

Figure 2. Physiographic diagram of the Upper Estuary showing the major features. Insert: Lower Estuary. Geographic locations: a: R.La Malbaie; b: R. du Gouffre; c: CapTourmente; d: Montmagny; e: Baie de Sainte-Anne; f: Rivière Ouelle; g: Kamouraska; h: Rivière-du-Loup; i: R.Manicouagan; j: R.Trois-Pistoles.

The two regions of the estuary differ considerably in depth. Approximately two-thirds of the Upper Estuary is shallower than 30 meters. Outliers of uplifted Paleozoic rocks form numerous northeast trending islands. These islands, the basement structural controls, as well as depositional or erosional features left by glaciers divide the Upper Estuary into a complex mosaic of physiographic sub-regions (Figure 2): -1) The North Channel, which consists of deeply incised interconnecting basins, with depths reaching 100 meters or more north of Ile-aux-Lievres; -2) the western platform between Ile d'Orleans and Ile aux Coudres, which is no deeper than 30 meters and is covered by migrating sand banks swept by strong tidal currents; -3) the English Banks, a broad central rise made up of moraine deposits; 4) the South Channel formed over Quaternary deposits; 5) the south shore inner shelf, a shallow 200 km^2 platform with MLLW depths of 5m or less. Physiographic contrasts between the North and South Channels create a noticeable assymetry in depth across the Upper Estuary, which exerts a major control both on the tidal penetration and on the mean tidal flow. This tends to be upstream over the North Channel and downstream over the South Channel.

The physiography of the Maritime Estuary is dominated by the Laurentian Trough, a long U-shaped glaciated valley which has regional depths above 300 meters with long axial depressions exceeding 400 meters. A precipitous wall with a 150 m drop marks the head of the Trough near the Saguenay. Steep-sided slopes occur along its entire length of 1500 km to Cabot Strait. In the estuary, the Laurentian Trough is flanked north and south by narrow shelves which do not exceed 15 km in width (Figure 1).

3. Late glacial and Holocene history

The present channel physiography is thought to be the product of ice movement during the Pleistocene glaciations acting on a preexisting Tertiary drainage system. During late Wisconsin, lobes from ice domes located to the north (Laurentian) and south (Appalachian) converged over the St. Lawrence valley (Dyke and Prest, 1987). The Saguenay fiord was a major corridor for the moving ice, which at its exit diverged both upstream and downstream along the valley (Loring and Nota, 1973). Laurentide ice also propagated south along other river channels now found on the north shore of the St. Lawrence Estuary (Le Gouffre, La Malbaie, Manicouagan). During ice retreat, these became the pathways for fluvioglacial sediment transport. The same applies to tributaries along the south shore draining the Appalachian highlands (Ouelle, Riviere-du-Loup, Trois Pistoles).

The northward retreat of the Wisconsin ice sheet led at about 14,000 y BP into a marine transgression over the region of the Maritime Estuary known as the Goldthwait Sea stage (Dionne, 1977). It was held back west of the Laurentian Trough near Riviere-du-Loup by a fragment of the retreating ice lobe. As this barrier disappeared near 12,800 y BP, the sea extended quickly over the present Upper Estuary to Quebec City. The subsequent retreat of the ice front to the west at about 12,000 y BP caused the sea to spread over the lower St. Lawrence valley. This submergence, known as the Champlain Sea stage, lasted for about 2000 years. It was followed by a dominantly regressive phase due to the isostatic rebound.

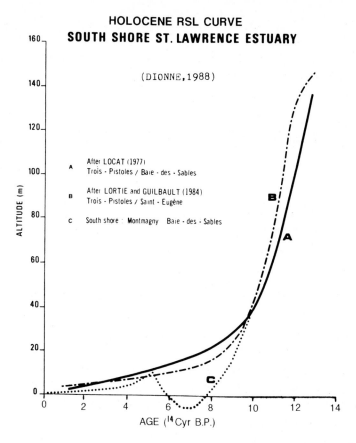

Figure 3. Relative sea level curve for the St. Lawrence Estuary during the last 14,000 years (from Dionne,1988).

The Holocene history of the St. Lawrence Estuary is dominated by complex fluctuations in relative sea level caused by the opposing and non-synchronous effects of the eustatic sea level rise and of the isostatic rebound. Various emergence curves have been documented from raised shorelines along the north and south shores of the estuary (Dionne, 1988; Figure 3). Marine maxima of 180 meters near Quebec City and of 100 meters near Matane have been established. Since deglaciation, the mean uplift rate decreased from 3 cm/year to 1mm/yr. Emergence was not continuous, and recent data obtained by Dionne (ibid) on the south shore near Montmagny provides evidence for a 8 m to 10 m transgression between 5,800 y BP and 4,400 y BP (Figure 3). Sea level reached its present stand about 3000 y BP. Based on recent erosion of the intertidal marshes, there is indication of a slight contemporaneous rise in relative sea level (Dionne, 1986).

4. Quaternary stratigraphy

The St. Lawrence Estuary Quaternary record was initially explored by boreholes (Lee, 1962; Simard, 1971). More comrehensive information has been obtained since by seismic reflection surveys (d'Anglejan and Brisebois, 1974; Syvitski and Praeg, in press).

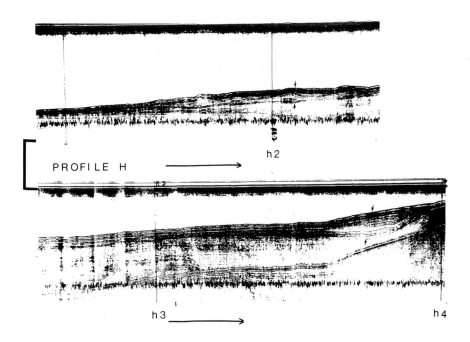

Figure 4. Sparker seismic reflection profiling of the North Channel off La Malbaie, Upper Estuary (see Fig. 2 for location): interval between arrows indicates the modern sandy deposits overlying accoustically transparent muds of the Goldtwaith Sea (d'Anglejan and Brisebois, 1978)

In the North Channel of the Upper Estuary, Syvitski and Praeg found more than 350 meters of glacial and postglacial sediments. Most of these consist of seismically transparent deep-water marine clays of Goldthwait age. These clays are overlain by about 30 meters of recent muds and lag deposits. Using a 4.5 KHz EG&G pinger probe, d'Anglejan and Brisebois (1974) have identified a regionally continuous reflector at subbottom depths between 10m and 30m marking the top of the transgressive clays. Near La Malbaie, sparker records show large buried bedforms representing reworked coarse fluvioglacial deposits supplied in the past by the La Malbaie River (d'Anglejan and Brisebois, 1978 Figure 4.a). There is evidence that lag deposits, presumably issued from sources on the south shore, extend conformably over the South Channel, while in the nearshore, erosion has exposed the older clays (d'Anglejan, 1981a). The English Banks, between the North and the South Channel, is now considered to be a front moraine left by a piedmont glacier readvance along the La Malbaie valley during the Laurentide ice sheet retreat (Poulin, 1976).

In the Lower Estuary, Quaternary sediments are 450 m thick (Syvitski and Praeg, in press). Seismic records over the Laurentian Trough indicate up to 250 meters of ice-contact deposits, representing basal tills and other sub-ice depositional features, overlain by wedge-shaped, coarse ice-proximal sediments reaching 200 meters near the head of the Trough. These record a stillstand of the retreating ice front at that site. The Goldthwait clays reach a maximum thickness of 167 meters in the thick western region of the Trough. Deltaic fans, the largest one offshore of the present Manicouagan delta being 90 meters thick, extend over the north slope. More than 60 meters of recent muds are found in the deep southern region of the Trough. On the shelves, these muds grade into lag deposits or merge into modern deltaic sediments in front of the local rivers.

5. Distribution of surficial sediments

The general distribution and composition of surficial sediments in the Upper Estuary (Figure 5) has been examined by d'Anglejan and Brisebois (1978), while Loring and Nota (1973) have reported on recent deposits in the Lower Estuary. These two regions are here reviewed separately.

5.1. Upper Estuary

Estuarine sediments supplied to the estuary under the present transportation regime form only minor deposits in the Upper Estuary, covering no more than 10% of the floor areas, as most river fine sediments are exported in suspension at a fast rate into the Lower Estuary and the Gulf of St. Lawrence.

These sediments are brownish grey soupy muds (pelites) consisting of silty clays, in which the clay-size fraction (d less than 0.002mm) is 20% to 30%, the sand fraction in the fine to very fine class (d between 0.063mm and 0.125 mm) ranges between 5% and 70%, and there are scattered gravel-size rock fragments. The silt-clay ratio is characteristically rather uniform (mean value about 2.3:1), being presumably controlled by the size composition of flocs in the water column. These modern muds form thin (less than 0.5 m) blankets over shallow depressions occupying the upper reaches of the North and South Channels seaward of Cap Tourmente and Montmagny (d'Anglejan and Brisebois, 1978; Silverberg and Sundby, 1979). These deposits mark the general area of the turbidity maximum, and tend to wax and wane seasonally in response to water discharge (Lucotte and d'Anglejan, 1986). As a result of the local circulation, a large lens of estuarine muds occupies the centre of Baie de Sainte-Anne on the south shore (d'Anglejan et al, 1981). Further downstream, the same sediments form thicker and more permanent accumulations over the central basin southwest of Ile-aux-Lièvres, which includes the Kamouraska banks and segments of the North and South Channels, as there is a reduction in flow intensity in this area compared to further upstream. Over the banks, these estuarine muds, less than 1m

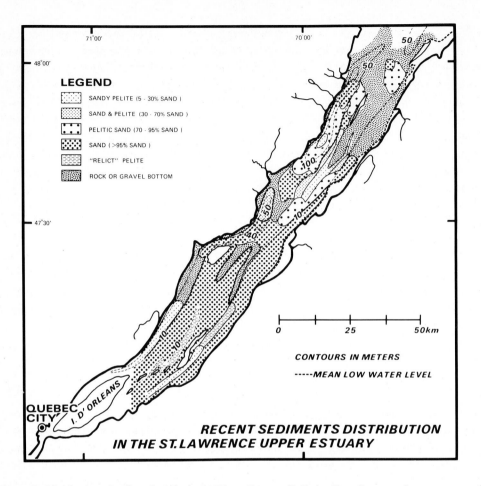

Figure 5. Distribution of surface deposits in the Upper Estuary. 'Relict' pelite refers to surface exposures of the Goldtwaith clays (d'Anglejan and Brisebois, 1978).

thick, are fixed by an abundant community of polychaetes (*Teribellidae* sp.) forming dense mats of anastomizing tubes. Further offshore, the muds are are underlain regionally by a 10 to 20 cm compact layer of mixed sands and gravels which sits unconformably over uniform light grey dominantly clay-size sediments. Based on textural and faunal evidence, these clays are interpreted as being Goldthwait or early post-Champlain in age (d'Anglejan and Brisebois, 1978). The intermediate coarse layer may record a brief period of intense shoreline erosion which followed a short transgressive phase at about 5800 yBP, as evidenced onshore by Dionne (1988). Although physical and biological mixing makes it impossible to determine absolute rates of deposition radiometrically, a consistent lack of ^{137}Cs activity whithin the surface portion of the modern muds suggests that net deposition is insignificant.

Surface sediments over the wide subtidal platform along the south shore are made up of mixed gravel and sand, with ice-rafted pebbles forming a thin compact layer over silty clays which on textural evidence do not appear to be recent. The coarse layer, very similar to the one discussed above and perhaps having the same origin, has a thin cover of settled muds, except in front of stream entrances, as off Rivière-du-Loup, where outflow deposits form broad lenses across the shelf (S. Lorrain, personal communication). There is also little deposition in the North Channel past Ile-aux-Coudres, where lag deposits consisting mainly of relict sands (see below) are abundant in the depressions, while the bedrock lies bare on the higher sills and on the steep sides.

Relict postglacial marine clays of Goldthwait or Champlain age, which underlie most of the St. Lawrence Lowlands, are found exposed or are buried under a light cover of modern sediments over many regions of the Upper Estuary. They mainly consist of a stiff light grey pelite characterized by an extremely fine grain size (more than 80% particles smaller than 0.002 mm) representing a deep water depositional facies. They outcrop at many localities in the intertidal zone or along stream channels. They occur under a thin layer of recent deposits in the South Channel (Loring and Nota, 1973) and in the Middle Traverse upstream of Ile-aux-Coudres (d'Anglejan and Brisebois, 1978). They are easily distinguished by color, particle size and plasticity from modern sediments.

At many places, the floor of the Upper Estuary is covered with an uneven layer of coarse lag deposits consisting of mixed gravel and sand, or well-sorted sands, often in a matrix of recent silts. These are either reworked glaciomarine sediments of Goldthwait age, or elements eroded from desintegrating moraines and raised beaches along the shores by proglacial streams following ice retreat. For instance, the downcutting during emergence of stream channels along the steep north shore, such as Le Gouffre and La Malbaie rivers, supplied large volumes of sands to the North Channel. The well-sorted sands forming large sand waves in the narrows north of Ile-aux-Coudres are most likely derived from the Le Gouffre river (d'Anglejan, 1971). Further downstream, thick sand lenses occupying the depression north of Ile-aux-Lièvres are a continuation of the St. Narcisse moraine, a large recessional moraine which intersects the North Channel there.

The degree of sorting of these sands depends on exposure to tidal currents and on the rates of settling. These sands are easily moved by the 50 cm/s to 100 cm/s near-bottom currents. This is consistant with theory and with observations of active bedload transport in all regions above 30m, where such velocities prevail. These processes are active over the platform to the northeast of Ile d'Orleans, a region covered with drifting sand banks and swept by particularly strong currents because of shoaling and increasing tidal convergence. Bottom sediment transport in that region leads to recurring siltation of the navigation channel, which needs to be maintained by frequent dredging to the required depth of 13m . The mobility of coarse sands under tidal motion is also well illustrated by the numerous fields of large sand dunes (amplitudes up to 15 meters; wave length up to 100 meters). These are particularly well developed in the North Channel near Cap Tourmente, in the Ile-aux-Coudres channel, over the English Banks and in the South Channel between Pointe-aux-Orignaux and Ile Verte.

4.2. Lower Estuary

The floor of the Laurentian Trough is covered with occasionally sandy pelites (5 to 30% sand; 30 to 55% clay-size fraction) which become progressively sandier along the slopes (Loring and Nota, 1973). Over the bottom of the Trough, the mean silt-clay ratio is close to 1:1. The sediments are composed essentially of detrital Shield minerals. The organic carbon content in a 45 cm box core obtained in typical sediments of the Trough varied between 1.8% and 1.3% (dry weight), with a very low inorganic carbon content (less than 0.04%) (Sundby et al, 1983). These pelites represent settling out over the Trough of the present supply from the St. Lawrence drainage system, or that being winnowed from shelf deposits. They tend to accumulate in the estuary due to the net landward flow along the bottom. They have maximum thicknesses of more than 60 meters in the southern section of the Trough over which the fresh water outflow is concentrated (Syvitski and Praeg, in press). Sedimentation rates ranging between 1.4 mm/a near the Gulf and 3.8 mm/a at the west end of the Trough have been measured using sediment traps, in close agreement with rates obtained by ^{210}Pb measurements in the surface sediments (Silverberg et al, 1986).

From submersible observations in the Lower Estuary, Syvitski et al. (1983) distinguished four benthic zones along the slope of the Trough on the basis of faunal community, sediment texture, current intensity and seston concentration. Although in places highly bioturbated by demersal fishes and sessile organisms, among which abundant polychaetes, the deeper bathyal zone has a sharp sediment-water interface. Increases in current energy and grain-size as well as changes in fauna define the infaunal and Ophiura zones on the slopes. Due to concentration of ice floes on the upper south slope, an ice-rafting zone is well developed only there, while the shallower wave base zone on the shelves is marked by intense sorting and linear features.

Sediments on the shelves are heterogeneous and patchy, with an abundance of lag deposits and ice rafted components. Muddy areas are related to local stream discharge. Extensive sandy patches at the edge of the shelf off Rimouski are out of equilibrium with present conditions and suggest a regressive-transgressive sequence (Silverberg, 1978).

5. Suspended sediment dynamics and composition

5.1. Physical factors controlling the suspended sediment distribution:

It is estimated that about 3.6×10^6 metric tons of fine sediments in suspension are exported annually by the St. Lawrence river (Milliman and Meade, 1983), nearly two orders of magnitude less than the Mississipi which carries not quite 50% more water. This represents the only major source of solids to the estuary. The solid input is small, in relation to either the fresh water discharge (11.9×10^3 m^3 s^{-1}; El Sabh, 1988) and the mean M$_2$ tidal transport (0.72×10^6 m^3 s^{-1}; Forrester, 1972), which should lead to rapid dilution of suspended particulate matter by turbulent mixing. However, the processes of dispersal and seaward transport are also controlled by the estuarine circulation.

According to Hansen and Rattray (1966)'s classification, the Upper Estuary falls in the category of estuaries where mixing is "appreciable", ranging from well mixed to moderately mixed in a seaward direction, while the Lower Estuary is partially mixed (El Sabh, 1988). In such estuaries, the downstream salt and water transport in the surface layer is balanced by a net landward motion along the bottom, which helps maintain a turbidity maximum in the null velocity zone at the upstream end of the salt intrusion (Dyer, 1973). Such a feature is common in most estuaries where gravitational flow at depth is required for salt and water balance (Meade, 1972; Peterson et al, 1975; Festa and Hansen, 1978). Besides the estuarine circulation, at least two other factors contribute to the turbidity maximum of the St. Lawrence Estuary: 1) the strong turbulence maintained by tidal currents, wind mixing and possibly breaking internal waves, which prevents deposition of fine particles and causes resuspension of silts and clays. The role of turbulence induced by tidal friction in maintaining high turbidity in the shallow headward regions has been noted in many estuaries (e.g., Gironde, Seine) with large tidal range (Avoine, 1981); 2) the increasing ebb-flood assymetry of the tidal wave upstream, due to shoaling and topographic convergence. In macrotidal estuaries such as the St. Lawrence, where the tidal range reaches nearly 5 m near the head, this causes upstream particle transport by the strong flood to dominate over the seaward motion during the weaker, longer ebb, contributing to form a tidal sediment trap (Allen et al, 1980). This process is particularly effective in the Gironde. In the St. Lawrence Estuary, its existence has been demonstrated by recent observations of Hamblin et al. (1988), and formalised by Hamblin (in press) in a one-dimensional model of particle transport. These authors concluded that the turbidity maximum is in part maintained by the Reynolds transport of particles associated with the effects of the tidal assymmetry on the turbulent field. In the St. Lawrence, because of intense turbulence and tidal exchanges, sediments are trapped either in suspension or on marginal tidal marshes and flats, rather than on the channel floor.

The core of the turbidity maximum (Figure 6) occurs northeast of Ile d'Orleans, within a broad region of brackish waters and more diffuse turbidity generally extending beyond Ile-aux-Coudres. In that region, concentrations of suspended particulate matter (SPM) ranging between 50 mg/L and 200 mg/L are commonly observed. These are highly variable in space and time, and are only loosely correlated with the ebb-flood cycle because of intense turbulent mixing. Peak concentrations are found near the upstream end of the North Channel, where values exceeding 400 mg/L have been recorded over the bottom near low water. Since concentrations in the river upstream of Quebec City are normally not above 20 mg/L, all brackish areas at the head of the estuary with SPM values exceeding this figure can be considered as part of the turbidity maximum zone.

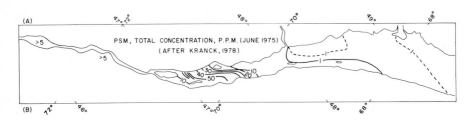

Figure 6. Distribution of suspended particulate matter (SPM) in the estuary in June 1975, showing the position and intensity of the turbidity maximum immediately following the spring freshet. From Kranck (1978).

The net cyclonic motion caused by deeper intrusion over the the North Channel and preferential fresh water outflow along the south shore causes turbid waters to spread further seaward over the South Channel. This results in significant transverse gradients in particle concentrations. The turbid plume advected along the south shore dissipates by tidal mixing along a frontal zone which is situated near Cap-aux-Orignaux during the low river stage and moves downstream about 30km to near Kamouraska during the spring freshette in May and early June (d'Anglejan, 1981b). The alternate trapping and discharge of suspended sediments in Baie de Ste Anne by the ebb and flood motion, modulated by the spring-neap cycle, causes the temporary build-up of mud flats inside this bay, followed by sediment export, a process which keeps reinforcing the turbidity front over the South Channel (d'Anglejan, 1981.b; d'Anglejan et al, 1981).

More generally, the lateral exchange of sediments between the channels and the tidal flats on both the north and south shores exerts a major control on the intensity of the turbidity maximum. In contrast to many partially mixed estuaries, where sediment deposition and storage in the zone occurs mainly over the channel areas (Meade, 1972), in the St. Lawrence these processes are characteristically linked to onshore-offshore transport. Such lateral exchanges are particularly important in the region of the North Channel fronting the tidal flats and marshes of Cap Tourmente. Studies carried out there by Serodes (1980) showed that 6.6×10^5 tons (dry sediment) were trapped over the brackish marshes during the growth season of the vegetation, which consists mainly of *Scirpus americanus* . These deposits were eroded and released to the estuary following the destruction of the plant cover in early October by flocks of the northward migrating greater goose (*Anower cerculescens atlantica*). Peak turbidity concentrations are found in the fall months over the fronting region of the North Channel. Sediment storage within the North Channel and north of Ile d'Orléans during the winter months, followed by resuspension, dowstream transport and dispersion in the water column during the spring freshette complete this annual cycle of onshore-offshore sediment exchanges (Lucotte and d'Anglejan, 1986).

Silverberg and Sundby (1979) noted that the peak values in suspended matter concentrations were landward of the salt intrusion during the spring freshette, and within the brackish waters in late fall.

Altogether, Serodes (1980) found that summer deposition in all the intertidal zones of the Upper Estuary reached 2.7×10^6 metric tons (dry sediment) in 1979. However, no net accumulation takes place, attesting to the efficiency of the physical and biological processes in controlling the cycle. Thus the temporary storage and release of particles withdrawn from the turbidity zone during that cycle involves a mass of sediments equivalent to about 65% of the St. Lawrence mean annual solid discharge.

The role of shore ice floes in releasing fine sediments deposited on the shorelines into the main stream during break-up , and in transporting this material seaward has been considered by Dionne (1969, 1984). A liquid mud layer 10cm to 20cm thick, which develops under the tidal flat ice cover, is returned to suspension in the spring. Mud incorporated within the ice may also be carried offshore at break-up. To what extent these sediments are brought into the channels by fragmenting ice floes is uncertain. The amount is limited by buoyancy forces acting on the floes during ice decay (Troude and Sérodes, 1988). Dionne's (1984) overall figure of 10×10^6 tons of sediments transported by ice drift seems to be an overestimate, as it is in no way supported by the observed extent and intensity of the early spring turbidity maximum.

Observations on the vertical distributions of the suspended matter in the Upper Estuary have been made by several authors (d'Anglejan and Smith, 1973; d`Anglejan and Ingram, 1976, 1984; Silverberg and Sundby, 1979; Kranck, 1979). There is generally a two-layer stratification with higher values near the bottom than at the surface in the turbidity maximum zone and the reverse downstream. Observations of local tidal fluctuations at various sites in the Upper Estuary (d'Anglejan and Ingram, 1976, 1984; Silverberg and Sundby, 1979) all indicate that suspended sediment concentrations are generally 90 degrees out of phase with respect to tidal velocities, with maxima occurring near the end of the ebb. This fact indicates that, generally speaking, advective transport and front migration control these concentrations, rather than current-induced local resuspension. Over the South Channel, a core of more turbid waters occurring near low tide in midwater was observed at several locations to correlate with a dominantly tranverse motion of the water mass, suggesting the importance of lateral exchanges with the south shore (d'Anglejan and Ingram, 1976, 1984). As expected in an estuary in which the circulation is dominated by tidal motion, the neap-spring cycle exerts a major control over these exchanges (d'Anglejan et al, 1981) and over the short term changes in turbidity (d'Anglejan and Ingram, 1984).

In the Lower Estuary, the SPM distributions are less dependant on tidal motion and more on water mass stratification as observed in open marine waters. The particle concentrations and characteristics differ in the three water masses observed during the summer months (Sundby, 1974). SPM concentrations in the 50m surface layer (0.1 mg/L to 2.9 mg/L) were much higher than in the cold intermediate layer (0.05 mg/L to 0.1 mg/L). The bottom 50 m, showed a linear increase to the floor of the Laurentian Trough, with concentrations up to 0.4 mg/L (d'Anglejan,1970). From a submersible, Syvitski et al (1983) observed abundant copepod fecal pellets in the top meter of the water column, with accumulation of *Oikopleura* houses near 20 m. These were most abundant near the temperature minimum of the intermediate layer where they were intermixed with the large aggregates of marine snow. Long bacterial filamentous stringers were also observed. The same authors recorded SPM concentrations up to 1.7 mg/L near the bottom. Particle aggregation in its various forms is seen as the cause for the rapid transfer of particles to the sediments of the Trough (Silverberg et al, 1986).

5.2. Particle size and composition

The suspended particle size distribution (Figure 7) in the water column of the St. Lawrence Estuary has been studied by several authors (d'Anglejan and Smith, 1973; Silverberg and Sundby, 1979; Kranck, 1979; Poulet et al, 1986). In the turbidity maximum, the size distribution is dominated by the inorganic components and is characteristically unimodal, with modal floc sizes between 10 μm and 20 μm in the more turbid near-bottom waters, and between 5 μm and 10 μm in the surface layer (Kranck, 1979). Particle collision within the mixing zone of flocs formed during river transport produces higher order aggregates which tends to settle out in the bottom layer. Seaward of the maximum, modal particle sizes in the surface decrease to a mean value of about 4 μm. The style of the particle size distribution also changes, and flat particle volume-frequency Coulter spectra, characteristic of open ocean waters, become dominant at the head of the Laurentian Trough, a region of upwelling. Kranck (1979) also found that the particle size-volume ratio increases seaward of the turbidity maximum, a fact which she explains by the retention of organic-rich larger aggregates in that zone. In a more recent study, Poulet et al (1986), using statistical analysis, identified five spectral types, groups of these being characteristic of different water masses, with possible continuity and discontinuity between types controlled by hydrological factors.

The seston present in the waters of the St. Lawrence Estuary is made up in different proportions of organic detritus and mineral matter. Tan and Strain (1983) measured mean POC concentrations ranging between 72 ug/L and 1056 ug/L in the Upper Estuary, and between 4 ug/L and 446 ug/L in the Lower Estuary. Based on its C/N ratios, the sources of the POC appear to be more terrigenous in the upper region, more from local production in the lower one (Pocklington and Leonard, 1979). The carbon content of settling material

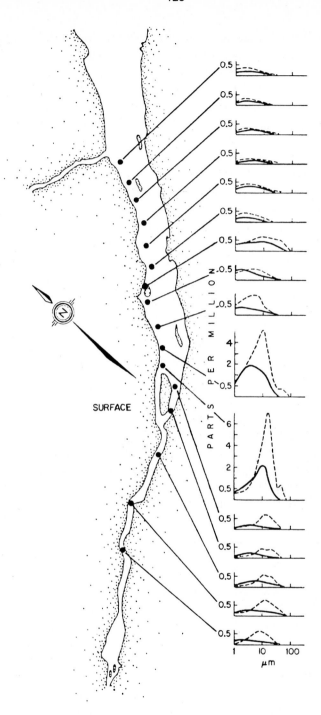

Figure 7. Coulter spectra of suspended particles in surface waters of the Upper Estuary, June 1978. Broken lines: total natural suspended matter; solid line:deflocculated inorganic grains. From Kranck (1979).

obtained with free-drifting sediment traps was found to have a mean value of 4.23% (n=32), with higher values in July than in May reflecting increased biological production throughout the summer (Silverberg et al, 1987). Although the carbon content of the SPM increased from spring to fall, the flux of organic matter to the sediments was found to decrease by a factor of 3 to 4 over the same period (Silverberg et al, 1985). Inorganic components of the suspended matter include, in order of abundance, plagioclase and orthoclase feldspars, quartz, chlorite and illite clays (in a ratio of 1:3), amphibole and amorphous iron oxihydrate colloids (d'Anglejan, 1970; d'Anglejan and Smith, 1973). At the head of the estuary, there are important changes in the major and minor element composition as the river SPM reaches the landward front of the turbidity maximum. These changes are attributed to both particle sorting during settling and resuspension and diagenetic alterations in recent mud deposits (Gobeil et al., 1981). Other changes which affect Fe colloids and trace elements are linked to flocculation in the freshwater-seawater transition zone (Bewers and Yeats, 1978, 1979; Lucotte and d'Anglejan, 1983). In the Lower Estuary, Sundby et al. (1981) have noted in the water column Mn oxide micronodules introduced by resuspension from surface sediments, where they accumulate through diagenesis. This process appears to be linked to the seaward transport of large quantities of Mn out of the estuary.

6. Sediment mass balance

The determination of a satisfactory mass balance for the suspended sediments in the St. Lawrence Estuary appears premature at this stage because of a lack of detailed information on the various entries, and because of contradictory evidence. Precise estimates on the means and variance of inputs and outputs are difficult to make because of the mere dimensions of the system, the year to year fluctuations in discharge, the lack of reliable relationships between run-off and suspended load and their respective rates of change, and finally because the assumption of steady state may not apply. Other complicating factors concern the magnitude of the near-bed load relative to the suspended load. More long term data at representative stations in the entrance and exit sections are also needed.

SPM budgets for the Gulf of St. Lawrence have been presented by Sundby (1974) and by Yeats (1988). Both put the input into the Gulf of St. Lawrence at 5×10^6 tonnes/a. This figure is 20% above the estimate of 3.6×10^6 tonnes/y given by Holeman (1968) and by Milliman and Meade (1983) for the river input. The increase accounts for all possible contributions from other sources to the estuary (shoreline and bottom erosion, tributaries, atmospheric input).

Major uncertainties concern the net rates of exchange with the estuary floor. In the Upper Estuary, sedimentation appears negligible under the present conditions. However, if bottom or nearshore erosion of exposed Goldthwait clays keeps pace with the residual uplift rate of 0.1 cm/a, the release of fine particles in the water column due to this source may be significant. If Goldthwait beds are exposed over 10% of the Upper Estuary surface this could amount to more than 0.5×10^6 tonnes/a.

A more serious discrepancy is found in the Lower Estuary where deposition appears to exceed river sediment input by at least a factor of 4. The determination of deposition rates from sediment traps gives a mean value of 0.51mg cm^{-2} d^{-1} (n=25), in general agreement with the results of ^{210}Pb measurements in sediments of the western Laurentian Trough (Silverberg et al, 1986). Given an area of 9000 km^2, this suggests a yearly deposition of above 1.6×10^7 tonnes/a compared to 3.6×10^6 tonnes/a for the river input. Either the steady state assumption is at fault, or there are major errors in our averaging of the river input, either because of internal recycling or because of large fluctuations through time.

Given the present information, an estimate of the mean passage time of the SPM through the Upper Estuary seems at least possible by comparing the above yearly input figure to the total sediment load in motion approximated from available data. At any instant, this consists of the SPM load in the water column, plus the seasonal deposits on intertidal mudflats and marshes. Total mud deposition during the summer was estimated as 2.7×10^6 tonnes (Serodes, 1980). Contouring of the summer SPM distribution averaged over selected depth intervals in the Upper Estuary yields a figure of 1.6×10^6 tonnes of sediment held in suspension at any time during that season. This suggests a total sediment load in motion of 4.3×10^6 tonnes, hence assuming steady state, a passage time slightly above one year. This compares with estimates of 3 to 4 months for the water itself (Forrester, 1972).

7. Summary and conclusions

Although deeply located within the North American landmass in contrast to most estuaries, the St. Lawrence Estuary has the characteristics of a large macrotidal system. In its shallow upper region, the dynamics of sediment transport is dominated by the fact that a relatively small river load is introduced into a widening basin subject to strong tidal motion. The influx of tidal energy is guided upstream along the steep sided, ice-gouged North Channel. Near the river end, shoaling increases both the tidal range and the frictional dissipation along the bottom. There, intense turbulence helps maintain a turbidity maximum. Semidiurnal and fortnightly tides control onshore-offshore transport of suspended sediment during the summer months, and shoreline erosion by sediment-laden ice floes in late winter, early spring. This, combined with the feeding activities of a migrant bird population, produces a seasonally modulated cycle of erosion and deposition on large tidal flats and brackish water marshes. An amount equivalent to more than one-half the annual river load is processed through this cycle. The instantaneous mass of sediment in suspension within the turbidity maximum is directly determined by it, and only indirectly by the river sediment input. The outflow of turbid water from the well-mixed turbid zone centres over the mainly sandy shallow South Channel and propagates seaward through complex tidal oscillations between the channel and a broad subtidal shelf along the south shore. Negligible permanent sedimentation takes place in the Upper Estuary, which is covered with coarse glacial lag deposits and is flushed of its solid input by the tidally modulated cyclonic circulation.

This situation contrasts with conditions in the Lower Estuary, dominated by the broad vertical circulation of the Gulf of St. Lawrence in which water masses carried landward over the bottom of the Laurentian Trough are returned seaward by surface flow after upwelling at the head of the Trough. This rather sluggish motion allows for fast settling and deposition of biologically aggregated terrigenous particles over the Trough floor, where up to 60 m of Recent deposits are found.

In its channel physiography as well as in the composition and stratigraphy of its surficial deposits, the estuary still carries many indications of its recent glacial past. In the Upper Estuary and over the shelves of the Lower Estuary, reworked moraines and fluvioglacial debris form a shallow cover over the massive ice-distal clay deposits of the Goldthwait Sea. The slow emergence up to the present of the land mass, due to the continuing recovery from the ice load, is recorded by subtidal platforms and raised shoreline terraces along the south shore.

There are still many questions to be answered, about both the Holocene evolution and modern sedimentary processes. Despite recent accoustical surveys, information on the subsurface stratigraphy of late glacial to recent deposits is still fragmentary, particularly in the not too easily accessible areas of the South Channel in the Upper Estuary. One would also wish to learn more about the depositional environments of the exposed Goldthwait beds, whether entirely marine, or whether in places fresh water, as suggested by Gadd (1971). Concerning the present, a more quantitative knowledge of the details of the throughput processes is desirable to arrive at a satisfactory understanding of the sediment mass balance. Further work is needed on the marshes and tidal flats of the Upper Estuary in order to assess their importance in the biogeochemical transformations of nutrients and other elements within the brackish zone. Finally, the response of the transport processes to changes in the intensity and timing of the spring run-off and of shore ice break-up would need to be documented over time.

REFERENCES

Allen, G.P., J.C. Salmon, P. Bassoulet, Y. Du Penhoat and C. De Grandpré (1980) Effects of tides on mixing and suspended sediment transport in macrotidal estuaries. Sedimentary Geology, 26, 69-90.

Avoine, J. (1981) L'estuaire de la Seine: sédiments et dynamique sédimentaire. Thèse 3eme cycle, Université Caen, 236 p.

d'Anglejan, B.F. (1970) Studies on particulate matter in the Gulf of St. Lawrence. Marine Sciences Centre, McGill University, Manuscript Report, 17, 0-31.

d'Anglejan, B. (1971) Submarine sand dunes in the St. Lawrence estuary. Canadian Journal of Earth Sciences, 8 (11), 1480-1486.

d'Anglejan, B. (1981a) Evolution post-glaciaire et sédiments recents de la plate-forme infra-littorale, Baie de Sainte-Anne, Estuarie du Saint-Laurent, Québec. Géographie Physique et Quaternaire, 2, 253-260.

d'Anglejan, B. (1981b) On the advection of turbidity in the St. Lawrence middle estuary. Estuaries, 4 (1), 2-15.

d'Anglejan, B. and E.C. Smith (1973) Distribution, transport and composition of suspended matter in the St. Lawrence estuary. Canadian Journal of Earth Sciences, 10, 1380-1396.

d'Anglejan, B. and M. Brisebois (1974) First subbottom acoustic reflector and thickness of recent sediments in the upper estuary of the St. Lawrence River. Canadian Journal of Earth Sciences, 11 (2), 232-245.

d'Anglejan, B. and R.G. Ingram (1976) Time-depth variations in tidal flux of suspended matter in the St. Lawrence Estuary. Estuarine Coastal Marine Science, 4, 401-416.

d'Anglejan, B. and M. Brisebois (1978) Recent sediments of the St. Lawrence Middle Estuary. Journal of Sedimentary Petrology, 48(3), 951-964.

d'Anglejan, B. and R.G. Ingram (1984) Near bottom variations of turbidity in the St. Lawrence estuary. Estuarine Coastal Shelf Science, 19, 655-672.

d'Anglejan, B., R.G. IIngram and J.P. Savard (1981) Suspended sediment exchanges between the St. Lawrence Estuary and a coastal embayment. Marine Geology, 40, 85-100.

Bewers, J.M. and P.A. Yeats (1978) Trace metals in the waters of a partially mixed estuary. Estuarine Coastal Marine Science, 7, 147-162.

Bewers, J.M. and P.A. Yeats P.A. (1979) The behaviour of trace metals in estuaries of the St. Lawrence basin. Naturaliste Canadien, 106, 149-161.

Dionne, J.C. (1969) Erosion glacielle littorale, estuaire du Saint-Laurent. Revue de Geographie de Montreal, 23(1), 5-20.

Dionne, J.C. (1977) La mer de Goldthwait au Québec. Géographie Physique et Quaternaire, XXX1(1-2), 61-80.

Dionne, J.C. (1984) An estimate of ice-drifted sediments based on the mud content of the ice cover of Montmagny, middle St. Lawrence Estuary. Marine Geology, 57, 149-166.

Dionne, J.C. (1986) Erosion recente des marais intertidaux de l'estuaire du St-Laurent, Quebec. Geographie Physique et Quaternaire, 40, 307-323.

Dionne, J.C. (1988) Holocene relative sea-level fluctuations in the St. Lawrence Estuary, Québec, Canada. Quaternary Research, 29, 233-244.

Dyer, K.R. (1973) Estuaries: A Physical Introduction. J. Wiley and Sons, 140 p.

Dyke, A.S. and V.K. Prest (1987) Late Wisconsinian and Holocene history of the Laurentide ice sheet in R.J. Fulton and J.T. Andrews, eds, The Laurentide Ice Sheet, Géographie Physique et Quaternaire, 41, 237-264.

El-Sabh, M.I. (1988) Physical oceanography of the St. Lawrence Estuary. In, Bjorne Kjerfve (ed), Hydrodynamics of Estuaries, II, 61-78.CRC Press, Bouca Raton, Florida.

Festa, J.F. and D.V. Hansen (1978) Turbidity maxima in partially mixed estuaries: a two-dimensional numerical model. Estuarine Coastal Marine Science, 7, 347-359.

Forrester, W.D. (1972) Courants et débits de marée dans le fleuve Saint-Laurent et son estuaire. Revue Hydrographique Internationale, XLIX(1).

Gadd, N.R. (1971) Pleistocene geology of the central St. Lawrence Lowland. Geological Survey of Canada, Memoire, 359, 1-112.

Gobeil, C., B. Sundby and N. Silverberg (1981) Factors influencing particulate matter geochemistry in the St. Lawrence Estuary turbidity maximum. Marine Chemistry, 10, 123-140.

Hamblin, P.F., K.R. Lum, M.E. Comba and K.L.E. Kaiser (1988). Observations of suspended sediment flux in the region of the turbidity maximum of the Upper St. Lawrence Estuary. Lecture notes on Coastal and Estuarine Studies, V. 29; D.G. Aubrey, L. Weishar (*Eds.*), Hydrodynamics and sediment Dynamics of Tidal Inlets. Verlag, New York, Inc. 246-254.

Hamblin, P.F. (in press). Observations and model of sediment transport near the turbidity maximum of the Upper St. Lawrence Estuary. Journal of Geophysical Research.

Hansen, D.V. and M. Jr. Rattray (1966) New dimensions in estuary classification. Limnology and Oceanography, 11, 319-326.

Holeman, J.N. (1968) Sediment yield of major rivers of the world. Water Resources Research, 4, 737-747.

Kranck, K. (1979) Dynamics and distribution of suspended particulate matter in the Saint-Lawrence Estuary. Le Naturaliste Canadien, 106, 163-173.

Lee, H.A. (1962) Surficial geology of Rivière-du-Loup - Trois Pistoles area. Geological Survey of Canada, Paper, 61-32, 2p. and map 43-1961.

Loring, D.H. and D.J.G. Nota (1973) Morphology and sediments of the Gulf of St. Lawrence. Fisheries Research Board of Canada, Bulletin, 182, 147 p.

Lucotte, M. and B. d'Anglejan (1983) Forms of phosphorus and phosphorus-iron relationships in the suspended matter of the Saint-Lawrence Estuary. Canadian Journal of Earth Sciences, 20, 1880-1890.

Lucotte, M. and B. d'Anglejan (1986) Seasonal control of the Saint-Lawrence maximum turbidity zone by tidal flat sedimentation. Estuaries, 9, 84-94.

Meade, R.H. (1972) Transport and deposition of sediments in estuaries. Geological Society of America, Memoire, 33, 91-120.

Milliman, J.D. and R.H. Meade (1983) World-wide delivery of river sediment to the oceans. Journal of Geology, 91, 1-21.

Peterson, D.H., T.J. Conomos, W.W. Broenkow and P.C. Doherty (1975) Location of the non-tidal current null zone in northern San Francisco Bay. Estuarine Coastal Marine Science, 3, 1-11.

Pocklington, R. and J.D. Leonard (1979) Terrigenous organic matter in sediments of the St. Lawrence Estuary and the Sagnenay Fjord. Journal Fisheries Research Board of Canada, 36, 1250-1255.

Poulet, S.A., J.P. Chanut and M. Morissette (1986) Etudes des spectres de taille de particules en suspension dans l'estuaire et le golfe du Saint-Laurent. I. Variations spatiales. Oceanologica Acta, 9 (2), 179-189.

Poulin, P. (1976) Le complexe morainique de Saint-Narcisse dans le secteur sud de la rivière Malbaie. Interpretation paléoclimatique par l'analyse pollinique. Québec, Université Laval, Dept. Géographie, thèse M.A. non publiée. 83 p.

Serodes, J.B. (1980) Etude de la sédimentation intertidale de l'estuaire moyen du Saint-Laurent. Environnement Canada, Direction Générale des Eaux Interieures; Region du Québec, 28 p.

Silverberg, N. (1978) Sediments of the Rimouski shelf region, Lower Saint Lawrence estuary. Canadian Journal of Earth Sciences, 15, 1724-1736.

Silverberg, N. and B. Sundby (1979) Observations in the turbidity maximum of the Saint-Lawrence estuary. Canadian Journal of Earth Sciences, 16, 939-950.

Silverberg, N., H.M. Edenborn and N. Belzile (1985) Sediment response to seasonal variations in organic matter input. *In* : A.C. Sigleo and A. Hattori (*eds*), Marine and Estuarine Geochemistry, Lewis Publishers Inc., Ch 5, 69-70.

Silverberg, N., H.V. Nguyen, G. Delibrias, M. Koide, B. Sundby, Y. Yokoyama and R. Chesselet (1986) Radio nuclide profiles, sedimentation rates, and bioturbation in modern sediments of the Laurentian Trough, Gulf of St. Lawrence. Oceanologica Acta, 9 (3), 285-290.

Silverberg, N., J. Bakker, H.M. Edenborn and B. Sundby (1987) Oxygen profiles and organic carbon fluxes in Laurentian Trough sediments. Netherlands Journal Sea Research, 21(2), 95-105.

Simard, L. (1971) Catalogue of St. Lawrence River soil and rock data. Department of Transport, Marine Hydrography Branch; St. Lawrence Ship Channel Directory; Ingineering Field Investigation Section, Montreal, Quebec, 2 vol., 339 p.

Sundby, B. (1974) Distribution and Transport of suspended particulate matter in the Gulf of St. Lawrence. Canadian Journal of Earth Sciences, 11, 1517-1533.

Sundby, B., N. Silverberg and R. Chesselet (1981) Pathways of manganese in an open estuarine system. Geochemica et Cosmochimica Acta, 45(3), 293-307.

Sundby, B., G.Bouchard, J. Lebel and N. Silverberg (1983) Rates of organic matter oxidation and carbon transport in early diagnenesis of marine sediments. *In* : Advances in Organic Geochemistry 1981. John Wiley & Sons Limited. 350-354.

Syvitski, J.P.M. and D.B. Praeg Quaternary sedimentation in the St. Lawrence Estuary and adjoining areas: an overview based on high resolution seismo-stratigraphy. Géographie Physique et Quaternaire (in press).

Syvitski, J.P.M., N. Silverberg, G. Ouellet and K.W. Asprey (1983) First observations of benthos and seston from a submersible in the Lower St. Lawrence Estuary. Geographie Physique et Quaternaire, 3, 227-240.

Troude, J.P. and J.B. Serodes (1988). Le rôle des glaces dans le régime morphosédimentologique d'un estran de l'estuaire moyen du St.Laurent. Revue Canadienne de Génie Civil.15, 348-354.

Tan, F.C. and P.M. Strain (1983) Sources, sinks and distribution of organic carbon in the St. Lawrence estuary, Canada. Geochimica et Cosmochimica Acta, 47, 125-132.

Yeats, P.A. 1988 Distribution and transport of suspended particulate matter. p. 15-28, *In*: Chemical Oceanography in the Gulf of St. Lawrence - Can. Bull. Fish. Aquatic Sciences 220.

Chapter 7

Nearshore Sediment Dynamics in the St. Lawrence Estuary

Georges Drapeau

INRS-Océanologie, Université du Québec, 310 des Ursulines, Rimouski, Québec, Canada, G5L 3A1

ABSTRACT

The principal agents of nearshore sediment dynamics for the St. Lawrence Estuary are tides, waves, and the formation of ice in winter. The Estuary is 400 km long, 70 km wide near the mouth and 15 km wide at the upper end. The dimensions of the Estuary are such that the hydrodynamic conditions prevailing in the lower portion are typically marine while those prevailing in the upper portion are typically estuarine. The tide is semi-diurnal and reaches 4.2 m at the mouth and 7.5 m at the head of the estuary and waves can exceed 4 m at the mouth of the estuary. Waves and tides have a definite gradient along the axis of the estuary; the wave energy diminishes while the tidal range increases landward. In the St. Lawrence Estuary ice forms in December and persists until April. The activity of sea ice is conspicuous along the shores of the estuary. Ice plays a major role on nearshore sediment dynamics; it transports sediments of all sizes from clay to boulders and it also erodes in some cases and in other cases protects the shore zone. Onshore-offshore exchange of sediment is an important process in the St. Lawrence Estuary. High sedimentation rates in harbors reveal the importance of the nearshore suspension load. The harnessing of large tributaries for the production of electric power has cut the supply of sediments to the littoral zone in adjacent areas. Seasonal variations are important in the estuary, even including the migratory stopover of Greater Snow Geese. Relative sea level changes are intricate in the St. Lawrence Estuary and the long-term trends of erosion and sedimentation are not uniform from one end of the estuary to the other.

Key Words: Boulder pavement, dredging, erosion, snow geese, harbors, sedimentation, ice rafting, littoral drift, littoral zone, maintenance dredging, nearshore, onshore-offshore transport, seasonal variations, sediment transport, sedimentation, tributaries.

1. Introduction

The dimensions of the St. Lawrence Estuary (Fig. 1) are such that the hydrodynamic conditions prevailing in the lower portion of the estuary are typically marine while those prevailing in the upper portion are typically estuarine. The sand beaches alternating with granite seacliffs of the north shore of the Lower Estuary have little in common with the salt marshes that predominate on both shores of the Upper Estuary. The St. Lawrence differs

Coastal and Estuarine Studies, Vol. 39
M. I. El-Sabh, N. Silverberg (Eds.)
Oceanography of a Large-Scale Estuarine System
The St. Lawrence
© Springer-Verlag New York, Inc., 1990

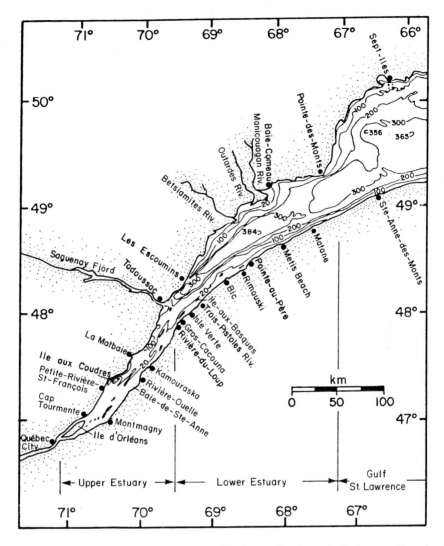

Figure 1. Toponymy of the St. Lawrence Estuary. The figure also shows the bathymetry (in meters) and the divisions between the Lower and the Upper Estuary.

from other large estuaries because the formation of ice in winter substantially modifies the sediment dynamics pattern. Tides, waves and ice are the preponderant and ubiquitous sedimentary processes in the littoral zone of the St. Lawrence Estuary but they are not the only ones. Locally, the exchange of sediments between the intertidal zone and the offshore, the presence of large tributaries and even the grazing of tidal marshes by large colonies of migratory birds have an impact on nearshore sediment dynamics. This chapter concentrates on outlining the mechanisms that control sediment dynamics along the shores of the St. Lawrence Estuary, rather than systematically reviewing the literature.

Some trends are directly related to the geometry of the estuary. The St. Lawrence Estuary is 400 km long. It is 70 km wide near the mouth and only 15 km wide at the upper end near Ile-d'Orleans (Fig. 1). The geometry and prevailing wind directions imply on the one hand, that the tidal range increases landward and, on the other hand, that the waves diminish in height and in period from the mouth of the estuary, opening on the Gulf of St. Lawrence, to the upper portion of the estuary, where fetches are tenfold shorter. The formation of ice is not systematically linked to the dimensions of the estuary although the drifting of ice in springtime is.

The principal mechanisms of nearshore sediment dynamics have been identified for the St. Lawrence Estuary, but most investigations are site specific. Ideally, it should be possible to integrate the different studies and come up with a model of the overall nearshore sediment dynamics. To achieve that task one would need to quantify each mechanism in relative terms, firstly with respect to the time scale and secondly, in relation to the geography of the estuary. For instance, siltation in harbours provides some measurement of the longshore suspended sediment load but, because the coverage is too spotty, it is not possible at this time to link these data into a global model.

The long term trends of erosion and accretion of the shorelines are not easily defined because variations of sea level have to be taken into account. The St. Lawrence valley has been subject to major crustal warping since the last glaciation. Relative sea level was some 140 m above the present level 12,000 yr. BP in the Rimouski area during the maximum extension of the Goldthwait Sea (Locat, 1977). The submersion increased landward, but the pattern of isostatic recovery does not necessarily coincide with the pattern of inundation, particularly during the contemporaneous stages of vertical crustal movement. An evaluation of the long term trends is then dependant not only on our ability to correctly interpret the role of the sedimentological processes that are currently taking place along the shores of the St. Lawrence Estuary, but also on our ability to forecast the trends of relative sea level variations for different portions of the Estuary shoreline.

The plan of the chapter is firstly to describe the general setting, secondly to outline the mechanisms of nearshore sediment dynamics and thirdly to examine the long term trends.

2. General Setting

2.1 Tides

The tide in the St. Lawrence Estuary is semi-diurnal. The amphidromic point of the semi-diurnal tidal constituent M2 is located near the Iles-de-la-Madeleine within the Gulf of St. Lawrence and the diurnal tidal constituent K1 is located near Sable Island on the Scotian Shelf. The phase lag of the tide is one hour between the mouth of the Estuary and the Saguenay Fjord and three hours from that point to Ile-aux- Coudres (Godin, 1979). The range increases as the tide propagates into the Estuary as shown in Table 1 (after Godin, 1979).

TABLE 1
Tidal ranges in the St. Lawrence Estuary (meters)

Locality	Mean	High
Ste-Anne-des-Monts	2.3	4.2
Pointe-au-Père	3.0	5.5
Rivière-du-Loup	3.8	6.7
Saint-François	5.0	7.5

2.2 Waves

The wave climate in the St. Lawrence Estuary is controlled by winds that blow predominantly from the western quadrant and having a tendency to align parallel to the axis of the Estuary and also by fetches that diminish in size in a landward direction. During the spring and summer seasons, winds blow more often from the eastern quadrant. Histograms of wave height versus wave period, recorded by the Marine Environmental Data Services (Pêches et Océans Canada, 1988), are shown in Figure 2 for the center of the

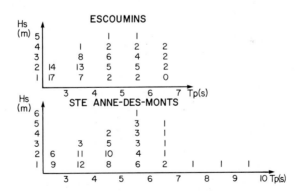

Figure 2. Wave climate at Escoumins and Ste-Anne-des-Monts. The two histograms show the relative occurences of significant wave heights (Hs) as a function of peak periods (Tp) for the two locations. From Pêches et Océans Canada (1988).

Estuary, 20 km seaward from the Saguenay Fjord (Escoumins) and off Ste-Anne-des-Monts. An examination of the two histograms clearly reveals the influence of fetch and the

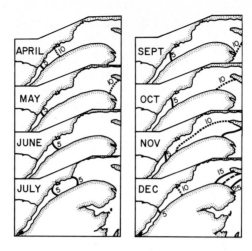

Figure 3. Percentage of occurences of 2-meter significant wave height. The 2- meter significant wave heights occurrences portray the wave climate for the ice free months in the Estuary. From Vigeant (1984).

resulting wave climate, which differs considerably at the two locations. Off Ste- Anne-des-Monts, the highest waves recorded between June and November, 1971 were 3.6 m. A climatological atlas of the St. Lawrence Estuary and Gulf (Vigeant, 1984) outlines the seasonal variability of the wave climate in the Estuary (Fig.3). In winter, the presence of ice impedes the formation of waves in the Estuary. Otherwise the highest waves are found in April and December and the lowest wave heights are recorded in August.

2.3 Ice

Ice plays an important role in all coastal areas of the eastern Canadian seaboard, as pointed out by McCann et al. (1981). Only the outer coasts of Nova Scotia and Newfoundland are ice-free. The role of ice is also important in the Bay of Fundy (Knight and Dalrymple, 1976) although different from the St. Lawrence Estuary because of the magnitude of the tide in the Bay of Fundy. Ice plays an even greater role in the Canadian Arctic where it dominates all other processes.

In the St. Lawrence Estuary, ice forms in December and persists until April (Brochu, 1960). The percentage of ice cover in the St. Lawrence Estuary is shown in Figure 4, which summarizes ice observations over the 11-year period from 1963 to 1973 (Markham, 1980). Ice forms rapidly during the second fortnight of December. By the end of January, and for the remainder of the winter, ice is most abundant on the southern shore of the Estuary. These conditions result from winds that are predominantly from the northwest in winter and the ice being drifted against the south shore. The combination of winds and residual currents finally clears the drifting ice out of the Estuary by April.

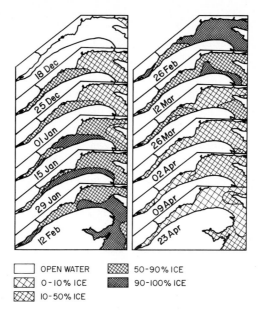

OPEN WATER   50-90% ICE
0-10% ICE   90-100% ICE
10-50% ICE

Figure 4. Ice cover in the St. Lawrence Estuary. The figure shows the ice cover for 0-10, 10-50, 50-90 and 90-100 percent coverage of the sea surface for different dates, based on observations between 1963 and 1973. From Markham (1980).)

3. Mechanisms of Nearshore Sediment Dynamics

3.1 Waves and tides

The specific role played by waves on sediment dynamics in the St. Lawrence Estuary has received relatively little attention. At the mouth of the estuary, wave energy levels are comparatively high for an estuarine environment. Beaches in this area are more characteristic of marine beaches than estuarine beaches, although the influence of the comparatively large tidal range is perceptible. Figures 2 and 3 show that wave energy diminishes progressively landward. The difference between the two histograms in Figure 2 is even more important than the numbers indicate, as only the highest waves can induce significant sediment movement. The specific contribution of waves to the transport of sediments has been analyzed for the Trois-Pistoles area (Drapeau and Morin, 1981). The threshold of motion of coarse sand at a depth of 5 m is reached only 4 per cent of the time in that area. In fact, wave action on sediments is limited to very shallow water depths except during rare storm events. A particularly interesting phenomenon is the erosion of backshore beach scarps in areas where the intertidal zone is otherwise under equilibrium. At Cap Tourmente, on the north shore of the Upper Estuary, Troude and Serodes (1988) report the retreat of a one meter high beach scarp at a rate of almost 2 m/yr, although the intertidal zone is essentially under equilibrium. Erosion of backshore beach scarps combined with the erosion of the upper portion of the tidal marsh has been monitored by

Dionne (1986) on the south shore of the St. Lawrence Estuary. At Montmagny the measured retreat averaged 1 m/yr and at Rivière-du-Loup the retreat was between 2 and 3 m/yr and reached 4 m/yr in 1985-86 at one location.

Tides play an important role because the waves can agitate a band of sediments that shifts across the littoral zone depending on the height of the tide. The result is that a broader zone of sediments is influenced by wave dynamics but, in consequence, the action is less intense because of the continuous shifting of the active zone. These conditions are typical of the St. Lawrence Estuary, where sediments in the intertidal zone are more widely spread and not as well sorted as on shorelines experiencing smaller tidal amplitude.

Waves and tides have a definite gradient along the axis of the Estuary; the wave energy diminishes while the tidal range increases landward. At the mouth, the Estuary is a wave-dominated environment and, at the head, it is tide-dominated. Sand beaches diminish in size and number landward, while tidal marshes increase in size. The first genuine tidal marshes are located at Pointe-au- Père and Rimouski but they are much smaller than those that have developed in the Upper Estuary.

3.2 Ice

The activity of sea ice is conspicuous along the shores of the St. Lawrence Estuary. Observations and descriptions date back to the 19th Century. For example, Dawson in 1886 described the action of ice on the shores of the Metis Beach area. The sediment dynamics in the intertidal zone are strongly influenced by ice-related processes. This section summarizes how ice contributes to the sedimentation, transport and erosion of sediments along the shores of the St. Lawrence Estuary.

3.2.1 Sediment transport by ice

There are different ways by which sediments are integrated within sea ice (Dionne, 1970). The phenomenon takes place at the ice foot but also over the whole intertidal zone. The rise and fall of the tide brings the bottom of the ice cover in contact with shallow surficial sediments twice daily. During winter cold spells the top layer of the sediment that comes in contact with the ice sheet freezes against it and large quantities of sediment are incorporated into the ice by this continuously renewed process (Fig. 5). Sediments are also spilled onto the upper surface through the many cracks that develop in the ice cover.

The volume of sediments caught in the ice is extremely variable. The sediment load frozen into the ice has been estimated by Dionne (1984), on the basis of visual observations of ice cross-sections in the Montmagny area, to average 4×10^6 tons over an area of 20 square kilometers. This corresponds to a sediment load of 200 kg/m^2. Troude and Serodes (1988) cored the ice cover in the Cap Tourmente area to evaluate its sediment load. They noted a high variability in the concentration of sediment frozen into the ice and obtained a mean value of 101 kg/m^2 with a 95 per cent confidence interval of 84 kg/m^2.

Figure 5. Photograph showing surficial sediments incorporated into the ice cover at Montmagny in April 1969. From Dionne (1971b).

The quantity of sediment exported from the intertidal zone by ice is highly variable depending upon whether the sediment-laden ice melts *in situ* or is drifted away. The fate of the ice is controlled by the direction of the prevailing winds (onshore or offshore) at the time of ice break-up. Dionne (1984,1987a) estimates that 85 per cent of the ice is drifted away in the Montmagny area. Other observations by Dionne (1987b) upstream of Quebec City lead him to estimate that the sediment frozen into the ice would average one million tons over a study area of 200 km^2. Troude and Serodes (1988) observed ice break up at Cap Tourmente and found that some ice floes drifted away and others were stranded at the high tide level. Troude and Serodes obtained ice core samples and have shown that the density of ice floes is increased by the sediment load, so that at the beginning of the thawing process ice floes are barely buoyant and strand in place. As thawing progresses, the ice floes shed the sediments that were frozen underneath and they then become more buoyant and prone to be carried away by winds and tides. It is estimated that 90 per cent of the ice floes finally drift away from the intertidal zone. At Cap Tourmente, the sediment load transported by ice away from the intertidal zone is estimated to be in the order of 50,000 tons per year.

The most spectacular action of drifting ice is the transport of boulders that are strewn over the intertidal zones. The capacity of ice to transport boulders in the St. Lawrence Estuary is impressive. Dionne (1988c) measured 176 ice-rafted boulders at 18 localities along the St. Lawrence Estuary. The largest boulder was 7.25 m long and weighed 176 tons, but most

Figure 6. Ice-rafted boulders paving the intertidal zone at Cap Marteau near Trois-Pistoles. From Dionne (1972a).

boulders measured between 1 and 2.5 m (Fig. 6). The median weight of boulders observed specifically at three localities is: Petite-Rivière-Saint-François, 57 tons, Bic, 16 tons and Rimouski, 17 tons (Dionne, 1981). These boulders are good indicators of the main trends of ice transport in the Estuary. The two shores of the St. Lawrence Estuary belong to different geological provinces. The south shore is composed of Appalachian sedimentary rocks while the north shore is bordered by the igneous Precambrian Shield. Most boulders in the intertidal zone of the south shore of the estuary are of Precambrian origin, which implies that they have been ice-rafted from the north. In the Rivière-du-Loup/Trois-Pistoles area, on the south shore of the Estuary, some 60 per cent of the boulders originate from the north shore (Dionne, 1972a). By contrast, the proportion of crystalline boulders is only 3 per cent in the glacial till, which emphasizes the ice-rafted origin of the boulders. The distribution of ice-rafted boulders is different in the Upper Estuary where Precambrian boulders make up only 20 per cent of the ice-rafted material (Dionne, 1987a).

Although ice-rafted boulders can be found at any location in the intertidal zone of the St. Lawrence Estuary, they are not distributed at random. In many locations, particularly in the vicinity of the headlands, boulders are concentrated at the mid-tide level. This type of accumulation of ice-rafted boulders has been observed in many cold-climate regions (Lyell, 1854; Tanner, 1939; Rosen, 1979; Lauriol and Gray, 1980). In the St. Lawrence Estuary the alignment of boulders in the intertidal zone has been observed at many locations (Brochu, 1961; Dionne, 1962, 1970, 1972a) and different explanations have been proposed. The phenomenon is particularly well developed in the Rimouski area (Guilcher, 1981) where tightly packed boulder pavements, 50 to 100 m wide and up to 2 km long, stretch along the mid-tide zone. Guilcher (1981) has explained this alignment of boulders at the mid-tide level on the basis of field observations and monitoring of the phenomenon. During springtime, when the ice blocks rafting the boulders are melting and dropping the boulders, they are pushed by the wind towards the shore, but are impeded from reaching the upper portion of the intertidal zone by the large ice foot that remains in place during the thawing period. The boulders are thus concentrated in the lower portion of the intertidal zone.

The distribution of pebble lithology has also been examined. A comprehensive analysis (175 sites visited and 34,234 pebbles identified) was carried out by Dionne (1971a), not only for the intertidal zone but also for tills, fluvio-glacial frontal moraines, outwashes, deltas, submerged tills and ancient raised beaches. The results show that the proportion of pebbles coming from the Precambrian shield is below 5 per cent except for the ancient beaches where the content reaches 12.1 per cent and the modern beaches where the proportion is much higher and reaches 22.6 per cent.

A gradient has been observed between the north- and south-shore provenances for sand-size sediments in the Isle-Verte/Trois- Pistoles area (Drapeau and Morin, 1985). Suites of minerals and rock fragments were associated specifically either with the sedimentary rocks of the south shore or with the Precambrian rocks located on the north shore of the St. Lawrence Estuary. Trend surface analyses were used to show that the sand-size sedimentary constituents diminish offshore, while the igneous constituents decrease shoreward (Fig. 7). These trends, drawn from two different grain size populations, independently confirm the provenance of the sand-size surficial sediments in the area. At Trois-Pistoles, the estuary is 30 km wide and 300 m deep (Fig. 1), which precludes bedload transport from one shore to the other. As the layer of sand-size surficial sediments blanketing Goldthwait Sea clays is only a few centimeters thick (Morin, 1981), it indicates that these sand grains are not relict. It implies that the "north shore" components of the surficial sandy sediments would be ice-rafted the same way the boulders are.

3.2.2 Erosion by ice

Erosion by ice takes place at two levels; that of the freeze-thaw cycles fracturing the bedrock and that of the mechanical action, that is the bulldozing and eroding effect of ice blocks (Dionne and Brodeur 1988a, 1988b). The freeze-thaw action on different types of bedrock encountered along the south shore of the St. Lawrence Estuary has been investigated by Trenhaile and Rudakas (1981) who found that shales were most susceptible to frost breakdown. Freeze-thaw action is one mechanism that contributes to the formation by cryoplanation of sub-horizontal intertidal platforms. Cryoplanation of the intertidal zone is particularly well developed on the south shore of the estuary where tightly folded Cambro-Ordovician shales outcrop almost vertically along headlands. Guilcher (1981) has described the process using examples in the Rimouski area, where he could observe the freeze-thaw action on shales outcroping in the intertidal zone. Guilcher pointed out that

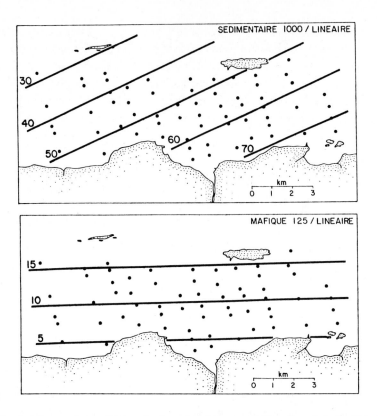

Figure 7. Trend surface analyses of mineral suites of nearshore surficial sediments in the l'Isle-Verte/Trois-Pistoles area. Upper: Linear trend surface of 1000-micron sedimentary grains. Lower: Linear trend surface of 125-micron mafic minerals. From Drapeau and Morin (1985).

cryoplanation alone could not build the intertidal bedrock platforms, but it is a process that enhances a morphological evolution that has a longer geological history.

Ice can produce some erosion by friction. The scouring action of ice is particularly apparent immediately after the ice has either melted or drifted away. Dionne (1971b, 1985c, 1988d) describes the erosion by ice of the tidal flats of the St. Lawrence Estuary (Fig. 8). This takes place mainly where ice blocks and floes are moved seaward by the combined action of offshore winds and tides. He identifies three types of erosional features: 1) grooves 15 to 125 cm wide, 10 to 45 cm deep and up to 2 km long; 2) circular and irregular depressions up to 45 cm deep and 400 cm in diameter; and 3) chaotic networks of basins and ridges. The scouring observed on tidal flats is also present on tidal marshes. The tidal marshes in the St. Lawrence Estuary are characterized the presence of numerous shallow pans (Dionne, 1972b, 1989). These pans are erosional scars, produced in the spring by ice blocks, that commonly measure 1 to 2 m in diameter although some are much larger (20-50 m). They are relatively shallow, that is 10-50 cm deep (Fig. 9).

Figure 8. Tidal flat erosion by ice at Montmagny in April 1969. From Dionne (1971b).

Figure 9. Large ice-made pans in a tidal marsh at Isle-Verte. From Dionne (1972b).

3.2.3 Protection by ice

The ice that forms in the intertidal zone in winter plays an important protective role. Ice protects the intertidal zone from wave action during the period of the year during which winds are strongest. The shores of the St. Lawrence Estuary would otherwise be severely eroded by wave action in winter time. The most severe erosion by waves takes place just before the ice begins to form in December and to a lesser extent in mid-April mid-May (cf. Figures 3 and 4).

3.3 Onshore-offshore sediment exchanges

Exchange of sediment between the intertidal zone and the water column is an important sedimentological process in the St. Lawrence Estuary. D'Anglejan et al. (1981) and Lucotte and d'Anglejan (1986) have studied this phenomenon at both extremities of the turbidity maximum zone; at Cap Tourmente on the north shore and at Baie de Ste-Anne on the south shore. Lucotte and d'Anglejan (1986) explain that the core of the turbidity maximum zone in the Upper St. Lawrence Estuary is located in the North Channel and oscillates in front of the large intertidal flats and marshes of Cap Tourmente. They also show that the intensity and position of the turbidity maximum are sustained by suspended sediment exchanges between the St. Lawrence Estuary channel and the intertidal zone. Fine sediments found seaward in late winter and early spring are advected onto the Cap Tourmente marshes by the tide during the summer months. Sedimentation over the tidal marshes is favored by the growth of vegetation during the summer months (Troude and Serodes, 1985). Greater White Geese, during their fall stopover in the area (see section on seasonal variations for more details), contribute to the intense (estimated to reach 4500 tons per tide) erosion of the tidal marshes by grazing down the vegetation. This material is transported upstream and deposited in the Ile d'Orléans channel (Lucotte and d'Anglejan, 1986). The yearly cycle is completed when the spring freshet carries this material back downstream.

At Baie-de-Ste-Anne the exchange of sediments is less intense . Baie-de-Ste-Anne is a 20 km long intertidal flat onto whose southern extremity flows the Rivière Ouelle. Current and suspended sediment measurements taken at different times of the year during flood and neap tides have led to an explanation of the tidal dynamics of the fine sediments in that area (d'Anglejan et al., 1981). An intensification of turbidity in the Ouelle river area provides the intertidal flats with a local source of suspended sediments and tidal exchanges between the intertidal flat and the adjacent subtidal platform contribute to maintaining high turbidity levels. This system seems to have reached a steady state, as witnessed by comparison of aerial photographs taken in 1927 and 1976, showing that the shoreline has remained at the same location during that period.

3.4 Seasonal variations

Waves and ice are two phenomena seasonally affecting the shores of the St. Lawrence Estuary. Sediment dynamics of large tidal flats and marshes, particularly in the Upper Estuary, are also seasonally controlled by the intertidal vegetation and by the passage of aquatic migratory birds. The seasonal sedimentary pattern of Upper Estuary tidal marshes is summarized by Serodes and co-workers (Serodes and Dube, 1983; Serodes and Troude, 1984; Troude and Serodes, 1985). The seasonal cycle of erosion-sedimentation in typical Upper Estuary tidal marshes is schematized in Figure 10.

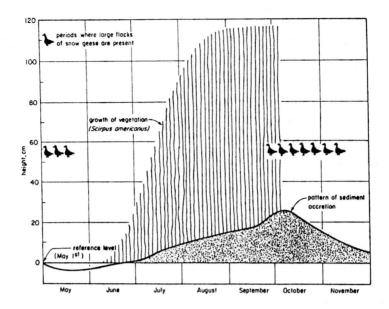

Figure 10. Schematic diagram describing the erosion-sedimentation cycle from May to November for the Cap Tourmente tidal marsh. From Serodes and Troude (1984).

During the spring, the removal of the ice cover either by melting *in situ* or drifting away leaves the tidal zone unprotected. At that time of the year, the spring thaw has considerably weakened the cohesiveness of the surficial sediments of the intertidal zone, which become particularly prone to erosion. The combination of waves and wind-driven currents during that period episodically erodes large volumes of sediment (Serodes and Troude, 1984).

As summer progresses, the growing vegetation in the upper half of the intertidal zone favors the retention of sediments. Summer is the period of lowest wave energy, which also favors the accumulation of sediments.

Many factors contribute to the erosion of sediments during the fall season. Higher waves, resulting from stronger winds, are more common than during the other seasons. Wave heights increase proportionally seaward as the fetches become wider and wave erosion is a widespread phenomenon over the whole estuary during the fall season. In the Upper Estuary tidal marshes are inhabited during the fall by colonies of Greater Snow Geese that interfere with sedimentary processes. Large colonies, in the order of 250,000 birds, feed in the marshes of the Upper Estuary during their migration from the Canadian Arctic (Reed, 1984). Their preferred food is the rhizome of American scirpus (*Scirpus americanus*), and during their migratory stopover in the Upper St. Lawrence Estuary, the geese peck the sediments to get at the *Scirpus* rhizomes. In some areas the geese are so active that Dionne

(1985b) could see the difference between bird hunting zones and those protected from hunting, the former showing some erosion of the upper tidal marsh, which had been deprived of its protective vegetation.

In winter, except for the sediments frozen into the ice, it is difficult to evaluate the sedimentary conditions prevailing under the ice. The mobility of sediments is mainly confined to the lower portion of the intertidal zone. Because of the protection that the ice cover offers, sedimentation is likely to take place but the circulation of water is restricted by the ice itself that is in contact with the bottom during part of the tidal cycle.

3.5 Tributaries

Rivers contribute a sediment load into the systems in which they flow and the sediment yield is proportional, for a given type of vegetation cover, to the size of the drainage basin and the annual precipitation (Langbein and Schumm, 1958). The larger tributaries of the St. Lawrence Estuary are important sediment sources that have a direct impact on nearshore sediment dynamics (Dubois, 1980; Lambert, 1985). The Saguenay Fjord is discussed in a separate chapter. What is particular to the St. Lawrence Estuary is that many tributaries have been harnessed to produce electricity. The building of dams on rivers results in the trapping of sediments that would otherwise feed the system into which they flow. This is the case for three large rivers on the north shore of the Estuary, the Betsiamites, Outardes and Manicouagan rivers.

The impact of the harnessing of the Outardes River has been analysed by Cataliotti-Valdina and Long (1984). Before the building of the dam network, the river flow reached 2,700 m^3/s during the spring freshet while it is now regulated to flow at a uniform yearly rate of 560 m^3/s. The spring freshet had been flushing large quantities of sediment into the St. Lawrence Estuary, which is now deprived of that sediment supply. Cataliotti-Valdina and Long (1984) have compared the sediment load of the harnessed Outardes River with that of an unharnessed river of comparable characteristics, the St-Jean River. Under flood conditions, the sediment load of the Outardes is 462 tons/day for a flow rate of 1,278 m^3/s, while for the Saint-Jean River the sediment load is 3,974 tons/day for a lower flow rate of 1,035 m^3/s. Nowadays large sand flats are developing in the estuary of Outardes River, which has become more a sediment sink than a sediment source (Long, 1983). The situation is complex because although the spring freshet has been suppressed, the mean flow velocity is now sustained at a higher level, which promotes the erosion of the banks of the Outardes Estuary (Hart, 1987). A state of equilibrium has not yet been reached and it is difficult to quantitatively foresee what will be the long term impact of river harnessing on the nearshore dynamics of the St. Lawrence Estuary (Hart and Long, 1990).

3.6 Sedimentation in harbors

Trapping of sediments in harbor basins is common in the St. Lawrence Estuary. Given a specific basin geometry and sediment load in the waters outside the basin, the nature of water circulation determines both the rate and the pattern of deposition in harbor basins (Mehta et al., 1984). Ouellet and Serodes (1984) have examined the sedimentation in small harbors located at Berthier-sur-Mer (near Montmagny), Ile-aux-Coudres, Cap-à-l'Aigle (near La Malbaie), Matane and Baie-Comeau. The authors indicate that all these small harbor basins are trapping enough sediments to require dredging.

Sedimentation in the ports of Rivière-du-Loup and Gros-Cacouna has been analysed more extensively. Troude and Ouellet (1987) studied the siltation in the harbor of Rivière-du-Loup. They measured an accumulation of 32 cm of silts and clays at the head of the harbor basin over a period of 6 months and determined that tidal currents are the principal vector for these sediments. The tidal excursion (ellipse) reaches 10 km and the residual displacement is of the order of 1,000 m at Rivière-du-Loup. The suspended sediment concentration in the nearshore waters varies considerably over a yearly cycle, but averages 20-40 mg/l.

Gros-Cacouna harbor is located 10 km seaward from Rivière-du-Loup in a similar setting. Drapeau and Fortin (1978) have monitored current velocities and sediment concentrations during tidal cycles in June and September 1976 to obtain sediment transport rates and patterns across the entrance of this harbor. They compared these results against the siltation in the harbor, that had averaged 31 cm/yr over a 8 year period. They measured a net transport of suspended sediments into the harbor at a rate of 13.1 tons per tidal cycle in June and 3.5 tons per tidal cycle in September. The higher transport rate in June is an indication of the seasonal variability of the suspended load in the estuary. However, bursts of higher sediment concentration than those measured are necessary to account for the 31 cm/yr accumulation rate measured between 1968 and 1976 (Fortin and Drapeau, 1979). This was corroborated by D'Anglejan and Ingram (1982), who determined higher suspended sediment transport rates for Gros-Cacouna harbor.

Maintenance dredging is that performed exclusively to locally maintain constant water depths. It does not include any development dredging, such as to increase the navigable area of a harbor. The sites of maintenance dredging in Rivière-du-Loup and Rimouski harbors are shown in Figure 11. At Rivière-du-Loup the approaches to a ferry berth are dredged systematically, while at Rimouski the maintenance dredging is more spotty and varies from year to year. The data for maintenance dredging listed in Table 2 are interesting because they show consistency from year to year. The repetitive dredging operations in Rivière-du-Loup harbor serve the same purpose as if a sediment trap had been built specifically to study the deposition of sediment into that harbor. The data in Table 2 indicate that siltation in Rivière-du-Loup harbor averaged 1.37 m/yr, with a standard deviation of 0.37 m/yr, over a 10- year period from 1979 to 1988. The sedimentation rates

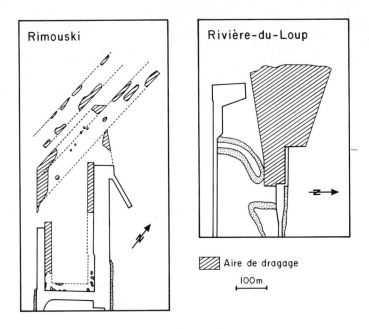

Figure 11. Maps showing the areas of maintenance dredging (cross-hatched) at Rimouski and Rivière-du-Loup harbors. The same area is dredged every year at Rivière-du-Loup while dredging is more spotty at Rimouski and varies from year to year. From Travaux Publics Canada (1989).

TABLE 2.
Maintenance dredging

	Rivière-du-Loup			Rimouski		
Year	Volume (m^3)	Area (m^2)	Rate (m/yr)	Volume (m^3)	Area (m^2)	Rate (m/yr)
1988	41,400	32,000	1.29	5,976	25,000	0.24
1987	48,475	32,000	1.51	13,134	12,000	1.09
1986	26,875	32,000	0.84	12,378	11,000	1.13
1985	26,322	32,000	0.82			
1984	67,563	40,000	1.69	52,096	63,000	0.83
1983	59,866	32,000	1.87			
1982	44,666	32,000	1.40			
1981	45,470	32,000	1.42			
1980	62,192	32,000	1.94	28,573	30,000	0.95
1979	28,907	32,000	0.90	23,372	32,000	0.73
1978				50,378	29,000	1.73
1977				57,873	65,000	0.89

From: Travaux Publics Canada (1989)

for Rimouski harbor are more variable, likely because of the spotty nature of maintenance dredging operations. Nonetheless, they are of the same order of magnitude as those obtained for Rivière-du-Loup harbor. These results indicate that large quantities of fine sediments are mobilized in the nearshore zone of the St. Lawrence Estuary. We have seen in the previous paragraphs that the tidal circulation provokes the onshore-offshore exchange as well as the longshore transport of suspended sediments. The flux of suspended sediment follows the cycles of the tide, but harbors interfere with the tidal circulation in the nearshore zone and trapping of suspended sediment results. The volume of sediments accumulating in Rimouski and Rivière-du-Loup harbors is then a measure of the nearshore tidal sediment flux in these two areas.

4. Long Term Trends: Erosional and Accretional Shorelines

Systematic sediment budgets are not available to evaluate the long term trends of the St. Lawrence Estuary shorelines. Available estimates are then based on site-specific observations, which lead to a bias because erosion is often more detectable than sedimentation. Evidences for erosion of the shores are readily apparent along the St. Lawrence Estuary, unlike the case for progradation, because the bulk of sediments transported are fine-grained and they do not form conspicuous features. They tend to be deposited in sheltered areas where they intermingle with the tidal marsh vegetation and deposition is not apparent unless markers are used to record it. The long term erosional and accretional trends are controlled by the cumulative effect of the hydrodynamic variables, including ice action, relative sea-level fluctuations and the harnessing of rivers for the production of hydroelectric power.

At the eastern end of Ile-d'Orléans, at the head of the Upper Estuary, Allard (1981) has surveyed an undisturbed section of a 640 m wide tidal marsh and provided a stratigraphic record of marsh evolution above a glacio-marine till dated at 11,200 years BP. The surficial stratigraphic unit is a 90cm-thick uniform mud, suggesting continued sedimentation at an accumulation rate of 0.8 mm/yr. Horizontal shifting of the different zones of the tidal marsh, observed on aerial photographs, indicates that the tidal marsh portion of the intertidal zone is prograding while the lower tidal flat portion is eroding. Only 20 km seaward, at Montmagny and Cap Tourmente, erosion of the upper portion of the tidal marsh has been observed on both shores of the Estuary. Dionne (1986) has monitored severe erosion of the upper tidal marsh at three locations in the Montmagny area, where annual measurements indicate a retreat of the order of 1 m/yr. On the north shore of the Estuary, at Cap Tourmente, Troude and Serodes (1985) have measured a retreat of the shoreline of almost 2 m/yr over a 17 year period. Erosion of the upper portion of the tidal marsh has also been reported to be active at Rivière-du-Loup (Dionne, 1986) where aerial photographs show a retreat of 2 m/yr between 1967 and 1985. It is worth noting, however, that at mid-distance between Montmagny and Rivière-du-Loup, various workers have shown that the shoreline is stable. D'Anglejan et al.(1981) report that a detailed comparison of two airborne photographic surveys in 1927 and 1976 over the Baie-de-Ste-Anne - Rivière-Ouelle area gives no indication of shoreline progradation or erosion. Field investigations by Serodes and Dubé (1983) near Kamouraska have also lead them to conclude that the net sedimentological budget is nil in that area and that the shoreline is stable.

The above examples emphasize the difficulty of outlining major trends of erosion and sedimentation along the shores of the St. Lawrence Estuary. One possible explanation for the lack of spatially continuous trends could be the irregular pattern of recent vertical crustal movement. The relation between the Wisconsin deglaciation and the relative sea-level record has been reconstituted by Quinlan and Beaumont (1982) who show that the readjustment is not uniform. Vertical crustal movement is estimated by Gale (1970) to have been 4.9 cm higher at La Malbaie than at Quebec City between 1919 and 1969. Patterns of recent vertical crustal movement determined by Vanicek (1976) show a relative gradient of the order of 30 cm per century between Pointe-au-Père (higher) and Kamouraska (lower). In addition, the relatively high seismic activity in the La Malbaie-Ste-Anne area is associated with a north-south rift fault system (Lamontagne, 1987) that can favor differential crustal warping. Recent investigations by Dionne (1985a, 1988a, 1988b, 1988e) demonstrate that the vertical crustal movements have been more intense than previously evaluated since the recession of the postglacial Goldthwait Sea. Dionne provides evidence that sea level was 5 m lower than present 7 000 yr BP and then transgressed to a height of 8 to 10 m above present some 4 800 yr BP.

Shoreline erosion related to anthropogenic activities is easier to explain. For instance, Dubois (Personal communication, 1987) measured an erosion rate of 60 cm/yr of the shoreline between the Outardes and Manicouagan rivers. This reflects the fact that, since these two rivers have been dammed to produce hydroelectric power, the adjacent Estuary shorelines are now deprived of their normal supply of sediments and more prone to erosion. Across the Estuary, a port built at Matane has triggered the erosion of the adjacent leeward shoreline , because waves are energetic enough in that area to generate substantial littoral drift.

It is common to find only a few centimeters of sandy sediments covering the clayey substratum deposited during the Goldthwait Sea marine transgression (Dionne, 1981). In the Trois-Pistoles area, Morin (1981) has shown that for half the bottom sediment sampling stations located between Ile-aux-Basques and the south shore of the Estuary, the sandy sediment layer covering the Goldthwait clays is less than 5 cm thick. This is the case even at the mouth of the Trois-Pistoles River where no delta has developed despite the fact that the river cuts through thick sections of older high-stand deltaic deposits, an abundant source of sandy sediment. The conclusion that can be drawn from these observations is that the actual interface between the Goldthwait clays and the surficial sediments results from a continued adjustment between the supply of sediments to the nearshore system and the capacity to transport it. The present profile of the nearshore zone is then a long term response to the nearshore sediment dynamics; the interface between the surficial mobile sediment layer and the relict Goldthwait clays adjusting to the changes of sea level.

5. Summary and Conclusions

Presently our understanding of the nearshore sediment dynamics of the St. Lawrence Estuary is essentially site specific. The situation is complicated by the fact that three types of processes interact and it is not possible in most cases to delineate their relative importance. In the St. Lawrence Estuary the nearshore sediment dynamics are controlled by: 1) the hydrodynamics of the estuary, 2) the ice that forms in winter and acts as an important sedimentological agent, and 3) the relative changes of sea level. .

Several chapters in this volume demonstrate that the hydrodynamics of the St. Lawrence Estuary are more complex than those of smaller estuaries and simple estuarine models are not applicable to the St. Lawrence (El Sabh, 1988). The same conclusions apply for the nearshore hydrodynamics. Nearshore waves and currents have been measured at a few sites but the coverage is not sufficient to generate a data base that could be used to build a model encompassing the whole estuary. Broad statements can be made on the basis of the geometry of the estuary. For instance, wave energy increases seaward as fetches become longer, while the tidal range increases landward because of the funnel effect of the estuary. This type of information partially explains the observation that tidal marshes increase in size and number as one progresses landward. The onshore-offshore circulation and the resulting sediment exchanges also have to be taken into account. For example, the role that the turbidity maximum plays in the Upper Estuary as a sediment sink-source for the intertidal zone is known to be important, but the sedimentological models remain open-ended at present.

By comparison with the hydrodynamic parameters the role of ice is even more difficult to quantify. While specific physical laws correlate the transport of sediments to the levels of turbulence and advection, the volume of sediments transported by ice is not linked to simple physical laws. As indicated earlier, all depends on whether ice forms within the intertidal zone or not, permitting the bottom sediments to be frozen within the lower ice interface. If this is the case, the rate of incorporation of sediments into the ice depends on the sediment grain-size and on the air and water temperatures and salinity. Ice formed offshore will be essentially free of sediments, while ice that formed in the intertidal zone will contain a substantial load of mud, sand and gravel and even large boulders. Another major difference is that the capacity of ice to transport sediments is almost unlimited. In the St. Lawrence Estuary sediment concentration in ice is not a limiting factor even for the transport of boulders (Drake and McCann, 1982). The movement of ice is difficult to predict because ice floes, in addition of being entrained by surface currents, also respond to wind stress. The distribution of boulders in the intertidal zone of the south shore of the Estuary has shown that the predominantly north-westerly winds strongly influence the movement of ice. The deposition of sediments transported by ice can be witnessed everywhere along the shores of the estuary, but this represents an unknown proportion of the total sediment load transported by ice because, in springtime, when the ice cover breaks up, all the ice covering the intertidal zone will drift away if winds are blowing offshore or strand and melt on place if the winds are blowing inshore at that time.

Sea level changes are relatively rapid in the St. Lawrence Estuary and the rate of change varies considerably along the axis of the estuary. When discussing present day sediment dynamics one has to take into account that the present geomorphology of the St. Lawrence Estuary results from adjustments to variations of sea level as well as to the hydrodynamics of the environment. The fact that the relative changes of sea level are not uniform along the St. Lawrence Estuary makes any attempt to integrate this parameter within a tentative model more difficult.

Another approach to summarizing the situation would be to identify and budget the sources and sinks of sediments that play a major role in the nearshore sediment dynamics, but the available data do not allow this. Exchanges of fine sediments between the intertidal zone and the Estuary have been studied, but it is not possible to determine what is the total exchange budget on the basis of measurements obtained only at a few locations. Sediment trapping in harbors provides interesting data. It would be prematurate, however, to use data from Table 2 and extrapolate these results beyond Rivière-du-Loup and Rimouski. The sources and sinks of sand-sized sediments are easier to track but they are not everywhere obvious in the St. Lawrence Estuary. For instance, we have seen how the Goldthwait clays are often covered by only a thin veneer of sand size sediments. At the mouth of Trois-Pistoles River, which is cutting through thick sand and gravel deposits, no accumulation of sand takes place in the intertidal zone. This implies that the sandy sediments are withdrawn from the nearshore at the same rate as they are supplied. The inshore-offshore transport is obviously an important process but, presently, they are no complete sets of data to quantify the phenomenon along the shores of the Estuary.

This chapter on nearshore sediment dynamics of the the St. Lawrence Estuary finally leads to similar conclusions as those of Fairbridge (1980), discussing the definition and geodynamic cycle of estuaries. He pointed out that, because of the multiple variables, every estuary is unique, and added that estuaries are ephemeral features in long-term, geologic history and must be regarded as dynamically evolving land-forms. His final conclusion was that accurate prediction of future estuary behaviour must await accurate analyses of estuarine histories. Seen in this perspective, we realize that our knowledge of the St. Lawrence Estuary nearshore sediment dynamics, although incomplete, is oriented towards fundamental questions leading to a better understanding of this complex system.

Acknowledgements

The author wishes to thank Drs. N. Silverberg and J.B. Serodes for reviewing the manuscript and Dr. J.C. Dionne for his review and helpful suggestions. This work was supported by the Natural Sciences and Engineering National Research Council of Canada, grant OGPIN 010.

REFERENCES

Allard, M. 1981. L'Anse aux Canards, Ile d'orléans, Québec: Evolution Holocène et dynamique actuelle: Géogr. Phys Quat. 35(2):133-154.

d'Anglejan, B. and Ingram, R.G. 1982. Investigation of natural sediment and dredge spoil movement in the vicinity of Gros-Cacouna harbour (st. Lawrence Estuary) DSS, contract 1 SV 80-000229, 50 p.

d'Anglejan, B., Ingram, R.G. and Savard, J.P. 1981. Suspended-sediment exchanges between the St. Lawrence Estuary and a coastal embayment: Marine Geology 40:85-100.

Brochu, M. 1960. Dynamique et caractéristique des glaces de dérive de l'estuaire et de la partie nord-est du golfe du Saint-Laurent: Min. Mines et relevés tech.,Ottawa, Etude géographique no 24, 93 p.

Brochu, M. 1961. Déplacement de blocs et d'autres sédiments par la glace sur les estrans du Saint-Laurent en amont de Québec: Mines et Relevés tech. Can. Etude géogr. 30, 16p.

Cataliotti-Valdina, D. and Long, B.F., 1984. Evolution estuarienne d'une rivière régularisée en climat sub-boréal; la rivière aux Outardes (côte nord du golfe du Saint-Laurent, Québec): Can. J. Earth Sci. 21(1):25-34.

Dawson, J.W. 1886. Note on boulder drift and sea margin at Little Metis, lower St. Lawrence: Can. Rec. Sci. 2(1):36-38.

Dionne, J.C. 1989. An estimate of shore ice action in a spartina tidal marsh, St. Lawrence Estuary, Québec, Canada. Jour. Coastal Res. 5(2):281-293.

Dionne, J.C. 1988a. Holocene relative sea-level fluctuations in the St. Lawrence Estuary, Québec, Canada: Quaternary Res. 29:233-244.

Dionne, J.C. 1988b. Note sur les variations du niveau marin relatif à l'Holocène, à Rivière-Ouelle, côte sud du Saint-Laurent. Géogr. Phys. Quat. 42 (1):83-86.

Dionne, J.C. 1988c. Ploughing boulders along shorelines with particular reference to the St. Lawrence Estuary: Geomorphology. 1(4):297-308.

Dionne, J.C. 1988d. Characteristic features of modern tidal flats in cold regions. In: Tide-influenced sedimentary environment and facies, P.L. Boer et al. eds., Riedel, Dordrech, p. 301-332.

Dionne, J.C. 1988e. Evidence d'un bas niveau marin durant l'Holocène à Saint-Fabien-sur-Mer, estuaire maritime du saint-Laurent. Norois 35(137): 19-34.

Dionne, J.C. 1987a. Lithologie des cailloux de la baie de Montmagny, côte sud du Saint-Laurent: Géogr. Phys. Quat. 41:161-169.

Dionne, J.C. 1987b. La charge sédimentatire glacielle des rivages du haut estuaire du Saint-Laurent. Conf. Can. Littoral 1987. CNRC p. 67- 96.

Dionne, J.C. 1986. Erosion récente des marais intertidaux de l'estuaire du Saint-Laurent: Géogr. Phys. Quat. 40(3):307-323.

Dionne, J.C. 1985a. Observations sur le Quaternaire de la rivière Boyer, côte sud de l'estuaire du Saint-Laurent, Québec. Géogr. Phys. Quat. 39(1):35-46.

Dionne, J.C. 1985b. Tidal marsh erosion by geese, St. Lawrence Estuary, Québec: Géogr. Phys. Quat. 39(1)99-105.

Dionne, J.C. 1985c. Formes, figures et faciés sédimentaires glaciels des estrans vaseux des régions froides. Palaeogeography- Paleoclimatology, 51:415-451.

Dionne, J.C. 1984. An estimate of ice-drifted sediments based on mud content of the ice cover at Montmagny, Middle St. Lawrence Estuary: Marine Geology 57:149-166.

Dionne, J.C. 1981. Observations sur le déplacement de méga-blocs par la glace sur les rivages du Saint-Laurent: Atelier sur l'action des glaces sur les rivages, CARERE-CNR, Rimouski, 5-6 avril 1981, 16 p.

Dionne, J.C. 1972a. Caractéristiques des blocs erratiques des rives de l'estuaire du Saint-Laurent: Rev. Géogr. Montréal 26(2):125-152.

Dionne, J.C. 1972b. Caractéristiques des schorres des régions froides, en particulier de l'estuaire du Saint-Laurent. Z.Geomorph.N.F. Suppl. Bd.13:131-162.

Dionne, J.C. 1971a. Nature lithologique des galets des formations meubles quaternaires de la région de Rvière-du-Loup/Trois-Pistoles, Québec. Rev. Géogr. Montr. 25(2):129-142.

Dionne, J.C. 1971b. Erosion glacielle de la slikke, estuaire du Saint- Laurent: Rev. Géomorph. Dyn. 20(1):5-21

Dionne, J.C. 1970. Aspects morpho-sédimentologiques du glaciel, en particulier des côtes du Saint-Laurent: Univ. Paris, Thèse doct. 412 p.

Dionne, J.C., 1962. Notes sur les blocs d'estran du littoral du Saint- Laurent: Can. Geographer 6(2):69-77.

Dionne, J.C. et Brodeur, D. 1988a. Erosion de plates-formes rocheuses littorales par affouillement glaciel. Z. Geomorph. N.F. 32(1):101-115.

Dionne, J.C. et Brodeur, D. 1988b. Frost weathering and ice action in shore platform development with particular reference to Quebec, Canada. Z. Geomorph. N.F. Suppl.-Bd. 71, p. 117-130.

Drake, J.J. and McCann, S.B. 1982. The movement of isolated boulders on tidal flats by ice floes: Can. J. Earth Sci. 19:748-754.

Drapeau, G. and Morin, R. 1985. Influence du glaciel sur la répartition minéralogique de la fraction sableuse de la zone littorale dans la région de Trois-Pistoles: Naturaliste Can. 112:51-56.

Drapeau, G. and Morin, R. 1981. Contributions des vagues au transport des sédiments littoraux dans la région de Trois-Pistoles, estuaire du Saint-Laurent: Géogr. Phys. Quat. 35:245-251.

Drapeau, G. and Fortin, G. 1978. Tidal sedimentation in Gros-Cacouna Harbour: Proc 16th Intrn. Conf Coastal Engr. ASCE : 1986-2000.

Dubois, J.M.M. 1980. Géomorphologie littorale de la côte nord du Saint- Laurent: in The Coastline of Canada (McCann, B. ed.) Geol. Sur. Can. Paper 80-10:215-238.

El-Sabh, M.I. 1988. Physical oceanography of the St. Lawrence Estuary: Hydrodynamics of Estuaries 2:61-78, CRC Press, Bouca Raton, Florida.

Fairbridge, R.W. 1980. The Estuary: Its Definition and Geodynamic Cycle: in Chemistry and Biochemistry of Estuaries (Olausson, E. and Cato, I. eds.) John Wiley and Sons, Chichester. p. 1-35.

Fortin, G. and Drapeau, G. 1979. Envasement du port de Gros-Cacouna situé sur l'estuaire du Saint-Laurent: Naturaliste Can. 106(1):175-188.

Gale, L.A. 1970. Geodetic observations for detections of vertical crustal movement: Can. J. Earth Sci. 7:602-606.

Godin, G. 1979. La marée dans le golfe et l'estuaire du Saint- Laurent: Naturaliste Can. 106(1):105-121.

Guilcher,A., 1981. Cryoplanation littorale et cordons glaciels de basse mer dans la région de Rimouski, Géogr. Phys. Quat. 35 (2):133-154.

Hart, B.S. 1987. The evolution of the Outardes Estuary: unpublished M.Sc. thesis, Université-du-Québec-à-Rimouski, 197 p.

Hart, B.S. and Long, B.F. 1990. Recent evolution of the Outardes Estuary, Quebec, Canada: Consequences of dam construction on the river: Sedimentology vol. 37, in press.

Knight, R.J. and Dalrymple, R.W. 1976. Winter conditions in macrotidal environments: Rev. Géogr. Montréal 30:65-85.

Lambert, N. 1985. Evolution de l'estuaire de la rivière Saint-Jean: unpublished M.Sc. thesis, Université-du-Québec-à-Rimouski

Lamontagne, M. 1987. Seismic activity and structural features in the Charlevoix region, Quebec: Can. J. Earth Sci. 24:2118-2129.

Langbein, W.B. and Schumm, S.A. 1958. Yield of sediment in relation to mean annual precipitation: Trans. Am. Geophys. Union 39:1076-84.

Lauriol, B. and Gray, J.T. 1980. Processes responsible for the concentration of boulders in the intertidal zone in Leaf Basin, Ungava, in The Coastline of Canada (McCann, S.B. ed) Geol. Surv. Can., Pap 80-10, p. 281-292.

Locat, J. 1977. L'émersion des terres dans la région de Baie-des- Sables\Trois-Pistoles: Géogr. Phys. Quat. 31(3-4):297-306.

Long, B.F. 1983. Evolution of the Outardes estuary after the Hydraulic power regulation: Proc. Canadian Coast. Conf. 83 (Holden,B.J. ed.) NRC p. 327-328.

Lucotte, M. and d'Anglejan, B. 1986. Seasonal control of the St.- Lawrence maximum turbidity zone by tidal-flat sedimentation: Estuaries 9(2):84-94.

Lyell,C. 1854. Principles of Geology, Appleton, N.Y. 847 p.

Markham, W.E. 1980. Atlas des glaces de l'est canadien: Environnement Canada, Service de l'Environnement atmosphérique 33p.

McCann, S.B., Dale, J.E. and Hale, P.B. 1981. Subarctic tidal flats in areas of large tidal range, Southern Baffin Island, Eastern Canada: Géogr. Phys. Quat. 35(2):183-204.

Mehta, J.A., Ariathurai, R., Maa, P.Y. and Hayter, E.J. 1984. Fine sedimentation in small harbors: Coastal and Oceanogr. Eng. Dept., U. of Florida, UFL-COEL-TR-051, 114 p.

Morin, G. 1981. Contribution à la sédimentologie de la région de Trois- Pistoles, estuaire du Saint-Laurent: unpublished M.Sc. thesis, Université-du-Québec-à-Rimouski, 118 p.

Ouellet, Y. and Serodes, J.B. 1984. Problèmes d'agitation et de sédimentation des ports de refuge: Assises ann. AQTE, p.113-141.

Pêches et Océans Canada 1988. Service des données sur le milieu marin.

Quinlan,G. and Beaumont,C. 1982. The deglaciation of Atlantic Canada as reconstructed from the postglacial relative sea-level record. Can. J. Earth Sci. 19:2232-2246.

Reed, A. 1984. Production of Scirpus Americanus and its use by greater snow geese at Cap-Tourmente National Wildlife Area. Draft of a paper presented at the 5th American Snow Goose Conf. Québec, Oct. 4-7, 1984.

Rosen, P.S. 1979. Boulder barricades in Central Labrador. J. Sed. Petrol. 49:1113-1124.

Serodes, J.B. and Troude, J.P. 1984. Sedimentation cycle of a freshwater tidal flat in the St. Lawrence Estuary: Estuaries 7(2):119- 124.

Serodes, J.B. and Dubé,M. 1983. Dynamique sédimentaire d'un estran à spartines (Kamouraska, Québec): Naturaliste Can. 110:11-26.

Tanner, V. 1939. Om de blockrika strandgordlana vid subarktiska oceankustar, forekomstatt og upkomst: Terra, 51:157-165.

Travaux Publics Canada, 1989. Dossier 3900-1.

Trenhaile, A.S. and Rudakas, P.A., 1981. Freeze-thaw and shore platform development in Gaspé, Québec: Géogr. Phys. Quat. 35(2):171- 181.

Troude, J.P. and Serodes, J.B. 1988. Le rôle des glaces dans le régime morpho-sédimentologique d'un estran de l'estuaire moyen du Saint- Laurent. Can. J. Civ. Eng. 15:348-354.

Troude, J.P. and Serodes, J.B. 1985. Régime morpho-sédimentologique d'un eastran à forte sédimentation dans l'estuaire du Saint-Laurent: Proc. Canadian Coastal Conf.'85, NRC, p.105-119.

Troude, J.P. and Ouellet, Y. 1987. Phénomènes contribuant à l'envasement du port de Rivière-du-Loup: Proc. Canadian Coastal Conf.'85, NRC, p. 297-311.

Vanicek, P. 1976. Pattern of recent crustal movements in Maritime Canada. Can. J. Earth Sci. 13:661-667.

Vigeant, G. 1984. Cartes climatologiques du Saint-Laurent: Environnement Canada, Service de l'Environnement atmosphrique.

Chapter 8

Reactivity and transport of nutrients and metals in the St. Lawrence Estuary

P. A. Yeats

Physical and Chemical Sciences, Department of Fisheries and Oceans, Bedford Institute of Oceanography, P.O. Box 1006, Dartmouth, Nova Scotia, Canada   B2Y 4A2

ABSTRACT

The St. Lawrence estuary provides a valuable natural laboratory for studying processes influencing the reactivity and transport of trace chemicals. The progression from control of chemical distributions by physical and inorganic processes in the Upper estuary to biological ones in the Lower estuary is illustrated using oceanic surveys of nutrient and trace metal distributions. Models of trace metal transport through the estuary are used to describe the trend from predominantly horizontal (advective) transport in the Upper estuary to increased vertical transport in the Lower estuary. Observed reactivity of metals and nutrients in the St. Lawrence estuary is compared with reactivities in other estuaries.

Key Words: Trace chemicals, nutrients, turbidity maximum, suspended particulate matter

1. Introduction

Numerous studies of the distributions of chemicals in the St. Lawrence estuary and the Gulf of St. Lawrence have shown the estuary and gulf to be an almost ideal natural laboratory for geochemical investigations. They have demonstrated that broad surveys of various chemical parameters can be used to clearly identify and characterize important chemical processes and determine chemical transports through the estuary and gulf. Chemical transports through this system have also been used as an analogue of the more general coastal environment in several trace metal transport models.

The St. Lawrence estuary is generally divided into an Upper and Lower estuary based on the abrupt change in bottom topography that occurs near the mouth of the Saguenay fjord. In the transition region at the head of the Laurentian Trough, the waters depths shoal from 300 m to ~50 m in <20 km. This is a region of intense tidal mixing and upwelling (El-Sabh, 1988; Gratton et al., 1988). In the Upper estuary the topography is uneven with numerous channels and troughs. In the Lower estuary the topography is dominated by the Laurentian Trough with depths >300m.

The extent of mixing in the Upper estuary changes from nearly vertically homogeneous in the region adjacent to Ile d'Orleans to partially mixed throughout most of the Upper estuary and moderately well stratified at the eastern end. The tidally averaged circulation has the surface outflow predominantly on the south side of the estuary and the landward flowing counter-current confined predominantly to the deeper northern channel, although the actual

Coastal and Estuarine Studies, Vol. 39
M. I. El-Sabh, N. Silverberg (Eds.)
Oceanography of a Large-Scale Estuarine System
The St. Lawrence
© Springer-Verlag New York, Inc., 1990

circulation will be more complicated than implied by this generalized picture as a result of the topography, tides, internal waves and other processes that have an effect on mixing and circulation. An important feature of the Upper estuary that is maintained by a combination of tidal and non-tidal circulation and resuspension of sediments is the turbidity maximum zone (TMZ). The TMZ is a constant feature of the Upper estuary whose location is surprisingly insensitive to changes in river discharge (Silverberg and Sundby, 1979). In the Lower estuary the water column is stratified. The general circulation pattern shows seaward flowing brackish water in the Upper layer predominantly confined to the south side of the estuary and landward flowing deep water. As in the Upper estuary the details of the circulation are considerably more complex, including two gyres in the surface circulation and complicated tidal currents.

These hydrographic features produce a very interesting environment for the study of chemical processes and transports. The system changes, over the length of the estuary, from a partially mixed Upper estuary with many shoals and a prominent turbidity maximum zone to a very deep, stratified Lower estuary. This is a progression from a fairly typical estuarine system to one that is much more oceanic in character within the 400 km length of the estuary. The St. Lawrence estuary is also very large, both in terms of the freshwater discharge and the length and depth of the estuarine system. This physical environment produces a chemical system in which physical mixing and suspended matter settling and resuspension determine the major transformations and transports of chemicals in the Upper estuary, while in the Lower estuary, more oceanic processes, predominantly those associated with biological cycling, become important. Because of the size of the system, and the demarcations between hydrographic regimes, effects of different chemical, physical and biological processes can often be identified from field survey data.

A number of chemical processes can affect the distributions of inorganic chemicals in estuaries and their transports through estuaries. These obviously include dissolved phase reactions that result from mixing of fresh and saline waters, plus exchanges between dissolved and particulate phases, and interactions with the sediments that are determined by or influenced by the TMZ. In addition, interactions with estuarine biota can actively or passively remove trace inorganics from the dissolved phase and subsequent decomposition reactions can reintroduce them. The importance of these processes in understanding the reactivity of the nutrients and trace metals in the St. Lawrence estuary can be investigated using oceanographic surveys for the chemical parameters and the generalized picture of the estuarine circulation as a model for the physical environment. Models of chemical transports based on the same generalized descriptions can be used both to augment the studies of chemical processes and to estimate the fluxes of riverborne materials through the estuary and gulf. Modelling of chemical transports through the coastal zone is generally a very difficult problem because the offshore transports are hard to estimate. Because the St. Lawrence estuary and gulf is such a large estuary/marginal sea system with reasonably well understood exchanges at its seaward boundary, it has provided us with a useful laboratory for studying coastal zone processes and exchanges.

2. Nutrients

The distributions of silicate, phosphate and nitrate in the St. Lawrence estuary (Table 1) clearly illustrate the transition from distributions dominated by mixing processes in the Upper estuary to those dominated by biological processes in the Lower estuary.

Conservative mixing of silicate in the Upper estuary is indicated by the linear silicate-salinity relationships observed by Subramanian and d'Anglejan (1976) and Yeats (1988a). The absence of any deviations from linearity through the TMZ is interesting since several reviews of silicate behavior in estuaries (Liss, 1976; Aston, 1978) have summarized mechanisms for interactions with estuarine particulates. In the St. Lawrence, the TMZ is well-developed and removal of other trace chemicals such as iron is observed (Bewers and Yeats, 1978) but silicate removal is not seen.

TABLE 1.
Nutrient distributions.

		Concentration Range uM	Behavior	Reference
Upper estuary				
Silicate	July '74	9-26	conservative	Subramanian and D'Anglejan,1976
	May '74	17-38	conservative	Yeats, 1988a
	Sept '74	17-20	constant	Yeats, 1988a
Nitrate	May '74	15-19	input	Yeats, 1988a
	Sept '74	9-16	slight input	Yeats, 1988a
	June '75	8-13	slight removal	Greisman and Ingram, 1977
Phosphate	May '74	0.7-1.5	slight input	Yeats, 1988a
	Sept '74	0.6-1.6	slight input	Yeats, 1988a
Lower estuary				
Nitrate		3-21	"oceanic" pro-	Greisman and Ingram, 1977
Silicate		2-45	files with bio-	Coote and Yeats, 1979
Nitrate		2-23	logical uptake	Coote and Yeats, 1979
Phosphate		0.2-2.2	and regeneration	Coote and Yeats, 1979

Another interesting feature of the silicate distributions reported by Yeats (1988a) is the marked upstream increase in silicate concentrations in the river water samples just upstream of salt intrusion. These trends are unrelated to changes in salinity and completely at odds with the conservative silicate-salinity relationships in the estuary. The observed variability in the freshwater regime could reflect temporal changes in the riverine silicate concentrations, however, in that case a linear relationship between silicate and salinity would not be expected in the estuarine regime. If similar variability in the silicate concentration upstream of the limit of salt intrusion occurs in other rivers, this feature could lead to the impression of non-conservative mixing unless the zero salinity intercept for the mixing curve is carefully established.

Greisman and Ingram (1977) compared observed surface nitrate concentrations in the estuary with those calculated assuming mixing between river water and upwelling Lower estuary water. These calculations showed that the deviations from conservative mixing were fairly small with some uptake of nitrate indicated for several parts of the estuary. Yeats (1988a) also found nearly conservative behavior of nitrate and phosphate through most of the Upper estuary. Some deviations from linearity were observed in the TMZ. These would appear, at least for phosphate, to result from the addition of phosphate from the sediments.

In the Lower estuary, the vertical nutrient distributions become much more oceanic, being characterized by relatively low concentrations in the surface waters and increasing concentrations with depth (Coote and Yeats, 1979). These profiles and the general distribution of nutrients in the Lower estuary and Gulf of St. Lawrence are determined by the estuarine circulation pattern and biological cycling. Nutrients are supplied to the surface water by river runoff and vertical mixing and upwelling in the Lower estuary and removed by incorporation into the biota, a portion of which sinks out of the surface layer. The estuarine circulation pattern then returns nutrients regenerated in the deeper waters to the surface farther landward.

The supply of nutrients to the surface waters of the Lower estuary is very important for phytoplankton growth, and as a result, a number of studies of the mechanisms of nutrient input to the surface water have been reported. River input supplies only a small fraction of the total input of nutrients to the surface waters. For example, Greisman and Ingram (1977) found, for a cruise in July 1976, that <25% of the nitrate originated in the river. The remainder, comes from the entrainment of deeper waters rich in nutrients into the surface layer. Several papers have identified the importance of tidally induced upwelling at the head of the Laurentian Trough as a means of supplying nutrients to the surface waters (Therriault and Lacroix, 1976; Greisman and Ingram, 1977; Levasseur and Therriault, 1987). The paper of Levasseur and Therriault has an interesting discussion of the relative amounts of nitrate and silicate that are upwelled. Although the silicate concentrations in the deep water are approximately twice those of nitrate, the amounts being upwelled are approximately equal since the upwelled water originates from relatively shallow depths (<100 m) where nitrate and silicate concentrations are similar. Other mechanisms for mixing deeper water into the surface are clearly also important and these will all tend to introduce nutrients to the surface water. The regulation of freshwater discharge is interesting in this context since changes in freshwater discharge will alter the entrainment of saline water and hence the entrainment of nutrients (Bugden et al., 1982). Finally, before leaving the discussion of nutrients, it should be pointed out that whatever the mechanism of supply, nutrients do not generally appear to be the critical parameter limiting the primary production in the estuary (Therriault and Levasseur, 1985).

3. Trace metals

3.1. Upper estuary

Conservative behavior, ie linear metal-salinity relationships, are found in the St. Lawrence estuary for a number of dissolved metals and other trace inorganics. These include species, such as lithium (Stoffyn-Egli, 1982), fluoride (Young, 1976), and iodate and total dissolved iodine (Takayanagi and Cossa, 1985) whose river concentrations are lower than those in seawater and metals such as copper and nickel (Yeats, 1987) whose river concentrations are higher. These observations mean that within the limits of the precision of the measurements, the distributions of these metals are unaffected by the TMZ, releases from the sediments or any other processes that would alter the metal-salinity relationship.

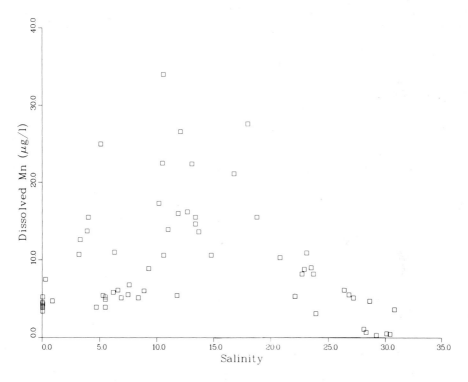

Figure 1. Dissolved manganese vs salinity relationship for the Upper estuary, BIO Cruise 79-024.

It is important to keep in mind, however, that the uncertainty in these relationships can be quite large and could easily mask some non-conservative behavior. Observations on dissolved manganese serve to illustrate this point. The first two observations of dissolved manganese distributions in the Upper estuary showed linear relationships with a lot of scatter (Subramanian and d'Anglejan, 1976; Bewers and Yeats, 1978), but in the latter case addition of samples from the Lower estuary improved the definition of the relationship and showed that the manganese-salinity relationship was actually curved. A more recent survey, at a time of lower freshwater discharge (BIO cruise 79-024, August 1979), shows a much more distinctly curved manganese-salinity relationship with a shape that suggests addition of Mn_d at low salinity and removal at higher salinity (Figure 1).

Metal-salinity relationships that clearly show losses of metals in the estuary have been seen for dissolved chromium (Campbell and Yeats, 1984), iron (Subramanian and d'Anglejan, 1976; Bewers and Yeats, 1978), mercury (Cossa et al., 1988) and zinc (Bewers and Yeats, 1978). For chromium and mercury the removal occurs very rapidly at low salinity in the region of maximum turbidity. In both cases interactions with particles in the TMZ, perhaps flocculation of organic-chromium and organic-mercury colloids, is proposed as the mechanism responsible for the removal. This mechanism is supported by the fact that flocculation of organic matter has been observed in the St. Lawrence estuary (Kranck, 1979), and dissolved iron removal, which is also thought to occur by organic flocculation (Boyle et al., 1977), is observed. This mechanism is also consistent with laboratory experiments (Li et al., 1984).

TABLE 2.
Dissolved trace metal distributions in the Upper estuary.

		Concentration range	Behavior	Reference
Li	Aug '79	2-160 ug/l	conservative	Stoffyn-Egli, 1982
B	May '76	0.73-3.7 mg/l	slight removal	Pelletier and Lebel, 1978
Cr	Aug '79	0.75-0.23 ug/l	removal	Campbell and Yeats, 1984
Mn	May '74	10-2 ug/l	conservative	Bewers and Yeats, 1978
	July '74	14-1 ug/l	conservative	Subramanian and D'Anglejan, 1976
	May '76	30-2 ug/l	input and removal	Yeats, 1987
	Aug '79	34-0.3 ug/l	input and removal	Yeats, 1987
Fe	May '74	55-14 ug/l	removal	Bewers and Yeats, 1978
	July '74	55-3 ug/l	removal	Subramanian and D'Anglejan, 1976
	May '76	190-2 ug/l	removal	Yeats, 1987
	Aug '79	50-2 ug/l	removal	Yeats, 1987
Ni	May '76	1.8-0.3 ug/l	conservative	Yeats, 1987
	Aug '79	1.4-0.2 ug/l	conservative	Yeats, 1987
Cu	May '74	5-1 ug/l	removal	Bewers and Yeats, 1978
	May '76	2.8-0.3 ug/l	conservative	Yeats, 1987
	Aug '79	1.6-0.3 ug/l	conservative	Yeats, 1987
Zn	May '76	3.3-0.5 ug/l	conservative	Yeats, 1987
	Aug '79	7.9-0.4 ug/l	removal	Yeats, 1987
Cd	May '76	18-130 ng/l	no trend	Yeats, 1987
	Aug '79	22-100 ng/l	input	Yeats, 1987
Hg	June '85	2.2-0.5 ng/l	removal	Cossa et al., 1988

Boron also shows a slight depletion in the TMZ (Pelletier and Lebel, 1978). In this case the deviation from linearity is very slight but the precision is such that the deviation can be detected. Table 2 summarizes the dissolved metal-salinity relationships for the Upper estuary.

Particulate metal distributions in the Upper estuary have been reported by Cossa and Poulet (1978) and Gobeil et al. (1981). Gobeil et al. found high particulate manganese and iron concentrations in the SPM in the St. Lawrence River that decreased markedly within the upstream part of the turbidity maximum. In the downstream part of the turbidity maximum, manganese concentrations tended to increase again but did not reach the levels observed in the river. In the eastern basin of the Upper estuary, intermediate to high concentrations were observed, with slightly higher concentrations in the deep water. Cossa and Poulet also described the distribution of particulate manganese as well as those of particulate zinc, lead and cadmium. Their manganese results were generally similar to those of Gobeil et al. although Cossa and Poulet found higher river concentrations and lower eastern basin concentrations. Particulate zinc, lead and cadmium concentrations were also found to decrease markedly in the upstream part of the estuary.The effects of two important processes can be seen in the dissolved and particulate manganese distributions. First, exchange of manganese from the particulate to dissolved phases can be inferred from the complimentary nature of the dissolved and particulate manganese distributions. While the decreases in Mn_p could result simply from dissolution of particulate manganese, resuspension of sediment particles in this region will complicate matters. Estuary sediments have lower particulate manganese concentrations than do the river particulates so resuspension of sediments will have the effect of decreasing the observed manganese content of TMZ particulates. Calculations based on observations made in August 1979 show that resuspension cannot totally account for the decrease in particulate manganese and some dissolution of particulate manganese must also be occurring.

Second, the dissolved manganese distribution on a TMZ tidal station (Figure 2) suggests release of dissolved manganese from the bottom sediments. Elevated near bottom concentrations on some other stations from this cruise are also consistent with releases from the sediments. The combination of these processes can account for the observed distributions including the scatter in the dissolved manganese distribution. The tidal cycle data for salinity, SPM and dissolved manganese (Figure 2) clearly shows that manganese concentrations do not covary with either salinity or SPM. Rather, manganese shows elevated near bottom concentrations on the falling tide that are propagated up into the surface water near low water. It is fairly easy to see that when these data are plotted on the manganese-salinity plot (Figure 1), they will contribute significantly to the scatter.

3.2. Lower estuary

In the Lower St. Lawrence estuary the bathymetry and oceanography change considerably. The depth increases to >250 m in most of the Lower estuary and the water column becomes stratified. Although a basically estuarine circulation with outflowing surface water and a subsurface inflow is maintained, the physical oceanography of the region is rather complex (see for example, El-Sabh, 1988). Estuarine characteristics are maintained for some elements as indicated by a continuation of Upper estuary element-salinity relationships into the Lower estuary. For others, particularly the nutrients, distributions become typical of

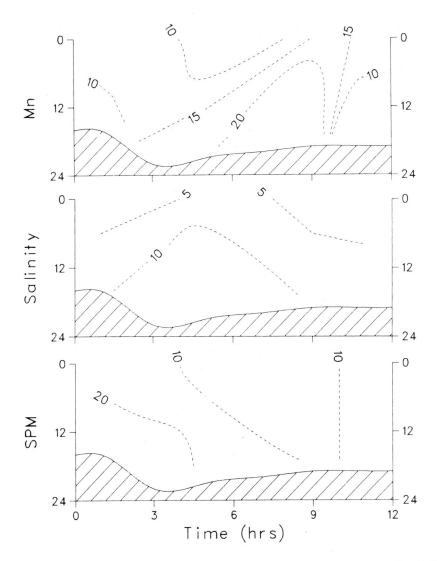

Figure 2. Dissolved manganese (ug/l), salinity and SPM (mg/l) distributions at a tidal station (47° 10'N, 70° 38'W) in the turbidity maximum zone, BIO Cruise 79-024.

coastal ocean conditions reflecting the increasing importance of biological processes that can affect the chemical distributions. An additional feature is the isolation of estuarine mixing processes in the surface waters from benthic processes. These similarities and differences result in interesting comparisons between the distributions in the Upper and Lower estuaries.

TABLE 3.
Dissolved trace metal distributions in the Lower estuary.

		Concentration range		Depth profile	Reference
B	May '76	3.0-4.2	mg/l	conservative	Pelletier & Lebel, 1978
Li	Aug '79	140-180	ug/l	conservative	Stoffyn-Egli, 1982
Cr	Aug '79	0.14-0.23	ug/l		Campbell & Yeats, 1984
Mn	May '74	0.07-20.3	ug/l	(large increases	Yeats et al., 1979
	Aug '79	0.02-7.4	ug/l	near bottom)	Yeats, 1988b
Fe	Aug '79	1.3-5.2	ug/l	increases near bottom	Yeats, 1988b
Ni	Aug '79	0.25-0.52	ug/l	decreases with depth	Yeats, 1988b
Cu	Aug '79	0.25-0.64	ug/l	decreases with depth	Yeats, 1988b
Zn	Aug '79	0.04-0.73	ug/l	increases with depth	Yeats, 1988b
Cd	Aug '79	26-59	ng/l	increases with depth	Yeats, 1988b

As already mentioned, biological processes become the dominant control on the nutrient distributions in the Lower estuary. While the estuarine circulation contributes to the buildup in nutrient concentrations in the deep waters of the Lower estuary, the surface depletion and subsurface increases in nutrient concentrations are basically a biological phenomenon. The importance of biological uptake and regeneration on several of the trace metals, notably cadmium, is also evident in the Lower estuary distributions (Table 3). The near surface distributions of a number of other trace metals are unaffected by the relatively high levels of biological activity in the Lower estuary. The Upper estuary metal-salinity relationships for dissolved lithium, boron, chromium, manganese, iron, nickel and copper continue through the Lower estuary without major deviations from the trends.

Processes occurring in the surface waters are relatively isolated from those occurring in the deep waters. As a result, the effects of benthic processes on the water column distributions can be observed in the deep waters of the Lower estuary, which flow sluggishly landward throughout most of the region. This benthic geochemistry of manganese has been studied most thoroughly. Within the water column, elevated near bottom levels that result from fluxes out of the sediments and precipitation of manganese to form suspended particles with extremely high manganese concentrations have been observed (Yeats et al., 1979). The overall cycling of manganese between sediments and water column has been described by Sundby et al., (1981). Other sediment geochemistry studies (Gendron et al., 1986; Gobeil et al., 1987) indicate that dissolved fluxes to the water column might also exist for Co and Cd. However, only for iron, and manganese of course, is there any indication of near bottom increases in dissolved metal concentrations in the water column (Yeats, 1988b).

4. Metal transport

An important aspect of estuarine trace metal studies is the estimation of the effect that processes occurring in estuaries have on the net transport of metals from rivers to coastal waters. It is easy to observe the qualitative effect of these processes on the distributions, but more difficult to estimate the magnitude of the effect on metal transports. In order to estimate the transport of metals through the estuary, the distributions of the metals and also the transport of both water and SPM must be understood. A lot of effort has been devoted to understanding the physical oceanography of the St. Lawrence estuary and Gulf. Knowledge of the general circulation pattern can be combined with the understanding gained from metal and SPM studies to develop first order estimates of the net transport of metals through the estuary.

A first attempt to estimate metal fluxes through the St. Lawrence estuary was based on a two year sampling of the St. Lawrence River (Yeats and Bewers, 1982) and a single cruise to the estuary in April 1974 (Bewers and Yeats, 1978, 1979). The results of this simple model (Bewers and Yeats, 1977; Yeats and Bewers, 1983) show that the net downstream SPM flux increases to 140% of the river flux in the turbidity maximum but then decreases to the extent that the net flux out of the Upper estuary is equal to the river flux. In the Lower estuary, the SPM flux is further reduced such that by Pte des Monts, the efflux is only 60% of the input. The particulate iron flux generally followed the SPM flux increasing to 110% in the TMZ, decreasing to 90% for the Upper estuary and 55% for the Lower estuary. Particulate manganese behaved differently since the increase in SPM flux through the TMZ was not matched by an increase in particulate manganese flux. The Mn_p flux decreased to 80%, 60% and 45% for the TMZ, Upper estuary and Lower estuary respectively. Dissolved manganese, as would have been expected from the discussion in the previous section, showed an increase in flux through the TMZ and no net loss for the rest of the system. The other dissolved metals all showed decreases in net flux to ~50% of the river flux by Pte des Monts.

This simple model can be used to compare horizontal transports of metals out of the estuary to those that are lost by sedimentation. Figure 3 shows the relative importance of horizontal versus vertical fluxes expressed as the ratio of vertical to horizontal transports for cruises in April 1974, April 1976 and August 1979 for both the Upper and Lower St. Lawrence estuaries. A positive value for the ratio of vertical to horizontal transport indicates a flux into the region from the sediments and a negative value, removal to the sediments. A ratio near zero indicates that horizontal transport dominates and one near 1 or -1 that horizontal and vertical transports are nearly balanced. In the Upper estuary, manganese and cadmium show evidence for release from the sediments so the transport ratio is generally small but positive since the fluxes are dominated by horizontal transport. Although releases from sediments are indicated for three different cruises, they cannot continue indefinitely. They must be balanced by removal at other times, or inputs from unidentified sources such as bedload transport. SPM shows little or no net sedimentation on any of these cruises so again the transport is almost wholly horizontal. While the reactivity of some metals in the Upper estuary is dominated by exchanges between dissolved and particulate phases and releases from sediments, the net transports are dominated in almost all cases by horizontal transports. This comment would also apply for the nutrients, which are largely advected through the Upper estuary without alteration.

In the Lower estuary, vertical removal to the sediments becomes more important such that vertical removal accounts for 40-80% of the input fluxes for all but manganese on one cruise. For manganese, releases from the bottom sediments, exchanges between dissolved and particulate phases and advection of enriched particles out of the estuary contribute significantly to the horizontal flux. This is reflected in the transport models that show the greatest relative importance of horizontal transport for manganese. The change from dominance by horizontal transport in the Upper estuary to dominance by vertical transport in the Lower estuary is clearly seen in Figure 3. Biogenic uptake and settling, and the settling of terrigenous inorganic particles can both contribute to the downward vertical transports in the Lower estuary. It is worthwhile to note here the dramatic change in behavior for cadmium that goes from net input in the Upper estuary to extensive removal by biological processes in the Lower estuary.

The transport calculations described in the previous paragraphs are based on a rather qualitative general type of model that helps to visualize some of the processes and their relative importance. Progress on estimating fluxes more accurately will depend on the closer integration of metal distribution measurements with more detailed investigations of the water transports and suspended sediment dynamics in order to relate observed variability in metal distributions to the variability in these other parameters. This integration of physical and chemical measurements is more advanced in studies of nutrient transport (e.g. Sinclair et al., 1976) than it is for metals. Nevertheless, a useful first order estimate of the metal transports and some information on metal reactivity can be determined from the very simple models described here. In fact, the extension of these models to the whole Gulf of St. Lawrence has been used to make some estimates of the overall removal of metals in the coastal environment (Bewers and Yeats, 1977; Yeats and Bewers, 1983). Because we were unable to estimate metal transport across the open shelf, observations on metal removal within the coastal zone and transport out of the coastal zone based on a Gulf of St. Lawrence model were used to revise global ocean residence times for trace metals (Bewers and Yeats, 1977) and potential anthropogenic effects in the North Atlantic (Yeats and Bewers, 1983).

5. Comparison to other estuaries

In broad terms, the Upper St. Lawrence estuary is a large but fairly typical partially mixed estuary with a prominent turbidity maximum zone. The nutrients and trace metals have distributions and reactivities that are also reasonably typical for this type of estuary and

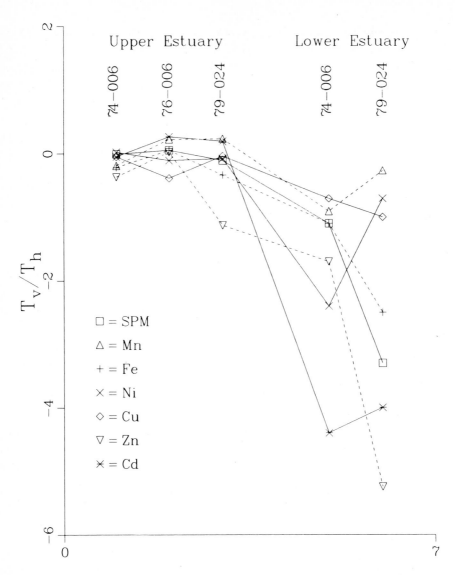

Figure 3. Ratios of vertical to horizontal transports (T_v/T_h) in the Upper and Lower estuaries for three different cruises.

consistent with ideas of estuarine behavior. For example, silicate shows conservative mixing which is consistent with the distributions and models presented by Peterson et al.(1978) that showed conservative silicate behavior at high flow rates and uptake at lower flow rates. Estuarine removal of dissolved iron is a fairly typical feature of most estuarine iron distributions as is the more complicated augmentation at low salinity and removal at higher salinity observed for dissolved manganese. Copper and zinc distributions are perhaps more interesting. Changes from estuarine removal at high river concentrations to conservative behavior at lower river concentration are observed. A comparison of these observations with those seen in other estuaries shows that observations in estuaries that at first appear to be rather divergent are in fact very consistent. The observed reactivity of copper, for example, seems to depend to a large extent on the river concentrations. So for very low river concentrations estuarine addition of copper is observed (e.g. Windom et al. 1983, Moore and Burton 1978) while conservative behavior is observed at intermediate river concentrations (e.g. Boyle et al., 1982) and removal at high concentrations (Duinker and Nolting, 1977). The river concentration for conservative behavior estimated from the results from the St. Lawrence estuary and elsewhere is ~1-2 ug/L. Similar changes would also appear to occur for zinc where again addition, conservation and removal are observed for different river concentrations.

The reactivity of nutrients and metals in the Upper estuary is governed by abiotic processes such as the flocculation of organo-metal colloids, desorption from estuarine particles and diagenetic releases from estuarine sediments. The distributions and transports are influenced by these processes but the major control on the distributions and transports is generally dilution of the river input with lower concentration seawater and horizontal advection out of the Upper estuary.

The Lower St. Lawrence estuary is more unusual. The large size of the estuary and its geometry allow for the observation of the estuarine behavior of trace chemicals over a much larger space scale than is normally possible. The great depth of the Lower estuary allows for the observation of interactions between the deep water and the bottom sediments in isolation from processes dominating the distributions in the intermediate and surface waters. These same features also make this system appropriate for the development of simple models of the chemical reactivity and transport. Most other large rivers may not be as appropriate for similar studies of reactivity, or models of transports, because they tend to flow directly onto the continental shelves rather than into an extended estuary and an enclosed marginal sea.

The descriptions of nutrient and metal distributions contained in this paper have mostly been based on large surveys of the distributions and rather general discussions of the processes influencing these distributions. These distributions have formed a basic description that has been used to model transports within the estuary and gulf. Further progress on understanding the reactivity of metals in the estuary and their transport through the estuary will depend on more contained experiments that are conducted in closer collaboration with studies of the water transport and suspended matter dynamics actually occurring at the time of the chemical surveys.

REFERENCES

Aston, S.R. 1978. Estuarine Chemistry. In Chemical Oceanography, 2nd edition Vol. 7, J.P. Riley and R. Chester editors, Academic Press, London, pp.361-440.

Bewers, J.M. and P.A. Yeats. 1977. Oceanic residence times of trace metals. Nature (London), 268, 595-598.

Bewers, J.M. and P.A. Yeats. 1978. Trace metals in the waters of a partially mixed estuary. Estuar. Coast. Mar. Sci., 7, 147-162.

Bewers, J.M. and P.A. Yeats. 1979. The behavior of trace metals in estuaries of the St. Lawrence basin. Nat. Can., 106, 149-161.

Boyle, E. A., J.M. Edmond, E.R. Sholkovitz 1977. The mechanism of iron removal in estuaries. Geochim. Cosmochim. Acta, 41, 1313-1324.

Boyle, E.A., S.S. Huested and B. Grant. 1982. The chemical mass balance of the Amazon Plume-II. Copper, nickel and cadmium. Deep-Sea Res., 29, 1355-1364.

Bugden, G.L., B.T. Hargrave, M.M. Sinclair, C.L. Tang, J.-C. Therriault and P.A. Yeats. 1982. Freshwater runoff effects in the marine environment: the Gulf of St. Lawrence example. Canad. Tech. Rep. Fish. Aquat. Sci. no 1078.

Campbell, J.A. and P.A. Yeats. 1984. Dissolved chromium in the St. Lawrence estuary. Estuar. Coast. Shelf Sci., 19, 513-522.

Coote, A.R. and P.A. Yeats. 1979. Distribution of nutrients in the Gulf of St. Lawrence. J. Fish. Res. Board Can., 36, 122-131.

Cossa, D. and S.A. Poulet. 1978. Survey of trace metal contents of suspended matter in the St. Lawrence estuary and Saguenay fjord. J. Fish. Res. Board Can., 35, 338-345.

Cossa, D., C. Gobeil and P. Courau. 1988. Dissolved mercury behavior in the Saint Lawrence estuary. Estuar. Coast. Mar. Sci., 26, 227-230.

Duinker, J.C. and R.F. Nolting. 1977. Dissolved and particulate trace metals in the Rhine estuary and the Southern Bight. Mar. Pollut. Bull., 8, 65-71.

El-Sabh, M.I. 1988. Physical oceanography of the St. Lawrence estuary. In Hydronamics of estuaries, B. Kjerfve, editor, CRC Press, Boca Raton, FL, USA, pp 61-78.

Gendron, A., N. Silverberg, B. Sundby and J. Lebel. 1986. Early diagenesis of cadmium and cobalt in sediments of the Laurentian Trough. Geochim. Cosmochim. Acta, 50, 741-747.

Gobeil, C., B. Sundby and N. Silverberg. 1981. Factors influencing particulate matter geochemistry in the St. Lawrence estuary turbidity maximum. Mar. Chem., 10, 123-140.

Gobeil, C., N. Silverberg, B. Sundby and D. Cossa. 1987. Cadmium diagenesis in Laurentian Trough sediments. Geochim. Cosmochim. Acta, 51, 589-596.

Gratton, Y., G. Mertz and J.A. Gagne. 1988. Satellite observations of tidal upwelling and mixing in the St. Lawrence estuary. J. Geophys. Res., 93, 6947-6954.

Greisman, P., and G. Ingram. 1977. Nutrient distribution in the St. Lawrence estuary. J. Fish. Res. Board Can., 34, 2117-2123.

Kranck, K. 1979. Dynamics and distribution of suspended particulate matter in the St. Lawrence estuary. Nat. Can., 106, 163-173.

Levasseur, M.E., and J.-C. Therriault. 1987. Phytoplankton biomass and nutrient dynamics in a tidally induced upwelling: the role of the $NO_3:SiO_4$ ratio. Mar. Ecol. Prog. Ser., 39, 87-97.

Li, Y.H., L. Burkhardt and H. Teraoka. 1984. Desorption and coagulation of trace elements during estuarine mixing. Geochim. Cosmochim. Acta, 48, 1879-1884.

Liss, P.S. 1976. Conservative and non-conservative behavior of dissolved constituents during estuarine mixing. In Estuarine chemistry, J.D. Burton and P.S. Liss, editors, Academic Press, London, pp. 93-130.

Moore, R.M., and J.D. Burton. 1978. Dissolved copper in the Zaire estuary. Nether. J. Sea Res., 12, 355-357.

Pelletier, E. et J. Lebel. 1978. Determination du bore inorganique dans l'estuare du Saint-Laurent. Can. J. Earth Sci., 15, 618-625.

Peterson, D.H., J.F. Festa and T.J. Conomos. 1978. Numerical Simulation of dissolved silica in the San Fransisco Bay. Estuar. Coast. Mar. Sci., 7, 99-116.

Silverberg, N., and B. Sundby. 1979. Observations in the turbidity maximum of the St. Lawrence estuary. Can. J. Earth Sci., 16, 939-950.

Sinclair, M., M. El-Sabh and J.-R. Brindle. 1976. Seaward nutrient transport in the lower St. Lawrence estuary. J. Fish. Res. Board Can., 33, 1271-1277.

Stoffyn-Egli, P. 1982. Conservative behavior of dissolved lithium in estuarine waters. Estuar. Coast. Mar. Sci., 14, 577-587.

Subramanian, V. and B. D'Anglejan. 1976. Water chemistry of the St. Lawrence estuary. J. Hydrolog., 29, 341-354.

Sundby, B., N. Silverberg and R. Chesselet. 1981. Pathways of manganese in an open estuarine system. Geochim. Cosmochim. Acta, 45, 293-307.

Takayanagi, K. and D. Cossa. 1985. Behavior of dissolved iodine in the Upper St. Lawrence estuary. Can. J. Earth Sci., 22, 644-646.

Therriault, J.-C., and G. Lacroix. 1976. Nutrients, chlorophyll and internal tides in the St. Lawrence estuary. J. Fish. Res. Board Can., 33, 2747-2757.

Therriault, J.-C., and M. Levasseur. 1985. Control of phytoplankton production in the Lower St. Lawrence estuary: light and freshwater runoff. Nat. Can. 112, 77-96.

Windom, H., G. Wallace, R. Smith, N. Dudek, M. Maeda, R. Dulmage and F. Storti. 1983. Behavior of copper in southeastern United States estuaries. Mar. Chem., 12, 183-193.

Yeats, P.A. 1987. Trace metals in eastern Canadian coastal waters. Can. Tech. Rep. Hydrog. Ocean Sci. No. 96: 49p.

Yeats, P.A. 1988a. Nutrients. In Chemical Oceanography in the Gulf of St. Lawrence, P.M. Strain (editor), Can. Bull. Fish. Aquat. Sci. 220, Chap. 3, p. 29-48

Yeats, P.A. 1988b. Trace metals in the water column. In Chemical Oceanography in the Gulf of St. Lawrence, P.M. Strain (editor), Can. Bull. Fish. Aquat. Sci. 220, Chap. 6, p. 79-98.

Yeats, P.A., and J.M. Bewers. 1982. Discharge of metals from the St. Lawrence river. Can. J. Earth Sci., 19, 982-992.

Yeats, P.A., and J.M. Bewers. 1983. Potential anthrogogenic influences on trace metal distributions in the North Atlantic. Can. J. Fish. Aquat. Sci., 40 (suppl. 2), 124-131.

Yeats, P.A., B. Sundby and J.M. Bewers. 1979. Manganese recycling in coastal waters. Mar. Chem., 8, 43-55.

Young, W. 1976. Fluoride to chlorinity ratios in the waters of the St. Lawrence estuary and the Saguenay fjord. Bedford Institute of Oceanography Report BI-R-76-1, 12p.

Chapter 9

Organic Geochemical Studies in the St. Lawrence Estuary

J.N. Gearing[1] and R. Pocklington[2]

[1] Fisheries and Oceans Canada, Maurice Lamontagne Institute, C.P. 1000, Mont-Joli, Qué., G5H 3Z4

[2] Fisheries and Oceans Canada, Bedford Institute of Oceanography, P.B. 1006, Dartmouth, N.S., B2Y 4A2

ABSTRACT

Organic material in the St. Lawrence Estuary is, as yet, incompletely understood. Much of the particulate organic matter is produced *in situ* , especially in upwelling areas. Terrestrially-derived material makes up as much as 60% of the POM (as indicated by both C/N ratios and stable isotope ratios). Inputs from freshwater runoff of the major rivers have been measured but not characterized; although many rivers are dammed,these waters provide a significant proportion of the estuarine POM in the spring. In the Upper Estuary, exchange with the adjacent salt marshes adds and removes organic material seasonally. Inputs from small streams, agricultural runoff, and municipal discharge include potentially harmful substances and have not yet been addressed in the St. Lawrence. Evidence from C/N ratios and from work with sediment traps indicates a rapid downwards transport of organic material on particulates.

Key Words: POC, PON, C/N, stable isotope ratios, hydrocarbons, fatty acids, biogeochemical markers

1. Introduction

Organic matter (OM) plays a large role in the functioning of a water body, especially in dynamic systems such as estuaries. Overall, it serves as the carbon source for organisms, while many individual biochemicals are important for growth, reproduction, osmo-regulation, and other aspects of life. Other organic compounds (and sometimes the same ones at higher concentrations) may be inhibitory or toxic. The presence of organic material may also affect the behaviour of chemicals in water and sediments. For example, dissolved organic carbon (DOC) may form complexes with, and thus increase the solubility of, some metals, consequently changing their routes and rates of transport in the ecosystem. The flocculation and precipitation of DOC in saline waters may facilitate the deposition and concentration of compounds in estuarine sediments. The oxidation of OM in sediments is instrumental in the processing of metals; its oxidation in the water column helps to regulate

Coastal and Estuarine Studies, Vol. 39
M. I. El-Sabh, N. Silverberg (Eds.)
Oceanography of a Large-Scale Estuarine System
The St. Lawrence
© Springer-Verlag New York, Inc., 1990

the oxygen concentration. Such processes are very important in understanding and predicting the effects of pollutants.

The St. Lawrence Estuary obtains terrestrial and anthropogenic OM from a wide variety of sources. At Quebec City (see Figure 2, Chapter 1), it receives the discharge of the St. Lawrence River, the second largest river system in North America. The material load of this river consists of (1) the combined contributions from the Great Lakes, the St. Lawrence Lowlands, the Canadian Shield, and the Northern Appalachians; (2) the products of such natural biogeochemical processes as occur within the river and its tributaries; and (3) the effluvia of human agriculture, human industry, and human intestines. The major tributary is the Saguenay River via the Saguenay Fjord. The Saguenay contains high levels of organic compounds from pulp and paper mills as well as waste products from aluminum refineries (see Chapter 17). There are also numerous towns and smaller tributaries along both coasts (400 km ea.) that release material to the Estuary. Finally, there are exchanges with the open Gulf of St. Lawrence as well as production of OM within the Estuary.

These inputs qualitatively and quantitatively influence the organic content of the waters and sediments of the Estuary. In the Lower Estuary, the disposition of deep water is a major determinant in the control of marine production, supplying nutrient-enriched saline waters to the head of the Laurentian Channel to sustain high levels of primary productivity (see Chapter by Therriault and Coote, and Yeats, 1979).

This chapter will summarize the general state of knowledge about the sources, concentrations, and behaviour of OM in the St. Lawrence Estuary. Our approach has been to examine first the total organic matter (dissolved and particulate) in order to put the St. Lawrence in global perspective and to model the annual cycle of OM. Total concentrations are important in understanding the extent to which OM affects the processing of other chemicals in the estuary. Knowledge of the OM within the estuary, and its changes in space and time, is largely limited to the bulk parameters of particulate organic carbon (POC), particulate organic matter (POM = POC x 1.7; factor derived from determinations of carbon and nitrogen in mixed plankton and organic detritus), DOC, dissolved organic matter (DOM), total organic carbon (C_{org} = POC + DOC), and particulate organic nitrogen (PON). In the case of sediments, organic carbon is converted to OM using the empirical multiplier 1.887.

Next are presented the results from studies focusing on parameters which serve to identify the source and state of the organic matter. These include the stable isotope ratios of carbon and nitrogen (tracers of bulk material) and individual compounds or groups of compounds which, though they make up only a small part of the total OM, may have disproportionate importance. These tracer compounds include essential nutrients, toxic pollutants, indicators of bacterial activity, etc. This chapter focuses primarily on lipids used as geochemical tracers, with some discussion of petroleum- and combustion-derived hydrocarbons.

Discussions of other compounds of biological interest and of pollutants may be found in other chapters of this volume by Cossa, Schafer et al., and Therriault et al.

2. Annual Cycles of Organic Matter

The sections dealing with bulk carbon (POC, DOC, C_{org}) are based primarily on work by the second author. Details of the methods used can be found in Gershey et al. (1979), MacKinnon (1978), Pocklington (1973), Pocklington and Hagell (1975), and Pocklington and Kempe (1983), Pocklington and Morash (1979).

2.1. General features of organic matter distribution within the Estuary

The data-base for OM is in certain months (Apr-Aug) extensive, in some (Jan, Nov) adequate, but in others, nonexistent. Vertical profiles of POC, PON and DOC have been taken at a number of stations, mostly within the Lower Estuary; surficial sediments have been sampled throughout the Estuary; and the discharge of the St. Lawrence River - the largest contributor of terrestrial OM to the Estuary - has been measured monthly over a recent five-year period (Pocklington & Tan, 1987). This data-base permits a description and preliminary quantification of the annual cycle of OM within the Estuary.

In the water column, systematic variation in OM concentration and composition occurs over much of the year and the whole of the Estuary (Table 1).

2.1.1. Surface-layer
In the spring (Apr-May), concentrations of POC in the surface-layer range from a low of 55 mg/m3 in the colder, more saline water along the south shore to a high of around 230 mg/m3 in the warmer, fresher water in the centre channel; PON behaves in similar fashion. Carbon to nitrogen ratios in the POM (C/N) range from (5.7) in water along the north shore to (12.6) in the fresher water in the centre channel, denoting terrestrial influence there. At this season, the concentration of POC and (especially) PON in the Estuary is lower than that in the Gulf as a whole (Pocklington, 1988), probably because primary production in the Estuary is inhibited by low water temperature and higher turbidity.

In early summer (May-June), POC concentrations range from a low of around 20 in the centre channel to a high of 200 in the most saline water entering the Estuary at Pte-des-Monts; PON behaves similarly. The C/N ratio is generally low, implying the predominance of primary marine production. The concentrations of POC and PON in the Estuary are still somewhat lower than those of the Gulf as a whole.

In summer (Jul-Aug), POC concentrations are at their highest for the year and range from a low of 90 in the centre channel to a high of above 400 in warm, fresher water leaving the

TABLE 1
Monthly ranges of suspended particle concentrations (POC, PON) and C/N ratios in each defined water layer of the Estuary.

Month	Depth range (m)	n	POC	PON	C/N
			(mg.m^{-3})		(atoms)
Jan	0 - 50	16	13 - 36	1.3 - 4.2	6.5 - 12.3
	75 - 150	7	7 - 22	1.1 - 2.4	7.1 - 11.5
	145 - 300	13	7 - 16	1.1 - 2.2	6.2 - 11.6
April/	0 - 50	17	55 - 234	7.6 - 26.7	5.7 - 12.6
May	45 - 200	28	16 - 34	1.5 - 6.1	6.5 - 12.9
	200 - 325	16	13 - 37	1.3 - 3.2	8.8 - 13.5
May/	0 - 50	21	23 - 200	3.8 - 31.6	6.5 - 8.8
June	40 - 200	20	12 - 23	1.5 - 3.3	7.0 - 10.6
	150 - 450	14	16 - 30	1.7 - 4.2	6.9 - 13.6
July/	0 - 50	13	90 - 412	12.3 - 58.6	6.5 - 9.6
Aug	25 - 200	22	17 - 32	2.3 - 4.5	7.9 - 10.3
	200 - 300	10	15 - 20	1.6 - 3.0	7.6 - 11.5
Nov	0 - 50	6	37 - 56	5.1 - 6.6	8.0 - 9.9
	50 - 100	4	15 - 34	1.9 - 4.4	9.1 - 9.5
	100 - 250	5	13 - 35	1.7 - 5.5	9.1 - 11.2

Estuary along the south shore; PON ranges from low to high between the same locations, but the range of C/N is low, showing little influence of terrigenous inputs at this time of low river flow. At this season, POC and PON concentrations in the Estuary are higher than those in the rest of the Gulf, and the C/N ratio is lower, which implies that the primary production of OM in the Estuary exceeds that of other regions, a conclusion in accord with the comprehensive production data of Steven (1975).

There are few data for fall (Nov). They show that the range of POC concentrations is toward the lower end of the ranges of spring and summer concentrations; PON concentrations are similar. The range of C/N ratios gives little evidence of fresh organic production at this season.

In winter (Jan), POC and PON concentrations are lower than the lowest values at other seasons, being similar to those found outside the Gulf on the Nova Scotian shelf in the same month and little higher than open ocean values (Pocklington, 1985). The range of C/N is broad and implies some fresh production (though not much: the mean POM in January is only 20% of that for Apr-Aug) and some terrigenous addition (possibly by ice-rafting; Dionne, 1981). In general, the seasonal pattern of POC and PON concentrations, and composition of POM as revealed by the C/N ratio, in the surface mixed layer of the Estuary conforms well to the cycle of primary production (Steven, 1975; Therriault and Lavasseur, 1985) and the sequence of terrigenous inputs described by Pocklington & Tan (1987).

2.1.2. Intermediate-layer ("cold-layer", T below 1.5 deg.C)

In spring, concentrations of POC and PON in the intermediate-layer are lower than those in the surface layer but the C/N ratios range somewhat higher. In early summer, POC and PON concentrations decrease and the C/N ratio is a little lower than earlier in the year. In late summer, concentrations increase again but the C/N ratio is little changed. In fall, there is little change in concentration, but the C/N ratios tend toward higher values. In winter, concentrations of POC and PON are minimal; C/N ratios are higher. This sequence of events in the intermediate layer is interpreted as a consequence of POM produced during the year replacing that of a previous annual cycle.

2.1.3. Deep-layer

Concentrations of POC and PON in the deep layer, other than being somewhat lower in winter and late summer, do not vary greatly during the year, but the C/N ratios are (except for winter) consistently higher than in the rest of the water column, implying a greater proportion of terrigenous (or more degraded) OM in the deep water. However, the presence of some C/N ratios as low as those in the surface layer also suggests rapid transfer of OM produced at the surface to the deep water. Additional evidence for this rapid transport is the seasonal change in organic carbon and nitrogen concentrations, as well as C/N ratios, shown by sinking particles. Silverberg et al. (1985) reported C/N ratios from free-drifting sediment traps suspended at approximately 150 m in the zone of minimal suspended particulate material over four years. More terrestrial values (>14) were found during spring when there was high freshwater runoff and more marine values (~7) in the summer, again indicating a progressive decrease in the influence of terrigenous OM in favour of marine OM during the year.

2.2. Autochthonous production of organic matter

A 2-dimensional box model has been constructed by Pocklington (1988) to give a quantitative description of the cycle of OM within the Gulf of St. Lawrence . Comparison with it shows that though the Estuary - defined as the region west of Pte.-des-Monts to Quebec City - has an area of 13×10^3 km^2 or only about 6% of the total area of the Gulf (226×10^3 km^2), autochthonous production of OM within the Estuary is disproportionately ($\times 2$) high. The concentrations of POC and PON in the surface layer of the Estuary in spring and summer (Apr-Aug) are as high as in the most productive regions of the World Ocean - e.g. off the northwest African coast during the months of intense upwelling (Pocklington & MacKinnon, 1982). Estimates of annual primary production in the Estuary have been derived from measurements of ^{14}C uptake made throughout the euphotic zone in all regions of the Gulf, from April to the end of November, by Steven (1975). In the Estuary, an estimate of 510 gC m^{-2} a^{-1} - close to the potential maximum attainable of 565 gC m^{-2} a^{-1} calculated from 200 mgC m^{-2} hr^{-1} (Steven, 1975) and 2826 hours of adequate insolation (~0.6 ly/hr) - which would place the Estuary among the most productive marine systems in the world, gives a total of 6.62 x 10^6 t C/a. In contrast, Therriault and Levasseur (1985) calculated only 134 gC m^{-2} a^{-1} for the production of the central part of the lower Estuary, which is little more than the global average oceanic production of 130 gC m^{-2} a^{-1}. Without entering the debate as to which of these two estimates is more likely correct, we shall use for present purposes the estimates of Steven (1975) as these were made consistently throughout the Gulf and permit comparison between the Estuary and the open Gulf, while treating his actual values for primary production as upper estimates.

In the Estuary, the mean DOC concentration of 1.4 mg/l in summer implies 805 x 10^3 t C in the surface layer. Even if all of this came from *in situ* production (which it most certainly does not), it could be accounted for by the primary production of one year on the assumption of only 13% of photosynthesate passing directly to the dissolved state (Aaronson, 1978). Therefore, the Estuary does not hold the equivalent of many years' accumulation of organic production, unlike some other estuarine systems e.g. the Baltic (Pocklington, 1986), probably because the surface waters have mean residence times of months rather than years.

2.3. Freshwater additions of organic matter

The largest single source of freshwater to the Estuary is the St. Lawrence River. Data for C_{org} and POM collected bimonthly at Quebec City over five recent years (1981 - 1985, Pocklington & Tan, 1987) permit the calculation of the fluvial contribution of OM to the Estuary. The annual discharge of the St. Lawrence River (413 ±7 km^3; years 1981 - 1985) contains OM in particulate (POC; 240 - 370 x 10^3 t) and dissolved (DOC; 1.29 - 1.72 x 10^6 t) form. The concentration of POC (and PON) is linearly correlated with discharge (increased during the spring flood and the fall enhancement of flow), but concentration of DOC is inversely related to discharge. In consequence, the C_{org} (1.66 - 2.09 x 10^6 t C/a, of which 12% is particulate) discharged by the river is relatively invariant. It is a substantial quantity of OM on a world scale - more than half that contributed by the Mississippi River to the Gulf of Mexico (3.40 x 10^6 t C/a, Dahm et al., 1981). It is of the same order as net primary production in the Estuary (1.65 x 10^6 t C/a, if net is 25% of the gross as *per* Wafar et al., 1984).

Other rivers (Table 2) bring an additional volume of water equivalent to 25% of the St. Lawrence River to the Estuary. If we assume that they contain OM in the same proportion as the St. Lawrence, then 457 x 10^3 t C/a (70 x 10^3 t C/a of which is POC) is contributed to the Estuary by these secondary fluvial sources. The contribution of local anthropogenic organic inputs, including sewage from coastal towns, is quantitatively small but may be of environmental significance (Sections 3 and 4).

TABLE 2.
Contribution of rivers other than the St. Lawrence to the St. Lawrence Estuary

Name	Basin area	Discharge				Years
		mean	max.	min	annual	
	$(10^3 km^2)$	(m^3/s)			(km^3/a)	
Saguenay	88.1	1470	9260	51	46.4	1914-1984
Manicouagan	45.8	871	6030	0	27.5	1947-1959
R. aux Outardes	19.0	399	2830	11	12.1	1923-1977
Betsiamites	18.7	331	1350	0	10.4	1964-1984

The C/N ratios in the fluvial POM ranged from 8.4 to 12.7 about a mean of 10.4, close to the mean C/N ratio of 9.9 calculated for all the major rivers of the world (Meybeck, 1982). The ratio was consistently highest for the year in late fall (Nov/Dec) and lowest in the summer (July). The high ratios in the fall coincided with the enhanced POC loads at that season when more POM of terrestrial origin (high C/N) is added to the river by runoff (30 - 60 x 10^3 t C, Pocklington & Tan, 1987). In contrast, the C/N ratio is lowest in summer when conditions favour within-river organic production (in the case of the Loire, 55% of total POM in summer is from *in situ* production; Billen et al., 1986).

2.4. Additions of organic matter from the atmosphere

The small addition of OM from the atmosphere was calculated from the concentration of C_{org} in rain falling over adjacent terrain and the precipitation volume over the Estuary to be 18 x 10^3 t C/a, of which 20% is particulate (Pocklington, 1986). It is quantitatively trivial relative to other inputs but may be important qualitatively (see Chapter 11).

2.5. Exchanges of organic matter with external waters

The greatest amount of OM leaving the Estuary is in the water that flows out in the surface layer through the section at Pointe-des-Monts. An estimate of the quantity of OM contributed to the open Gulf through this section was made by combining OM concentrations with geostrophic volume transports adjusted to satisfy the condition of zero net mean salt transport through the section (Table 3). The net export of C_{org} from the

Estuary (1.96×10^6 t C/a) is of the same order as the river input (see above) but the quantity in particulate form (128×10^3 t C/a) is less than half that contributed by the river to the Estuary. A marked decrease in the proportion of POM in the C_{org} as salinity increases is shown in other estuarine systems (Cadée, 1984).

TABLE 3.
Mass transports through the Pte-des-Monts section in different months of the year.

Month	OUT				IN			
	Water	DOC	POC	PON	Water	DOC	POC	PON
	(km^3)	$(10^3$ t$)$			(km^3)	$(10^3$ t$)$		
Jan	906	1075	17.4	2.36	839	793	9.8	1.40
July	710	800	36.1	5.53	667	638	17.2	2.47
Nov	1189	1296	42.6	5.61	1153	1279	39.1	5.03

Water volumes from El-Sabh (1975) as modified by Bugden (1981).

2.6. Losses of organic matter to the sediments

The mean OM content of sediments in the Estuary is generally low due to the large area of the bottom covered with coarse-grained sediments (concentration of OM being highly dependent upon the grain size of sediments; Bordovskiy, 1965). In Table 4 are listed the organic carbon and nitrogen concentrations and C/N ratios for different grain sizes of sediments in the two regions of the Estuary and in the Saguenay Fjord. Carbon concentrations in sands range from 0.6 to 8.0 mg/g sediment. In pelites (i.e. sediments of grain size <50 um), organic carbon varies over the range 10.3 to 37.8 mg/g. The higher concentrations can be found locally inshore near obvious OM sources, e.g. at the head of the Saguenay Fjord. Sands on the continental shelf range over 0.5 to 7 mg/g (Nigeria, Gulf of Mexico, Hudson Canyon, Atlantic off New York; Gearing, 1975; Gearing et al., 1976; Farrington and Tripp, 1977). Silts and clays from off the Mississippi River, off Washington State, in the Bay of Bengal, and off the Niger River range from 5 to 29 mg/g (Gearing, 1975; Gearing et al., 1976; Hedges and Van Geen, 1982).

The rate of organic carbon burial in the Estuary was calculated from a mean rate of (total) sedimentation of 0.51 mg cm-2 d-1 (Silverberg et al., 1986) and a mean C_{org} concentration of 1.65 % (Table 4) to be 30.7 g C m-2 a-1. Over the area of the Estuary covered by fine sediments (2.44×10^3 km2; Canadian Hydrographic Service, 1972), the burial of POC in the pelites is 75×10^3 t C/a. The contribution of the St. Lawrence River (Section 2.3) alone is more than adequate to account for this, even if only POC (300×10^3 t C/a) is considered as source (i.e. assuming conservative transport of DOM through the Estuary as per Mantoura and Woodward, 1983). This does not in itself prove that the OM in the

TABLE 4.
Means and ranges of sedimentary organic carbon and nitrogen and the C/N ratio of surficial sediments in the Estuary and in the Saguenay Fjord.

	Organic C (mg/g)		Organic N (mg/g)		C/N (atoms)	
			Upper Estuary			
True sands	1.0	(0.6 - 1.6)	0.13	(0.11-0.26)	8.9	(5.4-15.5)
Pelitic sands	7.4	(5.8 - 8.5)	0.73	(0.58-1.00)	11.9	(9.1-15.6)
Pelites	18.2	(10.3 - 29.4)	1.56	(1.00-2.66)	13.6	(11.9-15.9)
			Saguenay Fjord			
Sands	4.2	(0.5 - 8.0)	0.32	(1.00-0.63)	15.3	(7.5-21.2)
Pelites	25.7	(11.7 - 37.8)	1.42	(0.61-1.85)	21.1	(15.3-42.8)
			Lower Estuary			
Sands	4.1	(1.9 - 6.3)	0.35	(0.15-0.83)	13.8	(12.0-15.9)
Pelites	14.8	(11.8 - 19.3)	1.36	(0.92-1.61)	12.7	(9.5-16.4)

Data from Pocklington and Morash, 1979.

sediments of the Estuary derives from fluvial POM, only that the latter is quantitatively sufficient for the role. Carbon isotope ratios for sedimentary organic matter in the Estuary (see below) provide independent evidence for the important contribution of terrigenous carbon to the OM accumulating in the Estuary.

2.7. Removal of organic matter by fishing

Though quantitatively small, this is probably the most socially-desirable removal of OM from the system. The fishery in the Estuary is equivalent to 2-3 x 10^3 t C/a.

2.8. Conclusion with regard to annual cycle of organic matter

The fate of the rest of the OM within the Estuary is not to be exported or buried in the sediments, but to be mineralized and, as inorganic nutrients, returned to the surface layer. In order to produce observed increases in nutrients (Coote & Yeats, 1979) within the Estuary, POM must settle out of the surface layer and be oxidized in the deep water. This process consumes oxygen, reducing the oxygen concentration in the deep water by one half (6.8 to 3.5 ml/l) between Cabot Strait and the Saguenay (Fig.1). Were it not for the vigorous vertical exchange of water (see Chapter 5), deep water within the Estuary might become anoxic intermittently, as occurs in the Baltic (Pocklington, 1986).

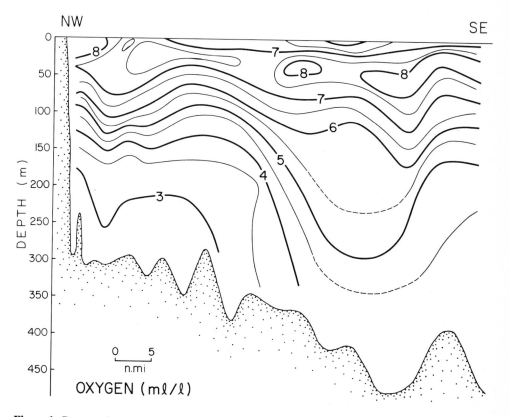

Figure 1. Decrease in deep-water oxygen concentration from Cabot Strait (SE) to the Saguenay.

3. Stable Isotope Ratios

One approach to estimating the relative contribution of organic matter from different sources is to examine the natural abundances of stable carbon and nitrogen isotopes that have been measured in the St. Lawrence Estuary. The method is based on the fact that many reactions involved in the biochemical production of organic carbon and nitrogen fractionate isotopes. The inorganic precursors may also have different isotope ratios. The net result is that, in general, terrestrial organic material can be distinguished from that produced in the marine environment; in some cases different types of terrestrial or marine organic matter can be differentiated (for example grasses versus trees and net plankton versus nanoplankton). The isotopic signals produced by plants are preserved with little change as this material moves through the biosphere and geosphere. Gearing (1988) gives a more complete discussion of the potential and limitations of stable isotopes as a research tool to distinguish land and marine production. Isotope ratios are reported as $\delta^{13}C$ and $\delta^{15}N$ (per mille), where more positive values indicate heavier samples (enriched in C-13 or N-15); analytical precision is ± 0.2 to 0.3 o/oo.

In order to use isotope ratios quantitatively, the end member values for terrestrial and marine organic matter should be determined at each location. Equations such as that given in Newman (1973) are used to calculate the fraction from each source or end member. Pocklington and Tan (1987) have made a three-year study of the carbon isotope ratios of surface POC entering the St. Lawrence Estuary at Quebec City. The values ranged from -23.6 to -27.0 (average = -25.6 o/oo), varying seasonally with water discharge and suspended load. Values for POC entering from the Saguenay range between -26 and -27 o/oo. From this we have assumed a terrestrial end member for carbon of -27.0 o/oo. The situation for the marine carbon value is less clear. Gulf of St. Lawrence sediments average -22.4 o/oo (Tan and Strain, 1979b) while surface POC averaged -24.9 in 1974 (Tan and Strain, 1979a) and -23.9 in 1979 (Tan and Strain, 1983). All of these values are more negative than POC measured at the mouth of the estuary which is presumably due to fresh planktonic production (-21.5 to -21.8 o/oo; Tan and Strain, 1979a, 1983). In this paper we assume a value for marine phytoplankton of -22.0 o/oo, in line with values of plankton measured elsewhere (see, for example, Gearing et al., 1984) while noting that the actual phytoplankton values will vary with species composition and growth conditions. In regions where nanoplankton forms a significant part of the *in situ* production, it is possible that a better estimate of the marine end member is around -23 o/oo (Gearing et al, 1984); thus the present estimates of terrigenous contribution should be considered maxima.

There have been very few studies of nitrogen isotopes on which to estimate end member values. For this chapter we will use +2.5 o/oo for terrestrial organic nitrogen (from values obtained for POC in the Saguenay Fjord, unpublished data) and +6.5 o/oo for marine organic nitrogen (from values obtained in the mouth of the estuary and from Macko, 1981 for the Gulf of Maine).

Within the estuary itself, four major isotope studies have been published on organic carbon (Tan and Strain, 1979a, 1979b, 1983, Lucotte, 1989). Also included here are some unpublished values for carbon and nitrogen (Gearing and Macko, methods as in Macko, 1981). The data are summarized in Figure 2.

3.1. Sediments

The situation for sediments is relatively straightforward. Values in the Upper Estuary average around - 25 o/oo and are relatively constant, indicating that up to 60% of the total organic matter is of terrigenous origin. These values are typical of values found for the most fluvial sections of estuaries, but most estuarine sediments become rapidly less

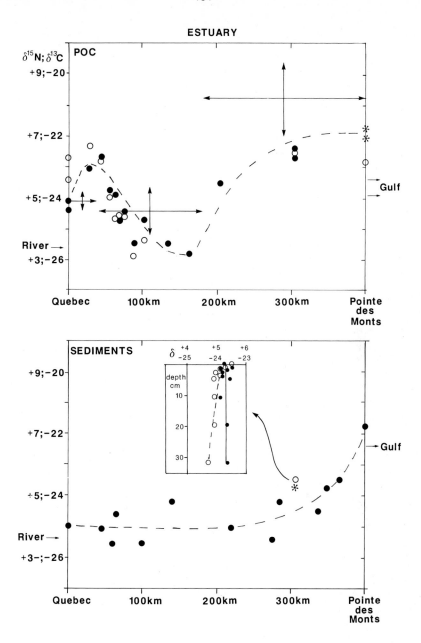

Figure 2. Summary of stable isotope ratios of suspended particulates (top) and sediments (bottom) in the St. Lawrence Estuary. Solid symbols are used for $\delta^{13}C$ and open symbols for $\delta^{15}N$ values. Arrows indicate regional averages for $\delta^{13}C$ with one standard deviation. Values for POC are from Pocklington and Tan (1987), river; Tan and Strain (1979a), star symbols; Tan and Strain (1983), regional averages; and unpublished data (Gearing and Macko) open and solid circles. Values in the bottom box are for surface sediments (collected in the middle of the channel when possible; the depth profile at one location (average of two cores) is shown as an insert. sedimentary data is from Tan and Strain (1979b) and unpublished work (Gearing and Macko).

negative with increasing distance from the head and salinity (for example Hunt, 1970 andGearing et al., 1977). St. Lawrence sediments do not become less negative until well into the Lower Estuary, where the expected gradient can be seen (Fig. 2). The constancy in the Upper Estuary may result from the extensive sediment reworking and exchange with older, salt marsh sediments (Gobeil et al., 1981; Lucotte and d'Anglejan, 1986).

In the more saline half of the Estuary, values in the sediments rise from -25 (60% terrestrial) off Tadoussac to -21.8 o/oo (100% marine) in the Laurentian Channel off Pte.-des-Monts. Surface sediments from the channel near Rimouski in the Lower Estuary contain 25 to 34% terrestrial organic matter based on nitrogen and carbon isotope ratios respectively. These are the first nitrogen values reported for the St. Lawrence. It is interesting to note that the carbon values become more positive and thus the relative amount of terrestrial material decreases with water depth across transverse sections. Sediments near both coasts are more "terrigenous" by up to 3 per mille compared with deep, channel sediments (Tan and Strain, 1979b); this corresponds to around 60% nearshore compared with virtually no terrestrial material in the trough. A shift in grain size may account for some of this difference if marine organic matter is associated with smaller sized particles than terrestrial material, some studies have found terrestrial plant waxes to be more concentrated on the sand-sized particles than on clays. The difference with depth may also result from the input of sewage from the coastal towns; sewage-contaminated sediments in New York, California, and Rhode Island have shown relatively negative $\delta^{13}C$ values of -24 to -26 per mille (Burnett and Schaeffer, 1980; Sweeney et al., 1980; Oviatt et al., 1987). Fatty acids also vary with depth along a transverse section (Rodier and Khalil, 1982 - see section 4.3.). Such changes may be more pronounced in the St. Lawrence than in most estuaries. Further examination should help map the movement of organic matter within the estuary.

Two box cores from the Lower Estuary show no change of $\delta^{13}C$ with depth over approximately 100 years (Fig. 2). The average for carbon in the sediments (-23.7 ±0.2o/oo, n=20) is equal to the average for material collected in sediment traps in the same location over several years (-23.7 ±1.3 o/oo, n=29). The situation for nitrogen is different. Sediment trap material (+6.1 ±1.5 o/oo) is more positive than the surface sediment (+5.5) and the ratio becomes less positive with depth (+4.7 o/oo at 30 cm). This possibly results from fractionation during degradation for nitrogen but not for carbon, as planktonic organic nitrogen is used up and is replaced by bacterially-produced organic nitrogen. Pocklington and Tan (1987) present some evidence for this possibility.

3.2. Suspended Particles

Carbon and nitrogen isotope ratios in both particulates and sediments of the St. Lawrence are significantly correlated (>99.99%), implying common sources. More nitrogen values are needed to interpret this with confidence. If suspended particulates are the major source for sedimentary organics, they should show the same general isotopic trends as sediments, representing a single point in time rather than the average found in the sediments. Certainly, on the whole, suspended particulates in the River and the Upper Estuary are more negative than in the Lower Estuary, both for carbon and nitrogen. However, little trend or reversed trends have been found at times in the Upper Estuary. Other transects in this region have found a small change in the expected direction for carbon (more positive values at higher salinities; Lucotte, 1989). The differences reflect an unstable situation in the Upper Estuary with (1) low planktonic productivity (possibly dominated by nanoplankton with some heterotrophic capacity) and (2) high levels of turbulence, bacterial activity, sediment resuspension, and macrophyte detritus from adjacent salt marshes.

Isotope ratios (POC and PON) in the Lower Estuary are considerably more positive than in the Upper Estuary. Values of -21 to -22 and +6 o/oo, found for diatoms isolated in the Lower Estuary, are typical of marine phytoplankton (Gearing, 1988). These values change seasonally due to changes in phytoplankton composition and growth conditions (Gearing et al., 1984). Sediment trap material from the Lower Estuary also ranged over ±2 (C) and ± 2.5 o/oo (N) during several years. The seasonal pattern is of more negative values in spring (due to freshwater inputs of terrigenous material), more positive values in summer (in situ diatom production), and a return to more negative values in fall. The fall values probably result from a shift in the phytoplankton assemblage to nano- and pico-plankton or to increased heterotrophic activity because C/N ratios still indicate little terrestrial influence. The smaller sized phytoplankton are known to have less C-13 (Gearing et al., 1984).

3.3. Conclusions

In summary, the stable isotope ratios of carbon and nitrogen tell similar stories in the St. Lawrence Estuary, indicating similar sources for these two elements. The isotope ratios give considerable insight into the relative importance of different sources of organic matter; this information is quite useful when used in conjuction with the concentrations from the budget (Section 2.).

Isotope ratios in the Lower Estuary show seasonal trends for POM in the surface waters consistent with the hypotheses made from POC, PON, and C/N data. Values in spring indicate up to 60% terrestrial material. Authochthonous production become dominent with time (up to 100% at times in summer). In fall and winter when concentrations decrease, isotope ratios point to a shift to production by nanoplankton while the C/N ratios remain at "marine" values. The little data available from intermediate and deep water are similar to those from surface water (Tan and Strain, 1983), compatible with the idea of rapid vertical transport.

The influence of freshwater runoff, in addition to the effect noted above on the seasonal pattern in the Lower Estuary, is noticable in the Upper Estuary. Trends here do not fit simple models, the isotope ratios being influenced by sewage from Quebec City, by production in the adjacent salt marshes, and possibly by exchanges with the waters from the Saguenay (very negative values).

There is no isotopic data which shed light on atmospheric inputs, exchanges with the Gulf, or changes due to fishing.

Ratios in the sediments indicate significant (60 to 80%) amounts of terrestrially-derived organic material over the upper 300 km. of the Estuary. Terrestrial influence in the central channel decreases to almost nothing by Pte.-des-Monts. In the Lower Estuary, ratios in the sediments mirror the average ratios in the surface particulates, indicating that these particles may be the principal source of sedimentary organic matter. Isotopic differences across the Estuary implies that much of this material may be coming from secondary tributaries or may be reworked within the estuary.

4. Molecular Tracers

In addition to the measurements of bulk organic matter discussed in the previous sections, there have been several studies of some individual chemicals such as lignin, hydrocarbons, and fatty acids. Such measurements are particularly sensitive for tracing specific organisms, reactions, or sources. Results from some compounds are covered in other chapters, for example chlorinated hydrocarbons as indicators of pollution (Chapter by Cossa) and chlorophyll-a and other pigments as indicators of primary production (Chapter Therriault).

4.1. Organic waste from pulp and paper mills

Lignin - an unequivocal indicator of terrigenous OM that contributes up to 30% of the weight of dry wood and is essentially undegraded in anoxic sediments - was the tracer used for over a decade to measure the impact of pulp mill wastes on nearshore waters (Pocklington and MacGregor, 1973; Pocklington, 1975, Pocklington and Roy, 1975). Table 5 lists the lignin concentrations found in some representative environments in the Estuary. Concentrations are highest in waters receiving relatively direct inputs of pulp mill waste e.g. the upper Saguenay Fjord. These high lignin concentrations are indicative of highly concentrated organic waste, rather than a general indicator of terrigenous OM. Much lower concentrations of lignin are found in regions more distant from these inputs of organic wastes and represent a terrigenous "background " signal.

TABLE 5.
Contribution of lignin to sediments of the Estuary and Saguenay. Table adapted from Pocklington, 1976.

Location	Lignin (mg/g)	Organic Matter (%)	Lignin (as % of OM)
Upper Saguenay	4.6	5.5	8.2
Middle Saguenay	9.0	7.1	12.7
Lower Saguenay	0.03	0.67	0.5
St. Lawrence Estuary	0.77	3.6	2.1
Laurentian Channel (outside estuary)	0	2.1	0

In the Saguenay Fjord, inputs of lignified material from pulp and paper plants have been traced not only longitudinally along the axis of the Fjord, but down the sedimentary column where they persist for times of the order of at least one hundred years (see Schafer et al., Chapter 17).

Dissolved phenols - degradation products of lignin - were examined by Gagné and Brindle (1985) in north-shore coastal areas around Baie-Comeau, Riviere Mistassini, Godbout, and Riviere Manicouagan. Values from the water column and in interstitial waters indicate high levels of vanillin and eugenol only near Baie-Comeau. Compared with levels in other Kraft Paper Mill effluents (Keith, 1977) vanillin is typical and eugenol is quite high in this one spot.

4.2. Ketones

Sediment cores and sediment trap material from one station in the Gulf and Maritime Estuary of the St. Lawrence was examined for the presence of long chain ketones and ketone alcohols (Nichols and Johns, 1986). These lipids, produced by *Emiliania huxleyi* and other Haptophycean algae, have been proposed as biological markers. In contrast to samples from the open Gulf, these compounds were not found in the estuary.

4.3. Fatty acids

Fatty acids, major components of complex lipids containing a hydrocarbon chain and a terminal carboxyl group, are major components of all living organisms and are nutritionally important. They are used by organisms for energy storage and in membranes. Terrestrial plants have a slightly different suite of fatty acids than algae. Some fatty acids are synthesized by animals, either *de novo* or from dietary chemicals. An organism's fatty acid makeup may also change with environmental conditions. These chemicals have been used to deduce feeding relationships, terrestrial versus marine influence, bacterial activity, etc. (for example Parker, 1967; Jeffries, 1972; Schultz and Quinn, 1972; Cranwell, 1978; Nishimura and Baker, 1987).

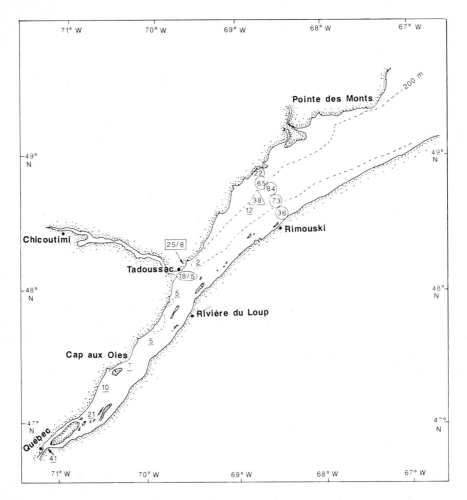

Figure 3. Summary of fatty acid concentrations in suspended particulates (numbers underlined), in surface water in November and then June (numbers in oval), in the surface microlayer (numbers in rectangle), and in surface sediments (numbers in circles and triangle) of the St. Lawrence Estuary. Values are from Marty and Choiniere (1979), oval and rectangle; Rodier and Khalil (1982), circles; and unpublished work by Tronczynski and Gearing, underlines and triangle.

In the St. Lawrence Estuary, there have been a few published reports of fatty acids (Marty and Choinière, 1979; Rodier and Khalil, 1982) and some, as yet, unpublished work by Tronczynski and Gearing (Tronczynski et al., 1987; Gearing and Tronczynski, 1988; and this work). The results are summarized in Figure 3. The data show large differences; values are too limited to define the limits or reasons for the variability but several trends are apparent: (1) The values in the water column vary greatly both seasonally and spatially. Marty and Choinière (1979) sampled the same station off Tadoussac (Fig. 3) in November, June and August (1974-75). Concentrations of fatty acids varied by almost one order of magnitude and were attributed to differences in production. The spacial trends noted by Tronczynski et al. (1987) along a transect from Quebec City to Rimouski (Fig. 3) in June, 1986 also reflect production, being highest at the two ends where there were phytoplankton blooms. (2) Values in the surface microlayer are, on average, 1.6 times higher that in the underlying water. These are typical enrichment factors for fatty acids, which are hydrophobic and surface active. (3) Values in the sediments vary with depth in the sediment due to degradation (Rodier and Khalil, 1982) as well as (for one transverse transect from Rimouski to the north shore, Fig. 3) with water depth and grain size, being highest in the pelites of the Laurentian Channel. There is also a difference in binding to the sediments. Approximately half the sedimentary fatty acids in the Channel are extractable with organic solvents as compared to the quantities extracted with base (38 versus 73 and 84 ug/g), these so called free fatty acids are presumably more recently added to the sediments.

Compared with those in diverse aquatic systems, fatty acid concentrations and distributions in the St. Lawrence Estuary fall within the range reported for estuarine systems. Tables 6 and 7 summarize values for sediments and suspended particulates respectively. Each table begins with more terrestrial and polluted regions, continues through various estuaries and ends with the open ocean. Of particular importance on these tables is the indication of method. The concentrations and ratios of different fatty acids are affected by the strength of the extraction (see Nishimura and Baker, 1987, for a recent discussion and other references). The two most used methods are extraction with organic solvents such as hexane or dichloromethane to obtain "free" fatty acids and extraction with a base (usually KOH) and organic solvents to obtain "total" fatty acids. The former are less tightly bound to the sediment and make up approximately half the "total" fatty acids, the actual fraction varying with particular locations. Correcting for this difference, the three studies of the St. Lawrence are in good internal agreement. The values are about the same as those reported for smaller bays such as Narragansett, Buzzards, Baffin (Texas), and Aransas and are slightly higher than nearshore, marine waters such as Rhode Island Sound, and the Gulf of Maine (Fig. 4A).

Also included on Tables 6 and 7 are various fatty acid parameters indicative of relative terrestrial contribution (n-C16:0 /n-C24:0), recent *in situ* production (n-C16:0 /n-C16:1), and bacterial activity (i-C16 /n-C16 and a+i-C15 /n-C15). Sedimentary fatty acid n-C16/n-C24 ratios from the Lower St. Lawrence near Rimouski indicate a relatively high proportion of terrestrial acids (Fig. 4B). Using values from the Great Lakes and Davis Strait as end members (see Table 6 for references), this ratio corresponds to approximately 90% terrestrial fatty acids. Because the end members are not well known and can vary

TABLE 6
Fatty acids reported from aquatic sediments.

Code[a]	Location	Ref[b]	Type[c]	OC[d]	TFA[e]	C16[f]	16/24[g]	S/U[h]	X16[i]	X15[i]
T1	Florida salt marsh	5	F	17.6	4600	1000	7.0	2.0	0.07	2.8
T2	Sewage treat. plant	13	T	10.0	4100	2180		25.1		2.1
F1	Providence River	13	T	3.0	155	57		3.5		4.0
F2	Houma River, LA	4	F	4.9	61	11.7	4.0	5.1	0.051	3.6
F3	Great Lakes	6	F				0.6			
E1	Offshore Houma, LA	4	F	2.0	13	2.7	13.5	4.5	0.037	3.0
E2	mid- Narragansett Bay	13	F	2.5	38	12.2		4.1		3.7
E3	mid- Narragansett Bay	13	T	2.5	93	31.6		4.1		4.1
E4	Buzzards Bay, Mass.	3	F	2.4	32	10.0		2.1		8.2
E5	Baffin Bay, Texas	7,10	F	1.6	42	11.0		2.0	0.118	2.4
E6	Aransas Bay, Texas	7,10	F	0.6	32	7.4		1.6	0.092	0.9
	Lower St. Lawrence Est.	12	F	1.0	38	4.0	1.2	0.8	0.075	4.3
	Lower St. Lawrence Est.	11	T	0.9	60	15.5	2.7	1.5		
S1	Rhode Island Sound	13	F	1.0	23	6.7		2.5		4.6
S2	Gulf of Maine	13	F		10	4.0		8.7		3.5
S3	Gulf of Maine	13	T		14	5.5		4.0		3.3
S4	Atlantic, shelf	2	T	0.3	22	7.7			0.17	
S5	Atlantic, shelf	15	T	1.6			2.4			
S6	Gulf of Mexico, shelf	4	F	0.7	11	1.6	5.3	2.7	0.063	1.3
S7	Gulf of Mexico, shelf	9	F	1.1	6		1.5	12.0		8.0
S8	Gulf of Mexico, shelf	7,10	F	0.5	11	2.4		1.4	0.108	1.2
O1	Atlantic, slope off NY	15	T	1.0			1.6			
O2	Gulf of Mexico, basin	8	F	0.8	19	7.3	11.1	14.9	0.012	0.4
O3	Davis Strait	9	F	0.4	15		18.0	0.4		
O4	Smith Sound	9	F	0.9	22		12.0	0.3		
O5	Antarctic Ocean	14	T	0.9		1.5	3.5	30.0	0.1	4.7
O6	Walvis Bay, diat. ooze	1	T				200.	4.4		1.8

[a] T=terrestrial, F=freshwater, E=estuarine, S= ocean shelf, O= open ocean
[b] 1 = Boon et al., 1975; 2 = Farrington & Quinn, 1973; 3 = Farrington et al., 1977; 4 = Gearing, 1975; 5 = Johnson & Calder, 1973; 6 = Leenheer & Meyers, 1983; 7 = Leo & Parker, 1966; 8 = Newman, 1973; 9 = Nishimura & Baker, 1987; 10 = Parker, 1967; 11 = Rodier & Khalil, 1982; 12 = This Study; 13 = Van Vleet & Quinn, 1979; 14 = Venkatesan & Kaplan, 1987; 15 = Venkatesan et al., 1987.
[c] F = free (organic solvent extraction); T = total (base extraction)
[d] Percent organic carbon
[e] Total fatty acids analyzed, concentration in $\mu g/g$ dry weight
[f] Concentration of n-C16:0, $\mu g/g$ dry weight
[g] Ratio of n-C16:0 to n-C24:0
[h] Ratio of n-C16:0 to n-C16:1 (saturated to unsaturated)
[i] Ratio of branched to straight-chain fatty acids C16 and C15

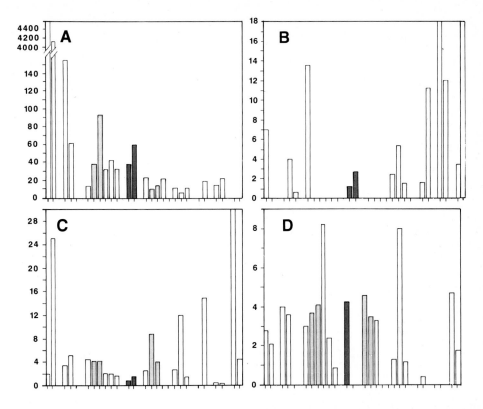

Figure 4. Comparison of fatty acid parameters from different marine sediments. In all cases the X-axis indicates location and corresponds to the references in Table 6 (from T1 on the left to O6 at right). From left to right the locations go from riverine through estuarine to oceanic. The St. Lawrence values (free followed by total fatty acids) are heavily shaded; light shading indicates free and total fatty acids from Narragansett Bay (to the left of the St. Lawrence) and acids from Narragansett Bay (right of the St. Lawrence). The sum of resolved fatty acids in ug/g is given by A; B shows the ratio of nC16:0 to nC24:0 fatty acids, C gives the ratio of nC16:0 to nC16:1; and D indicates the ratio of branched C15 (a-C15 + i-C15) to nC15:0.

differential degradation rates for the two compounds, but the ratio generally agrees with other indications of the importance of land-derived organic material throughout the estuary.

Analyses of cores with depth show decreases in concentrations due to biodegradation, but little alteration in the pattern of terrestrial dominance over the last 100 years. There is not enough fatty acid data from suspended particulates to determine the causes of large spacial and seasonal differences noted earlier, the variation in n-C16 to n-C24 seemingly being more closely related to primary production (both freshwater and marine) than to the amount of terrestrial material.

TABLE 7
Fatty acids on suspended particles.

Location	Ref[a]	Type[b]	TFA[c]	C16[d]	16/24[e]	S/U[f]	X16[g]	X15[g]
Loire River, France	7	F	22			1.1	0.02	
St. Lawrence River	8	F	41	4.2	13.4	0.72	0.028	3.4
Narragansett Bay, USA	6	T	41	14.0		1.7		
Rhone Delta	1	?	11					
Upper St. Lawrence Estuary	8	F	10	1.7	5.5	1.0	0.044	3.6
Loire turbidity zone, France	7	F	88		0.6	1.5	0.046	
Aransas Pass, Texas	2	F	1.8					
Villefranche Bay, France	3	F	2.4	1.1	13.5	6.7		
Mediterranean Sea	3	F	3.4	0.7		6.9		
St. Law. Est., off Tadoussac	8	F	2.2	0.5	13.2	0.93	0.023	2.0
St. Law. Est., off Tadoussac	5	F	7.9	2.1		1.9		
Lower St. Lawrence Estuary	8	F	12.0	1.9	33.0	0.81	0.007	1.1
Offshore Loire, France	7	F	5.5			2.8	0.016	
Gulf of Mexico shelf	4	F	<0.4					
Tasman Sea, Australia	9	F	1.0					

[a] 1=Denant & Saliot, 1987; 2=Entzeroth, 1982; 3= Goutx & Saliot, 1980; 4= Kennicutt & Jeffrey, 1981; 5= Marty & Choinière, 1979; 6= Schultz & Quinn, 1972; 7= Tronczynski, 1985; 8= this study; 9= Volkman et al., 1986
[b] F = free (organic solvent extraction); T = total (base extraction)
[c] Concentration of total fatty acids analyzed, ug/L
[d] Concentration of n-C16:0, ug/L
[e] Ratio of n-C16:0 to n-C24:0
[f] Ratio of n-C16:0 to n-C16:1 (saturated to unsaturated)
[g] Ratio of branched to straight-chain fatty acids C16 and C15

The ratio of n-C16:0 to n-C16:1 is relatively low in Lower Estuary sediments (Fig. 4C), indicating high levels of recent (undegraded) biological production compared to other estuaries. This may be a result of relatively rapid sedimentation and slow biodegradation in these cold waters. The values for particulates are similar or slightly lower than ratios from other estuaries (Table 7).

The ratios of branched to straight-chain fatty acids (Tables 6 & 7, Fig. 4D) increase with bacterial activity . They are similar to those reported for other estuaries, both in suspended particles and in the sediments. The suspended particulates also show a trend, high values in the Upper Estuary decreasing in the Maritime Estuary (Table 7), similar to the gradient found in the Loire.

4.4. Hydrocarbons

Hydrocarbons are primary components of crude oil and its refined products, entering the marine ecosystem from direct spills and as indirect runoff from sewage, used crankcase oil, etc. There are also suites of biologically-produced hydrocarbons, primarily aliphatic. Aromatic hydrocarbons are produced during petroleum formation as well as by combustion (e.g. forest fires, home heating, factories, cars). By examining the total quantity of hydrocarbons as well as the pattern of individual compounds and groups of compounds, researchers can estimate the relative importance of different sources (see, for example, Gearing et al., 1976 and Kennicutt et al., 1987). In general, fossil fuels have high concentrations of a complex group of chromatographically-unresolved compounds and approximately equal concentrations of n-alkanes with an even and with an odd number of carbon atoms. In unaltered biological hydrocarbons, compounds with an odd number of carbon atoms usually predominate. Marine plants usually produce relatively low molecular weight aliphatic hydrocarbons (up to C20), while terrestrial alkanes maximize at C25 to C31. The aromatic hydrocarbons have little or no biological contribution. Those from fossil fuel tend to be highly substituted because they were formed at low temperatures; formation at higher temperatures of combustion produces progressively fewer substituents.

Thus these hydrocarbons can be used as both an indicator of pollution and a measure of terrestrial versus marine production. They are particularly useful for studies over longer time periods because they are among the most stable biochemicals and are little altered in the food chain or in sediments.

As with the fatty acids, there are two factors that must be borne in mind when interpreting hydrocarbon results: terminology and methods. There are many hundreds of individual hydrocarbons found in the environment. Each has a slightly different behaviour (residence time, physical form, toxicity, etc.) and often several different names. For example, most papers speak to two general classes of hydrocarbons - aliphatic and aromatic. Aliphatic compounds may be straight-chain (n-alkanes), branched-chain (mostly isoprenoids) or cyclic; aromatic hydrocarbons are usually grouped by the number of rings and side chains. Different methods of extracting and analyzing hydrocarbons give different concentrations because they are measuring different things, but this is not always made clear. For example, extracted hydrocarbons are often measured gravimetrically and then by gas chromatography, the former method always giving a higher number because not all the compounds pass through the chromatographic column. Great attention must always be payed to methods when comparing results from different papers.

In the St. Lawrence Estuary a variety of methods have been used to characterize hydrocarbons in the water, sediment and organisms. Table 8 summarizes the water column results. The most extensive study has been that of Levy (1985) who, between 1971 and 1979, measured dissolved and dispersed hydrocarbons by fluorescence. This method measures all extractable compounds fluorescing at a certain wavelength, primarily aromatic hydrocarbons, and uses a calibration factor to convert to total hydrocarbons. It makes the assumption that the distribution of hydrocarbons is like that of an oil. This is a very effective method for monitoring oil spills of known composition because it is very quick and relatively inexpensive; it is less satisfactory when the actual composition is unknown and changing. Levy (1985) found that the overall levels in the estuary had declined, the levels in 1976-79 being 25-30% of 1971-73 values. There was a general trend to lower levels closer to the open Gulf and farther away from the major cities and rivers. Highest values (1976-79) were in the Upper Estuary near Quebec (about 2 ug/L); levels in the Saguenay (0.7 - 1 ug/L) were lower overall than in the Upper Estuary; and the Lower Estuary had very low levels (0.35 ug/L), indistinguishable from concentrations in the open Gulf and in clean areas such as Baffin Bay and Hudson Strait.

TABLE 8.
Summary of water column hydrocarbon concentrations (ug/L).

Location	Total hydrocarbons	n-alkanes	"oil"	PAHs (resolved)	benzo(a)pyrene
Upper Estuary					
average of 93[a]			1.1		
Quebec[b]	57.3	1.7		0.020	0.0006
turbidity zone[b]	24.6	1.2		0.037	0.0021
Rivière du Loup[b]	3.3	0.1		0.005	0.0002
Saguenay Fjord					
average of 64[a]			0.72		
Chicoutimi[b]				0.0046	0.0003
Baie de Ha! Ha![c]					0.0001
lower fjord[b]				0.0033	0.0001
Lower Estuary					
average of 36[a]			0.35		
Tadoussac[b]	2.2	0.05		0.001	0.00008
Tadoussac[d]		0.06	1.		
Tadoussac[e], Nov.		0.04			
Tadoussac[e], June		0.07			
Rimouski[b]	1.1	0.06		0.0012	0.00007
Pte-des-Monts[d]		0.07	0.7		

[a] Levy, 1985, [b] Tronczynski et al., 1987, [c] Primeau and Goulet, 1983
[d] Keizer et al., 1977, [e] Marty and Choinière, 1979

Keizer et al. (1977) collected water at two upper estuarine stations in 1976 and compared the fluorescence method with a gas chromatographic quantification of n-alkanes (Table 8). Fluorescence values agreed well with those of Levy (1985). Alkane concentrations also indicated little or no petroleum pollution when compared with other waters (Whittle et al., 1974). In fact they indicate some import of hydrocarbons from the open Gulf to the Lower

Estuary. The n-alkane values are consistent with analyses in 1974-75 near Tadoussac (Marty and Choinière, 1979): .04-.3 ug/L versus .07-.13 ug/L. Of particular note, however is the fact that n-alkane concentrations as measured by two laboratories were 4 to 10 times lower than oil-equivalents by fluorescence, again done in two independent laboratories (compare Saguenay and Tadoussac values in Table 8).

The same pattern (higher levels near Quebec decreasing toward the open Gulf) as observed for dissolved and dispersed hydrocarbons has been found in a 1986 study of particulate material (Tronczynski et al., 1987; Gearing and Tronczynski, 1988). These values were obtained by gas chromatography of suspended particulate extracts.

TABLE 9.
Concentrations of aliphatic hydrocarbons in surface marine sediments (ug/g).

Total	UCM	Location	Reference
25 - 1800		New York Bight	Farrington & Tripp,1977
730	725	San Pedro, Calif.	Venkatesan et al., 1980
43 - 165	28 - 140	Beaufort Sea	Wong et al., 1976
115		Narragansett Bay	Gearing et al., 1978
110	105	St. Lawrence Lower Estuary	Gearing & Tronczynski, 1988
71	60	Mediterranean Sea	Burns, 1986
1 - 87	0.5 - 81	Gulf of Mexico	Kennicutt et al., 1987
45	33	San Nicolas, Cal.	Venkatesan et al., 1980
1 - 5		Atlantic basin	Farrington & Tripp, 1977

Overall the levels found in these studies indicate elevated concentrations of hydrocarbons such as are found in many modern estuaries. This is best seen by a comparison of sedimentary concentrations because sediments show less variability than in the water column (Tables 9-11). The Lower Estuary has concentrations of most hydrocarbons similar to those in mid-Narragansett Bay (not considered highly polluted) and Massachusetts Bay. Levels are an order of magnitude lower than values from highly polluted areas such as the New York Bight and Boston's Charles River. However, the levels are about twice as high as in "unpolluted" shelf sediments such as the San Nicolas Basin and the continental shelves of the Gulfs of Maine and Mexico. It is also disturbing to note that the sediments show the levels of some anthropogenic compounds to have risen sharply over the last one hundred years (Gearing and Tronczynski, 1988).

TABLE 10.
Aromatic hydrocarbon concentrations in surface marine sediments (ug/g).

TOTAL PAHs	LOCATION	REFERENCE
5 - 600	New York Bight	Laflamme & Hites, 1978 Farrington & Tripp, 1977
230	San Pedro, Calif.	Venkatesen et al., 1980
30	Narragansett Bay, USA	Gearing et al., 1978
1 - 30	Beaufort Sea	Wong, et al., 1976
17	St. Lawrence Estuary	Gearing & Tronczynski,1988
1 - 5	Saguenay Fjord	Martel et al., 1986
1 - 6	Gulf Of Mexico Shelf	Gearing et al., 1976
0.2 - 2	Atlantic Abyssal Plain	Laflamme & Hites, 1978 Farrington & Tripp, 1977

TABLE 11.
Summary of benzo(a)pyrene concentrations in marine sediments (ug/g)

Concentration	Location	Reference
~10.	Charles River, Boston	Windsor & Hites, 1979
~1.	Boston Harbour, USA	Windsor & Hites, 1979
~0.5	Pettaquanscutt basin	Hites et al., 1980
~0.15	Saguenay Fjord	Martel et al., 1986
~0.1	Massachusetts Bay	Windsor & Hites, 1979
0.09	St. Lawrence Estuary	Gearing & Tronczynski, 1988
~0.05	Gulf of Maine	Windsor & Hites, 1979
0.01	North Atlantic Shelf	Venkatesan et al., 1987
0.004	Antarctic Ocean	Venkatesan & Kaplan, 1987

These results also confirm the previous findings that concentrations in the Saguenay Fjord are intermediate between those of the Upper and Lower Estuary. The relative role of the Saguenay Fjord as a source of organic contaminants is an important one since organic compounds have been proposed to be harming marine mammals in the fjord (see Cossa, Chapter 11. Aluminum processing plants, such as are found along the Saguenay River, are known to release PAHs. One survey of aromatic hydrocarbons in mussels from the estuary (Cossa et al., 1983; Picard-Berube et al., 1983) reported the levels of total aromatic hydrocarbons (by high performance liquid chromatography) and of benzo(a)pyrene to be very low (similar to background levels in the ocean) everywhere except very near factories and at the mouth of the Saguenay. Mussels transported into the Saguenay showed uptake of PAHs, but the uptake was no higher near industrial centres than near the Estuary. Measurements of sediments in the Saguenay (Primeau and Goulet, 1983; Martel, 1985; Martel et al., 1986; Cossa et al., 1988; Levy and Smith, 1988) show elevated levels of total PAHs and of individual compounds like benzo(a)pyrene near the industrial centres. The concentrations have also changed with time, the pattern being consistent with the history of aluminum plant discharges (Levy and Smith, 1988). Thus the factories seem to be an important source of PAHs, at least locally. On the other hand, in one core from the Lower Estuary the concentrations of these compounds have remained constant over approximately one hundred years (Gearing and Tronczynski, 1988), indicating that the influence of the Saguenay does not extend to the estuary. More work is needed to elucidate the sources of this pollution.

5. Summary

Considering its size and its importance, relativly little is known about the organic geochemistry of the St. Lawrence. Most existing studies have concentrated on bulk measurements and properties such as POC, PON, C/N ratios, and stable isotope ratios. This data has been used to construct a simple, semi-quantitative model of organic matter processing in the Estuary which can be used as a basis for future work, both on organic tracers of natural processes and on the sources and sinks of pollutants. What is needed now is more data; there are glaring holes in our information about dissolved organic matter, seasonal distribution of organic compounds, degradation processes in the water column and in the sediments, and the basic levels of many pollutants such as PCBs, PAHs, phenols, and dibenzofurans. While present data indicates the Estuary to be only slightly polluted, levels of some pollutants seem to increasing and possible effects are starting to be noted. Increasingly sophisticated measurements and understandings of the processes occurring in the Estuary are needed to protect the St. Lawrence.

Present understanding of these processes indicates that the surface waters show a seasonal cycle with terrestrial material from freshwater runoff being important in spring. This is confirmed by measurements of POC, PON, C/N ratios, and stable isotope ratios. Measurements of fatty acids and hydrocarbons show changes, but are not yet numerous enough (less than 10 total!) to pinpoint and quantify sources. A similar situation exists for the importance of *in situ* production in summer. Originally hypothesized based on concentration and C/N ratios, carbon and nitrogen isotope ratios show the summer particulate material to be virtually 100% autochthonous. They show, in addition a shift in fall that may be related to the change at this time to production by nanoplankton; this was not distinguished by C/N ratios. No comprehensive studies have yet been published on the seasonal changes in organic compounds, although they have great potential for elucidating specific sources (for example some hydrocarbons are characteristic of diatoms while another pattern can be seen where dinoflagellates dominate).

Measurements of POC, PON, and C/N suggest rapid vertical mixing of organic material through the water column. Isotope studies are consistent with this hypothesis as is the presence of relatively labile lipids in the underlying sediments. Detailed studies of this process are needed.

Calculations from concentrations indicate that POC and PON are sufficient to provide all the organic matter present in the sediments. Isotopic data and a limited amount of information on lipids shows there to be great similarity in composition between particulates and sediments, lending credence to particulates are the major source of sedimentary organic matter.

The model calculates there to be little change due to atmospheric inputs, exchange with other waters, or removal of organic material by fishing. Studies of the levels of polycyclic aromatic hydrocarbons in the Saguenay show that, while qualitatively small, these atmospherically-transported contaminants can be important in terms of potential adverse effects. More such studies need to done.

Organic geochemistry in the St. Lawrence Estuary is still in its infancy. Present studies show a very dynamic ecosystem with extensive seasonal changes due to changing freshwater and autochthonous production. There are spatial differences caused by exchanges with salt flats in the Upper Estuary and inputs of terrestrial material from the Saguenay. Spatial differences also include considerable variation from north to south across the Estuary, more than has been noted for most estuaries. Finally, vertical mixing and sedimentation appear to be quite rapid in this estuary. More work is needed in all these areas in order to understand the St. Lawrence Estuary.

REFERENCES

Aaronson, S., 1978. Excretion of organic matter by phytoplankton *in vitro* . - Limnol. Oceanogr. 23: 838.

Billen, G., G. Cauwet, S. Dessery, M. Meybeck & M. Somville, 1986. Origines et comportement du carbone organique dans l'estuaire de la Loire. - Rapp. P. V. Réun. Cons. Int. Explor. Mer 186: 375-391.

Boon, J.J., J.W. deLeeuw & P.A. Schenck, 1975. Organic geochemistry of Walvis Bay diatomaceous ooze - I. Occurrence and significance of the fatty acids. - Geochim. Cosmochim. Acta 39: 1559-1565.

Bordovskiy, O.K., 1965. Accumulation and transformation of organic substances in marine sediments, Parts I and II. -Mar. Geol. 3: 3-31.

Bugden, G.L. 1981. Salt and heat budgets of the Gulf of St. Lawrence. - Can. J. Fish. Aquat. Sci. 38: 1153-1167.

Burnett, W.C. & O.A. Schaeffer, 1980. Effect of ocean dumping on $^{13}C/^{12}C$ ratios in marine sediments from the New York Bight. -Estuar. Coast. Mar. Sci. 11: 605-611.

Burns, K.A. 1986. Evidence for rapid *in situ* oxidation rate of pollutant hydrocarbons in the open Mediterranean. -Rapp. P.-v. Réun. Cons. int. Explor. Mer 186: 432-441.

Cadée, G.C. 1984. Particulate and dissolved organic carbon and chlorophyll a in the Zaire River, estuary and plume. - Netherlands J. Sea Res. 17:426-440.

Canadian Hydrographic Service, 1972. Distribution of surface sediments - Gulf of St. Lawrence. - Chart 811-G, Ottawa, Canada.

Coote, A.R. and P.A. Yeats, 1979. Distribution of nutrients in the Gulf of St. Lawrence. - J. Fish. Res. Board Can. 36: 122-131.

Cossa, D., M. Picard-Berube and J.P Gouygou, 1983. Polynuclear aromatic hydrocarbons in mussels from the Estuary and northwestern Gulf of St. Lawrence, Canada. - Bull. Environ. Contam. Toxicol. 31: 41-47.

Cossa, D., A. Moinet & M. Picard, 1988. Nature et distribution d'hydrocarbures aromatiques polycycliques dans les sédiments du fjord du Saguenay, Québec. Unpublished manuscript.

Cranwell, P.A. 1978. Extractable and bound lipid components in a freshwater sediment. - Geochim. Cosmochim. Acta 42: 1523-1432.

Dahm, C.N., S.V. Gregory & P.K. Park. 1981. Organic carbon transport in the Columbia River. - Estuarine Coastal Shelf Sci. 13: 645-658.

Denant, V. & A. Saliot, 1987. Biogeochemistry of organic matter at the fresh/sea water interface in the Rhone Delta, Mediterranean Sea, France. -In: 8th International Symposium of Environmental Biogeochemistry, p. 63.

Dionne, J.-C. 1981. Données préliminaires sur la charge sédimentaire du couvert de glace dans la baie de Montmagny, Québec. - Géographie physique et Quaternaire 35: 277-282.

El-Sabh, M.I. 1975. Transport and currents in the Gulf of St. Lawrence. Bedford Inst. Oceanogr. Rep.BI-R-75-9: 180 p.

Entzeroth, L.C. 1982. Particulate matter and organic sedimentation of the continental shelf and slope of the northwest Gulf of Mexico. Ph.D. Diss., Univ. of Texas, Austin.

Farrington, J.W. & J.G. Quinn, 1973. Biogeochemistry of fatty acids in recent sediments from Narragansett Bay, Rhode Island. -Geochim. Cosmochim. Acta 37: 259-268.

Farrington, J.W. & B.W. Tripp, 1977. Hydrocarbons in western North Atlantic surface sediments. - Geochim. Cosmochim. Acta 41: 1627-1642.

Farrington, J.W., S.M. Henricks & R. Anderson, 1977. Fatty acids and Pb-210 geochemistry of a sediment core from Buzzards Bay, Massachusetts. -Geochim. Cosmochim. Acta 41: 289-296.

Gagné, J.-P. & J.-R. Brindle, 1985. Composés phenoliques en milieu cotier: contribution de la région de Baie-Comeau et du fjord du Saguenay. - Naturaliste can. (Rev. Ecol. Syst.) 112: 57-64.

Gearing, J.N., 1988. The use of stable isotope ratios for tracing the nearshore-offshore exchange of organic matter. -In B.-O. Jansson (ed.): Lecture Notes on Coastal and Estuarine Studies 22:69-101. Springer-Verlag

Gearing, J.N. & J. Tronczynski, 1988. Spatial and temporal changes in particulate lipids of the St. Lawrence Estuary. Third Chemical Congress of North America Abstracts of Papers (ISBN 8412-1444-1), American Chemical Society.

Gearing, J.N., P.J. Gearing, T.F. Lytle and J.S. Lytle, 1978. Comparison of thin-layer and column chromatography for separation of sedimentary hydrocarbons. -Anal. Chem. 50: 1833-1836.

Gearing, J.N., P. Gearing, D.T. Rudnick, S.G. Requejo & M.J. Hutchins, 1984. Isotopic variability of organic carbon in a phytoplankton-based, temperate estuary. - Geochim. Cosmochim. Acta 48: 1089-1098.

Gearing, P.J. 1975. Organic carbon stable isotope ratios of continental margin sediments. Ph.D. Diss., Univ. Texas, Austin.

Gearing, P., J.N. Gearing, T.F. Lytle & J.S. Sever. 1976. Hydrocarbons in 60 northeast Gulf of Mexico shelf sediments: a preliminary survey. -Geochim. Cosmochim. Acta 40: 1005-1017.

Gearing, P., F.E. Plucker & P.L. Parker, 1977. Organic carbon stable isotope ratios of continental margin sediments. -Mar. Chem. 5: 251-266.

Gershey, R.M., M.D. MacKinnon, P.J. LeB. Williams & R.M. Moore. 1979. Comparison of three oxidation methods used for the analysis of the dissolved organic carbon in seawater. Mar. Chem.7: 289-306.

Gobeil, C., B. Sundby & N. Silverberg, 1981. Factors influencing particulate matter geochemistry in the St. Lawrence Estuary turbidity maximum. -Mar. Chem. 10: 123-140.

Goutx, M. & A. Saliot, 1980. Relationship between dissolved and particulate fatty acids and hydrocarbons, chlorophyll a and zooplankton biomass in Villefranche Bay, Mediterranean Sea. -Mar. Chem. 8: 299-318.

Hedges, J.I. & A. Van Geen, 1982. A comparison of lignin and stable carbon isotope compositions in Quaternary marine sediments. -Mar. Chem. 11: 43-54.

Hites, R.A., R.E. Laflamme, J.G. Windsor, J.W. Farrington & W.G. Deuser. 1980. Polycyclic aromatic hydrocarbons in an anoxic sediment core from the Pettaquanscutt River (Rhode Island, U.S.A.) -Geochim. Cosmochim. Acta 44: 873-878.

Hunt, J.M., 1970. The significance of carbon isotope variations in marine sediments. - In G.D. Hobson & G.C. Speers (eds.): Advances in Organic Geochemistry, 1966, pp. 27-35. Pergamon Press, Oxford.

Jeffries, H.P. 1972. Fatty-acid ecology of a tidal marsh. -Limnol. Oceanogr. 17: 433-440.

Johnson, R.W. & J.A. Calder, 1973. Early diagenesis of fatty acids and hydrocarbons in a salt marsh environment. -Geochim. Cosmochim. Acta 37: 1943-1955.

Keith, L.H. 1977. GC/MS analyses of organic compounds in treated Kraft paper mill wastewaters. - In L.H. Keith (ed.): Identification and Analysis of Organic Pollutants in Water, pp. 671-707. Ann Arbor Science Publishers, Ann Arbor, Mich.

Keizer, P.D., D.C. Gordon & J. Dale, 1977. Hydrocarbons in eastern Canadian marine waters determined by fluorescence spectroscopy and gas-liquid chromatography. -J. Fish. Res. Board Can. 34: 347-353.

Kennicutt, M.C. & L.M. Jeffrey, 1981. Chemical and GC-MS characterization of marine particulate lipids. -Mar. Chem. 10: 389-407.

Kennicutt, M.C., J.L. Sericano, T.L. Wade, F. Alcazar & J.M. Brooks, 1987. High molecular weight hydrocarbons in Gulf of Mexico continental slope sediments. -Deep-Sea Res. 34: 403-424.

Laflamme, R.E. & R.A. Hites, 1978. The global distribution of polycyclic aromatic hydrocarbons. - Geochim. Cosmochim. Acta 42: 289-303.

Leenheer, M.J. & P.A. Meyers, 1983. Comparison of lipid compositions in marine and lacustrine sediments. -In: M. Bjoroy et al. (eds.): Advances in Organic Geochemistry 1981, pp.309-316. J.W. Wiley and Sons.

Leo, R.F. & P.L. Parker, 1966. Branched chain fatty acids in sediments. Science 152: 649-650.

Levy, E.M. 1985. Background levels of dissolved/dispersed petroleum residues in the Gulf of St. Lawrence, 1970-79. -Can. J. Fish. Aquat. Sci. 42: 544-555.

Levy, E.M. & J.N. Smith, 1988. A geochronology for PAH contamination in the sediments of the Saguenay Fjord. Third Chemical Congress of North America Abstracts of Papers. American Chemical Society (ISBN 8412-1444-1).

Lucotte, M. 1989. Organic carbon isotope ratios and implications for the maximum turbidity zone of the St. Lawrence Upper Estuary. - Est. Coastal Shelf Sci. 29: 293-304.

Lucotte, M. & B. d'Anglejan. 1986. Seasonal control of the Saint-Lawrence maximum turbidity zone by tidal-flat sedimentation. -Estuaries 9: 84-94.

MacKinnon, M.D. 1978. A dry oxidation method for the analysis of the TOC in seawater. -Mar. Chem. 7: 17-37.

Macko, S.A., 1981. Stable nitrogen isotope ratios as tracers of organic geochemical processes. Ph.D. Diss., Univ. Texas, Austin.

Mantoura, R.F.C. & E.M.S. Woodward, 1983. Conservative behaviour of riverine dissolved organic carbon in the Severn Estuary: chemical and geochemical implications. Geochim. Cosmochim. Acta 47:1293-1309.

Martel, L. 1985. Analyse spatio-temporelle des hydrocarbures polycycliques aromatiques (HPA) dans les sédiments du fjord du Saguenay, Québec. Memoire, Maitrise, Univ. du Québec à Chicoutimi.

Martel, L., M.J. Gagnon, R. Masse, A. Leclerc & L. Tremblay. 1986. Polycyclic aromatic hydrocarbons in sediments from the Saguenay Fjord, Canada. -Bull. Environ. Contam. Toxicol. 37: 133-140.

Marty, J.C. & A. Choinière, 1979. Acides gras et hydrocarbures de l'écume marine et de la microcouche de surface. -Naturaliste can. 106: 141-147.

Meybeck, M. 1982. Carbon, nitrogen, and phosphorus transport by world rivers. Amer.J. Sci. 282:401-450.

Newman, J.W. 1973. Quaternary deep sea sediments from the Gulf of Mexico: an organic geochemical study. Ph.D. Diss., Univ. of Texas, Austin.

Nichols, P.D. & R.B. Johns, 1986. The lipid chemistry of sediments from the St. Lawrence estuary. Acyclic unsaturated long chain ketones, diols and ketone alcohols. - Org. Geochem. 9: 25-30.

Nishimura, M. & E.W. Baker, 1987. Compositional similarities of non-solvent extractable fatty acids from recent marine sediments deposited in differing environments. -Geochim. Cosmochim. Acta 51: 1365-1378.

Oviatt, C.A., J.G. Quinn, J.T. Maughan, J.T. Ellis, B.K. Sullivan, J.N. Gearing, P.J. Gearing, C.D. Hunt, P.A. Sampou, & J.S. Latimer, 1987. Fate and effects of sewage sludge in the coastal marine environment: a mesocosm experiment. -Mar. Ecol. Prog. Ser. 41: 187-203.

Parker, P.L. 1967. Fatty acids in recent sediment. -Contrib. Mar. Sci. 4:135-142.

Picard-Berube, M., D. Cossa & J. Piuze, 1983. Teneurs en benzo-3,4-pyrene chez *Mytilus edulis* L. de l'Estuaire et du Golfe du Saint- Laurent. - Mar. Environ. Res. 10: 63-71.

Pocklington, R. 1973. Organic carbon and nitrogen in sediments and particulate matter from the Gulf of St. Lawrence. Bedford Inst. Oceanogr. Rep. BI-R-73-8: 16 pp.

Pocklington, R. 1975. Carbon, hydrogen, nitrogen and lignin determinations on sediments from the Gulf of St. Lawrence and adjacent waters. Bedford Inst. Oceanogr. Rep. BI-R-75-6: 12 pp.

Pocklington, R. 1976. Terrigenous organic matter in surface sediments from the Gulf of St. Lawrence. - J. Fish. Res. Board Can. 33: 93-97.

Pocklington, R. 1985. Organic matter in the Gulf of St. Lawrence in winter. - Can. J. Fish. Aquat. Sci. 42: 1556-1561.

Pocklington, R. 1986. The Gulf of St. Lawrence and the Baltic Sea: two different organic systems. - Dt. hydrogr. Z. 39: 65-75.

Pocklington, R. 1988. Organic matter in the Gulf of St. Lawrence. - In P.M. Strain (ed.): Chemical Oceanography in the Gulf of St. Lawrence. Can. Bull. Fish. Aquat. Sci. 220: 49-58.

Pocklington, R. & G.T. Hagell. 1975. The quantitative determination of organic carbon, hydrogen, nitrogen and lignin in marine sediments. Bedford Inst. Oceanogr. Rep. BI-R-75-18: 16 pp.

Pocklington, R. & S. Kempe. 1983. A comparison of methods for POC determination in the St. Lawrence River. - Mitt. Geol.-Palaont. Inst. Univ. Hamburg 55: 145-151.

Pocklington, R. & C.D. MacGregor. 1973. The determination of lignin in marine sediments and particulate form in seawater. - Intern. J. Environ. Anal. Chem. 3: 81-93.

Pocklington, R. & M.D. MacKinnon. 1982. Organic matter in upwelling off Senegal and The Gambia. - Rapp. P.-V. Reun. Cons. Int. Explor. Mer 180: 254-265.

Pocklington, R. & L. Morash. 1979. Organic carbon, nitrogen and lignin in sediments from the Gulf of St. Lawrence and adjacent waters. Bedford Inst. Oceanogr. Rep. BI-R-79-1: 14 pp.

Pocklington, R. & S. Roy. 1975. Utilisation de composés organiques comme traceurs de materiel d'origine terrestre dans l'océan. - Int. Coun. Explor. Sea Hydrography Committee CM 1975/c:10.

Pocklington, R. & F.C. Tan. 1987. Seasonal and annual variations in the organic matter contributed by the St. Lawrence River to the Gulf of St. Lawrence. - Geochim. Cosmochim. Acta 51: 2579-2586.

Primeau, S. & M. Goulet, 1983. Resultats d'echantillonnage de l'eau et des sediments de la baie des Ha! Ha! -Service de la qualite des eaux, ministere de l'Environnement du Quebec, document no 83-19, 61 pp.

Rodier, L. & M.F. Khalil, 1982. Fatty acids in recent sediments in the St. Lawrence estuary. - Estuar. Coast. Shelf Sci. 15: 473-483.

Schultz, D.M. & J.G. Quinn, 1972. Fatty acids in surface particulate matter from the North Atlantic. -J. Fish. Res. Bd. Can. 29: 1482-1486.

Silverberg, N., H.M. Edenborn & N. Belzile, 1985. Sediment response to seasonal variations in organic matter input. - In A.C. Sigleo & A. Hattori (eds.): Marine and Estuarine Geochemistry, pp. 69-85. Lewis Publ., Chelsea, MI.

Silverberg, N., H.V. Nguyen, G. Delibrias, M. Koide, B. Sundby, Y. Yokoyama & R. Chesselet, 1986. Radionuclide profiles, sedimentation rates, and bioturbation in modern sediments of the Laurentian Trough, Gulf of St. Lawrence. Oceanol. Acta 9: 285-290.

Steven, D.M. 1975. Biological production in the Gulf of St. Lawrence, p. 229-248. - In T.W.M. Cameron and L.W. Billingsley [ed.] Energy Flow-Its Biological Dimension, Royal Society of Canada, Ottawa, 319 pp.

Sweeney, R.E., E.K. Kalil & I.R. Kaplan, 1980. Characterization of domestic and industrial sewage in southern California coastal sediments using nitrogen, carbon, sulphur, and uranium tracers. - Mar. Environ. Res. 3: 225-243.

Tan, F.C. & P.M. Strain, 1979a. Organic carbon isotope ratios in recent sediments in the St. Lawrence estuary and the Gulf of St. Lawrence. - Estuar. Coast. Mar. Sci. 8: 213-225.

Tan, F.C. & P.M. Strain, 1979b. Carbon isotope ratios of particulate organic matter in the Gulf of St. Lawrence. -J. Fish. Res. Bd. Can. 36: 678-682.

Tan, F.C. & P.M. Strain, 1983. Sources, sinks, and distribution of organic carbon in the St. Lawrence estuary, Canada. -Geochim. Cosmochim. Acta 47: 125-132.

Therriault, J.-C. & M. Levasseur, 1985. Control of phytoplankton production in the lower St. Lawrence Estuary: light and freshwater runoff. -Naturaliste can. 112: 77-96.

Tronczynski, J. 1985. Biogéochimie de la matière organique dans l'estuaire de la Loire: origine, transport et evolution des hydrocarbures et des acides gras. These, Diplome de Docteur de 3ième Cycle, L'Univ. Pierre et Marie Curie, Paris.

Tronczynski, J., J.N. Gearing & S. Macko, 1987. Characterization of selected organics on particles and sediments in the St. Lawrence Estuary. 21st Annual Congress of the Canadian Meteorological and Oceanographic Society Programme with Abstracts, p.51.

VanVleet, E.S. & J.G. Quinn, 1979. Early diagenesis of fatty acids and isoprenoid alcohols in estuarine and coastal sediments. -Geochim. Cosmochim. Acta 43: 289-303.

Venkatesan, M.I. & I.R. Kaplan, 1987. The lipid geochemistry of Antarctic marine sediments: Bransfield Strait. -Mar. Chem. 21: 347-375.

Venkatesan, M.I., S. Brenner, E. Ruth, J. Bonilla & I.R. Kaplan, 1980. Hydrocarbons in age-dated sediment cores from two basins in the Southern California Bight. Geochim. Cosmochim. Acta 44:789-802.

Venkatesan, M.I., E. Ruth, S. Steinberg & I.R. Kaplan, 1987. Organic geochemistry of sediments from the continental margin off southern New England, U.S.A. - Part II. Lipids. -Mar. Chem. 21: 267-299.

Volkman, J.K., D.A. Everitt & D.I. Allen, 1986. Some analyses of lipid classes in marine organisms, sediments and seawater using thin-layer chromatography-flame ionisation detection. -J. Chromatogr. 356: 147-162.

Wafar, M., P. Le Corre & J.-L. Birrien. 1984. Seasonal changes of dissolved organic matter (C, N, P) in permanently well-mixed temperate waters. -Limnol. Oceanogr. 29: 1127 - 1132.

Windsor, J.G. & R.A. Hites, 1979. Polycyclic aromatic hydrocarbons in Gulf of Maine sediments and Nova Scotia soils. -Geochim. Cosmochim. Acta 43: 27-33.

Whittle, K., P.R. Mackie & R. Hardy, 1974. Hydrocarbons in the marine environment. -South Afr. J. Sci. 70: 141-144.

Wong, C.S., W.J. Cretney, P. Christensen & R.W. Macdonald, 1976. Hydrocarbon levels in the marine environment of the southern Beaufort Sea. -Beaufort Sea Technical Report No. 38, Fisheries and Oceans Canada.

Chapter 10

Sediment-Water Interaction and Early Diagenesis in the Laurentian Trough

Norman Silverberg and Bjørn Sundby

Maurice-Lamontagne Institute, Department of Fisheries and Oceans, Box 1000, Mont-Joli, Québec, Canada G5H 3Z4

ABSTRACT

This chapter summarizes the results of more than a decade of study of sediment-water interactions and early diagenesis in the St. Lawrence Estuary. A review is made of the role of organic matter (input to the sediment surface, inmixing through bioturbation, degradation by bacteria) and of the distributions of oxidants, nutrients, trace metals and other chemical substances either directly and indirectly involved in the cycling of organic matter. Manganese, in particular, is used to illustrate the development of our understanding about sediment geochemistry in the Laurentian Trough.

Key Words: geochemistry, vertical flux, organic matter cycle, trace metals, bioturbation, diagenetic modelling, benthic boundary layer,

1. Introduction

Diagenesis is the sum total of the processes that bring about changes in a sediment subsequent to deposition in water (Berner, 1980). It begins immediately after a sedimentary particle has arrived at the sediment-water interface, and it transforms and degrades many components of the sedimentary deposit well before rock formation processes become important. Strong interactions between the biosphere, the hydrosphere and the lithosphere characterize the early stages of diagenesis: Burrowing benthic organisms mix and ingest the sediment particles; bacteria mineralize energy-rich metabolizable organic matter; and inorganic sediment components dissolve or precipitate in response to the changing chemical environment that is created by the organisms.

The seafloor is thus more than simply a passive dump for debris falling from the water column. It is a complex, living biogeochemical factory, which, by continually interacting with the overlying water, controls the composition of seawater itself. Understanding the processes involved in diagenesis is necessary for a diversity of other scientific endeavors ranging from the reconstruction of paleo-environments to the assessment of the fate of toxic

Coastal and Estuarine Studies, Vol. 39
M. I. El-Sabh, N. Silverberg (Eds.)
Oceanography of a Large-Scale Estuarine System
The St. Lawrence
© Springer-Verlag New York, Inc., 1990

chemicals. The study of early diagenesis is a fascinating, stimulating and challenging subject of scientific research; it is also of great practical interest.

Studies of early diagenesis in the sediments of the Gulf of St. Lawrence, and particularly of the Lower (or Maritime) Estuary of the St. Lawrence, began in the late 1970's, inspired by the rapid development that was taking place internationally in this field. Leading up to these studies were extensive investigations of the morphology and sedimentology and the geochemistry of the bottom deposits, culminating in the monograph on the Gulf of St. Lawrence by Loring and Nota (1973). The study of the Maritime Estuary has profited from a particularly suitable conjuncture of large dimensions, relatively low sedimentation rates and long residence times of the water masses. In many of the world's estuaries, shallow depths, short distances, rapid mixing, short residence times, and, in some cases, repeated resuspension of bottom sediments, make it difficult to isolate and observe diagenetic phenomena and their interactions with the water column. In the Maritime Estuary of the St. Lawrence property gradients created by diagenesis have spatial and temporal dimensions that allow them to be readily observed and sampled. In the sediments of the Maritime Estuary, the depth zonation of redox reactions that is associated with the successive exhaustion of electron acceptors (Froelich et al, 1979), is readily discernable. In the bottom water of the Laurentian Trough, the influence of diagenesis can be observed as a progressive change in the chemical composition as the water overlying the sediment moves slowly landward. Although the large dimensions of the St. Lawrence may also complicate diagenesis research, the logistics are nevertheless simple in comparison to those required for deep-sea studies.

1.1. The Laurentian Trough environment

The Gulf of St. Lawrence joins one of the largest rivers in the world to the Atlantic Ocean. It must therefore play an important role in modifying the input of continental material to the sea. The Laurentian Trough (Fig. 1) connects the waters of the Upper St. Lawrence Estuary and the tributary Saguenay Fjord with those of the open Atlantic Ocean, 1200 Km to the east. The glacially modified, U-shaped valley dominates the Maritime Estuary of the St.Lawrence, where it is bordered by narrow shelves mantled with coarse-grained surficial deposits. Additional inputs of fresh water and terrigenous material arrive via a series of intermediate sized rivers, principally along the north shore. In the Gulf of St. Lawrence, broad sandy shelves isolate the Trough from land drainage before the valley cuts through the Cabot Strait and finally opens onto the continental slope south of Newfoundland.

The Trough, particularly the Maritime Estuary portion where the generally two-layer, complex, estuarine circulation (El-Sabh, 1979) acts as an efficient sediment trap, appears to be the preferred site of modern mud deposition (Loring and Nota, 1973). Little information is available about water motion very close to the bottom. Observations made from the submersible Pisces IV on the floor of the Trough (Syvitski et al., 1983) indicated weak

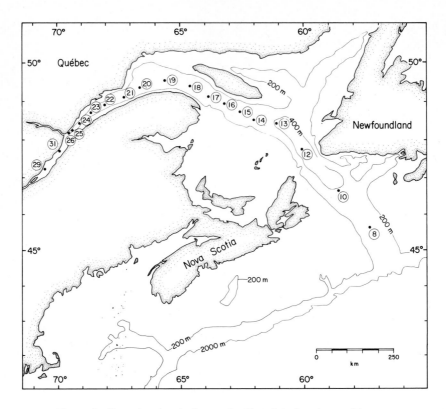

Figure 1. Chart showing the Laurentian Trough and station positions

tidal currents (up to 12 cm/sec), capable of transporting particles along the bottom but of much resuspension, most of which appeared to be effected by benthic organisms. Little if any fine grained sediment is presently accumulating in the Upper Estuary (see d'Anglejan, Chapter 6).

The Trough sediments consist of a brown colored, watery surface layer, 1-3 cm thick, grading downwards to an homogeneous unstratified olive grey mud containing about equal parts of silt and clay and less than 5-10% sand (Bouchard, 1983). The mineralogy, with quartz, feldspars, amphiboles and pyroxenes occurring in the clay fraction along with illite indicates that the sediments are immature. They are principally derived from the mechanical weathering of the crystalline rocks and the Quaternary deposits of the Canadian Shield to the north, as few important tributaries drain the Appalachian sedimentary terrain of the south shore (Loring and Nota, 1973).

The major element chemistry, established from grab samples of the surficial sediments, also reflects this source. The composition varies little throughout the Maritime Estuary (Si: 25%, Al: 7.5%, Fe: 5%, Ca: 2.5%, K: 2.5%, Na: 2%) (Loring and Nota, 1973). These authors also performed the first partition analyses of the sediments. They divided the bulk chemistry into a "detrital" or crystal-lattice bound component, soluble only in strong acids, and a "non-detrital" component, extractable in acetic acid solution and indicative of "loosely bound" elements. The non-detrital components of metals as well as organic matter were found to be preferentially associated with the finer grain sizes.

2. Diagenesis in the Laurentian Trough

2.1. Sedimentation rates

Although modern sedimentation was known to be occurring in the deep portion of the Maritime Estuary, direct measurements of deposition and accumulation rates were not available until the 1980's. Loring and Nota (1973) had dated certain horizons below the modern mud layer in the open Gulf of St. Lawrence, and, based upon the depth at which present-day foraminifera assemblages appeared, had inferred the onset of modern estuarine conditions at about 5000 years B.P. Sundby (1974) used this information to estimate a sediment accumulation rate of about 0.1 mm/yr. His mass balance estimates suggested that, if this rate was applicable to the entire Laurentian Trough, additional sources of fine-grained sediment other than the St. Lawrence River must exist.

TABLE 1.
Laurentian Trough sediment trap data (after Silverberg et al., 1986)

Station	Season	mg cm^{-2} d^{-1}	mm yr^{-1} [*]
10	06/81	0.20	0.99
12	06/81	0.10	0.49
	08/82	0.14	0.67
18	07/81	0.51	2.49
	08/82	0.03	0.16
20	07/81	0.31	1.51
21	08/82	0.28	1.39
23	08/80	0.46	2.27
	07/81	0.47	2.32
	mean of 25 measurements (1980-84)	0.51	2.50
24	08/80	1.18	5.82
	07/81	2.81	13.82
	08/82	0.43	2.13
25	07/81	0.62	3.04
	08/82	0.35	1.73

[*] Assuming 50% water content and a particle density of 2.65 g cm^{-3}

Silverberg et al. (1986) compiled a series of measurements of excess ^{210}Pb distributions in cores from several sites within the Trough, as well as several ^{14}C and ^{137}Cs profiles. The influence of bioturbation on the radionuclide profiles was examined, and direct, short-term measurements of sedimentation rates were obtained with free-drifting sediment traps. The data (Table 1) indicated that sedimentation rates in the Maritime Estuary were an order of magnitude greater than previously assumed from Loring and Nota's (1973) open Gulf data. They were shown to be unusually high in a zone near the head of the Trough (as much as 14 mm/yr), while elsewhere in the Maritime Estuary they were of the order of 1-3 mm/yr. The rates in the open Gulf were indeed of the order of tenths of a mm/yr.

During the 1980's more than 40 measurements were made with a free-drifting sediment trap, mostly at 150 m depth at a fixed site in the central portion of the Maritime Estuary, off Rimouski (Station 23, Fig. 1). Although limited to the ice-free months between April and October, the data indicate an average sedimentation rate of 0.48 mg cm^{-2} d^{-1}. This corresponds to about 2.4 mm yr^{-1}, assuming a density of 2.65 for the sediment particles and a water content of 50%, typically found below the top 10 cm of the accumulating sediment. The close agreement with the longer term accumulation rates estimated from ^{210}Pb measurements (Silverberg et al., 1986) suggests that sedimentation continues during the winter months.

The single-day sediment-trap measurements show that sedimentation is most intense during spring, and that it decreases by a factor of 3 during summer and fall (Fig. 2a.). There is considerable scatter in the data for the same period from one year to another, and even among measurements made on the same day using duplicate sediment traps (see section on Future Considerations). Data from other sites are much more limited, but reveal similar patterns.

2.2. Bioturbation

Bottom-dwelling organisms, besides directly influencing the composition of sediments, mix and irrigate the sediment over several cm depth, control particle-sizes distributions and micro-topography, and play an important role in stabilizing or pelletizing sediments (Rhoads, 1974). The intensity of bioturbation, inferred from biological structures in radiographs and visual observation of sediment boxcores, and from faunal types and abundances (Ouellet, 1982), tends to decrease in a seaward direction along the Trough.

It is difficult to quantify the combined effects on sediment mixing by complex assemblages of 40 or more species. However, if one assumes that the melange of burrowing, feeding and evacuation activities creates an essentially random movement of sediment particles, then an analogy to molecular diffusion can be entertained and a biological "mixing" coefficient can be proposed and applied to the transport of chemical substances in the sediment. This approach, based upon the work of Cochran and Aller (1979), Benninger et

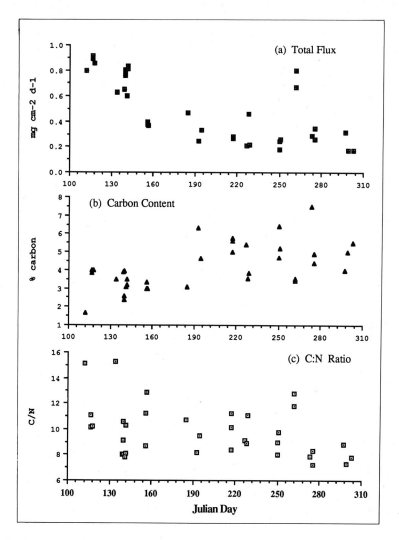

Figure 2. Sediment trap data from Station 23 in the Lower St. Lawrence Estuary during the ice-free months. (a) sedimentation rate (mg total dry weight cm-2 d-1); (b) carbon content (%); (c) C/N ratio.

al. (1979) and Aller et al. (1980), has been used to estimate the rates of particle reworking by benthic organisms in the Laurentian Trough (Silverberg et al., 1986). The depth distributions of the activity of the short-lived radionuclides, [234]Th (24 day half-life) and [228]Th (1.8 year half-life) were measured in boxcores from several sites. Since the activities of these isotopes would normally have been reduced to undetectable levels before they were buried to any significant depth in the sediment, their presence below the surface can be attributed to mixing. By applying a one-dimensional diffusion-reaction model to the depth distribution of the radionuclide activity, a mixing coefficient "D_b" can be determined. For the Laurentian Trough sediments, D_b's were in the order of 10^{-7} to 10^{-9} cm^2 sec^{-1}, and, as anticipated from the biological studies, decreased in a seaward direction. This appears reasonable for an environment with a benthic population intermediate between that of the dense nearshore communities and the rarified deep-sea communities.

2.3. Fluxes and degradation of organic matter

The mineralization of organic matter provides the energy to drive many of the other diagenetic reactions. Although there is some secondary production of living organic matter and detritus in the sediment, the bulk of the organic matter is derived from particles settling from the overlying water column.

2.3.1. Flux of organic matter to the sediment

The many sediment trap samples have provided an ideal way to examine the organic matter input to the sediment. A portion of each sample was examined under a binocular microscope immediately after recovery. There had perhaps been some additional aggregation (and dispersal) of the particles while in the collecting cylinder of the trap and during the recovery, but it was clear that a changing variety of naturally-produced aggregated particles up to several mm in size (e.g. stick-like and oval fecal pellets, stringy aggregates, and fluffy marine snow) made up the bulk of the settling material. The large size of some of these particles could be observed directly through the transparent walls of the collecting tube as the last ones settled quietly to the bottom of the trap. Settling velocities were timed at 0.3-1 cm s^{-1}. Subsequent observations, using a high-powered inverted microscope, confirmed the presence of the various kinds of aggregates, the dominance of mineral particles, and the inclusion of various kinds of empty or only slightly degraded diatom frustules and strings and other planktonic debris within a diffuse organic matrix (Fig.3). As in the deep-sea, it is apparent that "biological packaging" of particulate matter permits the rapid transfer of relatively fresh organic matter from the photic zone to the seafloor.

A portion of each trap sample was analyzed for its carbon and nitrogen content (Fig. 2 b,c). The trend of the carbon flux is similar to that of the total sedimentation flux, but the carbon content and the C:N ratios indicate that the quality as well as the quantity of settling organic matter vary with the time of year. The high flux, low organic carbon contents, and high C:N ratios in the spring are indicative of terrigenous material that arrives during the period of the highest river run-off. During the summer and fall, terrigenous material is less important, and in-situ production contributes more to the organic matter flux to the sediments. The 35 individual trap measurements obtained over 7 years at this site, give a mean total particulate flux of 482 µg cm^{-2} d^{-1}, a mean carbon content of 4.2 % and a mean carbon flux of 1.7 µmoles C cm^{-2} d^{-1}.

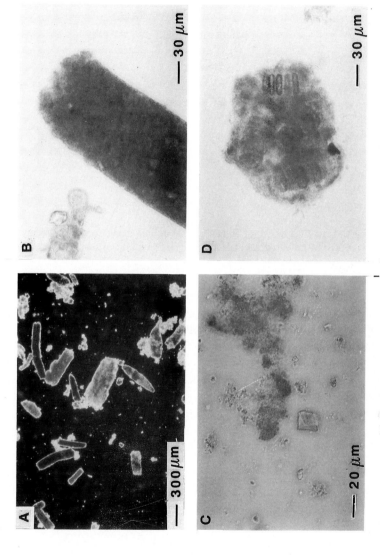

Figure 3. Photomicrographs (courtesy of Juan Carlos Colombo) of some particles from sediment trap samples: (A) Examples of oval and stick-like pellets and a few fluffy aggregates; (B) Close up of a stick-like pellet; (C) fragment of a fluffy aggregate (marine snow) and examples of the rarely observed individual mineral grains; (D) Fluffy aggregate containing relatively unbroken chain diatoms.

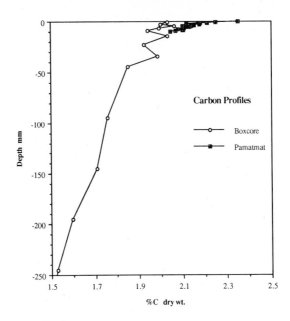

Figure 4. Typical vertical distributions of organic carbon in Laurentian Trough sediments. Note the very sharp gradient in the top few millimetres, obtained through sub-millimetre subsampling of Pamatmat cores.

2.3.2. Distribution of organic carbon in the sediment

Bouchard (1983) has shown that, within the Maritime Estuary, inorganic carbon makes up only about 0.03% dry wt. of the sediment. Hence the analysis of total carbon can be used directly as a measure of organic carbon.

Within the sediments, the distribution of organic carbon is much less variable than in the sedimenting particulate matter. In boxcores, the carbon content at the sediment surface is typically close to 2%, whereupon it diminishes almost linearly with depth, attaining between 15 and 45% of the surface value by about 35 cm depth (Fig. 4). Below this depth the concentration tends to be constant. In the top few cm the organic carbon distribution may be irregular due to bioturbation or grain-size differences (Bouchard, 1983). However, high-resolution subsampling (0.5 mm intervals) of undisturbed cores, collected with a multiple corer modelled after that of Pamatmat (1971), revealed carbon gradients over the top 3-10 mm of the sediment that are one to two orders of magnitude greater than the carbon gradient measured in boxcores (Silverberg et al., 1985, 1987).

2.3.3. Mineralization rates of organic carbon

There are several ways of examining the rate of decomposition of organic matter in these sediments. Bouchard (1983) noted that the organic carbon content in boxcores decreased quite regularly with depth. She calculated the average concentration gradient over the length

TABLE 2.
Downward fluxes of carbon in the water column and across the sediment-water interface (after Silverberg et al., 1985, used with permission of Lewis Publishers Inc.)

Date	Carbon gradient in top cm.	bioturbation flux (range) *	Water column flux
1984	$\%C\ cm^{-1}$	$\mu gC\ cm^{-2}\ d^{-1}$	$\mu gC\ cm^{-2}\ d^{-1}$
27/04	-1.31	5.2 - 52.5	43.9
06/06	-0.69	2.2 - 22.3	11.9
03/08	-1.44	5.8 - 57.7	15.2
01/09	-0.42	1.7 - 17.0	14.3

* $J = D_b\ (\partial C/\partial z)_{z=0}$ (uses $D_b = 10^{-2} - 10^{-3}\ cm^2 d^{-1}$; assumes 85% water content

of cores from different sites in the Trough to be 3-15 $\mu mol\ C\ cm^{-4}$. Since the sedimentation rates, except at st. 24, are nearly the same at these stations (0.2-0.3 cm y^{-1}) the average mineralization rates throughout the cores are 0.6-4.5 $\mu mol\ C\ cm^{-3}y^{-1}$, or 21-158 $\mu mol\ C\ cm^{-2}y^{-1}$, integrated over the top 35 cm. These are minimum estimates because they assume that carbon is transported by advection only.

Silverberg et al. (1987) calculated the mineralization rate over the top 35 cm in the sediment to be 405 $\mu mol\ C\ cm^{-2}y^{-1}$ by subtracting the burial rate of carbon below the 35 cm depth horizon (sedimentation rate times carbon concentration at 35 cm) from the flux of organic carbon to the sediment, measured with the the sediment traps. This calculation assumes that the degradation of organic carbon between 150 m depth, where the traps collected the settling particles, and the bottom at 325 m is insignificant. The assumption is reasonable in view of the high settling rates (>100 m d^{-1}), which leave little time for degradation in the water column, and the slow rates with which bacteria degrade organic matter, measured by incubations of fresh surface sediment using radio-labelled glucose (Silverberg et al., 1985).

The large difference between the estimates of Bouchard (1983) and Silverberg et al. (1987) can likely be attributed to the rapid inmixing by benthic organisms of freshly sedimented carbon-rich particles, ignored by Bouchard, and to the underestimation of the organic carbon gradient in the sediment surface layer. Silverberg et al. (1985) showed that the mixing across the steep carbon gradient in the top cm of the sediment accounts for a significant fraction of the vertical carbon transport (Table 2). Thus, most of the organic carbon flux is degraded in the top few millimeters of the sediment.

2.3.4. The importance of bioirrigation for the suboxic diagenesis of organic matter
Sediments accumulating in oxic basins are a mixture of reducing and oxidizing compounds diluted with stable, unreactive mineral components. At the moment of deposition, the reducing component of the mixture consists principally of organic matter. The oxidizing component consists mainly of manganese and iron oxides plus the oxygen, nitrate and sulfate that are dissolved in the sediment porewater. Upon burial, microbially mediated reactions lead to the oxidation and mineralization of labile organic compounds and the reduction of the oxidized forms of Mn, Fe, O, N and S.

The stoichiometry of organic matter oxidation and mass-balance calculations can be used to examine early diagenesis in the following way. Assuming that Redfield's (Redfield et al., 1963) formula for the elemental composition of marine plankton holds for the metabolizable fraction of sedimentary organic matter, the degree of mineralization of the organic carbon can be compared with the oxidizing capacity of the sediment at the moment of deposition. This establishes whether this oxidizing capacity is sufficient to account for the amount of organic carbon which has been mineralized in the sediment. Next the escape mechanism of mineralized carbon can be evaluated by comparing the flux of inorganic carbon out of the sediment, assuming molecular diffusion, with the rate of production of mineralized organic carbon within the sediment.

This approach has been applied to boxcores from a series of stations in the Laurentian Trough (Bouchard, 1983; Sundby et al., 1983). At St 23 (Fig. 1), the amount of carbon mineralized at depth is nearly four times greater than what can be accounted for by the oxidants present at the time of deposition. The sediment is therefore an open system with respect to oxidants. Accumulation of inorganic carbon is insignificant in these sediments and 98% of the mineralized carbon must have escaped into the overlying water. Transport out of the sediment by molecular diffusion of bicarbonate, which is the predominant dissolved inorganic carbon species at the pH of these sediments, can account, however, for only one tenth of the production rate of mineralized carbon. Thus, bioturbation and bioirrigation appear to be the dominant transport mechanisms for both oxidants and mineralization products during the early diagenesis of these sediments. The results imply that mineralization of sedimentary organic matter is more efficient and more complete than would be predicted from closed system considerations.

2.3.5. Oxygen fluxes
Oxygen is consumed by many of the reactions that take place during early diagenesis. Close to the sediment-water interface, aerobic microbial degradation of organic matter and respiration by the macro- and meiobenthos communities are the greatest consumers of oxygen. Within anoxic microenvironments in the surface layer of the sediment and in the underlying anoxic sediment, anaerobic processes produce reduced compounds which, upon transport to the oxygenated zone by diffusion, bioirrigation and bioturbation, are subjected to reoxidation. The oxygen required by all of these reactions can only be supplied through transport across the sediment-water interface.

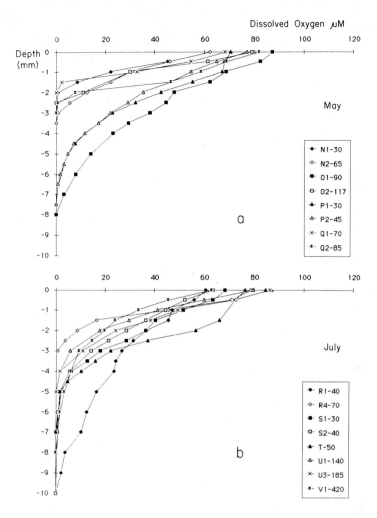

Figure 5. Vertical profiles of dissolved oxygen concentrations in cores from St. 23. Oxygen is exhausted within the top cm of the sediment. Note the variability in gradient and penetration depth.

Polarographic oxygen sensors (micro-electrodes) have been used to determine the distribution of dissolved oxygen in Laurentian Trough sediments (Silverberg et al., 1987). The results revealed a high degree of lateral variability, with oxygen penetration depths varying between 2 and 10 mm over distances as small as several cm (Fig. 5). The oxygen profiles were used to estimate a diffusive flux of oxygen across the sediment-water interface. This flux amounted to only 20% of the total flux necessary to account for the organic carbon that is mineralized in the sediment, implying that mechanisms other than molecular diffusion must dominate the transport of oxygen across the sediment-water interface in these sediments.

2.3.6. Sulfate reduction

Sulfate reduction rates in the sediments of the Laurentian Trough are among the lowest that have been reported for the coastal marine environment (Edenborn et al., 1987). Maximum rates measured ranged from 0.7 - 2.0 nmol cm^{-3} d^{-1}. These occurred between 5 and 15 cm depth in the sediment. The low rates are due to the relatively low sedimentation rates and the seasonally unchanging, low temperature (4°C) in the bottom waters of the Trough .

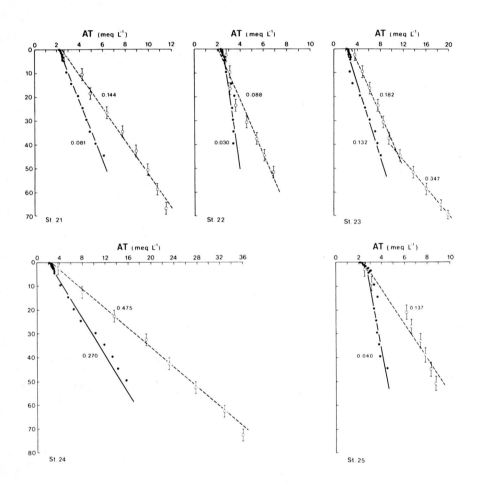

Figure 6. Near linear distributions of titration alkalinity in box and gravity cores from the Laurentian Trough (after Lebel et al., 1982).

Because of the low rates, sulfate persists to a considerable depth in the porewater of Laurentian Trough sediments (Edenborn et al., 1987; Lebel et al., 1982). Analysis of a piston core from the open Gulf station (St. 17, Fig. 1), for example, revealed 15 mmol sulfate at 8 m depth. Alkalinity profiles in gravity cores, corrected for core shortening

(Lebel et al., 1982) are virtually linear to depths of over 150 cm (Fig. 6), suggesting a deep source of alkalinity, possibly due to sulfate reduction by methane.

Sulfate reduction accounted for the mineralization of 26% of the organic matter flux to the sediment at station 23 in the Maritime Estuary, but only 5% of the flux at station 17 in the open Gulf. Unlike the absolute rate of sulfate reduction, the relative proportion of the carbon flux that is degraded via sulfate reduction is not directly correlated with the sedimentation rate, but is a function of the composition of the organic matter, the intensity of bioturbation, and the abundance of electron acceptors. The low proportion of carbon degradation via sulfate reduction in the open Gulf sediment is due to the low rates of sedimentation and the low rates of bioturbation. This combination allows a long residence time for organic matter near the sediment surface and, in consequence, a low flux of labile carbon into the sulfate reduction zone (Edenborn et al., 1987).

2.3.7. Nutrient salts

Many of the studies on early diagenetic processes in the Laurentian Trough have also collected data (Figs. 6-8) on the depth distributions of dissolved nitrate (actually nitrate + nitrite), phosphate, silicate and titration alkalinity in the sediment porewater (Lebel et al., 1982; Bouchard, 1983; Belzile, 1987, 1988; Gobeil, unpublished data) . The data indicate that the sediments may be an important source of nutrients for the overlying water, and that they probably contribute to the landward increase in deep-water nutrients observed by Coote and Yeats (1979). The results show considerable variability, which suggests that temporal and spatial variations in the bottom sediments are important for these substances as well.

The highest concentrations of nitrate in the porewater occur near the sediment-water interface. These concentrations are generally lower than nitrate concentrations in the overlying bottom water. There is thus no evidence to indicate that the sediment is a source of nitrate. On the contrary, the sharp concentration gradient below the interface suggests that the sediment is probably a sink for nitrate from the overlying water. In contrast, the few measurements that have been made of ammonia show steadily increasing ammonia concentrations from near detection limit at the interface to about 200 μM at 25-30 cm depth, suggesting that the sediment is a source of reduced nitrogen.

The concentration of phosphate in the porewater near the sediment-water interface is of the order of 5-6 μM in nearly all the profiles available, as opposed to about 1-2 μM in the overlying water. This concentration difference can support a diffusive flux out of the sediment. The phosphate concentration in the porewater increases with depth, with the most marked increases occurring within the zones of iron-oxide and sulphate reduction. Continuously produced during the mineralization of organic matter, phosphate also appears to be involved in the cycle of iron transformations (Belzile, 1988; Bouchard, 1983). The high concentrations of dissolved phosphate at St. 24 correspond to the highest

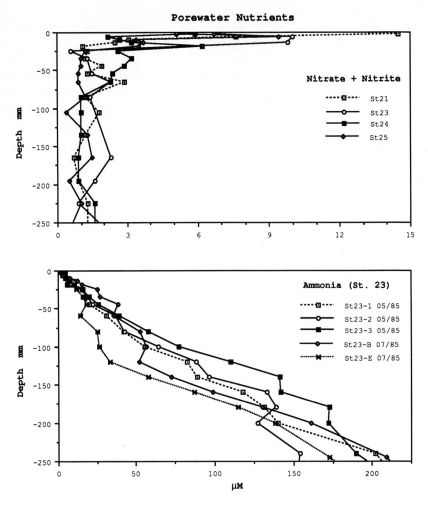

Figure 7. Examples of porewater profiles in Laurentian Trough sediments. (Top) nitrate plus nitrite concentrations; (Bottom) continuously increasing ammonia concentrations. The profiles suggest that the sediment act as a source for reduced nitrogen, and a sink for oxidized nitrogen.

concentrations of dissolved iron. There is also a trend of increasing phosphate concentrations in the landward direction, similar to that of dissolved iron.

The concentrations of dissolved silicate are very high in the top few mm of the sediment (130-270 µM) and increase rapidly to several hundred µM over the next few centimeters. The concentration of silicate in the overlying bottom water is only about 30 µM, so that diffusion into the overlying water is important. Deeper in the sediment the concentrations appear to approach constancy. There is no recognizable landward trend in the silicate concentrations, and it is not known whether the release of silicate from the sediment is

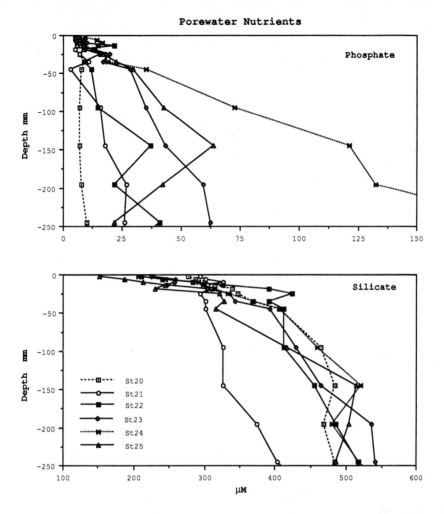

Figure 8. Examples of porewater profiles in Laurentian Trough sediments. (Top) phosphate distributions. (Note the restricted range of concentrations near the sediment-water interface and very high levels at depth at St. 24). (Bottom) silicate concentrations, with sharp gradients at the interface indicative of release to the overlying waters.

maintained by dissolution at depths greater than 40-50 cm, or within the upper part of the sediment column.

2.4. Biogenic trace metals

A number of elements, among them some trace metals, are extracted from sea water by living organisms, enter the sedimentary environment with the remains of these organisms, and are released again to sea water when these remains are destroyed. Cadmium, for example, appears to be closely tied to the carbon cycle. In the sea, particulate cadmium

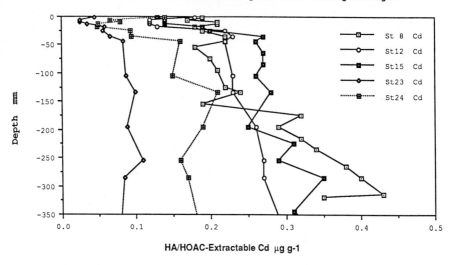

Particulate Cadmium profiles- along Trough

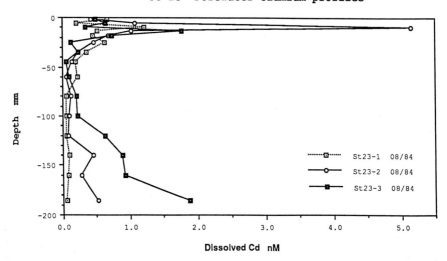

St 23   Porewater Cadmium profiles

Figure 9. Vertical profiles of cadmium concentrations in Laurentian Trough sediments (modified from Gendron et al., 1986; Gobeil et al., 1987). (Top) Particulate Cd - note the minimum in the top cm, and the trend towards increasing subsurface concentrations in a seaward direction. (Bottom) Dissolved Cd - note the very sharp maxima in the top cm.

occurs largely in association with biogenic particles, while dissolved cadmium follows the distribution of the nutrient salts.

In Laurentian Trough sediments the depth distribution of cadmium in all of the cores examined follows the same pattern: The concentration of particulate cadmium (Fig. 9 top) decreases initially, reaching a minimum within several millimeters of the sediment surface. This decrease is followed by a sharp increase over the next several millimeters or centimeters to concentrations which then remain constant or increase only slowly with further depth. Profiles of dissolved cadmium (Fig. 9 bottom) reveal a sharp subsurface concentration maximum, indicative that cadmium dissolution is taking place in the uppermost zone of the sediment (Gendron et al., 1986; Gobeil et al., 1987).

To explain the observed distributions of cadmium, the following mechanism was proposed: Particulate cadmium arrives at the sediment surface bound to biogenic material and is rapidly solubilized via an oxygen dependent reaction. This dissolution reaction could be the aerobic degradation of the fresh organic matter to which cadmium may be bound, or the dissolution of unstable cadmium phases (carbonate, for example) by the lowered pH associated with the intense oxidation of organic matter to CO_2. A portion of the solubilized cadmium then migrates downward into the reducing zone of the sediment and precipitates, possibly as a sulfide, enriching the deeper sediment column. The remainder migrates upward into the water column, where it is again available to biological processes.

Cadmium is efficiently recycled during early diagenesis. About 80% of the total cadmium flux to the sediment is returned to the water column through upward diffusion. About one quarter of the remaining 20% that is buried with the sediment is accountable through downward diffusion and precipitation below the oxygenated zone. Direct measurements of the release of cadmium from sediments have shown that the upward flux is dependent upon the availability of oxygen in the surface sediment (Westerlund et al., 1986).

In the Laurentian Trough, the cadmium content of sediment trap material was 3-5 times higher than that of the sediment. Thus it is remarkable that the highest concentration of cadmium within the cores was not found in the surface sediment, but at depth. Likewise, the highest concentrations were not found in the sediments closest to the head of the Trough, where anthropogenic input is most pronounced, but rather in the sediments in the open Gulf of St. Lawrence, farthest away from human influence. The distribution of elements such as cadmium, even when subject to strong anthropogenic influences, may nevertheless be controlled by early diagenetic redistribution processes. A portion of the lead in the sediments of the Maritime Estuary exhibits vertical profiles in the surface layer that are similar to those of cadmium (Gobeil and Silverberg, 1989). Other trace metals which may be at least partially biogenic are zinc and copper.

2.5. Manganese, iron, cobalt and associated elements

Unlike the biogenic trace metals, manganese, iron and cobalt are not involved directly with living organisms but respond to changes in the chemical properties of the sediment that are

Porewater manganese profiles

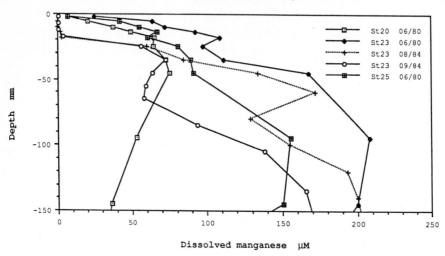

Porewater iron profiles

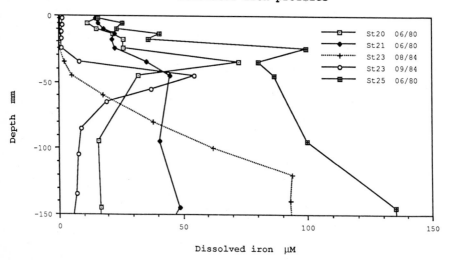

Figure 10. Examples of porewater profiles of manganese (top) and iron (bottom) in Laurentian Trough sediments. Note that concentrations of both metals tend to increase in a landward direction and that dissolved Mn increases at a shallower depth than Fe.

induced by organisms. The oxidized forms of manganese, iron and cobalt are insoluble in oxygenated water. They dissolve, however, below the sediment surface, when microbial activity has removed oxygen from the porewater and created reducing conditions.

In Laurentian Trough sediments, where the penetration of oxygen is limited to the top 3-10 mm (Silverberg et al., 1987), and the Eh decreases sharply over the first few cm (Belzile, 1988), reducing conditions exist close to the sediment surface. The effect of this on metal diagenesis can be seen in the porewater profiles of manganese and iron (Fig. 10) (Sundby and Silverberg, 1985; Gobeil et al., 1987; Belzile, 1988). From low levels at the sediment-water interface, the concentrations of dissolved manganese and iron increase sharply with depth to values that are orders of magnitude higher than in seawater. Porewater profiles of manganese collected at different times and places differ in shape near the sediment-water interface: Some profiles are concave down starting at the interface, indicating that manganese is dissolving in the surface layer of the sediment. Other profiles are concave up in the top several millimeters, an indication that precipitation is taking place in the surface layer. Where manganese and iron have been measured in the same cores, the profiles show that iron dissolution begins deeper in the sediment than manganese dissolution, consistent with the lower potential of the Fe(III)/Fe(II) redox couple than of the Mn(IV)/Mn(II) redox couple.

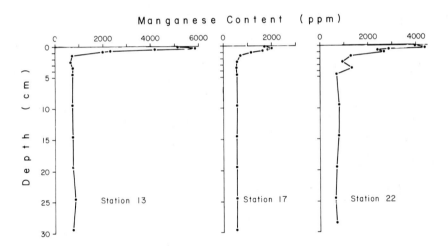

Figure 11. Typical manganese distributions in Laurentian Trough sediments (from Sundby et al., 1981).

The sharp concentration gradients in the porewater drive fluxes of manganese and iron back up across the respective redox boundaries and into the oxygenated layer of the sediment, and perhaps also into the bottom water. The precipitation which then takes place enriches the sediment surface layer in manganese and iron with respect to the subsurface sediment. This has been demonstrated most clearly for manganese. The sediment surface layer in the Laurentian Trough is everywhere enriched in manganese (Fig. 11), with concentrations as high as 7700 ppm, in contrast to the much lower and comparatively constant residual manganese values (500-900 ppm) in the subsurface sediment.

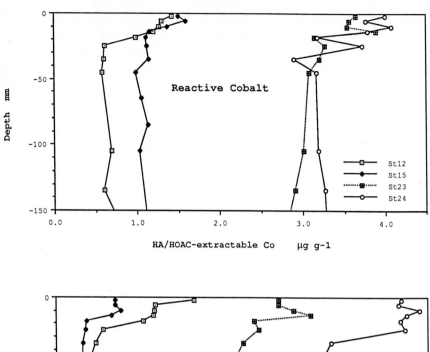

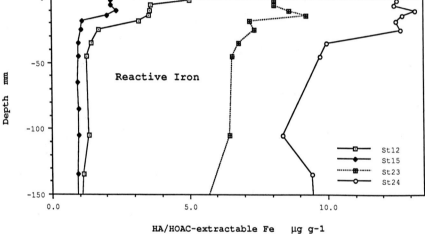

Figure 12. Vertical profiles of reactive cobalt (top) and iron (bottom). Note the surface enrichment and increasing concentrations in the landward direction (after Gendron et al., 1986; Belzile, 1988).

In contrast to manganese, the relative enrichment of iron and cobalt in the sediment surface layer cannot be demonstrated with bulk sediment measurements, because the fraction of these metals that participates in the process is a relatively small proportion of the total content of iron and cobalt. However, when selective extraction techniques are used to separate the unreactive forms of these metals from the more reactive, the surface enrichment of both cobalt and iron can be clearly seen (Gendron et al., 1986; Belzile, 1988). As a rule, the iron enrichment extends slightly deeper than the manganese enrichment

(Silverberg et al., 1982; Gobeil, unpubl. data), consistent with the different redox chemistry and porewater profiles of the two elements (Fig. 12).

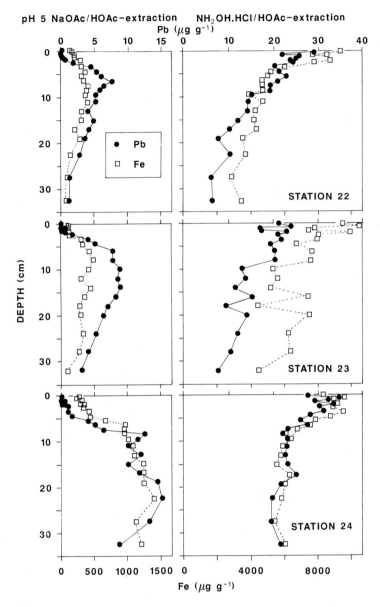

Figure 13. Parallelism in the vertical distributions of two extractable phases of Pb and Fe in sediments from the Laurention Trough (from Gobeil and Silverberg, 1989).

Characteristic of the hydrous oxides of iron and manganese, the form these elements have in the sediment surface layer, is their ability to absorb or coprecipitate other elements. Because of the transient nature of the hydrous oxides, elements associated with them will have distributions in the sediment that parallel iron and manganese, i.e. surface enrichment in the solid phase and subsurface enrichment in the porewater. A number of elements that have been studied in Laurentian Trough sediments do in fact show this parallelism. They include arsenic (Edenborn et al., 1986; Belzile and Lebel, 1986; Belzile, 1988), selenium (Takayanagi and Belzile, 1988; Belzile and Lebel, 1988), lead (Fig. 13) (Gobeil and Silverberg, 1989), and boron (Bergeron, 1980; Bergeron and Lebel, 1984; Marchand, 1984).

3. Manganese, a case study in diagenesis

In the case study of diagenesis that is presented below, pathways, fluxes, mass-balances, reaction rates, turn-over times and mathematical modeling will be dealt with in detail, using manganese as an example. Considering the close parallel between the observations involving manganese and those that involve iron and cobalt, there is reason to believe that many of the conclusions that are derived for manganese also hold true for iron and cobalt.

3.1. The role of continental margin sediments in transforming and modifying the continental input of the elements to the ocean

Remarkably little of the enormous quantities of sediment which rivers discharge from the continents ever reaches the open ocean, for the overwhelming proportion settles out along the continental margins, forming estuarine, shelf and slope deposits. For this reason, many calculations of the continental input of weathering products to the ocean have ignored the fraction carried as suspended sediments.

The rate of accumulation of non-lithogenic manganese in oceanic sediment is an order of magnitude greater than what can be accounted for by the river input of dissolved manganese (e.g. Boström, 1967; Elderfield, 1976). Additional sources proposed to explain the "excess" manganese include manganese diffusing upwards from deep reducing strata of oceanic sediments, manganese discharged into the ocean by submarine volcanism, and manganese carried by the fine fluvial particles that escape sedimentation along the continental margins.

The analysis of particulate material from the deep water of the Laurentian Trough revealed that this suspended matter contained very high manganese concentrations, as high as 1.7% of the bulk composition (Sundby, 1977). This observation suggested that continental margin sediments could be an important source of manganese to the deep ocean. The following mechanism was proposed: A portion of the manganese that is carried by the river-borne particles that settle out over the continental margins dissolves in the reducing environment of the sediment, diffuses out of the sediment, and precipitates in the bottom water to form fine-grained slow-settling particles which are then exported to the open ocean. A relatively small proportion of river-borne particulate manganese exported in this way would be sufficient to balance the oceanic manganese budget.

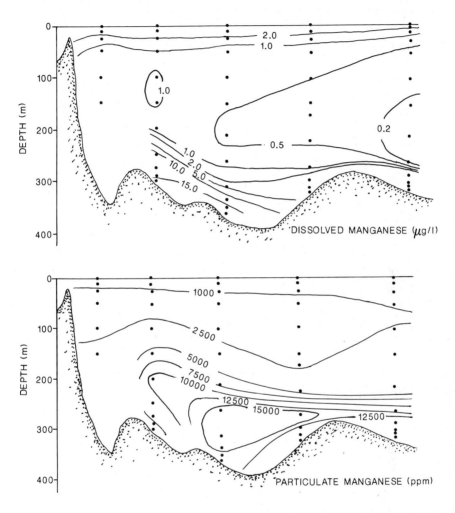

Figure 14. Distributions of dissolved Mn (top) and Mn concentrations on suspended particles (bottom) in the waters of the Laurentian Trough (after Yeats et al., 1979).

The autochtonous nature of the manganese rich suspended particulate matter was established by Yeats et al. (1979) who found that the bottom water of the Laurentian Trough was enriched not only in particulate manganese but also in dissolved manganese. Maximum concentrations of dissolved manganese were found close to the bottom while particulate matter with the strongest manganese enrichment was found 30-100 m above the bottom (Fig. 14). The observations implied that manganese is released from the bottom sediments and subsequently precipitated in the water column.

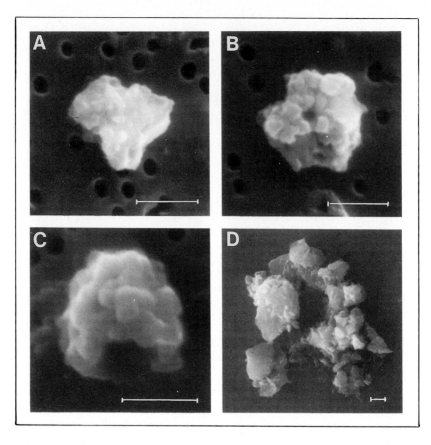

Figure 15. Photomicrographic examples of Mn-rich particles in suspension (top) and in the oxidized layer of the sediment (bottom). (A) Typically bright Mn-enriched grain: Mn - 33%, Fe - 10%, Si - 5%, Al - 3%; (B) Metal-oxide covered grain with exposed aluminosilicate nucleus at bottom: Mn - 25%, Fe - 6%, Si - 11%, Al - 6%. (C) Sediment particle: Mn - 11.9%, Fe - 26.6%, Si - 5.7%, Al - 4.5%. (D) Stringy aggregate containing enriched grains: Mn - 28%, Fe - 2%, Si - 15%, Al - 3%.

The application of electron microscopy to the study of the suspended matter and sediment particles made it possible to go beyond the bulk analysis and examine the composition of individual particles (Sundby et al., 1981). This analysis revealed that manganese is not distributed uniformly among the particulate matter. In fact, easily distinguishable, highly enriched particles are the principal carriers of manganese (Fig. 15). Prominent among these are individual grains, typically in the 1-2 micrometers size range, which contain 20-40% manganese, 1-15% iron and varying amounts of aluminum and silicon in a proportion of roughly 1:2. In addition to these particles, which were called micro manganese nodules, larger aggregate grains with 1-20% manganese are also common, but with increasing height above the bottom the abundance of aggregates decreases until, at 50-100 m above the bottom, almost no other form of manganese enrichment than the micronodules is detectable. The progressive change in in the relative proportion of aggregates and micronodules away from the bottom is the result of resuspension of particles from the manganese-enriched sediment surface-layer, resettling of the coarsest particles, and precipitation of manganese onto fine-grained, already enriched nuclei.

While the observations just described established that diagenetic remobilization is an important process in the coastal environment, they did not demonstrate that any of this manganese was indeed exported to the ocean. For that one had to turn to mass balance considerations. If a significant quantity of diagenetically mobilized manganese were indeed exported to the open ocean, the sediments that accumulate along the continental margins should be depleted in manganese relative to the river-borne particulate matter that constitutes their source. An examination of a large number of boxcores collected along the length of the Laurentian Trough showed that the sediments that accumulate in the Gulf of St. Lawrence are, indeed, depleted in manganese with respect to the particulate matter carried in suspension by the St. Lawrence River (Sundby et al., 1981). Only some 40% of the dissolved and particulate manganese brought in by the St. Lawrence River is retained in the sediments of the Estuary and Gulf of St. Lawrence. This means that this large coastal system is an inefficient trap for the manganese that enters it. An independent study of the manganese balance of the Mississippi River and the Gulf of Mexico (Trefrey and Presley, 1983) has lead to similar conclusions.

3.2. Pathways of manganese in the estuarine environment

Because of the processes in bottom sediments that transform, rather than simply bury, the material that arrives there, the pathway an element such as manganese follows from land to sea can be much more complex and intricate than it would otherwise be. For manganese, the combination of chemical reactions and physical transport that constitutes the pathways has been described schematically (Fig. 16).

The general circulation pattern of the water masses in the Laurentian Trough is estuarine with net seaward flow in the surface layer and landward flow in the deeper layers. Because of its large dimension, complex secondary patterns develop. At all depths seaward flows are locally and temporally superimposed on the primary estuarine circulation. The residence time of the surface water is about four months whereas the bottom water may take several years to traverse the length of the Trough. The landward moving bottom water, in contact with the sediment throughout its entire 1200 Km long journey, receives a continuous input of diagenetically remobilized manganese in both dissolved and resuspended form. This causes the concentration of both phases of manganese in the bottom water to increase in the landward direction. Although there is repeated resettling of suspended particles to the bottom, and precipitation of dissolved manganese onto suspended particles which may also settle out, the net effect is a sweeping of manganese owards the head of the Trough. This mechanism has been referred to as the "broom effect". The total landward near-bottom transport of manganese is considerable; across a section in the landward part of the Trough

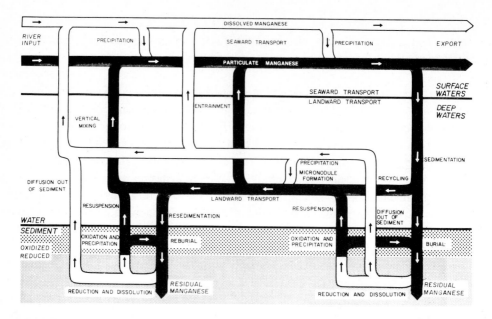

Figure 16. Pathways of Mn in an estuary (black - solid phase; white - dissolved phase; from Sundby et al., 1981)). A portion of the Mn that sediments is stripped from particles during early diagenesis and returned to the landward-moving bottom waters. Export of Mn can occur when fine-grained, slow-settling, Mn-rich particles are carried up into the seaward-moving surface waters.

the estimated transport is of the same order of magnitude as the total input of manganese by the St. Lawrence River.

Above and beyond the broom effect in the bottom 50 m of the water column there is also a transport of manganese at intermediate depth. Water at this depth moves landward more rapidly, entrains water from the bottom layer, and is in turn entrained in the seaward moving surface water. In addition to being entrained and mixed upward into the surface layer, the manganese enriched bottom water is also transported vertically in the upwelling area created by the steep sill at the head of the Trough. Once the mobilized manganese enters the surface circulation, the probability of escape to the open ocean is greatly enhanced, given the large exchange of both water and suspended particulate matter between the Gulf of St. Lawrence and the Atlantic Ocean. However, some of this manganese will resediment and go through the internal cycle many times before finally escaping the system.

The proposed pathways of manganese are based on data from the St. Lawrence System, but the mechanism of long-term residual two-layer flow, entrainment, and eventual mixing of bottom water and surface water is common to many estuaries. There is reason to believe, therefore, that the pathways proposed for the St. Lawrence are a general phenomenon of estuarine systems.

3.3. Manganese fluxes in the benthic boundary layer

The pathways outlined above describe qualitatively the movement of manganese within an estuary. With the information available about the St. Lawrence it is possible to go further, however, and obtain a quantitative description of the movement of manganese and its cycling within and between sediment and water column (Sundby and Silverberg, 1985).

The water and sediment column was divided into five reservoirs, two in the water column and three in the sediment (Fig. 17). The upper reservoir of the water column represents the primary source of new particulate manganese, mainly of terrigeneous origin. The settling of particles maintains the downward flux of manganese through the bottom water, where the flux is augmented by the precipitation of dissolved manganese which has diffused out of the sediment. In the first sediment reservoir, particulate manganese is received from the water column and the precipitation of dissolved manganese diffusing upward from the fourth reservoir. These two sediment reservoirs thus correspond to the zones where particulate manganese is produced and destroyed, respectively. They are separated by the depth below which MnO_2 is unstable and goes into solution. This depth has been observed to be as little as 2 mm below the sediment surface. The deepest reservoir is the site of burial of the manganese which filters through the zone of manganese dissolution.

Assuming steady state, diagenetic modeling and mass-balance calculations were applied to boxcore and sediment trap data from the Laurentian Trough. This yielded the fluxes in and out of the five reservoirs at three different locations, differing mostly in the intensity of biological mixing near the sediment surface. The results of these calculations showed that the cycling of manganese between the oxidizing and the reducing zone of the sediment was quantitatively more important than the cycling between the sediment and the water column. The fluxes across the redox boundary were 3-50 times greater than either the rates of sedimentation or the rates of accumulation of manganese.

By comparing the inventory of total manganese in the enriched surface layer of the sediment with the production rates of dissolved manganese in the sediment, turnover times of 45-200 days were calculated. The shortest turnover times correspond to the location with the higher rate of bioturbation. The calculations also showed that increasing the rate of bioturbation increased the rate of manganese cycling between precipitated and dissolved forms within the sediment column more than it increased the rate of cycling across the sediment-water interface. Thus it appears that the manganese contained in the surface layer of the sediment in the Laurentian Trough goes through a complete cycle of burial, dissolution, migration and reprecipitation on a time scale of a few months to less than a year. This vividly illustrates the dynamics involved in the early diagenesis of marine sediments.

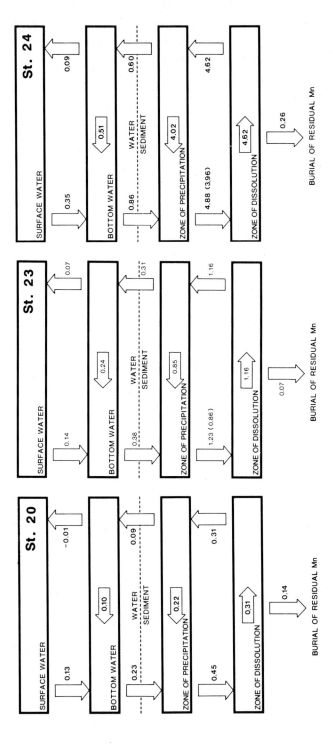

Figure 17. Modelled fluxes of Mn (mmol m^{-2} d^{-1}) between reservoirs (after Sundby and Silverberg, 1985). Note the much greater exchanges across the redox boundary in the sediment at St. 24, where bioturbation is much more intense.

3.4. Mathematical modeling of manganese diagenesis

The first attempt to account for the distribution of both dissolved and solid-phase manganese in sediments with a mathematical model was made by Burdige and Gieskes (1983). This model was primarily intended for use on sediments in which manganese reduction/oxidation takes place below the zone of bioturbation and does not included terms for the biological transport of manganese. To describe mathematically the diagenesis of manganese in Laurentian Trough sediments, Gratton et al. (1989) extended the model to incorporate bioturbation and bioirrigation.

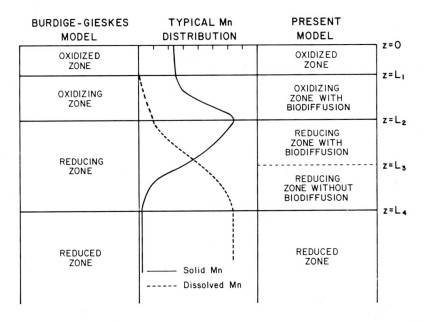

Figure 18. Sediment depth-zonation used in the diagenetic model of Gratton et al. (1990).

The sediment column was divided into four distinct zones (Fig. 18): the oxidized zone, the oxidizing zone, the reducing zone and the reduced zone. In the oxidized zone, the concentration of dissolved manganese is zero, and the concentration of solid manganese is constant with depth. It is assumed that the supply of solid manganese from the water column is constant. In the oxidizing zone, upward diffusion of dissolved manganese from the reducing zone and consumption via oxidation (precipitation) cause the concentration of both porewater manganese and solid manganese to increase with depth. Below the redox boundary, solid manganese is reduced (dissolved). The oxidizing zone and the upper layer of the reducing zone are bioturbated.

By modeling the precipitation and dissolution processes as first order reactions with depth-independent rate constants, and the biological mixing processes as analogous to diffusion, a pair of equations expressing the effects of physical, biological and chemical processes on the mass conservation of solid and dissolved manganese could be written for each layer and solved for the steady state case, using appropriate boundary and matching conditions.

The model was first applied to data from Station 23 in the Laurentian Trough, for which estimates of all the parameters required by the model, except the two rate constants, were available. By fitting the model to measured distributions of manganese in three replicate cores, values of 5×10^4 y^{-1} for the precipitation rate constant and 2.5×10^2 y^{-1} for the dissolution rate constant were obtained. For sediments from the same station, Edenborn et al. (1985) had obtained values for the pseudo-first-order oxidation rate constant that ranged from 10^4 to 10^5 y^{-1}. The agreement between the model result and the measurement is surprisingly good, considering the many simplifying assumptions that went into the model.

Although the precipitation reaction was modeled as a first-order reaction, the actual reaction is not first-order in dissolved manganese, but depends on pH, oxygen, and surface area of catalyst, i.e. MnO_2, as well. This dependence can be seen clearly in the measurements of Edenborn et al. (1985), which show that the pseudo-first-order rate constant decreases with depth in the sediment in nearly perfect parallel to the decreasing concentration of extractable manganese, presumably MnO_2. For this reason, the value of the precipitation constant obtained in the St. Lawrence cannot be compared with rate constants obtained by modeling pelagic sediments and other coastal sediments unless local conditions of pH, oxygen and reactive solid are known.

The diagenetic model was put to another test by asking it to reproduce the range of values of dissolved manganese that had been observed in sediment porewater at three stations in the Laurentian Trough (Sundby and Silverberg, 1985). The order-of-magnitude range of the maximum concentration of dissolved manganese was difficult to explain by invoking the equilibrium with a solid manganese phase as the control of the dissolved manganese concentration. The model calculations were in good to excellent agreement with the observations (the decreasing concentration of dissolved manganese with depth below the level of the maximum could not be reproduced by the model since it ignores reactions in the reduced layer).

The success of the model in reproducing the field observations is a strong indication that the profiles of dissolved manganese in the sediments of the Laurentian Trough represent a dynamic balance between the rate of production and the rate of removal of dissolved manganese. Hence it is not necessary to invoke solubility control to explain the observations. The production rate of dissolved manganese within sediments depends on the rate at which bioturbation transports dissolvable manganese down into the reducing zone. The higher the rate of bioturbation, the higher the production rate of dissolved manganese. As a result, the balance is shifted towards higher concentrations of dissolved manganese.

4. Considerations for Future Research

The many observations made on the diagenesis of St. Lawrence estuary sediments have brought with them a growing awareness of the complexities of the natural environment and have made it apparent that we are still in the embryonic stage in our understanding of the role of bottom sediments in the estuarine ecosystem. In this final section we would like to discuss the implications of spatial heterogeneity and temporal variability for early diagenesis, the interaction between sediment chemistry and the physics and biology of the overlying water, and the need for new approaches to the study of early diagenesis.

4.1. Spatial Heterogeneity

By coastal standards the Laurentian Trough is very deep and quiet. Near the bottom, tidal currents are attenuated with respect to the surface, and the net displacement of the bottom water is only 1-2 cm s^{-1} (El-Sabh, 1979). The temperature and salinity are virtually constant, and oxygen appears to be always present. The sedimentation rates are moderate, and tube-dwelling polychaetes dominate the fauna. Although these rather uniform, quiet conditions are reflected in the appearance and the grain-size distribution of the modern mud deposits of the Maritime Estuary, they are not reflected in the chemistry and biology of the sediments: The thickness of the Mn-enriched surface layer in replicate cores taken at St. 23 varied by a factor of three. In the same cores, the maximum concentration of dissolved cadmium varied by a factor of three within the surface layer and by an order of magnitude at 20 cm depth (Gobeil et al., 1987). The maximum concentration of dissolved phosphate in a core from St. 24 was as high as 425 µM whereas it only attained 275 µM in two other cores from the same station and less than 75 µM in cores from other stations (Bouchard, 1983). Some boxcores showed a predominance of 1-2 cm-thick Maldanid polychaete tube structures, whereas neighbouring cores did not (Ouellet, 1982, Bouchard, 1983). Profiles of dissolved oxygen, measured centimeters apart in the same core, showed penetration depths varying between 3 and 10 mm. Sediment traps, separated by less than a kilometer, sometimes showed total particle fluxes differing by as much as 1/3, and individual cylinders on the same frame sometimes collected as little as 1/4 of the flux collected by neighbouring cylinders, about 50 cm distant. These differences are not attributable to experimental error, but to heterogeneities in the environment, and they cast serious doubt on the representativity of an individual sample. Indeed, these observations point to the need to develop better sampling strategies that those presently used.

4.2. Temporal variability

In cores taken at the same station, but at different times, the shape of concentration depth profiles can be quite variable (Silverberg et al., 1985). The concentration of dissolved manganese in the porewater close to the sediment-water interface is sometimes much higher than in the bottom water. At other times, the profiles reveal little or no difference between the concentrations in the bottom water and in the porewater of the upper few millimeters of the sediment. It is possible that these differences reflect the variable arrival of organic matter from the overlying water, for the many sediment trap measurements at St. 23 reveal a pattern of seasonal change in both the bulk sedimentation rate and the quality and quantity of organic carbon fluxing toward the sediment surface.

There are indications that the thicknesses of the different redox zones in the sediment respond to changes in the organic matter flux, but the trends are confounded by spatial and temporal variability (Silverberg et al., 1985). Sediment traps deployed on successive days show factor of two differences in particle fluxes, and year to year differences on similar dates can be very large. It is therefore doubtful that the temporal pattern in the flux of particulate matter and organic matter content can be well-described by occasional field sampling. Since the degradation of freshly arrived organic matter drives most of the chemical transformations during early diagenesis, it is likewise doubtful that occasional sediment sampling would reveal temporal patterns in sediment chemistry. We are thus not yet in a position to state <u>when</u> specific diagenetic phenomena occur, or whether they are steady, oscillatory, or intermittent.

4.3. Spatial and Temporal Flux Patterns : Examples involving manganese and iron

In the absence of data about time and space variations, it has often been assumed that the profiles observed in the sediment are in steady state. As data from the St. Lawrence have accumulated, however, the steady state notion has had to be abandoned. For example, the profiles of dissolved manganese in the three cores investigated by Gobeil et al. (1987) are not in steady state with respect to the profiles of solid-phase manganese in the same cores: the former show that manganese is being precipitated in the uppermost sediment layer, whereas the latter profiles require that manganese is dissolving in the same layer. Other observations at the same station have shown, within the depth resolution of the measurements, that dissolution of manganese can occur throughout the sediment column up to the sediment-water interface.

Time- and space-averaged distributions of sediment properties are required for purposes such as mass balance calculations. Using this approach, it has been established that a large fraction of the manganese (Sundby et al., 1981), and possibly also the iron (Silverberg et al., 1982), that sediments in the Laurentian Trough does not accumulate, but escapes into the overlying water column. Elevated bottom water concentrations of dissolved manganese and dissolved iron (Yeats et al., 1979; Yeats, 1988), and the occurrence of individual manganese-rich and iron-rich particles in suspension in the water column (Sundby et al., 1981), also demonstrate that manganese and iron escape from the bottom sediments. The mechanism of escape remains unknown, but it is clear that it must somehow be related to the time and space variation of the porewater composition near the sediment-water interface.

Several mechanisms, driven by variations in the depth of oxygen penetration in the porewater, can be envisaged to explain an escape of manganese and iron into the overlying water. One could imagine, for example, that the sediment surface goes completely anoxic for brief periods of time, such as after the arrival of a blanket of fresh organic matter produced by an intense phytoplankton bloom in the surface water. During such anoxic periods, both iron and manganese oxides would dissolve at the very surface of the sediment, and the dissolved ions could easily escape into the bottom water. Instead of a continous blanket, one could imagine a heterogeneous organic matter distribution with "hotspots" along the sediment surface where particularly large amounts of organic matter have just arrived (the carcass of a fish would be an extreme example). Intense microbial activity at these hotspots could locally raise the depth of zero oxygen until it intersected the sediment-water interface. The hotspot would then serve as a chimney through which manganese and iron could escape for a brief time until the activity subsided. New chimneys would replace old ones as new large organic particles arrived. In this way, there would be escape of manganese and iron as long as organic matter arrived from the water column, but the escape would be confined to spots on the sediment surface.

These mechanisms would allow manganese and iron to escape the sediment in significantly different ways. One mechanism (temporary anoxia) would let them escape from the entire sediment surface, but only during certain brief periods; the other (local chimneys) would let the metals escape continuously, but only from small areas of the sediment at any given time. The mechanisms remain speculations, however, and can neither be confirmed nor eliminated without more information. Furthermore, whole-sediment mixing (bioturbation) and replacement of burrow water with fresh bottom water (bio-irrigation) also influence the depth of penetration of oxygen, and must play an important but largely unknown role in the mechanisms that let manganese and iron escape. The distribution patterns of the fauna (species composition and density) and the variable intensity of their activities, will induce additional spatial and temporal heterogeneity. Key questions about early diagenesis thus remain unanswered: How do solutes escape the sediment water interface? When do they escape? And where do they escape?

4.4. New Approaches to the Study of Sediment-Water Interactions

The processes that control the geochemistry at the sediment-water interface are clearly not such that they can be revealed by the usual approach to the study of sediment chemistry, i.e. collect the occasional core when shiptime is available and hope that the porewater composition is typical of the site and representative of the ensemble of processes that might influence it. Even if luck would have it that the porewater profile observed did represent an average distribution, that average might not be as relevant to the escape mechanisms of trace metals as the fluctuation about the average. Future research approaches must deal with the fact that the realm of early diagenesis involves both the hydrosphere and the biosphere, with all of their complexities. Multidisciplinary field studies, aiming at finding the link between physical and biological events in the surface waters and biological responses in the bottom sediments are needed. This will necessarily mean more thorough, more intense and probably more costly multidisciplinary sampling of the natural environment, using statistical approaches, and increased use of well-designed laboratory experiments.

REFERENCES

Aller, R.C., Benninger, L.K. and Cochran, J.K., 1980. Tracking particle-associated processes in nearshore environments by use of 234Th/238U disequilibrium. Earth and Planetary Science Letters 47: 161-175

Belzile, N., 1988. The fate of arsenic in sediments of the Laurentian Trough. Geochimica et Cosmochimica Acta 52: 2293-2302.

Belzile, N., 1987. Etude géochimique de l'arsenic et du sélénium dans les sédiments du chenal Laurentien. Ph.D. thesis, Université du Québec à Rimouski, 251 p.

Belzile, N. and Lebel, J., 1986. Capture of arsenic by pyrite in near-shore marine sediments. Chemical Geology 54: 279-281.

Belzile, N. and Lebel, J., 1988. Selenium profiles in the sediments of the Laurentian Trough (Northwest North Atlantic). Chemical Geology 68: 99-103.

Benninger, L.K., Aller, R.C., Cochran, J.K. and Turekian, K.K., 1979. Effects of biological sedimernt mixing on the 210Pb chronology and trace metal distribution in a Long Island Sound sediment core. Earth and Planetary Science Letters 43: 241-259.

Bergeron, M., 1980. Contribution à la géochimie du bore dans l'estuaire du St-Laurent. M. Sc. thesis, Université du Québec à Rimouski, 82 pp.

Bergeron, M. and Lebel, J., 1984. On the boron content in the sediments from the St. Lawrence estuary off Rimouski. Chemical Geology 42: 77-83.

Berner, R.A., 1980. Early diagenesis. A theoretical approach. Princeton University Press, 241 p.

Boström, K., 1967. The problem of excess manganese in pelagic sediments. In: Researches in Geochemistry, Vol. 2. P.H. Abelson (ed.). J. Wiley and Sons, p. 421-452.

Bouchard, G., 1983. Variations des paramètres biogéochimiques dans les sédiments du chenal Laurentien. M. Sc. thesis, Université du Québec à Rimouski, 161 p.

Burdige, D.J. and Gieskes, J.M., 1983. A porewater/ solid phase diagenetic model for manganese in marine sediments. American Journal of Science 283: 29-47.

Coote, A.R. and Yeats, P.A.,1979. Distribution of nutrients in the Gulf of St. Lawrence. Journal of the Fisheries Research Board of Canada 36: 122-131.

Cochran, J.K. and Aller, R.C., 1979. Particle reworking in sediments from the New York bight apex: evidence from 234Th/238U disequilibrium. Estuarine and Coastal Marine Science 9: 739-747.

Edenborn, H.M., Mucci, A., Silverberg, N. and Sundby, B., 1987. Sulfate reduction in deep coastal marine sediments. Marine Chemistry 21: 329-345.

Edenborn, H.M., Belzile, N., Mucci, A., Lebel, J. and Silverberg, N., 1986. Observations on the diagenetic behavior of arsenic in a deep coastal sediment. Biogeochemistry 2: 359-376.

Edenborn, H.M., Paquin, Y. and Chateauneuf, G., 1985. Bacterial contribution to manganese oxidation in a deep coastal sediment. Estuarine, Coastal and Shelf Science 21: 801-815.

El-Sabh, M.I., 1979. The Lower St. Lawrence Estuary as a physical oceanographic system. Naturaliste canadien 106: 55-73.

Elderfield, H., 1976. Manganese fluxes to the oceans. Marine Chemistry 4: 103-132.

Froelich, P.N., Klinkhammer, G.P., Bender, M.L., Luedtke, N.A., Heath, G.R., Cullen, D., Dauphin, P., Hammond, D., Hartman, B. and Maynard, V., 1979. Early oxidation of organic matter in pelagic sediments of the eastern equatorial Atlantic: suboxic diagenesis. Geochimica et Cosmochimica Acta 43: 1075-1090.

Gendron, A., Silverberg, N., Sundby, B. and Lebel, J., 1986. Early diagenesis of cadmium and cobalt in Laurentian Trough sediments. Geochimica et Cosmochimica Acta 50: 741-747.

Gobeil, C. and Silverberg, N., 1989. Early diagenesis of lead in Laurentian Trough sediments. Geochimica et Cosmochimica Acta 53: 1889-1895.

Gobeil, C., Silverberg, N., Sundby, B. and Cossa, D., 1987. Cadmium diagenesis in Laurentian Trough sediments. Geochimica et Cosmochimica Acta 51: 589-596.

Gratton, Y., Edenborn, H., Silverberg, N. and Sundby, B., 1990. A mathematical model for manganese diagenesis in bioturbated sediments. American Journal of Science. In press.

Lebel, J., Silverberg, N. and Sundby, B., 1982. Gravity core shortening and porewater chemical gradients. - Deep-Sea Research 29: 1365-1372.

Loring, , D.H. and Nota, D.J., 1973. Morphology and sediments of the Gulf of St. Lawrence. Bulletin of the Fisheries Research Board of Canada 182: 147p.

Marchand, B., 1984. Distribution et diagénèse du bore dans les sédiments du chenal Laurentien. M. Sc. thesis, Université du Québec à Rimouski, 159 p.

Ouellet, G., 1982. Etude de l'interaction des animaux benthiques avec les sédiments du chenal Laurentien. M. Sc. thesis, Université du Québec à Rimouski, 188 p.

Pamatmat, M.M., 1971. Oxygen consumption by the seabed. IV. Shipboard and laboratory measurements. Limnology and Oceanography 31: 305-318.

Redfield, A.C., Ketchum, B.H. and Richards, F.A., 1963. The influence of organisms on the composition of sea-water. In: The Sea, Vol. II. M.N. Hill (ed.). Interscience Publishers, New York, p. 26-77.

Rhoads, D.C., 1974. Organism-sediment relations on the muddy sea floor. Oceanography and Marine Biology Annual Review 12: 263-300.

Silverberg, N. and Sundby, B., 1979. Observations in the turbidity maximum of the St. Lawrence estuary. Canadian Journal of Earth Sciences 16: 939-950.

Silverberg, N., Bakker, J., Edenborn, H. and Sundby, B., 1987. Oxygen profiles and organic carbon fluxes in Laurentian Trough sediments. Netherlands Journal for Sea Research, 21: 95-105.

Silverberg, N., Nguyen, H.V., Delibrias, G., Koide, M., Sundby, B., Yokoyama, Y. and Chesselet, R., 1986. Radionuclide profiles, sedimentation rates, and bioturbation in modern sediments of the Laurentian Trough, Gulf of St. Lawrence. - Oceanologica Acta 9(3): 285-290.

Silverberg, N., Edenborn, H.M. and N. Belzile, 1985. Sediment response to seasonal variations in organic matter input. In: Marine and Estuarine Geochemistry. A.C. Sigleo and A. Hattori (eds.). Lewis Publ. Inc., Chelsea, MI., Ch.5, p. 69-80.

Silverberg, N., Gobeil, C., Sundby, B. and Lambert, C.E., 1982. Early diagenetic behaviour of reactive iron in muddy estuarine sediments. Abstract. AGU/ASLO Joint Meeting, San Antonio, Texas, Feb. 9-13.

Staresinic, N., Rowe, G.T., Shaughnessey, D. and Williams, A.J. III, 1978. Measurement of the vertical flux of particulate matter with a free-drifting sediment trap. Limnology and Oceanography 23: 559-563.

Sundby, B., 1977. Manganese-rich particulate matter in a coastal marine environment. Nature 270: 417-119.

Sundby, B.,1974. Distribution and transport of suspended particulate matter in the Gulf of St. Lawrence. Canadian Journal of Earth Sciences 11: 1517-1533.

Sundby, B. and Silverberg, N., 1985. Manganese fluxes in the benthic boundary layer. Limnology and Oceanography 30: 374-382.

Sundby, B., Bouchard, G., Lebel, J. and Silverberg, N., 1983. Rates of organic matter oxidation and carbon transport in early diagenesis of marine sediments. Advances in Organic Geochemistry 1981: 350-354.

Sundby, B., Silverberg, N. and Chesselet, R., 1981. Pathways of manganese in an open estuarine system. Geochimica et Cosmochimica Acta 45: 293-307.

Syvitsky, J.P.M. and Praeg, D.B., 1989. Quaternary sedimentation in the St. Lawrence and adjoining areas: An overview based on high-resolution seismo-stratigraphy. Géographie physique et Quaternaire, in press.

Syvitski, J.P.M., Silverberg, N., Ouellet, G. and Asprey, K.W., 1983. First observations of benthos and seston from a submersible in the Lower St. Lawrence Estuary. Géographie physique et Quaternaire 37: 227-240.

Takayanagi, K. and Belzile, N., 1988. Profiles of dissolved and acid-leachable selenium in a sediment core from the Lower St. Lawrence Estuary. Marine Chemistry 24: 307-314.

Trefrey, J. and Presley, B., 1982. Manganese fluxes from the Mississippi Delta sediments. Geochimica et Cosmochimica Acta 46: 1715-1726.

Westerlund, S., Anderson, L., Hall, P., Iverfeldt, A., Rutgers van der Loeff, M. and Sundby, B., 1986. Benthic fluxes of cadmium, copper, nickel, zinc, and lead in the coastal environment. Geochimica et Cosmochimica Acta 50: 1289-1296.

Yeats, P.A., 1989. Trace metals in the water column. In: Chemical Oceanography of the Gulf of St. Lawrence. P.M. Strain (ed.). Canadian Bulletin of Fisheries and Aquatic Sciences 220: 79-98.

Yeats, P.A., Sundby, B. and Bewers, J.M., 1979. Manganese recycling in coastal waters. Marine Chemistry 8: 43-55.

Chapter 11

Chemical Contaminants in the St. Lawrence Estuary and Saguenay Fjord

D. Cossa

Institut Français de Recherche pour l'Exploitation de la Mer, B.P. 1049, F-44037 Nantes, cedex 01, France

ABSTRACT

Metallic (mercury, cadmium, lead, zinc, etc.) and organic (PCB, Mirex, PAH, etc.) contaminants of the St. Lawrence Estuary and of its main tributary, the Saguenay Fjord, are assessed by examining data on water, suspended particles, sediments and biota. Past and present contamination levels and fluxes in various biogeochemical reservoirs are given. Sources are explored.

After a serious mercury contamination problem, mainly due to chloralkali plants, a large decrease of mercury inputs from liquid effluents has occurred in the St. Lawrence Basin. At the present time the remaining mercury inputs to these waters from anthropogenic sources originate from the atmosphere and sediments. Studies on sediments also show the presence of significant fractions of other metals of anthropogenic origin (e.g. Pb). However, their concentrations do not attain levels considered hazardous for marine life. Organic contaminants constitute a more worrying problem for marine life in the Estuary and Saguenay Fjord. High concentrations of PCBs and PAHs have been measured in different compartments of the St. Lawrence ecosystem and sediments. Such compounds have been suggested to be responsible for diseases or disorders observed in marine mammals living in the area.

Key Words: Mercury, cadmium, lead, zinc, PCB, Mirex, PAH, suspended particles, sediment, biota, pollution

RESUME

La contamination métallique (mercure, cadmium, plomb, zinc, etc.) et organique (PCB, Mirex, PAH, etc.) de l'estuaire du St-Laurent et du fjord du Saguenay, son principal affluent, est évaluée sur la base des données disponible sur l'eau, la matière en suspension, les sédiments et les organisms vivants. Les niveaux de contamination passés et actuels ainsi que les flux entre les différents réservoirs géochimiques sont explorées.

Après la diminution des rejets de mercure dans les affluents des usines d'électrolyse des chlorures alcalins qui ont causé par le passé une grave contamination du bassin du St-Laurent, l'atmosphère et les sédiments constituent aujourd'hui les sources majeures de mercure anthropique pour les eaux. L'étude des sédiments a aussi montré qu'une proportion significative de plusieurs autres métaux, parmi lesquels le plomb, provenait de sources anthropiques. Cependant les niveaux de concentration rencontrés n'atteignent pas les tenures considérées comme dangereuses pour la vie marine. Les contaminants organiques constituent un problème beaucoup plus grave. Des concentrations élevées en BPC et HAP ont été mesurés dans plusieurs compartiments des ecosystèmes du St-Laurent et dans ses sédiments. Ces composés ont été mis en cause dans l'apparition des affections ou maladies chez les mammifères marins vivant dans la région.

Coastal and Estuarine Studies, Vol. 39
M. I. El-Sabh, N. Silverberg (Eds.)
Oceanography of a Large-Scale Estuarine System
The St. Lawrence
© Springer-Verlag New York, Inc., 1990

1. Introduction

The review written by Trites (1972), entitled "The Gulf of St. Lawrence from a Pollution Viewpoint" was a sign of the growing interest within the scientific community in the problems of the preservation of an environment hitherto regarded as an inexhaustible resource and an "unfillable" dumping area. This paper made an inventory of the potential sources of chemical contaminants and of the physical mechanisms likely to govern their distribution (dilution, dispersion, transit time).

Research carried out in this field over the last 15 years has produced analytical results on the levels, distribution and fluxes of potentially toxic chemical substances; research on the behaviour of these substances is scarcer. The desire to synthesize the findings acquired in these fields was expressed on several occasions. In 1981 a conference on the pollution of the North Atlantic was organized in Halifax (cf. Can. J. Fish. Aquat. Sci., 40 suppl. N° 2); in 1984 the federal ministries of the Environment and of Fisheries and Oceans published a book entitled "Health of the Northwest Atlantic" (Wilson and Addison, 1984). In both these cases the geographical entities surveyed were too vast to allow a detailed approach towards the contamination of the St. Lawrence Estuary, although several articles published at the time of these events dealt with some particular aspects of the contamination of the St. Lawrence Estuary.

In this chapter a review of knowledge acquired over the past 15 years concerning metal and organic contamination is attempted. The sources, levels and distribution of contaminants in the various sections of the St. Lawrence Estuary and of the Saguenay Fjord are described and the current fluxes between the various related biogeochemical reservoirs and their variation in the past are assessed. Metals will be considered first (mercury, cadmium, lead, zinc, etc.) then organics (PCB, Mirex, PAH, etc.). The effects of these contaminants on marine life are raised in the conclusion section.

This survey is focused on non-radioactive chemical compounds and does not include persistent litter, microbiological problems, such as those related to toxic plankton, or the effects of the regulation of freshwater inputs.

2. Metals

The scientific literature offers signs of the presence of several elements of anthropogenic origin in the St. Lawrence Estuary. The most widely-documented case is undoubtedly that of mercury, but cadmium, lead and zinc have also been mentioned as potential contaminants. A large proportion of these metals are conveyed by the Saguenay Fjord and contamination of this tributary will be treated extensively. Mercury, given its particular importance, is dealt with first. Cadmium is then developed in a second section. A third section deals with other elements.

2.1. Mercury

In this section after some brief information on sources and fluxes of mercury in the Canadian environment, especially in the St. Lawrence Basin, the case of the contamination of the Saguenay Fjord and St. Lawrence Estuary are described in turn.

According to a 1977 report of the Canadian Academy of Science, anthropogenic mercury discharged into the Canadian environment comes from three main sources : fossil fuel combustion, chloralkali factories and mining industries. Airborne waste released by the combustion of coal and petroleum ranks first in quantity and reaches the aquatic environment in a relatively diffuse state, although concentrated around the major industrial centres, including those along the shores of the Great Lakes and the banks of the St. Lawrence River. In 1975, this source accounted for 30 x 10^3 Kg of mercury. Since then, emissions from combustion do not seem to have decreased. The second major source was, until 1975, chloralkali plants. In this case the waste, being mostly in liquid effluents and therefore more localized, reached as much as 67 x 10^3 Kg of mercury in 1970. The enforcement of 1972 federal regulations governing waste emission made it possible to bring this figure down to 6 x 10^3 Kg.a^{-1} as early as 1973, then to less than 500 Kg.a^{-1} in the early 1980's. Environment Canada established that in 1985 there were 130 Kg.a^{-1} of waste mercury in the liquid effluents of the chloralkali plants in operation in Canada (McBeath, 1986). In 1975, waste produced by mining industries, namely zinc, lead and mercury extraction, represented in all likelihood around 10 % of the total mercury discharged into the environment, i.e. approximately 7 x 10^3 Kg.a^{-1} (Canadian Academy of Science, 1977). In 1985 there seemed to be no grounds for an upward revision of this last estimate.

The St. Lawrence Basin was particularly affected by the presence of several chloralkali plants. Without taking into account those located around the Great Lakes, four plants were in existence until 1976 : in Cornwall, Beauharnois, Shawinigan and Arvida. In 1985 the Cornwall plant discharged only 42 g of mercury a day. The Shawinigan plant stopped production in 1978. Between 1984 and 1987 the plant in Beauharnois discharged approximately 200 g of mercury a day ; since then these amounts are reported to have been brought down to 30 g.d^{-1} (Environnement Québec, pers. comm.). The biggest plant, at Arvida on the Saguenay River, ceased production in 1976. It is reported to have released over 300 x 10^3 Kg of mercury into the environment during its existence (Loring and Bewers, 1978). During its 30 years of operation the Arvida plant was the greatest source of mercury contamination for the Saguenay Fjord and the St. Lawrence Estuary.

2.1.1. Contamination of the Saguenay Fjord

In the 70's high mercury contents were measured in the organisms of the Saguenay Fjord. Concentrations were specially high in organisms whose habitat or diet is linked to the sediment. Levels in the edible part of shrimp exceeded 10 µg.g^{-1} (wet weight) (Cossa and

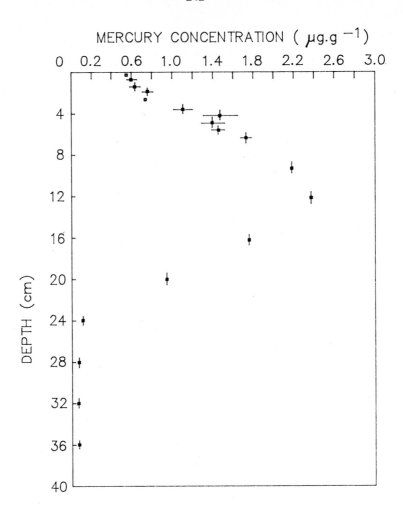

Figure 1. Profile of the mercury concentration levels in a sediment core sample taken from the inner basin of the Saguenay Fjord in 1978. From Gobeil and Cossa (1984).

Desjardins, 1984), attained 7.2 µg.g.$^{-1}$ (dry weight) in cod muscle, 3.8 and 0.7 µg.g^{-1} (dry weight) in soft parts of whelk and clams, respectively (Cossa, unpublished results). Several studies of the Fjord sediments revealed the importance of this contamination (Loring, 1975 ; Loring, 1978a; Loring and Bewers, 1978 ; Smith and Loring, 1981; Barbeau et al., 1981a; Loring et al., 1983; Gobeil and Cossa, 1984; Pelletier et al., in press). Mercury concentration levels in surficial sediments taken from the Fjord between 1964 and 1976 ranged from 0.2 to 12 µg.g^{-1} (Loring and Bewers, 1978). The quantities of mercury accumulated in the Fjord sediments were estimated at 20 x 10^3 Kg by Barbeau et al. (1981 a), while Loring (1978 a) estimated that accumulated in the Saguenay System at almost 120 x 10^3 Kg. In core samples taken in 1978 from the inner basin of the Fjord, mercury concentrations varied from 0.05 µg.g^{-1} in the deep sedimentary layers (> 25 cm) to more than 2 µg.g^{-1} a few centimeters beneath the surface of the sediment (Fig. 1).

A geochronological study based on excess ^{210}Pb distributions (Smith and Loring, 1981) provided the most convincing evidence as to the origin of the anthropogenic mercury in the area. Mercury contamination of the Fjord sediment began around 1948, which coincides with the building of the chloralkali plant at Arvida along the Saguenay River. The input of mercury to the sediment reached a peak in the 60's and then decreased over the period 1971-1976. The ages of the sediment where the mercury concentration decreases coincide with the dates when the regulations governing the discharge of effluents were imposed on the chlorine industry (1971-1972) until the final shutdown of the Arvida plant in 1976.

The rapid variations in the mercury concentrations in the sediment at the head of the Fjord bear witness to a very brief residence time for mercury in the water column in this area, whereas further downstream, it may reach 10 to 15 years (Loring et al., 1983). On the basis of of variations in mercury concentrations in the water column, Gobeil et al. (1984) estimate the residence time in the inner basin at between 3.5 to 7.5 years.

In 1983, more than 7 years after the plant implicated in the pollution was shut down, mercury concentrations in surface sediments (0 to 1 cm) varied with sampling site between 0.2 and 0.8 μg.g^{-1}. This is 4 to 17 times the natural (pre-industrial) concentration found deep in the sediment (Gobeil and Cossa, 1984). The same authors pointed out that mercury in the interstitial waters of the sediments was enriched in comparison with the overlying waters. It should be noted that mercury remobilization takes place in the layer of the sediments directly below the surface oxidizing zone. Given the concentration gradient measured at the sediment-water interface, a diffusive mercury flux towards the water column of 6×10^{-7} ng.cm^{-2}.sec^{-1} was calculated. However, if one takes into account the presence of the abundant benthic fauna in the inner basin of the Fjord, a larger flux can be predicted : up to five time the diffusive flux according to Silverberg et al. (1987). Assuming a residence time of about 3 months for the waters in the inner basin, such a "biologically mediated flux" can generate a 2.5 ng.L^{-1} rise in concentration in the deep waters of the Fjord. Measurements made in 1983 and 1984 (Table I) are consistent with this hypothesis ; they show that the mercury concentrations in the deep waters of the Fjord are higher (2.2 -2.9 ng.L^{-1}) than the surface and intermediate waters of the Lower Estuary (< 0.3 - 0.9 ng.L^{-1}) from which they originate.

The rapid decrease in mercury levels in the 70's and their subsequent stabilization has also been observed in biota from the Fjord. The measurements carried out on the edible part of the shrimp before and after the polluting plant was shut down show a dramatic decrease in mercury concentrations (Fig. 2). However, they reveal the persistance of a level higher than in those areas which had never suffered from contamination. This tendency has been confirmed by recent measurements made by Pelletier and Canuel (1988) on both surface sediment and shrimp sampled in 1986 and 1987. In summary, one can conclude that: (i) measurements have clearly identified the most important source of mercury contamination for the Fjord, the Arvida chloralkali plant ; (ii) federal regulations adopted in 1972 and the

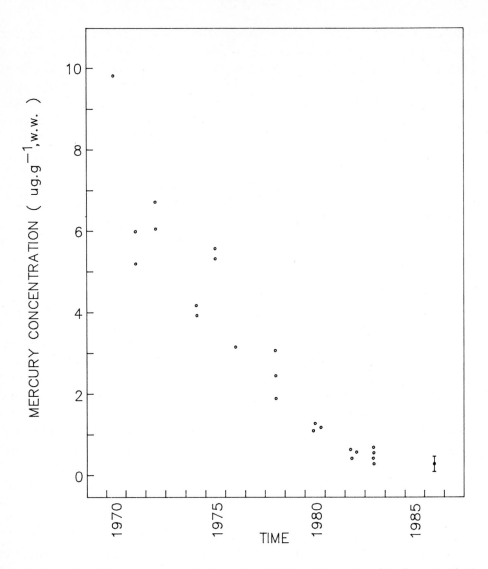

Figure 2. Evolution of the mercury concentration in the edible part of the shrimp of the Saguenay Fjord. From Cossa and Desjardins (1984) and Pelletier et al. (in press).

plant shutdown in 1976 have permitted the Fjord's partial rehabilitation ; (iii) the sediment remains a reservoir where considerable amounts of mercury are still stored, and that seems to be responsible for a persistant and still significant level of contamination of organisms whose diet is linked to the sediments. Research on mercury mobility at the sediment-water interface is needed to predict the time required for a complete decontamination of the Fjord's ecosystem. In the meantime, it can be firmly recommended that resuspension of the contaminated sediment with fishing equipement be avoided.

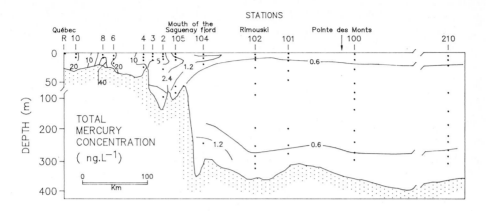

Figure 3. Longitudinal distribution of the mercury concentrations in unfiltered waters of the St. Lawrence Estuary. Data gathered between 1980 and 1984 (Gobeil et al., 1983 ; Cossa, unpublished).

2.1.2. Estuarine contamination

The first results on mercury concentrations in the waters of the St. Lawrence Estuary were published by Piuze and Tremblay (1979). Then Gobeil et al. (1984) indicated that total mercury concentrations in the Upper Estuary varied between 4 and 7 ng.L^{-1} and that those of dissolved mercury were never higher than 6 ng.L^{-1}. Since then additional results were obtained by means of "ultraclean" sampling and analytical methods described elsewhere (Cossa and Noel, 1987). The latter results are combined with those of Gobeil et al. (1984) to provide a description of the mercury distribution throughout the Estuary (Table 1 and Fig. 3). The longitudinal distribution of the total (unfiltered sample) mercury concentrations is shown in Figure 3. It appears clear that the highest concentrations are associated with the

TABLE 1
Mercury concentration (ng.L^{-1}) in unfiltered sea-water samples collected in the inner basin of the Saguenay Fjord and in the Estuary (Cossa and Gobeil, unpublished).

Depth (m)	Saguenay Fjord (1983)	Lower Estuary (1984)
1	4.3	0.7
5	3.1	0.9
10	2.8	0.4
25	1.1	<0.3
50	1.5	<0.3
100	1.1	0.4 (80 m)
150	2.2	--
200	2.9	0.4
250	2.5	0.4
300	--	0.7

maximum turbidity zone of the Upper Estuary, with levels as high as 40 ng.L^{-1} (Station 8). Seaward, the concentrations decrease to the detection limit (0.3 ng.L^{-1}) in the intermediate waters of the Laurentian Trough.

At the mouth of the Estuary and in the Laurentian Trough (seaward of station 3 ; Fig. 3) there is a surface-layer enrichment, which corresponds to the higher levels of suspended particulate matter. At stations 102, 101, 100 and 210 a slight tendency towards an increase in concentration within the bottom nepheloid layer can be seen (Fig. 3) ; results of analyses of some filtered samples collected off Rimouski (station 102) in the Laurentian Trough in June 1985 indicate that sub-surface and near-bottom mercury enrichment is linked with higher turbidity in these zones (Table 2).

TABLE 2.
Total mercury concentration at station 102 (off Rimouski, Fig. 3). (F) filtered samples, (NF) unfiltered samples. Turbidity has been measured with a Hach turbidimeter and expressed as nephelometic turbidity units (NTU) (Cossa, unpublished data from a cruise on the R/V L. Lauzier in June 1985).

Depth (m)	Salinity (.10^{-3})	Turbidity (NTU)	Total mercury (ng.L^{-1})	
			NF	F
1	24.14	0.4	0.7	0.7
7	24.46	1.8	0.9	0.6
200	33.86	0.1	<0.6	<0.6
325	34.46	0.5	0.9	0.8

During a 1983-85 monitoring programme of the St. Lawrence River (Tremblay, 1985) five samples were collected for particulate mercury determinations : the average concentration in the samples was 0.4 µg.g^{-1}. On the basis of a 5.2 x 10^9 Kg.a^{-1} input of solid material from the River (Tremblay, 1985), an influx of particulate mercury into the Estuary of 2.0 x 10^3 Kg.a^{-1} can be calculated, i.e. four times the amount given by Cossa et al. (1988 a) for transport in solution.

The analysis of the surface sediments in the Laurentian Trough (Loring, 1975) has shown that its mercury contamination was important, particularly in the centre of the Lower Estuary. More recent research (Gobeil and Cossa, 1988) noted a decrease in contamination of surface sediments in the Lower Estuary : while surface sediment concentrations reached values exceeding 0.5 µg.g^{-1} before 1975 (Loring, 1975)) they stood at 0.16-0.18 µg.g^{-1} in samples taken in 1987. The total quantity of anthropogenic mercury accumulated in the sediments of the Trough would be on the order of 180 x 10^3 Kg (Gobeil and Cossa,

1988). The same authors concluded that, owing to the low concentration gradient at the sediment-water interface one should not expect a important diffusive flux of mercury out of the sediment. However, if biologically mediated, mercury mobilization may be more significant.

The "Mussel Watch" type program carried out in 1977 along the intertidal zone of the Lower St. Lawrence Estuary and Gulf failed to point out any mercury "hot spots" in the area covered (Cossa and Rondeau, 1985). The distribution of mercury levels in mussel soft tissue reflected the distribution patterns of freshwater and of turbidity : the highest concentration levels (on the order of 0.3 μg.g^{-1}, d.w.) were observed in the landward part of the Lower Estuary and in the southern part of the Baie des Chaleurs.

2.1.3. Conclusions

It can be stated that the mercury contamination of the St. Lawrence Estuary and Saguenay Fjord has been considerably reduced since the 70's. However, sediments remain highly contaminated and organisms living in contact with the bottom present high mercury concentrations. Elsewhere the levels encountered are similar to those found in other industrialized areas (GESAMP, 1986). From the data presented here, the riverine mercury (particulate and dissolved) input from the St. Lawrence and Saguenay rivers are estimated to be approximately 3×10^3 Kg.a^{-1}. Current official inventories indicate that less than 200 Kg of mercury from localized anthropogenic sources are still discharged each year (Mc Beath, 1987). This is only 0.2 % of the anthrogenie flux of 20 years ago. It must be noted, however, that there has been a 40 % rise in mercury consumption by chloralkali plants between 1980 and 1985 which makes necessary a very strict control of waste discharges. Moreover, the mobilization of mercury from sediments and its possible transfer into the ecosystem calls for more study. Lastly, because of the well known volatility of this element, an evaluation of inputs of airborne mercury and of its anthropogenic components are required.

2.2. Cadmium

In the late 70's a series of papers reported relatively high cadmium contents in the waters of the St. Lawrence River and Estuary and of the Saguenay Fjord (Bewers and Yeats, 1977, 1978, 1979 ; Yeats and Bewers, 1976; 1983. High concentration levels, averaging 110 ± 290 ng.L^{-1} (dissolved) in the River, 85 ± 35 ng.L^{-1} (total) in the Estuary and ranging from 60 to 270 ng.L^{-1} (total) in the Fjord, suggested distribution anomalies of possibly anthropogenic origin. Upstream in the St. Lawrence basin (Lakes Superior, Michigan and Erie) the concentration levels measured did not exceed 41 ng.L^{-1} (Muhlkaler et al., 1979 ; Lum and Leslie, 1983 ; Rossman, 1986) and were 10 ng.L^{-1} in Lake Ontario (Lum and Callaghan, 1986). On the other hand, the highest concentrations encountered in the North Atlantic Ocean are lower than 40 ng.L^{-1} (e.g. Yeats and Campbell, 1983).

However, such a situation has been reflected neither in the scattered cadmium measurements conducted on plankton (Cossa, 1976 ; Dubé, 1982 ; Poirier and Cossa, 1981) nor in the sediments of the Estuary (Gendron et al., 1986 ; Gobeil et al., 1987). The only "hot spot" displaying a real cadmium contamination problem in the area is that of Belledune Harbour in the Baie des Chaleurs whose influence, however, is remote with respect to the Estuary itself (Uthe and Zitko, 1980 ; Bewers et al., 1987). Thus, a reassessment of the cadmium concentration levels found in the waters has proven necessary. The recent measurements made on St. Lawrence River waters (Lum, 1987 ; Lum et al., 1988) and data on estuarine waters (Cossa, 1988) - presented in the following sections - concur to show that the first measurements probably overestimated the actual concentrations, as was the case for a large number of analytical results on trace metals in the past decade.

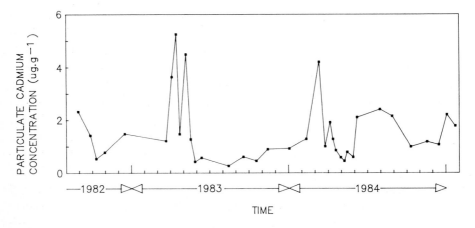

Figure 4. Temporal variations of the cadmium content of suspended particles collected in the St. Lawrence River near Quebec City (Cossa, unpublished).

2.2.1. The St. Lawrence River

Within the framework of its Marine Analytical Chemistry Standards Program, the National Research Council of Canada devised a riverine water reference standard based on filtered water drawn from the St. Lawrence in 1984 a few kilometers upstream from Quebec City. The analyses, carried out by several independent analytical methods, led to an average cadmium concentration of 15 ± 2 ng.L[-1]. Lum et al. (1986) and Lum (1987) measured dissolved and particulate cadmium in water samples from Lake Ontario to Quebec City in 1985. The average concentration of dissolved cadmium was 14 ± 8 ng.L[-1] in May and 16 ± 12 ng.L[-1] in October. Although the highest levels were found in the vicinity of Montreal and are probably the result of anthropogenic inputs (3.6×10^3 Kg.a[-1] according to the Quebec Department of the Environment, pers. comm.) such levels are within the range of not very contaminated river waters (Boyle et al., 1982 ; Shiller and Boyle, 1983 ; Elbaz-

Poulichet et al., 1987). Particulate cadmium averaged 1.88 ± 0.76 $\mu g.g^{-1}$ for the two periods. These levels which are very consistent with previous measurements made in 1974 (Cossa and Poulet, 1978) and with data presented in Figure 4, which cover two and a half seasonal cycles between August 1982 and January 1985. The temporal variation in the cadmium content of the particulate matter collected in the River at Quebec City shows maxima during the spring flood period (Fig. 4). These data allow a relatively precise estimate of the annual riverine flux of particulate cadmium to the Estuary of 8×10^{-3} Kg per year. On the basis of the available data, a dissolved flux of 6×10^3 $Kg.a^{-1}$ can be estimated (15 $ng.L^{-1} \times 405.10^{12}$ $L.a^{-1}$). According to figures of the Quebec Department of the Environment, 25 % of this 14×10^3 $Kg.a^{-1}$ total flux is of anthropogenic origin.

2.2.2. The Estuary

Samples have been collected during two cruises - October 1983 in the Upper Estuary and October 1984 in the Lower Estuary - using teflon coated Niskin or Go-Flo bottles on Kevlar hydrowire. Ultra-clean techniques were used dsuring sample filtration, storage and analysis. Differential pulse anodic stripping voltametry was used for the cadmium determinations (Gobeil et al., 1987). A standard reference sea water (NASS-1, National Research Council of Canada) was used to check the accuracy of the method. The analytical precision and detection limit were 15 % and 5 $ng.L^{-1}$, respectively. Particulate cadmium was determined by graphite furnace atomic absorption spectrophotometry after total solubilization of the matter collected on acid-cleaned Nuclepore membranes as described by Eggiman and Belzer (1976).

In the Upper Estuary, dissolved cadmium concentrations vary between 11 $ng.L^{-1}$ in the River to 25 $ng.L^{-1}$ at the mouth of Saguenay Fjord (Table 3). The relationship between dissolved cadmium and salinity shows no real departure from a dilution line. The distribution pattern of particulate cadmium results from the mixing of particles from three different sources : (i) River and (ii) marine waters, with cadmium concentrations higher than 0.5 $\mu g.g^{-1}$ in suspended particles, and (iii) the turbidity maximum zone, where concentrations range between 0.2 and 0.5 $\mu g.g^{-1}$. In addition, particles flowing out in the surface waters of the Upper Estuary are impoverished in cadmium compared to particles entering the Estuary. The pattern probably reflects the solubilization of cadmium associated with the degradation of particulate organic matter in the turbidity maximum zone. Due to the lower cadmium concentration in River water than in seawater, the relatively low level of suspended particulate matter in the St. Lawrence Estuary compared to other macrotidal estuaries, and the relatively short residence time of the water, this solubilization of cadmium from riverborne particles does not create a noticeable increase in dissolved cadmium, as is generally observed in other estuaries (Elbaz-Poulichet et al., 1987). In addition, the seasonality in the particulate cadmium input (Fig. 4) probably induces a seasonal variation in the cadmium behaviour in the Upper Estuary which is not reflected by the momentary distribution presented.

TABLE 3.
Dissolved cadmium concentration in the surface waters of the Upper St. Lawrence Estuary (R/V L.M. Lauzier cruise, October 1984 ; Cossa, 1988).

Station	Salinity $(x\ 10^{-3})$	Cd $(ng.L^{-1})$
46°45.5'N	< 0.2	11
71°15.3'W	< 0.2	14
	< 0.2	12
47°04.5'N	0.2	16
70°44.6'W	0.7	11
	1.4	13
	2.4	14
	5.0	13
47°13.1'N	5.8	19
70°36.4'W	7.2	14
	7.4	14
47°21.5'N	14.1	20
70°29.0'W		
47°27.0'N	15.9	18
70°15.0'W		
47°30.1'N	21.9	21
70°10.0'W		
47°55.0'N	25.8	23
69°46.8'W	28.0	25

In the Lower Estuary, the concentration range is similar to Atlantic Ocean waters (18-34 $ng.L^{-1}$, Fig. 5) and their distribution tends to show a subsurface minimum and an enrichment of deep water, specially near the head of the Laurentian Trough. As in other oceanic and coastal waters, a significant relationship exists between cadmium and phosphate concentrations with a regression coefficient of 0.13, i.e. similar to those found in cold high-latitude waters (Olafsson, 1983 ; Campbell and Yeats, 1982). This situation results from the implication of cadmium in the cycling of organic matter in the waters of the Laurentian Trough : the conjuction of the recycling of dissolved cadmium in the deep water and the estuarine circulation, produces a distribution similar to that described for observed for nutrient salt enrichment (Coote and Yeats, 1979).

2.2.3. The Saguenay Fjord
Samples collected during two cruises (1978 and 1979) were analyzed for dissolved and particulate cadmium. Results are given in Table 4. The concentration levels measured are close to those found in the Lower Estuary, from which the deep waters of the Fjord

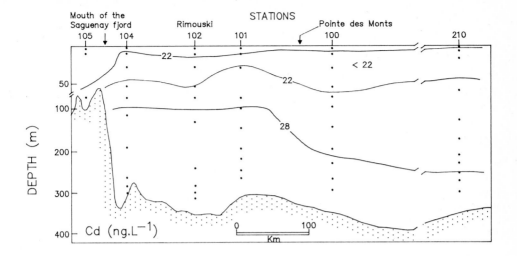

Figure 5. Longitudinal distribution of the total cadmium concentrations in the water of the Lower St. Lawrence estuary (Cossa, 1988).

originate (cf. Chapter 17, this volume) ; the Fjord therefore does not seem to represent an important source of cadmium, as might have been suspected following the publication of

TABLE 4.
Cadmium concentrations found in the waters of the Saguenay Fjord during two cruises of the Champlain Center for Ocean Sciences and Surveys in 1978 and 1979. (n) : the number of determinations (Piuze, Tremblay and Cossa, unpublished).

	Unfiltered sample $(\mu g.L^{-1})$		Particulate matter $(\mu g.g^{-1})$	
	range	(n)	range	mean ± s.d. (n)
Surface (0 - 10 m)	<0.01	(23)	0.8-2.7	1.7 ± 0.7 (14)
Deep water (25 - 250 m)	<0.01-0.03	(58)	0.4-5.4	2.2 ± 1.5 (23)

somewhat high cadmium concentration levels (0.2 to 0.5 $\mu g.g^{-1}$) measured in a core sample taken at the head of the Fjord (Loring et al., 1983). This contamination does not seem to be very widespread, since the levels measured downstream in the most of surface sediments of the Fjord are lower than 0.2 $\mu g.g^{-1}$ (Pelletier and Canuel, 1988).

2.2.4. Conclusions

There does not seem to be any important cadmium contamination in the St. Lawrence Estuary. Concentrations measured in the water and sediment are close to the levels of uncontaminated areas of other coastal regions of the world (Cossa and Lassus, 1988). The results of the mussel survey conducted in the Estuary and the Gulf confirm this (Cossa, 1980 ; Cossa and Bourget, 1980). The only "hot spot" mentioned by these authors is in the Belledune area in the Baie des Chaleurs where Uthe and Zitko (1980) found mussel cadmium contents as high as 100 μg.g^{-1} (d.w.).

On the basis of the data presented, some of the fluxes can be compared : the riverine inputs have been assessed at 14 x 10^3 Kg.a^{-1} (25 % of which are probably anthropogenic). Calculated on the basis of data from Gendron et al. (1986) and Silverberg et al. (1986), the estimated rate of cadmium burial in the sediments of the Laurentian Trough is about 14.3 x 10^3 Kg.a^{-1}. Given this approximate balance, the possible export of cadmium to the Atlantic Ocean would be dependent on the atmospheric inputs which have not as yet been measured. Knowledge concerning seasonal variations of the riverine inputs of dissolved cadmium by the St. Lawrence would contribute to a better assessment. It is not unlikely that these inputs display a marked seasonal pattern, as has been demonstrated for other elements in the St. Lawrence River (Cossa and Tremblay, 1983; Cossa et al., 1988). There is a slight indication of sporadic cadmium inputs from the melting of snow contaminated by airborne waste.

2.3. Other Metallic Contaminants

An assessment of the contamination will be arrived at *via* the examination of the metal contents in sediments, water and mussels (*Mytilus edulis*). The only elements examined here are those for which a possible contamination of the area has been pointed out.

TABLE 5.
Average and standard deviation of the concentrations (μg.g^{-1}) of some potential metallic contaminants in surface sediments of the St. Lawrence System, collected in 1972-74. From Loring (1978b, 1979).

Location	Pb	Zn	Cu	V
Saguenay Fjord	47 $\pm$ 12.2	131 $\pm$ 8	27 $\pm$ 2.5	121 $\pm$ 22
Upper Estuary	34 $\pm$ 4.5	185 $\pm$ 32	36 $\pm$ 8.5	97 $\pm$ 4
Lower Estuary	30 $\pm$ 8.0	115 $\pm$ 20	24 $\pm$ 7.2	110 $\pm$ 20

2.3.1. Sediments

Values for lead, zinc, copper and vanadium concentration in the sediments of the St. Lawrence Estuary and Saguenay Fjord have been provided by the work of Loring (1978a, b ; 1979). From the concentrations found in these sediments (summarized in Table 5) it is apparent that the Saguenay Fjord is the most contaminated area for lead. The distribution of "non-detrital" lead (i.e., according to Loring, lead which is not linked to the crystalline matrix of the sediment) and, to a lesser degree, also that of zinc and copper, would appear to be dependent on the dispersion of the industrial sources of mercury-rich organic matter from the Saguenay River. Barbeau et al. (1981a) have shed more light on these observations.

The analyses of numerous core samples of Fjord sediment (Table 6) reveal a tendancy towards an increase in zinc and lead concentrations since 1940. They also show that while mercury inputs into the Fjord had decreased since 1970, lead and zinc inputs had continued to increase. Such findings are confirmed by recent observations of Pelletier and Canuel (1988) who calculate a 20 % increase in zinc concentrations between 1972 and 1986. For Barbeau et al. (1981 a) the atmosphere seems to be a significant source of lead for the sediments of the Fjord's western basin. Gobeil and Silverberg (1989) have described the vertical profiles of lead concentration in some box cores from the Laurentian Trough. The concentration increases from the base of the cores towards the surface. According to the authors this indicates that the sediment of this area have been affected by the increased used of lead over the last century. In the surface sediment, two-thirds of the lead would be of anthropogenic origin. They also suggest that a subsurface maximum in lead concentration probably reflects the decrease in anthropogenic input over the last decade, associated with the increasing substitution of unleaded for leaded gazoline in North America.

TABLE 6.
Range of concentrations of Pb, Zn and Cu in the sediment of the Saguenay Fjord for the periods before and after the installation of plants in Arvida. From Barbeau et al. (1981 a).

	Pb (μg.g-1)	Zn (μg.g-1)	Cu (μg.g-1)
Pre - 1940	5 - 38	24 - 116	5 - 30
Post - 1940	5 - 72	45 - 201	8 - 39

For Loring (1979) the Lower Estuary appears to exhibit a relative vanadium enrichment of the fine sediments, probably from the Saguenay Fjord. The author also concludes that there is zinc contamination of the sediments in the Upper Estuary. The arsenic contents of the Saguenay Fjord surface sediments (Gobeil, pers. comm.) are twice as high as those found at similar depths in the sediments of the Laurentian Trough (Belzile, 1988). The anthropogenic origin of such a difference should not be ruled out *a priori*.

2.3.2. Water

The most recent data on concentration levels in water for some potential metallic contaminants are presented in Table 7. Lead concentrations in the River and Upper Estuary seem to be higher than those considered representative of a "pristine" region. The Quebec Department of the Environment estimates the industrial input of lead to the St. Lawrence River and its tributaries to be around 2×10^3 Kg.a^{-1}. On the other hand, according to Lum et al. (1987), the zinc concentration levels coincide with the range found in systems displaying few disturbances. The same applies tocopper contents in the Lower Estuary and the Fjord. It is difficult, however, to form a judgement on the apparently high zinc levels reported in the Lower Estuary and the Saguenay Fjord.

TABLE 7.
Average concentration (ng.L^{-1}) $\pm$ standard deviation (number of determinations) of potential metallic contaminants in the waters of the St. Lawrence System. (1) Filtered samples, Lum et al. (1987a) ; (2) unfiltered samples, Bewers and Yeats (1979) ; (3) filtered samples, Yeats (personnel communication).

Location	Pb	Zn	Cu	Reference -
St Lawrence River	89 ± 64(19)	362 ± 141(19)	--	1
Upper Estuary	146 ± 60(32)	486 ± 99(32)	--	1
Lower Estuary				
- surface	--	2110 ± 810(16)	840 ± 430(18)	2
- intermediate	--	1340 ± 410(17)	400 ± 130(15)	2
- deep	--	2290 ± 1010(15)	500 ± 180(15)	2
Saguenay Fjord				
- surface	--	1900 ± 800(16)	840 ± 330(17)	3
- intermediate	--	800 ± 300(13)	400 ± 120(7)	3
- deep	--	1000 ± 400(7)	500 ± 310(3)	3

2.3.3. The Blue Mussel (*Mytilus edulis*)

The zinc and lead contents in the soft tissues of mussels from the Lower Estuary range from 49 to 146 µg.g^{-1} and 0.5 to 4.1 µg.g^{-1}, respectively (Cossa, 1980). These levels do not indicate heavy contamination. After normalizing the concentration values to exclude the effects of biotic factors, no definite geographical distribution pattern emerges from the data that is indicative of a concentration gradient from a particularly important source.

Significant contamination was detected only for lead on the south shore of the Baie des Chaleurs, where the average lead content is on the order of 10 μg.g^{-1} (Cossa, 1980).

2.3.4. Conclusions

Analyses of sediments and water reveal a level of lead contamination in the St. Lawrence Estuary and Saguenay Fjord which seems, except in the Fjord, to have decreased in the last decade. Zinc contamination is even less significant. As for copper, the data offer no reason to suspect important anthropogenic input. It should nevertheless be stressed that little recent data is available for these two metals in the waters of the Estuary. Such data, as well as information about the distribution of other potentially toxic elements (arsenic, nickel, vanadium, tin, etc.) would be necessary to make a more comprehensive assessment of the metal contamination of the St. Lawrence Estuary and Saguenay Fjord.

3. Organics

There is fairly little geochemically-oriented literature on the nature and the degree of contamination by organic compounds in the St. Lawrence Estuary. Available data chiefly concern concentration levels found in living organisms and to a lesser extent in sediments. These will be dealt with in two main sections, one dealing with organohalogenated compounds, the other with hydrocarbons, and more specifically with polyclyclic aromatic hydrocarbons.

3.1. Halogenated Hydrocarbons

The seriousness of contamination by this type of substance is mainly the result of their lipophilic nature, which enables them to accumulate in aquatic organisms to levels as great as 10 000 times that found is the ambient water. In this category, the polychlorobiphenyls (PCBs), which clearly are major contaminants in the St. Lawrence River and Estuary will be reviewed first. We shall then review the less extensive data on Mirex, DDT and volatile organohalogenated substances.

3.1.1. Polychlorinated Biphenyls (PCBs)

PCB's are chlorinated hydrocarbons which concentrate in lipid matter. They are man-made compounds which began to be produced in the 1920's. For over 40 years they were put to a large variety of uses, which favoured their dissemination into the environment. Since 1970, their proven toxicity has led to a reduction in their production and to a control of their utilization.

At present they are only authorized as dielectrics in closed installations. In North America they are marketed under the brand name of 'Arochlor'. Arochlor 1254 is a mixture of substituted biphenyls averaging 54 % by weight chlorine. It was long used as a reference to quantify PCBs in environmental samples. Analysts now agree that a more representative quantification is obtained by means of a few selected individual compounds (congeners), which can now be isolated thanks to recent improvements in chromatography. Most of the results presented here were obtained during the Seventies and are therefore expressed as Arochlor 1254 equivalents.

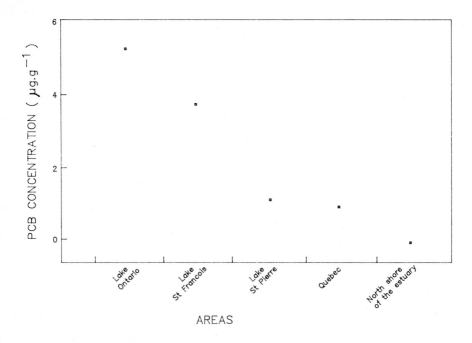

Figure 6. PCB concentration in sedentary eels in the St. Lawrence System. From Desjardins et al. (1983a).

In a review paper Couillard (1982) compiled a great deal of information on PCB concentrations in the sediments and living organisms of the St. Lawrence River, Estuary and Gulf. Contamination affects the whole area. The St. Lawrence near Montreal, Lakes St. Louis and St. François, are the most contaminated areas, with PCB levels in the sediments up to 0.4 μg.g^{-1}, and over 6 μg.g^{-1} in some species of fish. Sloterdijk (1986) reports that the young fish of the year, fairly good indicators of the ambient contamination level, are already highly contaminated in Lake St. François, where the stations closest to Lake Ontario are located. Fish in the Outaouais River are much less contaminated. The PCB concentrations in sedentary eels of the St. Lawrence Basin (Fig. 6) (Desjardins et al., 1983a) offer a more precise distribution pattern, tracing back to the upstream source mainly responsible for the contamination. Concentration levels are higher in sedentary eels from Lake Ontario than in those of the St. Lawrence Estuary.

Concentration levels in the Estuary are much lower than in the River, except for a geographically limited area, the Baie des Anglais (near Baie Comeau) where PCB discharges of industrial origin (Pydraul 230 oil), now stopped, significantly contaminated the sediments and organisms found there (Table 8). In this area, the consumption of shellfish can be harmfull to human health. In the Estuary, average concentration levels (depending on species and fishing areas) range from 0.02 to 1.6 $\mu g.g^{-1}$ (Couillard, 1982). In the oils of fish, concentrations are on the order of 10 $\mu g.g^{-1}$.

TABLE 8.
PCB concentrations ($\mu g.g^{-1}$, dry weight) in sediments and marine organisms (dry weight) from Baie des Anglais. Data from Canada and Quebec Departments of the Environment (pers. comm.).

	Min	-	Max
Sediment	<0.01	-	27.4
Mussel	0.4	-	11.0
Whelk	6.0	-	23.0

Recent studies have been concerned with the presence of PCBs in beluga whales (*Delphinapterus leucas*) of the St. Lawrence Estuary. Massé et al. (1986) and Martineau et al. (1987) report levels ranging from 10 to 312 $\mu g.g^{-1}$ in the blubber, the tissue displaying the highest concentrations. The results indicate a considerable variation in tissue concentration according to the age and sex of these whales. Concentrations are higher among males and increase with age. The lower levels found in females can be explained by a massive transfer during lactation. As a result concentrations of polychlorinated biphenyls among young animals are equal to those encountered in adult males. Martineau et al. (1987) suggest that the high contents of chlorinated hydrocarbons could be a major factor in the nonrecovery of the St. Lawrence beluga population over the last decades.

In order to assess the tendency over time of PCBs fluxes into the sediments of the Laurentian Trough, a sediment box-core sample was taken in the Lower Estuary off Rimouski during a cruise of the Champlain Center for Ocean Science and Surveys and analysed at the Institut Français de Recherche pour l'Exploitation de la Mer. The levels of PCBs increase from the sediment surface, where they are close to the detection limit (0.01 $\mu g.g^{-1}$), to 0.33 $\mu g.g^{-1}$ at a depth of 8 cm (Fig. 7). Concentrations then decrease again with the depth, down to the detection limit between 20 and 25 cm. The concentrations measured are of the same order of magnitude as those found in other coastal sediments near industrialized areas : 0.0002 to 0.035 $\mu g.g^{-1}$ in the Mississippi Delta (Giam et al., 1978), 0.002 to 0.16 $\mu g.g^{-1}$ in the Irish Sea (Dawson and Riley, 1977), 0.008 to 0.011 $\mu g.g^{-1}$ in the Kiel Bight (Osterroht and Smetacek, 1980).

258

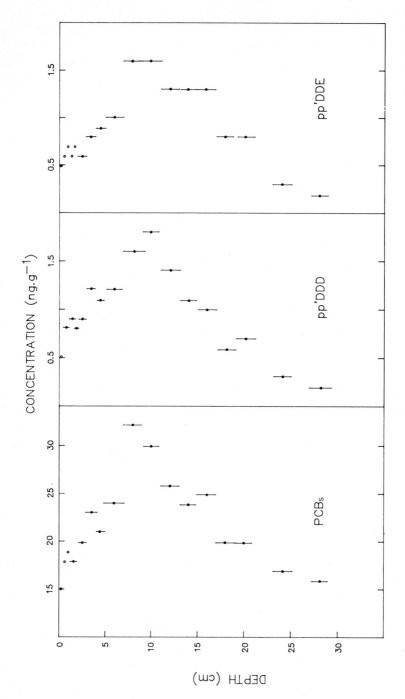

Figure 7. Chlorinated hydrocarbons distribution in a core sample collected in the Laurentian Trough near Rimouski in 1985 (Gobeil, Cossa and Luçon, unpublished results).

At the same station where the core sampling was carried out, Silverberg et al. (1986), using sediment traps and ^{210}Pb dating, assessed the sedimentation rate at 0.51 μg.cm^{-2}.d^{-1}, i.e. approximately 0.25 cm.a^{-1}. Thus the sedimentary layer displaying the highest PCBs concentration levels (7-9 cm) corresponds to the early 50's. Since then concentrations appear to have tended downwards. This interpretation of the profiles rests on the assumptions of a relative immobility of PCBs in the sediments and of a bioturbated layer of less than 8 cm. The latter assumption is verified by a number of core samples from the area (Silverberg et al., 1986).

3.1.2. Mirex

Mirex (dodecachloropentacyclo 5,2,1,0,0,0, decane) is a relatively insoluble and highly persistent insecticide the use of which has been banned since 1972 due to its high toxicity. Large quantities appear to be stored in the sediments of Lake Ontario (Holdrinet et al., 1978). Desjardins et al. (1983b) called attention to the contamination of eels by Mirex in the basin of the St. Lawrence. All eels from Lake Ontario are contaminated with Mirex (mean concentration 0.18 μg.g^{-1}). In the St Lawrence River the sedentary eels are only very slightly contaminated while the opposite is true for migratory eels.These authors conclude that Mirex contamination in St. Lawrence River eels originates from Lake Ontario. Based upon these observations, a technique for identifying fish stock has been suggested (Dutil et al., 1985).

Lum et al. (1987b) have assessed the relative importance of migratory eels and of suspended matter in the transfer of Mirex between Lake Ontario and the St. Lawrence Estuary. Migratory eels appear to carry 2.27 kg.a^{-1}, approximately twice as much as in suspended particulate matter (1.37 kg.a^{-1}). A mass balance calculation indicates that the combination and persistance of these two transport processes could lead to the decontamination of Lake Ontario's surface sediments in 100 years at the most.

3.1.3. DDT and Volatile Halogenated Hydrocarbons

--- DDT and metabolites:

pp'DDT (dichlorodiphenyltrichloroethane) and its degradation by-products (pp'DDD - pp'DDE) have been measured in the fish of the St. Lawrence Basin. According to Couillard (1982) the most contaminated fish (up to 0.5 μg.g^{-1}) are to be found in the upstream section of the River. Elsewhere in the River, with the exception of the Sorel area, levels in fish are below 0.2 μg.g^{-1}. The fact that the use of these products has been discontinued in Canada and the United States undoubtedly accounts for this situation.

Besides the inputs from the St. Lawrence River, the Estuary may have been subjected to inputs of DDT when it was used intensively in forests. In Quebec and New-Brunswick total discharge rates of up to 500 x 10^3 Kg a year had been attained between 1950 and

1960. DDT and metabolite levels in fish of the Estuary range from < 0.008 to 0.6 $\mu g.g^{-1}$; in certain invertebrates from < 0.05 to 0.8 $\mu g.g^{-1}$. As in the case of PCBs, the highest contents are found in the oils of fish (herring and cod) and of seals : 7.2 to 15.8 $\mu g.g^{-1}$ (Couillard, 1982).

The total DDT levels in the blubber of belugas vary from a few $\mu g.g^{-1}$ among females to 22.5 $\mu g.g^{-1}$ in a one-year-old male whose carcass was recovered in 1982 (Martineau et al., in press). The authors make the same comments concerning the behaviour of pp'DDT and its metabolites in belugas as were mentioned earlier in connection with PCB's (viz differences according to size and sex and transfer from mother of calf).

In one sediment core from the Lower Estuary (Fig. 7) surface levels of the metabolites pp'DDE and pp'DDD) are approximately 0.4 $ng.g^{-1}$. Levels increase with depth up to 7-11 cm where they reach a peak of 1.6 to 1.8 $ng.g^{-1}$ and then decline. The subsurface maximum corresponds to the period when DDT dusting in the forests of Quebec was at its height. One the other hand, throughout the entire sedimentary layer under study, the pp'DDT concentrations are below the detection limit.

TABLE 9.
Concentration range of volatile halogenated hydrocarbons (VHH) in the waters of St. Lawrence River (ng. L^{-1}). Data from Lum and Kaiser (1986).

VHH	Concentration
Carbon tetrachloride	10 - 100
Tetrachloroethylene	50 - 500
Chlorofluoromethane	20 - 100
1,1,1, - Trichloroethane and Trichloroethylene	0 - 50
Chloroform	20 - 100

--- Volatile Halogenated Hydrocarbons:
Lum and Kaiser (1986) measured a few volatile halogenated hydrocarbons in the St. Lawrence River (Table 9). The presence of this type of compound in the Estuary has not as yet been the subject of a published report. The same applies to dioxins and benzofurans, whose sources (e.g. incinerators, wood processing industry) are being classified on a Canadian level (Sheffield,1985). The high toxicity of these types of molecule makes them a subject for priority research.

3.2. Polycyclic Aromatic Hydrocarbons (PAHs)

PAHs are mainly generated by incomplete combustion, both natural and anthropogenic (forest and fossil fuel). Their occurrence in the aquatic environment has been estimated to result mainly from anthropogenic sources (Hase and Hites, 1977).

A broad study of the background levels of dispersed/dissolved petroleum residues was carried out in the Gulf of St. Lawrence during the period 1970-79 (Levy, 1985). The measurements were performed using the fluorescence of a carbon tetrachloride extract of the seawater samples and provide information about the relative concentration of polycyclic aromatic compounds. The results show that a significant decrease of contamination occurred in the mid 70's. The most recent levels are similar to those in the areas off the east coast of Canada, regarded as unpolluted. The author also shows that the waters from the Atlantic Ocean entering through the Cabot Strait are the main source of petroleum residues in the Gulf. However, at the upper reaches of the St. Lawrence Estuary and the Saguenay Fjord, inputs of petroleum residues from landbased sources is perceptible.

Using mussels as sentinel organisms, Cossa et al. (1983) pointed out that PAH levels present in most of the mussels of the Estuary reflect the natural background. These findings are consistent with seawater measurements made by Keizer et al. (1977). These authors suggest a biogenic origin for the hydrocarbons measured by fluorescence in the area. However, two mussel beds, located at the mouth of the Saguenay Fjord, present higher PAH levels. In those samples, the examination of the mass spectra of chromotographically separated compounds allowed the identification of several unsubstituted PAHs (Cossa et al., 1983). Such compounds are thought to be tracers of pyrolytic origin.

In a further study, Picard-Berubé et al. (1983) analyzed blue mussels taken in the Estuary and Gulf for their 3, 4 benzopyrene contents. Mussels from the mouth of the Saguenay Fjord exhibit high concentrations, up to 28 ng.g^{-1}. A mussel transplantation experiment performed in 1978 showed that the entire Fjord was contaminated by 3,4 benzopyrene. The authors suggest that the aluminium smelter in Arvida, which at the time still used the Soderberg process (baking coke for making carbon electrodes), was probably responsible for the contamination of the Fjord.

In order to determine the extent of PAH contamination, Martel et al. (1986) studied the distribution of different PAHs in the sediments of the Saguenay Fjord. Their results indicate that the PAH concentrations are 9 to 80 times higher than the background level of uncontaminated sediments. The spatial distribution showed that PAH concentrations in surface sediments reach a maximum near the aluminium smelting plant and decrease with distance from this zone. In addition, the ratios between substituted and unsubstituted PAHs suggest a pyrolytic origin for the sedimentary PAHs.

In a subsequent study, Martel et al. (in press) described the temporal variations and estimated the fluxes of PAHs to the sediments of the Fjord. The general increase in concentration towards the top of the sediment cores indicates that the inputs of PAHs have increased considerably since 1930. However, a decrease in the concentration in the Fjord sediments was noted for the last decade. Martel et al. (in press) estimated the total PAHs fluxes to the Fjord sediments to be 170 kg.a^{-1} in 1900 and 868 kg.a^{-1} in 1980 with a peak up to 1 373 kg.a^{-1} around 1968. The use of precooked electrodes instead of unbaked ones in the aluminium smelter is probably responsible for the substantial reduction of the PAHs emission of the plant and consequently the PAH input in the Fjord.

Troczynski et al. (1988) analyzed the PAH composition of suspended particulate matter (SPM) collected in 1986 in the Saguenay Fjord and St. Lawrence Estuary. In the Fjord, total PAH concentrations decreased from 3.4 µg.g^{-1} upstream near the industrial center to 0.6 µg.g^{-1} seaward. The SPM in the St. Lawrence Estuary have a similar concentration range. The levels correspond to what it is reported for other nearshore waters close to industrialized areas (Puget Sound, Loire Estuary, Mediterranean).

The extension of this contamination to the Laurentian Trough deserves study. The only results available indicate that 90 % of particulate hydrocarbons in the Lower Estuary exhibit patterns typical of anthropogenic pollutants (Tronczynski et al., 1987). This finding questions the conclusion of Levy (1985), that the inputs of petroleum residues from land-based sources are highly localized.

4. Conclusions

The mercury contamination of the St. Lawrence Estuary and of the Saguenay Fjord caused a significant concern in the 70's. The contamination by this metal has decreased considerably over the past 15 years. However, the mercury trapped in the sediments still represents a source of contamination for the neighbouring ecosystem, and is particularly perceptible in the Saguenay Fjord.

The levels and inputs from other potentially toxic metals (e.g. lead) denote an anthropogenic influence but have apparently never reached critical thresholds for marine life.

Contamination by organic substances, more particularly by halogenated substances is poorly documented. However, available data tend to show a significant contamination by organochlorines and polycyclic aromatic hydrocarbons in the Estuary and Saguenay Fjord. Few attempts have been made to assess the biological effects of those compounds at the concentration levels found in the area. In the late 60's it appeared that the pp'DDE levels in the Gulf of the St. Lawrence were high enough to contribute to an abnormally low productivity of Bonaventure gannets (Pearce et al., 1973). Since then, in spite of an apparent decrease in DDT, PCB and PAH inputs, it is clear that the levels in some sectors remain high and that concentration levels for an appreciable number of the most toxic organic contaminants remain unknown. The consequences of this situation are difficult to assess. Some authors have expressed the opinion that organochlorines might be contributory to observed diseases of the St. Lawrence beluga whales (Martineau et al. 1985, 1986, in press). Martineau et al. (in press) also suggest that PAHs probably contaminate the beluga food web and have to be considered as a possible aetiological factor for the high frequency of tumors seen in this population. However, Geraci et al. (1987) have pointed out that it is difficult to conclude that a single substance is to be incriminated and that a variety of other tumor-promoting and enhancing substances do exist. It is important that presence of such substances and their concentration levels in the St.Lawrence System be determined.

Monitoring of some biochemical indices in sedentary fish (e.g. EROD - ethoxyresorufine deethylas - and acetylcholinesterase activities) might constitute another step toward a better assessment of the impact of contaminants on marine life in the St. Lawrence Estuary and Gulf. Whatever the methodological choice, further research to assess the possible links between contamination levels and biological effects are needed.

Acknowledgments

Thanks are due to Y. Dansereau, C. Desjardins, E. Fedida, C. Gobeil, D. Green, M. Pelletier and M. Riverain for providing unpublished information very useful to the present paper. The help of L. Giboire is acknowledged for drawing the illustrations and M. Uhr for the typing. Thanks are also due to Drs. J. Gearing, E. Pelletier and N. Silverberg for their comments on the early manuscript.

REFERENCES

Barbeau, C., R. Bougie and J.E. Coté. 1981 a. Temporal and spatial variations of mercury, lead, zinc and copper in sediments of the Saguenay fjord. Can. J. Earth Sci., 18: 1065-1074.

Barbeau, C., R. Bougie and J.E. Coté. 1981 b. Variations spatiales et temporelles du césium 137 et du carbone dans les sédiments du fjord du Saguenay. Can. J. Earth Sci., 18 : 1004-1011.

Belzile, N. 1988. The fate of arsenic in sediments of the Laurentian trough.Geochim. Cosmochim. Acta, 52: 2293-2302.

Bewers, J.M. and P.A. Yeats. 1977. Oceanic residence times of trace metals. Nature, Lond., 268 : 595-598.

Bewers, J.M. and P.A. Yeats. 1978. Trace Metals in the Waters of a Partially Mixed Estuary. Est. Cstl. Mar. Sci., 7 : 147-162.

Bewers, J.M. and P.A. Yeats. 1979. The behavior of trace metals in estuaries of the St. Lawrence basin. Naturaliste can., 106 : 149-161.

Bewers, J.M., B. Sundby and P.A. Yeats. 1976. The distribution of trace metals in the western North Atlantic off Nova Scotia. Geochim. Cosmochim. Acta, 40 : 687-696.

Bewers, J.M., D.H. Loring, K. Kranck, G.H. Seibert, R.L. Charron, J.F. Uthe,C.L. Chou and D.G. Robinson. 1987. Cadmium pollution associated with coastal lead-smelting plant. Chap. II, p. 117-132. In : Oceanic Processes in Marin Pollution. Vol. 2. O'Connor, Burt and Duedall (eds.). R.E. Krieger Pub. Comp., Malabar, FL (USA).

Boyle, E.A., S.S. Huested and B. Grant. 1982. The chemical mass balance of the Amazon Plume - II. Copper, nickel, and cadmium. Deep Sea Res., 29 A : 1355-1364.

Canadian Academy of Science. 1977. Vue d'ensemble de la contamination par le mercure au Canada. Suppl. Québec Science, Oct. 1977.

Campbell, J.A. and P.A. Yeats. 1982. The distribution of manganese, iron, copper and cadmium in the waters of Baffin Bay in the Canadian Artic Archipelago. Oceanol. Acta, 5: 161-168.

Coote, A.R. and P.A. Yeats. 1979. Distribution of Nutriments in the Gulf of St. Lawrence. J. Fish. Res. Board Can., 36: 122-131.

Cossa, D. 1976. Sorption du cadmium par une population de la diatomée *Phaeodactylum tricornutum* en culture. Mar. Biol., 34: 163-167.

Cossa, D. 1980. Utilisation de la moule bleue comme indicateur du niveau de pollution par les métaux lourds et les hydrocarbures dans l'estuaire et le golfe du Saint-Laurent. Rapport INRS-océanologie (Université du Québec) SI - 43600/00. 74 pp.

Cossa, D. 1988. Cadmium behavior in the waters of the St. Lawrence estuary. International Symposium on Fate and Effects of Toxic Chemicals in Large Rivers and Estuaries. Québec (Canada), Oct. 10-14th, 1988.

Cossa, D. and S.A. Poulet. 1978. Survey of Trace Metal Contents of Suspended Matter in the St. Lawrence Estuary and Saguenay Fjord. J. Fish. Res. Board Can., 35 : 338-345.

Cossa, D. and E. Bourget. 1980. Trace element in *Mytilus edulis* L. from the estuary and gulf of St. Lawrence, Canada : lead and cadmium concentrations. Environ. Pollut. Serie A, 23 : 1-8.

Cossa, D. and G. Tremblay. 1983. Major ions composition of the St. Lawrence River : Seasonal Variability and Fluxes. In : Transport of Carbon and Minerals in Major World Rivers, Part 2. Degens, E.T., S. Kempe and H. Soliman (eds.). Mitt. Geol.- Palaeont. Inst. Univ. Hamburg, SCOPE/UNEP Sonberbd., 55 : 253-259.

Cossa, D. and C. Desjardins. 1984. Evolution de la concentration en mercure dans les crevettes du fjord du Saguenay (Québec) au cours de la période 1970-83. Rapp. Techn. Can. Hydrogr. Sci. océan. n° 32. 8 p.

Cossa, D. and J.G. Rondeau. 1985. Seasonal, geographical and size-induced variability in mercury content of *Mytilus edulis* in an estuarine environment : a re-assessment of mercury pollution level in the estuary and gulf of St. Lawrence. Mar. Biol., 88 : 43-49.

Cossa, D. and J. Noel. 1987. Concentrations of mercury in near shore surface waters of the Bay of Biscay and in the Gironde Estuary. Mar. Chem., 20: 389-396.

Cossa, D. and P. Lassus. 1988. Le cadmium en milieu marin : biogéochimie et écotoxicologie. Rapp. Sci. Techn. IFREMER n° 16 ; SDP, Brest (France). 112 pp.

Cossa, D., M. Picard-Bérubé and J.P. Gouygou. 1983. Polynuclear Aromatic Hydrocarbons in Mussels from the Estuary and Northwest Gulf of St. Lawrence, Canada. Bull. Environ. Contam. Toxicol., 31 : 41-47.

Cossa, D., C. Gobeil and P. Courau. 1988a. Dissolved Mercury Behaviour in the St. Lawrence Estuary. Est. Cstl. Shelf Sci., 26: 227-230.

Cossa, D., G.H. Tremblay and C. Gobeil. 1988 b. Variation saisonnière des apports fluviatiles en Al, Fe et Mn à l'estuaire du St-Laurent. International Symposium on Fate and Effects of Toxic Chemicals in Large Rivers and Estuaries. Québec (Canada), Oct. 10-14th, 1988.

Couillard, D. 1982. Evaluation des teneurs en composés organochlorés dans le fleuve, l'estuaire et le golfe Saint Laurent, Canada. Environ. Pollut. (Serie B), 3 : 239-270.

Dawson, R. and J.P. Riley. 1977. Chlorine-containing pesticides and polychlorinated biphenyls in British coastal waters. Est. Cstl. Mar. Sci., 4 : 55-69.

Desjardins, C., J.D. Dutil and R. Gélinas. 1983 a. Contamination de l'anguille (*Anguilla rostrata*) du bassin du fleuve Saint-Laurent par les biphényles polychlorés. Rapp. can. Ind. Sci. halieut. 144, v + 56 pp.

Desjardins, C., J.D. Dutil and R. Gélinas. 1983 b. Contamination de l'anguille (*Anguilla rostrata*) du bassin du fleuve Saint-Laurent par le Mirex. Rapp. can. Ind. sci. halieut. aquat., 141 : v + 52 pp.

Dubé, J. 1982. Etude de la distribution de quelques métaux dans le zooplancton de deux écosystèmes du Saint-Laurent. Mémoire de Maîtrise ès-science (océanographie). Université du Québec à Rimouski. 48 pp. + annexes.

Dutil, J.D., B. Légaré and C. Desjardins. 1985. Discrimination d'un stock de poisson, l'anguille (*Anguilla rostrata*), basée sur la présence d'un produit chimique de synthèse, le mirex. Can. J. Fish. Aquat. Sci., 42 : 455-458.

Elbaz-Poulichet, F., W.W. Huang, J.M. Martin and J.X. Zhu. 1987. Dissolved cadmium behaviour in some selected french and chinese estuaries. Mar. Chem., 22: 125-136.

Eggimann, D.W. and P.R. Betzer. 1976. Decomposition and Analysis of Refractory Oceanic Suspended Materials. Anal. Chem., 48: 886-890.

Gendron, A., B. Sundby, N. Silverberg and J. Lebel. 1986. Early diagenesis of cadmium and cobalt in sediments of the Laurentian trough. Geochim. Cosmochim. Acta, 50: 741-748.

Geraci, J.R., N.C. Palmer and D.J. St Aubin.1987. Tumors in cetaceans : analysis and new findings. Can. J. Fish. Aquat. Sci., 44: 1289-1300.

GESAMP. 1986. (IMO/FAO/UNESCO/WMO/WHO/IAEA/VN/UNEP. Joint Group of Experts on the Scientific Aspects of Marine Pollution). Review of Potentially Harmful Substances : Arsenic, Mercury and Selenium : Rep. Stud. GESAMP No 28.

Giam, C.S., H.S. Chan, G.S. Neff and E.F. Atlas. 1978. Phtalate ester plasticizers : a new class of marine pollutant. Science, 199 : 419-420.

Gobeil, C. and D. Cossa. 1984. Profils des teneurs en mercure dans les sédiments et les eaux interstitielles du fjord du Saguenay (Québec) : données acquises au cours de la période 1978-83. Rapp. Techn. Can. Hydrogr. Sci. océan. No 53 : vi + 23 p.

Gobeil, C. and D. Cossa. 1988. Mercury in sediment porewater of the Lower St Lawrence Estuary. International Symposium on Fate and Effects of Toxic Chemicals in Large Rivers and Estuaries. Québec (Canada). Oct. 10-14 th, 1988.

Gobeil, C. and N. Silverberg. 1989. Early diagenesis of lead in Laurentian trough sediments. Geochim. Cosmochin. Acta, 53: 1889-1895.

Gobeil, C., D. Cossa and J. Piuze. 1983. Distribution des concentrations en mercure dans les eaux de l'estuaire moyen du Saint-Laurent. Rapp. Techn. Can. Hydrogr. Sci. océan. n° 17 : iii + 14 pp.

Gobeil, C., D. Cossa and J. Piuze. 1984. Niveaux des concentrations en mercure dans les eaux du fjord du Saguenay (Québec) en octobre 1979 et septembre 1983. Rapp. Tech. Can. Hydrogr. Sci. océan. n° 36 : v + 14 pp.

Gobeil, C., N. Silverberg, B. Sundby and D. Cossa. 1987. Cadmium diagenesis in Laurentian trough sediments. Geochim. Cosmochim. Acta, 51 : 589-596.

Hase, A. and R.A. Hites. 1977. Identification of organic pollutants in water. Keith, L.H. (ed.). Ann Arbor Sci. Publish., Ann Arbor (USA).

Holdrinet, M., R. Franck, R.L. Thomas and L.J. Hetling. 1978. Mirex in the sediments of Lake Ontario. J. Great Lakes Res., 4 : 69-74.

Keizer, P.D., D.C.Jr. Gordon and J. Dale. 1977. Hydrocarbons in Eastern Canadian Marine Waters Determined by Fluorescence Spectroscopy and Gas-Liquid Chromatography. J. Fish. Res. Board Can., 34: 347-353.

Levy, E.M. 1985. Background Levels of Dissolved/Dispersed Petroleum Residues in the Gulf of St. Lawrence, 1970-79. Can. J. Fish. Aquat. Sci., 42: 544-555.

Loring, D.H. 1975. Mercury in the sediments of the gulf of St. Lawrence. Can. J. Earth Sci., 12: 1219-1237.

Loring, D.H. 1976a. The distribution and partition of zinc, copper and lead in the sediments of the Saguenay fjord. Can. J. Earth Sci., 13: 960-971.

Loring, D.H. 1976b. The distribution and the partition of cobalt, nickel, chromium and vanadium in the sediments of the Saguenay fjord. Can. J. Earth Sci., 13, 1706-1718.

Loring, D.H. 1978a. Industrial and natural inputs, levels, behavior, and dynamics of biologically toxic heavy metals in the Saguenay fjord, gulf of St. Lawrence, Canada. In : Proceedings of the IIIth Intern. Symp. Environ. Biogeochem. Vol. 3 ; p. 1025-1040. Krumbein, W.E. (ed.) Ann Arbor Sci. Publish., Ann Arbor (USA).

Loring, D.H. 1978b. Geochemistry of zinc, copper and lead in the sediments of the estuary and open gulf of St. Lawrence. Can. J. Earth Sci., 15 : 757-772.

Loring, D.H. 1979. Geochemistry of cobalt, nickel, chromium and vanadium in the sediments of the estuary and open gulf of St. Lawrence. Can. J. Earth Sci., 16 : 1196-1209.

Loring, D.H. and J.M. Bewers. 1978. Geochemical mass balances for mercury in a Canadian fjord. Chem. Geol., 22 : 309-330.

Loring, D.H., R.T.T. Rantala and J.N. Smith. 1983. Response Time of Saguenay Fjord Sediments to Metal Contamination. Environ. Biogeochem., 35 : 59-72.

Lum, K.R. 1987. Cadmium in Fresh Waters: The Great Lakes and St. Lawrence River. p. 35-50. In : Cadmium in the Aquatic Environment. Nriagu, J.O.and J.B. Sprague (eds.) J. Wiley and Sons New-York (USA).

Lum, K.R. and J.K. Leslie. 1983. Dissolved and particulate metal chemistry of the central and eastern basins of Lake Erie. Sci. Total Environ., 30 : 99-109.

Lum, K.R. and M. Callaghan. 1986. Direct determination of cadmium in natural waters by electrothermal atomic absorption spectrometry without matrix modification. Anal. Chim. Acta, 187 : 157-162.

Lum, K.R. and K.L.E. Kaiser. 1986. Organic and inorganic contaminants in the St. Lawrence River : some preliminary results on their distribution. Water Poll. Res. J. Canada, 21 : 592-602.

Lum, K.R., M. Callaghan and P.R. Youakim. 1986. Dissolved and particulate cadmium in the St. Lawrence River. In : Proceedings 2 nd Intern. Conf.Environ. Contam. Amsterdam (NL). Sept., 1986 pp. 189-191.

Lum, K.R., P.R. Youakim and C. Jaskot. 1987 a. Geochemical Availability of Lead and Zinc in the Upper Estuary of the St. Lawrence River. 6th Intern. Conf. on Heavy Metals in the Environment. 15-18 Sept. 1987, New-Orleans (USA).

Lum, K.R., K.L.E. Kaiser and M.E. Comba. 1987 b.Export of Mirex from Lake Ontario to the St. Lawrence Estuary. The Sci. Total Environ., 67 : 41-51.

McBeath, I. 1986. Status Report on Compliance with the Chlor-Alkali Mercury Regulations 1984-85. Reports EPS 1/CI/1. Environment Canada. 41 pp.

Martel, L., M.J. Gagnon, R. Massé, A. Leclerc and L. Tremblay. 1986. Polycyclic Aromatic Hydrocarbons in Sediments from the Saguenay Fjord,Canada. Bull. Environ. Cont. Toxicol., 37 : 133-140.

Martel, L., M.J. Gagnon, R. Massé, and A. Leclerc. (in press). The spatio-temporal variations and fluxes of polycyclic aromatic hydrocarbons in the sediments of the Saguenay fjord, Québec, Canada. Water Res. (in press).

Massé, R., D. Martineau, L. Tremblay and P. Béland. 1986. Concentrations and chromatographic profiles of DDT metabolites and polychlorobiphenyl (PCB) residues in stranded Beluga whales (*Delphinapterus leucas*) from the St. Lawrence Estuary, Canada. Arch. Environ. Contam. Toxicol., 15: 567-579.

Martineau, D., A. Lagacé, R. Massé, M. Morin and P. Béland. 1985. Transitional cell carcinoma of the urinary bladder in a Beluga whale *Delphinapterus leucas* . Can. Vet. J., 26: 297-302.

Martineau, D., A. Lagacé, P. Béland and C. Desjardins. 1986. Rupture of a dissecting aneurysm of the pulmonary trunk in a Beluga whale (*Delphinapterus leucas*). J. Wildlife Disease, 22: 289-294.

Martineau, D., P. Béland, C. Desjardins and A. Lagacé. 1987. Levels of organochlorine chemicals in the tissues of Beluga whales (*Delphinapterus leucas*) from the St. Lawrence Estuary, Québec, Canada.Arch. Environ. Contam. Toxicol., 16: 137-147.

Martineau, D., A. Lagacé, P. Béland, R. Higgins, D. Armstrong and L.R. Shugart (in press). Pathology of stranded beluga whales (*Delphinapterus leucas*) from the St. Lawrence estuary, Québec, Canada. J. Comp. Pathol.

Moreau, G. and C. Barbeau. 1982. Les métaux lourds comme indicateurs d'origine géographique de l'anguille d'Amérique (*Anguilla rostrata*). Can. J. Fish. Aquat. Sci., 39: 1004-1011.

Muhlkaler, J., C. Stevens, D. Graczyk and T. Tisue. 1982. Determination of cadmium in Lake Michigan by mass spectrometry following electrodeposition. Anal. Chem., 54: 496-499.

Nriagu, J.O. 1980. Cadmium in the Environment. Part I : Ecological Cycling. J. Wiley and Sons. New-York (USA). 682 pp.

Olafsson, J. 1983. Mercury concentrations in the North Atlantic in relation to cadmium, aluminium and oceanographic parameters. In : Trace Metals in Sea Water. Wong, C.S. et al. (eds). NATO Conf. Ser. Plenum Press, New York (USA). 920 p.

Osterroht, C. and V. Smetacek. 1980. Vertical transport of chlorinated hydrocarbons by sedimentation of particulate matter in Kiel Bight. Mar. Ecol. Progr. Series, 2 : 27-34.

Pearce, P.A., I.M. Gruchy and J.A. Keith. 1973. Toxic Chemicals in Living Things in the Gulf of St. Lawrence. Manuscript Report n° 24. Canadian Wildlife Service. Presented at the joint CSWFB/CSZ symposium: Renewable Resource Management of the Gulf of St. Lawrence; Jan., 5th, 1973, Halifax, N.S.

Pelletier, E. and G. Canuel. 1988. Trace metals in surface sediment of the Saguenay Fjord, Canada. Mar. Pollut. Bull., 19: 336-338.

Pelletier, E., C. Rouleau and G. Canuel. (in press). Niveau de contamination par le mercure des sédiments de surface et de crevettes du fjord du Saguenay en 1985-86. Revue Sci. de l'Eau.

Picard-Berube, M., D. Cossa and J. Puize. 1983. Teneurs en benzo 3, 4 pyrène chez *Mytilus edulis* de l'estuaire et du golfe du Saint-Laurent.Mar. Environ. Res., 10: 63-71.

Piuze, J. and M. Tremblay. 1979. Mercury in the waters of the St. Lawrence Estuary, Québec, Canada. Intern. Council Explor. Sea. C.M. 1989/E: 49, 13 pp.

Poirier, L. and D. Cossa. 1981. Distribution tissulaire du cadmium chez *Meganyctiphanes norvegica* (Euphausiacée) : état naturel et accumulation expérimentale de formes solubles. Can. J. Fish. Aquat. Sci., 38: 1449-1453.

Rossmann, R. 1986. Trace metal concentration in the offshore water of Lake Superior. Draft Report. Great Lake Research Division ; University of Michigan, Ann Arbor (USA). 105 pp.

Shiller, A.M. and E.A. Boyle. 1983. Variability of dissolved trace metals in the Mississippi River. Geochim. Cosmochim. Acta, 51: 3273-3277.

Shafer, C.T., J.N. Smith and D.H. Loring. 1980. Recent sedimentation events at the head of the Saguenay fjord. Environ. Geol., 3: 139-150.

Sheffield, A. 1985. Polychlorinated dibenzo-p-dioxins (PCDDs) and polychlorinated dibenzofurans (PCDFs): sources and releases. Report EPS 5/HA/2. Environment Canada, Environmental Protection Service. Ottawa (Ont.). 47 pp.

Silverberg, N., H.V. Nguyen, G. Delibrias, M. Koide, B. Sundby, Y. Yokoyama and R. Chesselet. 1986. Radionuclides profiles, sedimentation rates and bioturbation in modern sediments of the Laurentian trough, Gulf of St. Lawrence. Oceanol. Acta, 9: 285-290.

Sloterdijk, H. 1986. A propos du fleuve et des substances toxiques. Milieu, 33 : 15-18.

Smith, J.N. and D.H. Loring. 1981. Geochronology for mercury pollution in the sediments of the Saguenay fjord, Québec. Environ. Sci. Technol., 15 : 944-951.

Tremblay, G.H. 1985. Variations temporelles des concentrations en ions majeurs du fleuve Saint-Laurent et évaluation de l'apport fluvial dans la zone estuarienne. Mémoire de Maîtrise (océanographie). Université du Québec à Rimouski, Québec. 165 p.

Trites, R.W. 1972. The gulf of St. Lawrence from a pollution point of view. In : Marine Pollution and Sea Life. pp. 59-72. Ruivo, M. (ed.) Fishing News Books, Lond. (UK). 624 pp.

Tronczynski, J., J.N. Gearing and S. Macko. 1987. Characterization of selected organics on particles and sediments in the St. Lawrence estuary. 21st Annual Congress Canadian Meteorological and Oceanographic Society, June 16-19, 1987. St John's, NFL (Canada).

Tronczynski, J., J.N. Gearing, S. Ouellet and E. Pelletier. 1988. Hydrocarbon distribution in the Saguenay Fjord, Québec, Canada. International Symposium on Fate and Effects of Toxic Chemicals in Large Rivers and Estuaries. Québec (Canada), Oct. 10-14 th, 1988.

Uthe, J.F. and V. Zitko (eds). 1980. Cadmium pollution of Belledune Harbour, New Brunswick, Canada. Can. Tech. Rep. Fish. Aquat. Sci., 963 v + 107 p.

Wilson, R.C.H. and R.F. Addison. 1984. Health of the Northwest Atlantic. A Report to the Interdepartmental Committee Issues. Department of the Environment/Department of Fisheries and Oceans/Department of Energy, Mines and Resources. Dartmouth (Canada). 174 pp.

Yeats, P.A. and J.M. Bewers. 1976. Trace metals in the waters of the Saguenay fjord. Can. J. Earth. Sci., 13: 1319-1327.

Yeats, P.A. and J.M. Bewers. 1982. Discharge of metals from the St. Lawrence River. Can. J. Earth Sci., 19: 982-992.

Yeats, P.A. and J.M. Bewers. 1983. Potential Anthropogenic Influences on Trace Metal Distribution in the North Atlantic. Can. J. Fish. Aquat. Sci., 40 (suppl. 2): 124-131.

Chapter 12

Oceanography and Ecology of Phytoplankton in the St.Lawrence Estuary [3]

Jean-Claude Therriault[1], Louis Legendre[2] and Serge Demers[1]

[1] Institut Maurice-Lamontagne, Ministère des Pêches et des Océans, 850 Route de la Mer, Mont-Joli, Québec, Canada, G5H 3Z4

[2] Département de biologie, Université Laval, Québec, Québec, G1K 7P4

ABSTRACT

This paper examines, in turn, the influence of a number of oceanographic and ecologic factors on the control of phytoplankton production and biomass in estuarine and coastal environments, with particular attention to studies conducted in the St. Lawrence Estuary. These factors are then invoked to explain the differences between the known subregions of the St. Lawrence ecosystem, which have been defined principally on the basis of phytoplankton biomass and primary production criteria.

Key Words: oceanography, ecology, phytoplankton, St. Lawrence Estuary, freshwater runoff, turbidity, nutrients, temperature, mixing

RESUME

Cet article examine l'influence de plusieurs facteurs écologiques et océanographiques sur le contrôle de la biomasse et de la production phytoplanctonique dans les milieux côtiers et estuariens, avec une attention toute spéciale apportée aux travaux sur l'estuaire du Saint-Laurent. Ces facteurs sont ensuite utilisés pour expliquer les différences entre les sous-régions reconnues dans l'écosystème du Saint-Laurent qui ont été définies principalement en se basant sur des critères de biomasse et productivité phytoplanctonique.

[3] Contribution to the programs of the Maurice Lamontagne Institute (Department of Fisheries and Oceans) and of GIROQ (Groupe interuniversitaire de recherches océanographiques du Québec) .

Coastal and Estuarine Studies, Vol. 39
M. I. El-Sabh, N. Silverberg (Eds.)
Oceanography of a Large-Scale Estuarine System
The St. Lawrence
© Springer-Verlag New York, Inc., 1990

1. Introduction

In the marine environment, phytoplankton biomass and production are generally controlled by several factors such as temperature or salinity which directly affect the organisms, while other factors, such as vertical mixing and turbidity, influence phytoplankton through the proximal agency of light and/or nutrients (Legendre and Demers, 1984). A distinction should be made, however, between oceanic (offshore) and coastal (estuarine and inshore) environments, since each factor will have a different significance or relative importance for phytoplankton, depending on the area under study. For example, freshwater runoff, salinity, turbidity, topography and tides will have a much more pronounced influence on phytoplankton inshore, in an estuary or a coastal embayment, than offshore, in the open ocean. Conversely, wind is bound to play a relatively much more important role offshore than inshore.

In this paper, we examine in turn a number of oceanographic factors which we believe are important for the control of phytoplankton production and biomass in inshore areas. These factors are listed in Table 1, along with our interpretation of their potential influence on phytoplankton dynamics. To illustrate these effects, we shall consider examples from the literature, with special attention paid to studies conducted in the St. Lawrence Estuary. We shall also explain how the physical, chemical and biological factors interact with the topography, so as to determine a number of identifiable sub-regions in the St. Lawrence ecosystem (see Figures 1 to 3) based on phytoplankton biomass and primary production.

2. Environmental Control of Phytoplankton Biomass and Production

2.1. Freshwater runoff

Freshwater runoff is important in estuaries, especially at the time of spring discharges. Several studies have attributed to high runoff (flushing) the low phytoplankton biomasses usually observed in estuaries during the spring. Flushing time was invoked, for example, as a factor regulating phytoplankton growth and occurrence in the Cochin Blackwater, India (Wyatt and Quasim, 1973), the Lower Hudson River (Malone, 1977), the Saguenay Fjord (Côté and Lacroix, 1979), the Lower St. Lawrence Estuary (Levasseur et al., 1984; Therriault and Levasseur, 1985) and several other fjords (Sinclair et al., 1981). The importance of flushing time for the maintenance of phytoplankton populations within estuaries has been examined theoretically by Ketchum (1954) and Margalef (1967).

Therriault and Levasseur (1985; 1986) showed that freshwater runoff exerts a strong influence on the spatial and temporal distributions of phytoplankton at the scale of the whole Lower St. Lawrence Estuary. From their data, we can extrapolate and suggest that at high discharge rates, the whole St. Lawrence system forms a single freshwater plume

TABLE 1.
Environmental factors and their effects on phytoplankton in estuarine and inshore waters.

ENVIRONMENTAL FACTORS	EFFECTS ON PHYTOPLANKTON
Freshwater runoff	Flushing of biomass Seeding of the photic layer Dinoflagellate blooms Spatio-temporal location of bloom
Water temperature	Growth rate Species succession
Salinity	Species distribution Osmotic stress
Turbidity	Biomass Production
Nutrients	Biomass Growth rate
Vertical mixing	Physiology Photosynthesis Growth
-Buoyancy	Production
-Vertical shear	Biomass Photosynthesis Production
-Topography	Distribution of biomass
-Internal waves and tides	Production Biomass Photosynthesis
-wind mixing	Production Distribution of biomass

characterized by an increasing salinity gradient (a typical estuary) which extends from upstream at Ile d'Orléans (freshwater) to the Gulf, and even much farther on the Atlantic Coast (Bugden et al., 1982). Because of high flushing rates, phytoplankton biomass accumulation during this period is restricted to the downstream end of the salinity gradient (Gulf region) as shown by Steven (1974), de Lafontaine et al. (1981) and de Lafontaine et al. (1984). During periods of flow reduction (summer), the impact of freshwater runoff is much more geographically localized, which contributes to the high spatial heterogeneity of phytoplankton biomass and production observed, for example, by Therriault and Levasseur (1985) in the Lower Estuary. Therriault and Levasseur (1986) have pointed out that freshwater plumes have both detrimental and beneficial effects on phytoplankton in the St. Lawrence system, depending on the time and/or region considered. Indeed, because of varying flushing rates, detrimental effects on phytoplankton growth (biomass accumulation) are observed in the Lower Estuary at the same time as beneficial effects (stratification) much farther in the Gulf or the Atlantic coast, and *vice-versa* . On a seasonal time scale, displacement of the spring bloom of phytoplankton is therefore expected to follow the recession of freshwater in the Estuary due to lower discharge rates. This could be one of the reasons why the Lower Estuary blooms late as compared to the Gulf region. This phenomenon has been discussed by Sinclair et al. (1981).

Another effect of increased freshwater runoff (not independent from flushing rate) is decreased eddy exchange between surface and deeper waters (Bugden, 1981). In the spring, this could prevent the seeding of the photic layer by marine diatoms, since their growth depends upon resuspension of resting spores that have sedimented in the fall (Durbin, 1978), and/or advection of cells from an upstream region, and/or upstream transportation by deep currents of cells produced downstream (Margalef, 1958). In such cases, diatom blooms would be delayed by high freshwater runoff. This mechanism has been invoked, for example, to explain late diatom blooms in Puget Sound (Winter et al., 1975), the Lower Hudson River (Malone, 1977), Buzzards Bay (Roman and Tenore, 1978), the plume of the Hudson River (Malone and Chervin, 1979) and the Lower St. Lawrence Estuary (Levasseur et al., 1984; Therriault and Levasseur, 1985).

Several studies have suggested a positive effect of freshwater runoff on the growth of dinoflagellates due to low salinity, high temperature and high levels of nutrients and/or humic substances (see reviews by Prakash, 1975 and Provasoli, 1979). Margalef et al. (1979) have also suggested that freshwater runoff is important for the growth of dinoflagellates, not necessarily for its content in nutrients or humic substances, but for its higher potential of stabilization of the water column. Studies in the St. Lawrence Estuary by Therriault and Levasseur (1985), Therriault et al. (1985), Cembella and Therriault (1988) and Cembella et al. (1988) support this hypothesis for development of blooms of the toxic dinoflagellate *Protogonyaulax tamarensis* and associated shellfish poisoning events in the St. Lawrence Estuary. Their data indicate that developments of *Protogonyaulax* blooms have always been essentially restricted to the region under the direct influence of the freshwater plumes of the Manicouagan and Aux-Outardes rivers and the Gaspé current. It is interesting to note that environmental conditions in the St. Lawrence system are such that on a per cell basis, toxicity of natural *Protogonyaulax* populations is among the highest ever reported for this species (Cembella et al.,1988; Cembella and Therriault, 1988).

In another context, Gilmartin (1984) was able to show that freshwater runoff was entirely driving the seasonal dynamics of phytoplankton by affecting the vertical stability of the waters from Indian Arm, a fjord from western Canada.

2.2. Water temperature

There exists a well-known relationship between phytoplankton growth and water temperature (Eppley, 1972) according to which growth rate almost doubles for each 10°C increase. This is only true, however, below an optimum temperature, which varies from species to species and above which growth rate drops rapidly. Accordingly, abundances of some phytoplankters (e.g. μ-flagellates) in estuarine and inshore waters are often associated with year-round variations in water temperature (e.g., Briant, 1975; Durbin et al., 1975; Holligan and Harbour, 1977; Malone and Chervin, 1979; Walting et al., 1979; Malone and Neale, 1981). For the Lower St. Lawrence Estuary, Sinclair (1978) has examined the relationship between surface temperature and phytoplankton dynamics. He found that although no temperature effect could be detected in the field, a marked effect of temperature could be observed during shipboard incubation experiments to determine the potential growth rate (μ_{pot}) and assimilation number (P/B) of the phytoplankton. Generally, higher growth rate and assimilation number values were associated with higher temperatures. In another study, Levasseur et al. (1984) found a significant relationship between surface temperature and μ-flagellate numbers. Their results also suggested that long-term temperature variations have an important effect on the control of diatom succession. This led them to propose a hierarchical model of the relative importance of the different physical factors for the control of phytoplankton succession. In this model, the most important factor is the frequency of destabilization of the water column, which determines nutrient limitation in the mixed layer and consequently selects a range of growth rates. Then, the mean light intensity in the mixed layer, which is dependent upon solar radiation and density stratification cycles, determines the presence or absence of diatoms in the mixed layer. Finally, the temperature cycle sets the conditions for optimal metabolic activities, to which the phytoplankton community responds by a shift or a succession of species within the range of possible growth rates.

On the other hand, different species, as suggested above, exhibit maximum growth at different temperatures, which is also true for the formation of resting spores that occurs at different temperatures for different species. It follows that seasonal temperature variations in estuarine and inshore waters may be critical for species succession, as suggested for diatoms by Malone and Neale (1981) for different coastal environments, and by Levasseur et al. (1984) for the Lower St. Lawrence Estuary.

2.3. Salinity

Estuaries are the site of strong salinity gradients, both vertically and horizontally. However, salinity generally co-varies in estuaries with other environmental factors such as temperature, turbidity, stratification, etc. Therefore, it is not easy to demonstrate specific responses of phytoplankton to salinity gradients. Gessner and Schramm (1971), Hellebust (1976) and Yancey et al. (1982) have reviewed the effects of salinity on marine plants, including phytoplankton. There exist typically freshwater and salt water phytoplankton communities, but a typically brackish water community does not seem to exist. In general, brackish waters are mostly colonised by marine species, which suggests that increasing salinity probably has a detrimental effect on freshwater phytoplankton species. It is not clear, however, whether freshwater species in estuaries disappear seaward due to increasing salinity (osmotic stress), competition with marine phytoplankton (ecological effect) or sedimentation processes. For example, Morris et al. (1978) observed in the Tamar Estuary that minima of O_2, chlorophyll fluorescence and maxima of dissolved organic carbon coincided with salinities ≈ 1 $^o/oo$. They concluded that a mass mortality of freshwater halophobic phytoplankton wasdue to osmotic changes. Other examples of a shift from freshwater to marine phytoplankton species when passing from salinity 0 to about 10 $^o/oo$ are abundant in the literature. In the St. Lawrence Estuary, such a community shift was observed by Cardinal and Bérard-Therriault (1976) and Cardinal and Lafleur (1977) in the Upper St. Lawrence Estuary. Painchaud and Therriault (1985, 1989) and Painchaud et al. (1987) have also observed drastic changes in chlorophyll concentrations, from typically freshwater to typically marine waters in the St. Lawrence Estuary. They also found that the limit of salinity intrusion in the Upper Estuary (0 to 10 $^o/oo$) constitutes a sharp ecological barrier between two distinct bacterial and phytoplankton communities: the freshwater community on one side, and the estuarine community on the other side. From empirical measurements and laboratory experiments, Painchaud and Therriault (1985) and Painchaud et al. (1987) also showed that the growth and activity of freshwater bacteria was seriously impeded by exposure to increasing salinity. However, no such work has been carried out to look experimentally at the direct effect of salinity on phytoplankton in the St. Lawrence Estuary. It is of interest to note, however, that as with the phytoplankton, the brackish waters were colonised mostly by marine species. It seems, therefore, that marine phytoplankton as well as bacterial organisms are more tolerant to decreasing salinities than freshwater species to increasing salinities.

2.4. Turbidity

As mentioned above, hydrodynamics generally acts on phytoplankton through the agency of light and nutrients. In estuarine and inshore waters, one of the major determinants for light attenuation in the upper water column is turbidity. According to the now classical model of Riley (1942) and Sverdrup (1953), a bloom occurs only when the surface mixed layer becomes shallower than a certain "critical depth", above which irradiance is high enough for carbon fixation by phytoplankton to exceed respiration per unit area. This model assumes that nutrients are nonlimiting at the time of the bloom, and that seed cells are present in the upper water column. Given these assumptions, there are three factors that set the average irradiance ($<I>$) in the mixed layer: solar irradiance, depth of the surface mixed layer, and turbidity of the surface waters. When $<I>$ exceeds a critical threshold (e.g., Riley 1957), a bloom occurs. Variations in solar irradiance will not be discussed here since this factor is under meteorological control. Vertical mixing, on the other hand, will be discussed later in this review.

Turbidity is a very important factor in the phytoplankton dynamics for the whole St. Lawrence Estuary. It is the dominant factor in the Upper Estuary where a turbidity maximum is maintained by tidal current asymmetry and flocculation (Kranck, 1979). The dynamics of this turbidity maximum (or turbidity front) was studied by d'Anglejan and Smith (1973), d'Anglejan and Ingram (1976) and Ingram and d'Anglejan (1977). Kranck (1979) showed that preferential deposition of organic as opposed to inorganic particles occurs in this turbidity maximum, depleting the particulate matter of its organic fraction and of the associated potential pollutants. This turbidity front plays a significant role in the phytoplankton and bacterial distribution dynamics of the Upper St. Lawrence Estuary (Cardinal and Bérard-Therriault, 1976; Cardinal and Lafleur, 1977; Painchaud and Therriault, 1985; Painchaud et al., 1987). In the Lower Estuary, Sinclair (1978) as well as Fortier and Legendre (1979) have invoked long-term (seasonal and fortnightly) changes in turbidity to explain differences in phytoplankton biomass and production. Levasseur et al. (1984) have used the critical depth concept, with particular reference to turbidity, to explain the seasonal variations of phytoplankton biomass and production at one station in the Lower Estuary. Therriault and Levasseur (1985) were also able to determine spatial pattern in water masses characteristics using turbidity as one of the most important factors.

2.5. Nutrients

The growth rate of phytoplankton can be limited, in addition to light, by nutrient concentrations. These can also limit the final yield of phytoplankton. In estuarine and inshore waters, an important aspect of nutrients concerns their sources, since continental waters do not generally contain the same element(s) and in the same proportion as marine waters. It is often stated that continental waters are phosphorus-limited while oceanic waters are nitrogen-limited. A possible explanation of this phenomenon, based on molybdenum availability, was proposed by Howarth and Cole (1985). It follows that nutrient limitation in estuaries will not be caused by the same element, according to the hydrodynamical mechanism responsible for nutrient replenishment (i.e., advection of surface waters from continental sources versus upwelling or vertical mixing of underlying waters of marine origin). In waters dominated by upwelling, phytoplankton would most likely be limited by concentrations of nitrate or silicic acid (when diatoms are dominant), as shown by Levasseur and Therriault (1987) for the Lower St. Lawrence Estuary and by several authors for upwelling areas (references in the latter paper). On the other hand, limitation by phosphate concentrations would be expected in waters dominated by advection from the continent, as was observed by Gosselin et al. (1985) for ice algae growing in plume waters of the Grande Rivière-de-la-Baleine, Hudson Bay.

Steven (1974) was the first to recognize the importance of the supply of dissolved nutrients in the surface waters of the St. Lawrence ecosystem in controlling the productivity of phytoplankton. He discussed the mechanisms by which surface waters of the Estuary could be enriched, and estimated that the runoff of the St. Lawrence river could probably contribute about 10% of the nutrients in the surface waters at the head of the Estuary, while the remaining 90% would be supplied by entrainment into the surface layer of nutrient rich sub-surface waters by shear forces and mixing associated with large vertical oscillations due to internal tides. This led him to propose the concept of a continuously operating "nutrient pump" at the head of the Lower Estuary, largely responsible for the high productivity in the Gaspé current and in a large part of the Southern Gulf. Therriault and Lacroix (1976) later confirmed the occurrence of large vertical oscillations associated with internal tides which, linked to intense mixing at the head of the Laurentian channel, allowed nutrient enrichment of surface waters and eventual advection in the seaward direction.

Greisman and Ingram (1977) further studied nutrient distributions in the St. Lawrence Estuary, and estimated that nutrient supply in the diluted freshwaters represented < 25% of the total nutrient concentration in the surface waters of the Lower Estuary and in some respects, confirmed the "nutrient pump" hypothesis at the head of the Estuary. However, Sinclair et al. (1976) studied the volume transport and nutrient distribution at different seasons along a transect across the St. Lawrence Estuary and estimated that estuarine transport supplies a negligible percentage of the nutrients required to support the suggested primary production of the Gulf (by Steven, 1974), implying that other mixing processes and overturning events were much more important. They therefore concluded that the Estuary does not function as a "nutrient pump" of the magnitude and scale suggested by Steven (1974). It should be noted here that an error was picked up by Bugden (1981) in the original paper by Sinclair et al. (1976), but the same general conclusion can be reached after correction. This conclusion was in some way confirmed, in another context, by Levasseur and Therriault (1987) who studied the nutrient dynamics following an important tidal mixing event (referred to as "tidally induced upwelling") at the head region of the St. Lawrence Estuary, and found a rapid decrease of all nutrients as phytoplankton biomass increased exponentially up to ≈ 17 µg chl $\underline{a}$. L^{-1}. This indicates that, at least during the summer growing season, nutrients are consumed locally and cannot consequently contribute to enhancement of primary productivity in other part of the St. Lawrence ecosystem.

In their study, Levasseur and Therriault (1987) also found a decrease of the $NO_3:PO_4$ and $NO_3:SiO_4$ ratios during the growing periods, indicating that nitrate was the growth limiting factor at the time of their study. They also observed that silicic acid fell below levels required for diatom growth. Their data, as well as data from other upwelling areas, world-wide (see references in Levasseur and Therriault, 1987), support the hypothesis that the initial concentration of nitrate and silicic acid in the newly upwelled (or mixed) waters is the factor which mainly determines which nutrient will become exhausted (limiting) first.

Interestingly, Levasseur and Therriault (1987) observed an important trapping mechanism for silicic acid in the Estuary. An important increase in silicic acid concentrations below 100 m had been observed before by Coote and Yeats (1979) in the Estuary and Gulf of St. Lawrence. It had been attributed to the estuarine circulation and its ability to trap silicic acid more efficiently than nitrate: while nitrate (and phosphate) are flushed out rapidly from the estuary due to their longer residence time in the surface layer where most of the regenerative processes are occurring, the incorporation of silicic acid into fast-sinking particles (diatom theca and fecal pellets) results in silicic acid accumulation in bottom waters where the residual circulation is in the upstream direction (Ingram 1979).

2.6. Vertical mixing

The effects of vertical mixing on phytoplankton were reviewed in a recent paper (Demers et al., 1986), where numerous references are given. In the laboratory, there are a few studies showing that turbulence directly affects phytoplankton growth, but these small-scale effects are poorly documented. In the natural environment, on the other hand, there are numerous studies demonstrating larger-scale effects of vertical mixing on phytoplankton, through the agency of light and nutrients. Phytoplankton photosynthesis is often inhibited by high irradiance near the surface and limited by low irradiance at depth. So, there is a narrow range of depths where light favours phytoplankton growth. The downward light flux cannot be vertically mixed, but phytoplankton cells can be vertically redistributed relative to the light gradient. By contrast, the upward flux of nutrients can be greatly increased by vertical mixing. When phytoplankton cells are taken as the fixed reference, instead of the actual physical space, vertical mixing can thus results in homogenizing both nutrient and light responses of cells on the vertical.

A variety of mechanisms, reviewed by Demers et al. (1986), can lead to an increase in photosynthetic activity of vertically mixed phytoplankton, as a result of exposure to changing light intensities. Several authors have suggested that this could influence the succession of phytoplankton species, as was shown for the Lower St. Lawrence Estuary by Levasseur et al. (1984). On the other hand, there are numerous observations of positive phytoplankton responses to nutrient increases at various time and space scales. After reviewing such cases, Legendre (1981) concluded that a phytoplankton burst most often occurs upon stabilization of previously destabilized (and nutrient replenished) water so that, as far as nutrients are concerned, the potential for phytoplankton production depends mainly on the frequency of stabilization-destabilization of the water column. Detailed responses of phytoplankton to changes in light and nutrients, caused by vertical mixing, are described by Demers et al. (1986).

Demers and Legendre (1982) have shown that the intensity of vertical mixing due to tidal variations in water column stability covered a broad range of scales in the St. Lawrence Estuary. They suggested that these variations in stability of the water column have important effects on the photosynthetic capacity of phytoplankton. Levasseur et al. (1984) discussed the importance of stability and stratification in their model of the hierarchical control of phytoplankton succession in the Lower St. Lawrence Estuary. Demers and Legendre (1981) and Fortier and Legendre (1979) showed that fluctuations of biomass and photosynthetic activity over a semi-diurnal cycle in the Upper and Lower Estuary were related to changes in vertical stability. Sinclair (1978) also showed that biomass and species composition appeared to be function of the vertical circulation and, in particular, the changes in mixing characteristics. He showed that the weekly temporal variations were related to vertical stratification characteristics. In tank studies, where large volumes of natural water were isolated from environmental variations, Legendre et al. (1985) showed that variations in phytoplankton photosynthetic activity, which are possibly endogenous, can be phased on semi-diurnal variations in vertical tidal mixing (variations of the mean light conditions in the mixed layer).

There are several factors which influence vertical mixing in estuaries. These have been discussed by Bah and Legendre (1985), in their study of the distribution of phytoplankton biomass *versus* tidal mixing in the Upper St. Lawrence Estuary. The most important factors are: buoyancy, vertical shear in the horizontal velocity field, topography, internal waves and tides, and wind mixing. Let us now examine briefly each one of these factors.

2.6.1. Buoyancy
Buoyancy and static stability are two terms which are often employed when considering the resistance of a water column to vertical mixing. Buoyancy is therefore a fundamental factor which has a great influence on phytoplankton dynamics. In estuaries, the influx of freshwater, as well as temperature differences (due to warming) between the surface layer and deeper waters of marine origin, contribute to the maintenance of high buoyancy. High buoyancy has a stabilizing influence on the vertical structure of the water column, which opposes vertical mixing and which results in various degree of stratification of the water column.

In estuarine and inshore areas where this effect dominates, summer phytoplankton production may be either enhanced or depressed (see section on freshwater runoff above). If conditions are such that nutrients are replenished or regenerated in the photic layer, stabilization of the upper water column will favour primary production by maintaining the cells above the critical depth (see Turbidity); on the contrary, if nutrient replenishment is impeded by vertical stratification, summer production will be low. Therriault and Levasseur (1985) have presented a number of cases where these opposite effects of buoyancy play a major role in the Lower St. Lawrence Estuary. Let us just mention, for example, features such as the freshwater plumes of the Saguenay, Manicouagan and Aux-Outardes rivers and the Gaspé current to stress the importance of buoyancy related processes in the St. Lawrence Estuary.

2.6.2. Vertical shear

Dynamic stability of the water column is often estimated by the Richardson number (e.g., Krauss 1981), which compares the stabilizing effect of buoyancy to the destabilizing influence of vertical shear in the horizontal velocity field, over given depth intervals (Turner, 1973). Bah and Legendre (1985) have shown the relationship that exists between the Richardson number and the stratification parameter of Simpson and Hunter (1974; see Topography, below), which is often used to characterize vertical stability of the water column in coastal areas. In waters where vertical shear is strong, due to freshwater runoff and/or tidal currents, relationships between Richardson number and phytoplankton production are expected. Such relationships have been shown in the St. Lawrence Estuary for time series of phytoplankton biomass and production (Fortier and Legendre, 1979), and in the Gulf of St. Lawrence on vertical profiles of biomass and photosynthetic characteristics (Vandevelde et al., 1987). On the other hand, vertical shear may be an important mechanism for nutrient enrichment by entrainment (Neu, 1970; Bugden et al., 1978).

2.6.3. Topography

In shallow areas, the energy required to vertically mix the water column is derived from the dissipation of the tide and from wind stress on the water surface. Since the effect of wind mixing is relatively uniform spatially, differences in vertical mixing are mainly caused by the distribution of tidal energy dissipation relative to water depth (or topography). In this context, Simpson and Hunter (1974) proposed the characterization of vertical stability in shallow areas by the stratification parameter $S = \log_{10} h / CDu^3$, which depends on water depth (h), mean tidal stream velocity (u) and a drag coefficient on the bottom (CD). This coefficient was used in many shelf areas, including the northwest European Shelf (Pingree and Griffiths, 1978), the Bay of Fundy and the Gulf of Maine (Garrett et al., 1978), the Gulf of St. Lawrence (Pingree and Griffiths, 1980), Greater Cook Strait (Bowman et al., 1980), Hudson Bay (Griffiths et al., 1981) as well as the shallow waters of Long Island Sound (Bowman and Esaias, 1980). It is generally considered that values of the stratification parameter $S < 1$ characterize vertically well mixed waters and $S > 2$, stratified waters; between the well-mixed and well-stratified areas, fronts are often found in association with the $S = 1.5$ contour. Loder and Greenberg (1986) and Bowers and Simpson (1987) have further developed the Simpson-Hunter model by including wind mixing for the Gulf of Maine region and the European-shelf seas, respectively.

In order to study the combined influence of vertical stability and water column illumination, Pingree et al. (1978) introduced the S-kh diagram in which phytoplankton abundances are plotted as a function of both the stratification parameter (S) and the optical depth of the water column (k: coefficient of light attenuation). It should be noted that both axes of the s-kh diagram include topographic effects (water depth h). In estuarine waters, this diagram was successfully used by Bowman et al. (1981) in Long Island Sound and by Bah and Legendre (1985) in the Upper St. Lawrence Estuary to characterize the horizontal distributions of phytoplankton biomasses.

2.6.4. Internal waves and tides

The propagation of internal waves also influences the vertical structure of the water column. Their frequencies (Pollard, 1977) range between the local inertial frequency (a function of latitude) and the Brunt-Vasälaa frequency (a function of vertical stratification), and often include the M_2 tidal frequency (internal tide). According to Pond and Pickard (1983), causes for internal waves include strong tidal flow through passes or over bottom irregularities, passage of low atmospheric pressure systems and related short-period variations in wind stress and nonlinear transfer from surface waves.

Demers et al. (1986) have reviewed the effects of internal tides on phytoplankton using, in particular, many examples from the St. Lawrence Estuary. There is no use in repeating this review here, but let us just mention that internal tides affect almost all the characteristics that are influenced by changes in temperature and/or density in the water column, including the living organisms and the chemical and physical properties of the water. Near the surface, it is expected that vertical motions in the water column will exert a strong effect on the phytoplankton, primarily by modifying the conditions of irradiance and of nutrient supply. Several studies have reported variations in phytoplankton biomass, photosynthesis and production related to internal waves in the St. Lawrence (e.g., Denman and Platt, 1975; Steven, 1975; Therriault and Lacroix, 1976; Greisman and Ingram, 1977; Ingram, 1975, 1979; Lafleur et al., 1979; Sinclair et al., 1980; Fréchette and Legendre, 1982; Bah and Legendre, 1985; Vandevelde et al., 1987).

2.6.5. Wind mixing

Wind is another destabilizing agent of the water column in the vertical. In areas where factors such as those described above are not dominant, wind may become the most important source of vertical mixing that will influence phytoplankton. Aperiodic blooms of phytoplankton during the summer have been linked to nutrient replenishment of the upper water column at times of high winds, followed by stabilization of the water column (e.g., Iverson et al., 1974; Takahashi et al., 1977; Legendre et al., 1982). In addition, Therriault and Platt (1980) and Therriault et al. (1978) have demonstrated that spatial distribution of phytoplankton production and biomass is indeed controlled by wind stress induced turbulence, for wind velocities > 4-5 m s^{-1}. For the St. Lawrence, Levasseur et al. (1983) showed that wind greatly influenced the spatial organization of phytoplankton in surface waters. They showed that phytoplankton patch dimensions and shapes were dependent on interactions between tidal and wind conditions. Higher winds first have a tendency to increase the small-scale structure of the environment by breaking up larger patches into smaller patches, before structures are completely eliminated. In another study, Demers et al. (1987) found that, on a short time scale, chlorophyll a concentration in the littoral zone was highly variable and closely related to variations in the wind field. They attributed the higher chlorophyll values observed aperiodically at the sampling stations to mechanical resuspension of benthic diatoms due to wind, or wind-induced wave mixing in the littoral zone. Wind velocities > 4 m s[-1] were shown to cause significant resuspension of particulate organic matter in the water column.

3. Regional Divisions of the St. Lawrence Estuary

Most oceanographers (e.g., Brunel 1970) divide the St. Lawrence Estuary into three distinct areas : the Riverine Estuary, the Upper (or Middle) Estuary and the Lower (or Maritime) Estuary (see Fig. 1A). The Riverine Estuary extends upstream from Ile d'Orléans and is characterized by fresh waters under tidal influence; the Upper Estuary, between Ile d'Orléans and the mouth of the Saguenay Fjord, has partially mixed waters with salinities ranging from 0 to 25 °/oo. The Lower Estuary is characterized by salinities > 25 °/oo and shows pronounced stratification at least during the summer months. Since boundaries of these three zones were established after topographical and hydrological characteristics (Brunel, 1970; Neu, 1970; d'Anglejan and Smith, 1973; Ingram, 1976; El-Sabh, 1979, 1988; Kranck, 1979; Pingree and Griffiths, 1980; Ingram,1985), one may wonder whether these divisions also correspond to well defined biological characteristics.

According to Able (1978), Powles et al. (1984) and de Lafontaine (this book), the zones defined above represent natural divisions for the spatial distribution of ichthyoplankton. Concerning phytoplankton, it is interesting to note that several species very seldom cross the boundary between the Upper and the Lower estuaries. For example, Cardinal and Lafleur (1977) reported that 27 species found in the Lower Estuary almost never occurred in the Upper Estuary, whereas 10 other species were much more abundant upstream from the Saguenay Fjord than downstream. This is also reflected in phytoplankton cell numbers (Fig. 1B). On the other hand, Figure 1B also shows, at least for the July period, a strong cell number gradient within the Upper Estuary: much higher cell numbers being observed in the River and Lower parts of the Estuary than in the Upper one. This spatial pattern is paralleled by the distribution of chlorophyll a concentrations, as shown by Painchaud and Therriault (1985) and Painchaud et al. (1987). The Upper Estuary seems, therefore, to correspond to a zone of significant population and community changes in phytoplankton.

Using other characteristics in the distribution of phytoplankton, the Upper Estuary, between Ile d'Orléans and the Saguenay Fjord can be divided into three sub-regions: the **Freshwater,** the **Turbidity** and the **(Truly) Estuarine** Regions (Fig. 2). These divisions are based on phytoplankton data from Cardinal and Bérard-Therriault (1976), and also on a model proposed by Painchaud and Therriault (1985) to explain the distribution of heterotrophic bacteria in the St. Lawrence estuary. Using the distribution of phytoplankton production and biomass, Therriault and Levasseur (1985) have distinguished four main regions in the Lower Estuary. These are identified in the present paper as the **Outflow** (I), **Upwelling** (II), **Plume** (III) and **Near-Gulf** (IV) Regions (Fig. 3). Let us now examine each of these sub-regions in more detail.

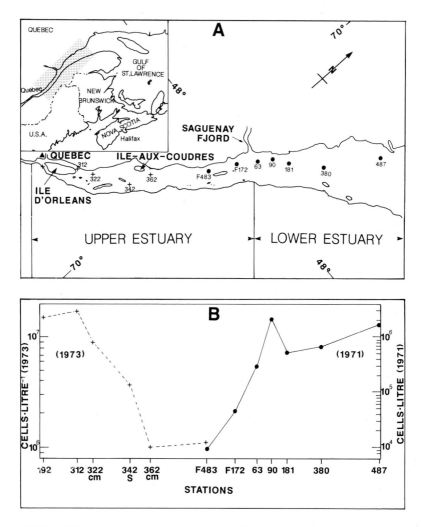

Figure 1. A) Map of the St. Lawrence Estuary showing Upper and Lower Estuaries. B) Numbers of phytoplankton cells in July, at 12 selected stations shown on the map (A). Data for 1971 from Cardinal and Lafleur (1977) and Fortier et al. (1978), and for 1973 from Cardinal and Bérard-Therriault (1976).

3.1. Upper Estuary

3.1.1. Freshwater Region

The Freshwater Region comprises the area around Ile d'Orléans and is characterized by surface salinity < 1º/oo. During the summer, this region shows high numbers of phytoplankton cells (Fig. 1B), and concentrations of chlorophyll $\underline{a}$ in excess of 5 mg.m^{-3} (Painchaud and Therriault, 1985; Painchaud et al., 1987). All phytoplankton in this region are typically freshwater species (Cardinal and Bérard-Therriault, 1976). This is the region that has received the least attention from oceanographers and limnologists to date, and given the importance of organic matter and pollutant discharges in this system, it should certainly receive more attention in the future.

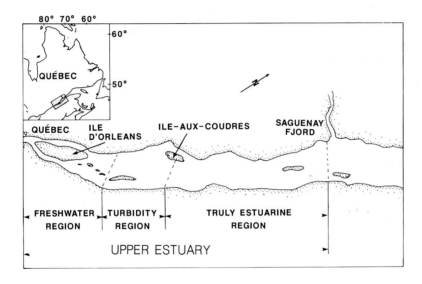

Figure 2. Map showing the biological sub-regions in the Upper Estuary.

3.1.2. Turbidity Region

The Turbidity Region corresponds to a well-developed, tidally and seasonally variable turbidity maximum (or front), observed at the transition from fresh to salt waters between Ile d'Orléans and Ile-aux-Coudres (d'Anglejan and Smith, 1973; Krank, 1979; Silverberg and Sundby, 1979; d'Anglejan, 1981; Gobeil et al., 1981). Salinities in this region vary between 0 and 10º/oo. The numbers of phytoplankton cells and chlorophyll $\underline{a}$ concentrations are much lower than in the Freshwater Region, but species composition is the same as upstream (Cardinal and Bérard-Therriault, 1976; Painchaud and Therriault, 1985; Painchaud et al., 1987), and very high ratios of particulate organic carbon to chlorophyll $\underline{a}$ (in excess of 1000: Painchaud and Therriault, 1985) indicate abundant detrital material. This suggests that the phytoplankton community in the Turbidity Region is composed of deteriorating freshwater cells, passively advected from upstream.

The major factors which influence the distribution, composition and physiology of phytoplankton cells and other biological organisms in the Turbidity Region are: (1) circulation processes, which are under the control of freshwater runoff and tidal excursion, (2) light availability, which is strongly influenced by the high turbidity, and (3) strong salinity gradients, which may cause important osmotic stress for biological organisms.

The extent of the Turbidity Region varies greatly with the semi-diurnal and fortnightly tidal cycles as well as with the seasonal and annual changes in freshwater discharge. Accordingly, the zone of influence of freshwater in the Turbidity Region will be variable in extension (Silverberg and Sundby, 1979). For example, in May, the 0 to 5 °/oo salinity range extends from Ile d'Orléans to 50 km downstream, whereas in November, this zone is just about 10 km wide (Silverberg and Sundby, 1979; D'Anglejan, 1981). Consequently, the turbidity front extends down to Ile-aux-Lièvres in May, whereas it is confined to the surroundings of Ile d'Orléans in August-October (Lucotte and d'Anglejan, 1986). These changes have major effects on the distribution of particulate matter in the Turbidity Region (Silverberg and Sundby, 1979; Lucotte and d'Anglejan, 1986).

Concerning phytoplankton, Cardinal and Bérard-Therriault (1976) showed that the cells are typically freshwater species in June, whereas in July and September, the community is composed of mixed populations that include freshwater and brackish water species. Secchi depths in this region are less than 1 m, suggesting limiting light conditions for primary production. Therefore, the high chlorophyll biomasses and cell numbers sometimes observed in the upstream part of this region are probably the result of direct advection from the Freshwater Region. The strong salinity gradients cause rapid disappearance of most of the freshwater species, which are gradually replaced by estuarine water species (Cardinal and Bérard Therriault, 1976; Cardinal and Lafleur, 1977; Painchaud and Therriault, 1985).

Nevertheless, even if phytoplankton production is probably negligible in this area, the high concentrations of particulate organic matter may not be lost from the food chain, since Painchaud and Therriault (1985, 1989) and Painchaud et al. (1987) have found high bacterial activity in the Turbidity Region. The freshwater bacterial community in this zone is gradually replaced, in the 1 to 3°/oo salinity range, by estuarine-type bacteria which thrive on undegraded cellular material released from freshwater organisms. In this region of the St. Lawrence Estuary, energy transfer from the primary to the secondary producers is thus probably mainly based on bacterial production instead of phytoplankton production (Painchaud et al., 1987).

3.1.3. (Truly) Estuarine Region
The (Truly) Estuarine Region lies downstream from Ile-aux-Coudres and is characterized by salinities varying between 10 and 25 °/oo. Cell numbers are very low (Fig. 1; see also

Demers et al., 1979; Lafleur et al., 1979) and concentrations of chlorophyll a are generally < 2 mg.m^{-3} (Demers and Legendre, 1981; Painchaud and Therriault, 1985; Painchaud et al., 1987). Species assemblages comprise both freshwater and marine taxa (Cardinal and Bérard-Therriault, 1976; Cardinal and Lafleur, 1977; Lafleur et al., 1979), but the dominance is by marine species. Phytoplankton in this area are affected by variations in freshwater runoff (flushing) and by strong tidal activity (Demers and Legendre, 1982; Bah and Legendre, 1985). This is the region of the St. Lawrence Estuary with the most typical estuarine conditions for biological organisms.

In the Estuarine Region, there is a high level of biological variability associated with various physical processes. Factors such as waves, semi-diurnal, fortnightly and internal tides, were invoked to explain the observed phytoplankton variability. In this part of the St. Lawrence Estuary, nutrients are rarely limiting (Demers and Legendre, 1981), such that light, as influenced by physical forcing mechanisms, is the main factor controlling phytoplankton dynamics.

In the Truly Estuarine Region of the Upper Estuary, turbidity is high (photic layer always < 5 m) and the surface mixed layer is continuously deeper than the photic layer. Bah and Legendre (1985) have demonstrated that turbulent eddies can reach maximum vertical dimension as high as 48 m with a mixing time of the order of 6 min in the upper layer of the water column. They also showed that materials other than phytoplankton account for more than 90% of light attenuation in the photic layer. Therefore, phytoplankton cells are always mixed to depths greater than the photic layer, where light becomes limiting, and respond to these changes by adjusting their photosynthetic capacity to the mean light intensity in the mixed layer (Savidge, 1979; Falkowski, 1980). Demers and Legendre (1979; 1981) found that, in these conditions, phytoplankton respond mainly to a 24-h cycle, since vertical mixing is too fast for the phytoplankton cells to continuously adjust their physiological responses to the variations in vertical stability.

Fortier et al. (1978) and Lafleur et al. (1979) observed semi-diurnal periodicities of biomass and species composition of phytoplankton, which they ascribed to advective processes. This led Fortier et al. (1978) to conclude that, on a short-time scale, the phytoplankton community in this part of the Estuary reacts as a conservative property of the environment such as temperature or salinity. Lafleur et al. (1979) found neap-spring tidal variability in the proportion of freshwater/marine taxa, which they also ascribed to advective processes. They further suggested that variations in the physiological status of the cells were associated with these changes, which would indicate that phytoplankton might also respond to short-term changes in the physical environment.

Changes in vertical mixing and advection associated with fortnightly tidal variations in the Estuarine Region have been the object of some intensive studies. Demers et al. (1979) found that increased tidal mixing during spring tides is paralleled by a decrease in chlorophyll concentration per unit cell. Primary production and photosynthetic capacity were also found to vary in response to fortnightly changes in vertical mixing (Demers et al., 1979: Demers and Legendre, 1979, 1981). It was suggested that phytoplankton cells were in better physiological condition during neap tidal periods than during spring tidal periods. Poor physiological condition of phytoplankton during spring tides would be linked to shortened residence time in the photic layer, due to increased vertical mixing. This was accompanied by a decrease in the amplitude of circadian variations in photosynthetic capacity (Demers and Legendre, 1981), which indicates that variations on the fortnightly scale may have effects on a shorter time scale. Over a two-month period, Demers and Legendre (1982) showed that changes in water column stratification characteristics were reflected in both the photosynthetic capacity and the variance of chlorophyll a.

3.2. Lower Estuary

The Lower Estuary extends from the Saguenay Fjord, at the head of the Laurentian channel (with depths > 300 m) to Pointe-des-Monts some 250 km downstream (Fig. 3). Because of its large vertical and horizontal dimensions and of the complex character and strong spatio-temporal variability of the circulation (El-Sabh, 1979; El-Sabh et al., 1982), the Lower Estuary cannot be considered as a typical estuary: it resembles more a shelf environment under the influence of numerous physical forcing mechanisms.

As mentioned above, Therriault and Levasseur (1985) have divided the Lower Estuary into 4 sub-areas (Fig. 3) which were defined on the basis of recurrent patterns in the spatial distribution of phytoplankton production and biomass. Discriminant analysis, using the physico-chemical variables, indicated a clear separation along two axes: 1) the first axis discriminates between region I, which is characterized by high turbidity (shallow photic zone), low temperatures and high nutrient concentrations, and region IV which had the least turbid waters (deep photic zone), the highest mean temperatures and the lowest nutrient concentrations; regions II and III showing intermediate characteristics; and 2) the second axis discriminates regions I and III from regions II and IV, on the basis of differences in mean temperatures and salinities. These results suggest that, on an annual/seasonal basis, the 4 regions identified by biological factors are also well separated by their physical-chemical characteristics. Each region corresponds to one of the major mesoscale features described by El-Sabh and Ingram (this book), and is dominated by a different hydrodynamical forcing mechanism. Other distribution patterns have been observed in the Lower estuary, probably corresponding to transient physical conditions at certain times of the year and/or of the tidal cycle, as suggested by Ingram and El-Sabh (this volume). However, the division into 4 regions is certainly characteristic of the summer productive period.

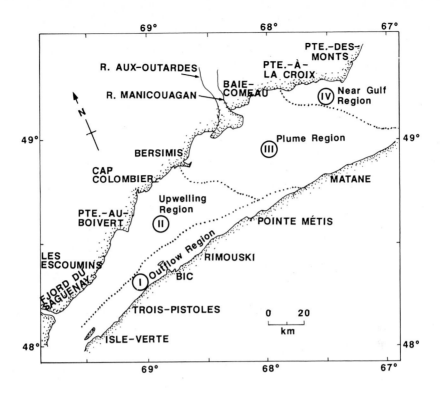

Figure 3. Map showing the division in 4 sub-regions of the Lower Estuary (After Therriault and Levasseur, 1985).

3.2.1. Outflow Region (I)

Region (I) corresponds to a zone where freshwaters from the combined plume of the Saguenay and the Upper Estuary are rapidly flushed along the south shore of the Lower Estuary (El-Sabh et al. 1982). Higher turbidities are associated with this freshwater input, as is also the case on the north shore of the Lower Estuary for the Manicouagan and Aux-Outardes river runoffs (Plume Region). In spite of the stabilizing potential (buoyancy) of this large flow of freshwater, this region is nevertheless characterized by high instability of the water column due to strong tidal mixing (Therriault and Lacroix, 1976; Fortier and Legendre, 1979; Pingree and Griffiths, 1980; Sinclair et al., 1981). This instability and the high flushing rates as well as the high turbidity are the main reasons that explain reduced phytoplankton production in this region. Therriault and Levasseur (1985) have estimated the mean residence time of waters in this region to be of the order of 2-3 days, which is comparable to the mean doubling time of phytoplankton. Therefore, it cannot be expected to find accumulations of biomass in this region, especially with a shallow photic layer. For those reasons, the Outflow Region is the least productive (ca. 30 gC m^{-2}.a^{-1}), with the shortest duration of growth.

During the summer, chlorophyll a concentrations are generally of the same order of magnitude as in the (Truly) Estuarine Region (< 2 mg.m-3). Because of logistic reasons (i.e., proximity of research laboratories), it is also the region which has been the most intensively studied in the Lower Estuary (e.g., Steven, 1974; Fortier et al., 1978; Sinclair, 1978; Fréchette and Legendre, 1978; Fortier and Legendre, 1979; Sinclair et al., 1980; Auclair et al.,1982; Levasseur et al., 1983; Legendre et al., 1984; Levasseur et al., 1984).

3.2.2. Upwelling Region (II)

The Upwelling Region corresponds to an area of higher mean salinity and nutrient values, and of lower mean temperature. Strictly speaking, this is not a real upwelling area since the introduction of sub-surface waters into the surface mixed layer results from frequent isopycnal shoaling, shear instabilities and tidal mixing (Greisman and Ingram, 1977) due to the generation of large amplitude internal tides (50-100 m) near the head of the Laurentian Channel (Forrester, 1974; Ingram, 1975; Therriault and Lacroix, 1976). There exist, however, indications that true upwelling occurs along the north shore (Simard et al., 1986). Since there is a neap-spring relationship in the amplitude of the internal tides (Therriault and Lacroix, 1976; Greisman and Ingram, 1977), destabilization of the water column associated with these mixing events (often referred to as tidal induced upwelling in the literature) occurs during the spring tides, while stability of the water column may be reestablished during neap tides, with the relaxation of upwelling. The stabilization of the surface waters following nutrient enrichment events is the reason for a higher productivity of the Upwelling Region compared to the Outflow Region, in agreement with the stabilization-destabilization model of Legendre (1981). Nutrient uptake dynamics of phytoplankton following such events was studied by Levasseur and Therriault (1987), and the main results have been reported above.

The Upwelling Region shows an annual cycle of primary production very similar to the mean cycle in the whole Estuary, with phytoplankton activity mostly occurring between June and September, and a decrease of biomass and production in August (Therriault and Levasseur, 1985). Annual production is of the order of 90 gC.m2, and mean chlorophyll concentrations in the photic zone during the summer are usually < 5 mg.m-3. Simard et al. (1986) have suggested that upwelling processes along the north shore can have large impacts on the aggregation and accumulation of secondary producers.

3.2.3. Plume Region (III)

The Plume Region is an area that is directly associated with the freshwater plume from the Manicouagan and Aux-Outardes Rivers (Fig.3), particularly during the summer period. The eastern boundary of this plume forms a sharp density front which has been described by Tang (1980). The location of this density front closely corresponds to the delimitation between the Plume and the Near Gulf Regions, and upwelling of cold intermediate waters has been associated with this front. Tidal mixing in this region is less intense than upstream, and the internal tides have lost most of their energy (Forrester, 1974; Therriault and Lacroix, 1976), so that the limited freshwater inputs from rivers are sufficient to stabilize the surface waters over a large area. Since these freshwaters flow over, and are continuously and moderately vertically mixed with nutrient-rich waters which are advected from the Outflow and Upwelling Regions, nutrient replenishment of surface waters is continuously insured, with the consequence of relatively high phytoplankton production (Therriault and Levasseur, 1985). The Plume Region shows a summer production cycle similar to the Upwelling Region, but more phytoplankton growth can be observed during late fall and winter in the former. Annual primary production is high (> 130 mgC.m^{-3}.a^{-1}), and mean concentrations of chlorophyll during the summer usually exceed 5 mg.m^{-3}. Therriault et al. (1985) and Cembella et al. (in press) have stressed the importance of the Plume Region for blooms of the toxic dinoflagellate *Protogonyaulax tamarensis* (see section on freshwater runoff, above).

3.2.4. Near Gulf Region (IV)

The Near Gulf Region is located at the lower end of the turbidity range, with the least turbid waters and the deepest photic zone. This region is also the most vertically stable of all four sub-areas of the Lower Estuary, as suggested by lower nutrient values and higher temperatures observed (Therriault and Levasseur, 1985). This is also the region where tidal mixing has the least influence. Temporal variations of the various biological parameters in this region are more typical of conditions generally encountered in the Gulf of St. Lawrence, where a massive phytoplankton bloom occurs early in the spring (April-May: Steven, 1974; Sévigny et al, 1979; de Lafontaine et al, 1981; de Lafontaine et al., 1984), followed by the classical (for temperate latitudes) nutrient-limited reduction of growth during the summer months with formation of sub-surface chlorophyll maxima (e.g., Vandevelde et al., 1987).

Therriault and Levasseur (1985) have compared the phytoplankton production of the Lower Estuary with different environments. Their production values were in general of the same order of magnitude as in many other estuarine environments, such as the Hudson River (Malone, 1977), the Bristol Channel (Joint and Pomeroy, 1981) or the Eastern Scotian Shelf (Mills and Fournier, 1979), but were much lower than in some other so called productive regions such as the adjacent Gaspé Current (Sévigny et al., 1979), the Long Island Sound (Riley, 1956), the Puget Sound (Winter et al., 1975), or St. Margaret's Bay (Platt, 1971) and the Bedford Basin (Platt, 1975) on the Atlantic Coast.

REFERENCES

Able, K. W. 1978. Ichthyoplankton of the St. Lawrence estuary: composition, distribution and abundance. J. Fish. Res. Board Can. 35: 1517-1531.

Auclair, J. C., S. Demers, M. Fréchette, L. Legendre and C. L. Trump. 1982. High frequency endogenous periodicities of chlorophyll synthesis in estuarine phytoplankton. Limnol. Oceanogr. 27: 348-352.

Bah, A. and L. Legendre. 1985. Biomasse photosynthétique et mélange de marée dans l'estuaire moyen du Saint-Laurent. Naturaliste Can. 112: 39-49.

Bowers D. G. and J. H. Simpson. 1987. Mean position of tidal fronts in European-shelf seas. Cont. shelf Res. 7: 35-44.

Bowman, M. J. and W. E. Esaias. 1980. Fronts, stratification, and mixing in Long Island and Block Island Sounds. J. Geophys. Res. 86: 4260-4264.

Bowman, M. J., W. E. Esaias and M. B. Schnitzer. 1981. Tidal stirring and the distribution of phytoplankton in Long Island and Block Island Sounds. J. Mar. Res. 39: 587-603.

Bowman, M. J., A. C. Kibblewhite and D. E. Ash. 1980. M_2 tidal effects in Greater Cook Strait, New Zeland. J. Geophys. Res. 85: 2728-2742.

Briant, F. J. P. 1975. Seasonal variations and associations of southern Californian nearshore phytoplankton. J. Ecol. 64: 821-835.

Brunel, P. 1970. Les grandes divisions du Saint-Laurent: 3ème commentaire. Revue Géogr. Montréal 24: 291-294.

Bugden, G. L. 1981. Salt and heat budgets for the Gulf of St. Lawrence. Can. J. Fish. Aquat. Sci. 38: 1153-1167.

Bugden, G. L., B. T. Hargrave, M. M. Sinclair, C. L. Tang, J.-C. Therriault and P. A. Yeats. 1982. Freshwater runoff effects in the marine environment: the Gulf of St. Lawrence example. Can. Tech. Rep. Fish. Aquat. Sci. 1078: 1-89.

Cardinal, A. and L. Bérard-Therriault. 1976. Le phytoplancton de l'estuaire moyen du Saint-Laurent en amont de l'Ile-aux-Coudres (Québec). Int. Rev. Ges. Hydrobiol. 61: 639-648.

Cardinal, A. and P. E. Lafleur. 1977. Le phytoplancton estival de l'estuaire maritime du Saint-Laurent. Bull. Soc. Phycol. France 22: 150-159.

Cembella, A. D. and J.-C. Therriault. 1988. Population dynamics and toxin composition of *Protogonyaulax tamarensis* from the St. Lawrence Estuary. In: Red Tides: Biology, Environmental Science and Toxicicology. pp. 81-84, T. Okaichi, D. M. Anderson, T. Nemoto (eds.), New York, N. Y. Elsevier, Science Publishing Co. Inc.

Cembella, A. D., J.-C. Therriault and P. Béland. 1988. Toxicity of cultured isolates and natural populations of *Protogonyaulax tamarensis* from the St. Lawrence Estuary. J. Shellfish Res. 7: 611-621.

Côté, R. and G. Lacroix. 1979. Influence des débits élevés et variables d'eau douce sur le régime saisonnier de production primaire d'un fjord subarctique. Oceanol. Acta 2: 299-306.

Coote, A. R. and P. A. Yeats. 1979. Distribution of nutrients in the Gulf of St. Lawrence. J. Fish. Res. Board Can. 36: 122-131.

d'Anglejan, B. 1981. On the advection of turbidity in the St. Lawrence middle estuary. Estuaries 4: 2-15.

d'Anglejan, B. and E. C. Smith. 1973. Distribution, transport and composition of suspended matter in the St. Lawrence estuary. Can. J. Earth Sci. 10: 1380-1396.

d'Anglejan, B. F. and R. G. Ingram. 1976. Time-depth variations in tidal flux of suspended matter in the St. Lawrence estuary. Estuar. Coastal Mar. Sci. 4: 401-416.

de Lafontaine, Y., M. Sinclair, S. N. Messieh, M. I. El-Sabh and C. Lassus. 1981. Ichthyoplankton distribution in the northwestern Gulf of St. Lawrence. Rapp. P.-v. Réun. cons. Int. explor. Mer, 178: 185-187.

de Lafontaine, Y., M. Sinclair, M. I. El-Sabh, C. Lassus and R. Fournier. 1984. Temporal occurrence of ichthyoplankton in relation to hydrographic and biological variables at a fixed station in the St. Lawrence estuary. Estuar. Coast. Shelf Sci. 18: 177-190.

Demers, S., P. E. Lafleur, L. Legendre and C. L. Trump. 1979. Short-term covariability of chlorophyll and temperature in the St. Lawrence estuary. J. Fish. Res. Board Can. 36: 568-573.

Demers, S. and L. Legendre. 1979. Effets des marées sur la variation circadienne de la capacité photosynthétique du phytoplancton de l'estuaire du Saint-Laurent. J. Exp. Mar. Biol. Ecol. 39: 87-99.

Demers, S. and L. Legendre. 1981. Mélange vertical et capacité photosynthétique du phytoplancton estuarien (estuaire du Saint-Laurent). Mar. Biol. 64: 243-250.

Demers, S. and L. Legendre. 1982. Water column stability and photosynthetic capacity of estuarine phytoplankton: long-term relationships. Mar. Ecol. Prog. Ser. 7: 337-340.

Demers, S., L. Legendre and J.-C. Therriault. 1986. Phytoplankton responses to vertical tidal mixing. In: Tidal mixing and plankton dynamics, pp. 1-40. J. Bowman, C. M. Yentsch and W. T. Peterson (eds.), Springer-Verlag, Berlin.

Demers, S., J.-C. Therriault, E. Bourget and A. Bah. 1987. Particulate matter resuspension in the littoral zone of a tide-dominated estuarine environment: Wind influence. Limnol. Oceanog. 32: 327-339.

Denman, K. L. and T. Platt. 1975. Coherences in the horizontal distributions of phytoplankton and temperature in the upper ocean. Mém. Soc. R. Sci. Liège 7: 19-30.

Durbin, E. G. 1978. Aspects of the biology of resting spores of *Thalassiosira nordenskioldii* and *Detonula confervacea* . Mar. Biol. 45: 31-37.

Durbin, E. G., R. W. Krawiec and T. J. Smayda. 1975. Seasonal studies on the relative importance of different size fractions of phytoplankton in Narragansett Bay (USA). Mar. Biol. 32: 271-287.

El-Sabh, M. I., 1979. The Lower St. Lawrence Estuary as a physical oceanographic system. Naturaliste Can. 106: 44-73.

El-Sabh, M. I. 1988. Physical oceanography of the St. Lawrence Estuary. In: Hydrodynamics of estuaries, Vol II Estuarine cases studies. pp. 61-78, B. Kjerfve (ed.). CRC Press Inc., Boca Raton, Florida.

El-Sabh, M. I., H.-J. Lie and V. G. Koutitonsky 1982. Variability of the near-surface residual current in the lower St. Lawrence Estuary. J. Geophys. Res. 87: 9589-9600.

Eppley, R. W. 1972. Temperature and phytoplankton growth in the sea. Fish. Bull. U. S. 70: 1063-1085.

Falkowski, P. G. 1980. Light and shade adaptation in marine phytoplankton. In: Primary productivity in the sea. pp. 99-119, P. G. Falkowski (ed.), Plenum Press, New York.

Forrester, W. D. 1974. Internal tides in the St. Lawrence Estuary. J. Mar. Res. 32: 55-66.

Fortier, L. and L. Legendre. 1979. Le contrôle de la variabilité à court terme du phytoplancton estuarien: stabilité verticale et profondeur critique. J. Fish. Res. Board Can. 36: 1325-1335.

Fortier, L., L. Legendre, A. Cardinal and C. L. Trump. 1978. Variabilité à court terme du phytoplancton de l'estuaire du Saint-Laurent. Mar. Biol. 46: 349-354.

Fréchette, M. and L. Legendre. 1978. Photosynthèse phytoplanctonique: réponse à un stimulus simple, imitant les variations rapides de la lumière engendrées par les vagues. J. Exp. Mar. Biol. Ecol. 32: 15-25.

Fréchette, M. and L. Legendre. 1982. Phytoplankton photosynthetic response to light in an internal tide dominated environment. Estuaries 5: 287-293.

Garrett, C. J. R., J. R. Keeley and D. A. Greenberg. 1978. Tidal mixing versus thermal stratification in the Bay of Fundy and Gulf of Maine. Atm. Ocean 16: 403-423.

Gessner, F. and W. Schramm. 1971. Salinity. Plants. In: Marine ecology, Vol. I, Part 2. pp. 705-820, O. Kinne (ed.). Wiley-Interscience, London.

Gobeil, C., B. Sundby and N, Silverberg. 1981. Factors influencing particulate matter geochemistry in the St. Lawrence Estuary turbidity maximum. Mar. Chem. 10: 123-140.

Gosselin, M., L. Legendre, S. Demers and R. G. Ingram. 1985. Responses of sea-ice microalgae to climatic and fortnightly tidal energy inputs (Manitounuk Sound, Hudson Bay). Can. J. Fish. Aquat. Sci. 42: 999-1006.

Greisman, P. and R. G. Ingram. 1977. Nutrient distribution in the St. Lawrence Estuary. J. Fish. Res. Board Can. 34: 2117-2123.

Griffiths, D. K., R. D. Pingree and M. Sinclair. 1981. Summer tidal fronts in the near-Arctic regions of Foxe Basin and Hudson Bay. Deep-Sea Res. 28: 865-873.

Hellebust, J. A., 1976. Osmoregulation. Ann. Rev. Plant Physiol. 27: 485-505.

Holligan, P. M. and D. S. Harbour. 1977. The vertical distribution and succession of phytoplankton in the western English Channel in 1975 and 1976. J. Mar. Biol. Ass. U. K. 57: 1075-1093.

Howarth, R. W. and J. J. Cole. 1985. Molybdenum availability, nitrogen limitation, and phytoplankton growth in natural waters. Science 229: 653-655.

Ingram, R. G. 1975. Influence of tidal induced mixing on primary productivity in the St. Lawrence estuary. Mém. Soc. R. Sci. Liège 7: 59-74.

Ingram, R. G. 1976. Characteristics of a tide induced estuarine front. J. Geophys. Res. 81: 1951-1959.

Ingram, R. G. 1979. Water mass modification in the St. Lawrence estuary. Naturaliste Can. 106: 45-54.

Ingram. R. G. 1985. Frontal characteristics at the head of the Laurentian Channel. Naturaliste Can. 112: 31-38.

Ingram, R. G. and B. F. D'Anglejan. 1979. On the importance of cross channel suspended sediment flux in the Upper St. Lawrence Estuary. Proc. Symp. Modelling of transport Mechanisms in Oceans and Lakes, Burlington, Ontario. Environment Can. MS. 43: 149-159.

Iverson, R. L., H. C. Curl Jr., H. B. O'Connors Jr., D. Kirk and K. Zakar. 1974. Summer phytoplankton blooms in Auke Bay, Alaska, driven by wind mixing of the water column. Limnol. Oceanogr. 19: 271-278.

Joint, I. R. and A. J. Pomeroy. 1981. Primary production in a turbid estuary. Estuar. Coastal Shelf Sci. 13: 303-316.

Ketchum, B. H. 1954. Relation between circulation and planktonic populations in estuaries. Ecology 35: 191-200.

Kranck, K. 1979. Dynamics and distribution of suspended particulate matter in the St. Lawrence Estuary. Naturaliste can. 106: 163-173.

Krauss, W. 1981. The erosion of a thermocline. J. Phys. Oceanogr. 11: 415-433.

Lafleur, P. E., L. Legendre and A. Cardinal. 1979. Dynamique d'une population estuarienne de diatomées planctoniques: effet de l'alternance des marées de morte-eau et de vive-eau. Oceanol. Acta 2: 307-315.

Legendre, L. 1981. Hydrodynamic control of marine phytoplankton production: The paradox of stability. In: Ecohydrodynamics. pp. 191-207, J. C. J. Nihoul (ed.). Elsevier, Amsterdam.

Legendre, L., S. Demers, J.-C. Therriault and C. A. Boudreau. 1985. Tidal variations in the photosynthesis of estuarine phytoplankton isolated in a tank. Mar. Biol. 88: 301-309.

Legendre, L., R. G. Ingram and Y. Simard. 1982. Aperiodic changes of water column stability and phytoplankton in an Arctic coastal embayment, Manitounuk Sound, Hudson Bay. Naturaliste Can. 109: 775-786.

Levasseur, M. and J.-C. Therriault. 1987. Phytoplankton biomass and nutrient dynamics in a tidally induced upwelling: the role of the $NO_3:SiO_4$ ratio. Mar. Ecol. Prog. Ser. 39: 87-97.

Levasseur, M., J.-C. Therriault and L. Legendre. 1983. Tidal currents,wind and the morphology of phytoplankton spatial structures. J. Mar. Res. 41: 655-672.

Levasseur, M., J.-C. Therriault and L. Legendre. 1984. Hierarchical control of phytoplankton succession by physical factors. Mar. Ecol. Prog. Ser. 19: 211-222.

Loder, J. W. and D. A. Greenberg. 1986. Predicted positions of tidal fronts in the Gulf of Maine region. Cont. Shelf Res. 6: 397-414.

Lucotte, M. and B. F. D'Anglejan. 1986. Seasonal control of the St. Lawrence maximum turbidity zone by tidal-flat sedimentation. Estuaries 2: 84-94.

Malone, T. C. 1977. Environmental regulation of phytoplankton productivity in the lower Hudson River. Estuar. Coast. Mar. Sci. 5: 157-171.

Malone, T. C. and M. B. Chervin. 1979. The production and fate of phytoplankton size fractions in the plume of the Hudson River, New York Bight. Limnol. Oceanogr. 24: 683-696.

Malone, T. C. and P. J. Neale. 1981. Parameters of light-dependent photosynthesis for phytoplankton size fractions in temperate estuarine and coastal environments. Mar. Biol. 61: 289-297.

Margalef, R. 1958. Temporal succession and spatial heterogeneity in phytoplankton. In: Perspective in marine biology. pp. 323-349, A. A. Buzzati-Traverso (ed.). Univ. California Press, Berkeley.

Margalef, R. 1967. Some concepts relative to the organization of plankton. Oceanogr. Mar. Biol. Ann. Rev. 5: 257-289.

Margalef, R. 1978. Life-forms of phytoplankton as survival alternatives in an unstable environment. Oceanol. Acta 1: 493-509.

Margalef, R., M. Estrada and D. Blasco. 1979. Functional morphology of organisms involved in red tides, as adapted to decaying turbulence. In: Toxic Dinoflagellate blooms, pp. 89-94, D. L. Taylor and H. H. Seliger (eds.), Elsevier, North Holland.

Mills, E. L. and R. O. Fournier. 1979. Fish production and the marine ecosystem of the Soctian shelf, eastern Canada. Mar. Biol. 54: 101-108.

Morris, A. W., R. F. C. Mantoura, A. J. Bale and R. J. M. Howland. 1978. Very low salinity regions of estuaries: important sites for chemical and biological reactions. Nature 274: 678-680.

Neu, H. J. A. 1970. A study on mixing and circulation in the St. Lawrence Estuary up to 1964. Atlantic Oceanogr. Lab., Bedford Inst. Oceanogr., Dartmouth, Nova Scotia, AOL Rep. 1970-9, 31p.

Painchaud, J. and J.-C. Therriault. 1985. Heterotrophic potential in the St. Lawrence Estuary: Distribution and controlling factors. Naturaliste Can. 112: 65-76.

Painchaud, J. and J.-C. Therriault. 1989. Relationships between bacteria, phytoplankton and particulate carbon in the Upper St Lawrence Estuary. Mar. Ecol. Prog. Ser. 56: 301-311.

Painchaud, J., D. Lefaivre and J.-C. Therriault. 1987. Box model analysis of bacterial fluxes in the St. Lawrence Estuary. Mar. Ecol. Prog. Ser. 41: 241-252.

Pingree, R. D. and D. K. Griffiths. 1978. Tidal fronts on the shelf seas around the British Isles. J. Geophys. Res. 83: 4615-4622.

Pingree, R. D. and D. K. Griffiths. 1980. A numerical model of the M_2 tide in the Gulf of St. Lawrence. Oceanol. Acta 3: 221-225.

Pingree, R. D., P. M. Holligan and G. T. Mardell. 1978. The effects of vertical stability on phytoplankton distributions in the summer on the northwest European Shelf. Deep-Sea Res. 25: 1011-1028.

Platt, T. 1971. The annual production by phytoplankton in St. Margaret's Bay, Nova Scotia. J. Cons. Int. Explor. Mer 33: 324-333.

Platt, T. 1975. Analysis of the importance of spatial and temporal heterogeneity in the estimation of annual production by phytoplankton in a small, enriched, marine basin, J. Exp. Mar. Biol. Ecol. 18: 1-11.

Pollard, R. T. 1977. Observations and theories of Langmuir circulations and their role in near surface mixing. In: Voyage of discovery: George Deacon 70th Anniversary Volume. pp. 235-251, M. V. Angel (ed.). Pergamon Press, Oxford.

Pond, S. and G. L. Pickard. 1983. Introductory dynamical oceanography, 2nd ed. Pergamon Press, Oxford. 329 p.

Powles, H., F. Auger and G. J. Fitzgerald. 1984. Nearshore ichthyoplankton of a north temperate estuary. Can. J. Fish. Aquat. Sci. 41: 1653-1663.

Prakash, A. 1975. Land drainage as a factor in "red tide" development. Environment Letters 9: 121-128.

Provasoli, L. 1979. Recent progress, an overview. In: Toxic dinoflagellate blooms. pp. 1-14, D. L. Taylor and H. H. Seliger (eds.), Elsevier, North Holland.

Riley, G. A. 1942. The relationship of vertical turbulence and spring diatoms flowerings. J. Mar. Res. 5: 67-87.

Riley, G. A. 1956. Oceanography of Long Island Sound, 1952-1954. IX. Production and utilization of organic matter. Bull. Bingham Oceanogr. Coll. 15: 324-334.

Roman, M. R. and K. R. Tenore. 1978. Tidal resuspension in Buzzards Bay, Massachusetts. 1. Seasonal changes in the resuspension of organic carbon and chlorophyll a. Estuar. Coast. Mar. Sci. 6: 37-46.

Savidge, G. 1979. Photosynthetic characteristics of marine phytoplankton from contrasting physical environments. Mar. Biol. 53: 1-12.

Sévigny, J.-M., M. Sinclair, M. El-Sabh, S. Poulet and A. Coote. 1979. Summer plankton distributions associated with the physical and nutrient properties of the northwestern Gulf of St. Lawrence. J. Fish. Res. Board Can. 36: 187-203.

Simard, Y., R. de Ladurantaye and J.-C. Therriault. 1986. Aggregation of euphausiids along a coastal shelf in an upwelling environment. Mar. Ecol. Prog. Ser. 32: 203-215.

Simpson, J. H. and J. R. Hunter. 1974. Fronts in the Irish Sea. Nature 250: 404-406.

Sinclair, M. 1978. Summer phytoplankton variability in the lower St. Lawrence estuary. J. Fish. Res. Board Can. 35: 1171-1185.

Sinclair, M., M. El-Sabh and J. R. Brindle. 1976. Seaward nutrient transport in the lower St. Lawrence estuary. J. Res. Board Can. 33: 1271-1277.

Sinclair, M., J. P. Chanut and M. El-Sabh. 1980. Phytoplankton distributions observed during a 3 1/2 days fixed-station in the lower St. Lawrence estuary. Hydrobiologia 75: 129-147.

Sinclair, M., D. V. Subba Rao and R. Couture. 1981. Phytoplankton temporal distributions in estuaries. Oceanol. Acta 4: 239-246.

Silverberg, N. and B. Sundby. 1979. Observations in the turbidity maximum of the St. Lawrence Estuary. Can. J. Earth Sci. 16: 939-950.

Steven, D. M. 1974. Primary and secondary production in the Gulf of St. Lawrence. McGill Univ. Mar. Sci. Centre MS Rep. 26: 1-116.

Steven, D. M. 1975. Biological production in the Gulf of St. Lawrence. In: Energy flow - Its biological dimensions. A Summary of the IBP in Canada, pp. 229-248. W. M. Cameron and L. W. Billingsley (eds.) Royal Society of Canada, Ottawa.

Sverdrup, H. U. 1953. On conditions for the vernal blooming of phytoplankton. J. Cons. Perm. Int. Explor. Mer 18: 287-295.

Takahashi, M., D. L. Siebert and W. H. Thomas. 1977. Occasional blooms of phytoplankton during summer in Saanich Inlet, B. C., Canada. Deep-Sea Res. 24: 775-780.

Tang, C.L. 1980. Mixing and circulation in the northwestern Gulf of St. Lawrence. J. Geophys. Res., 85, 2787-2796.

Therriault, J. C. and G. Lacroix. 1976. Nutrients, chlorophyll and internal tides in the St. Lawrence Estuary. J. Fish. Res. Board Can. 33: 2747-2757.

Therriault, J.-C., D. J. Lawrence and T. Platt. 1978. Spatial variability of phytoplankton turnover in relation to physical processes in coastal environment. Limnol. Oceanogr. 23: 900-911.

Therriault, J.-C. and M. Levasseur. 1985. Control of phytoplankton production in the lower St. Lawrence Estuary: light and freshwater runoff. Naturaliste Can. 112: 77-96.

Therriault, J.-C. and M. Levasseur. 1986. Freshwater runoff control of the spatio-temporal distribution of phytoplankton in the lower St. Lawrence Estuary (Canada). In: Proceedings of the NATO Freshwater/Sea Workshop, Bodo, Norway, pp. 251-260. S. Skreslet (ed.). NATO ASI Series, Vol. G 7, Springer-Verlag, New York.

Therriault, J.-C., J. Painchaud and M. Levasseur. 1985 . Factors controlling the occurrence of *Protogonyaulax tamarensis* and shellfish toxicity in the St. Lawrence estuary: freshwater runoff and the stability of the water column. In: Toxic dinoflagellates, pp. 141-146. D. M. Anderson, A. W. White and D.C.Baden eds.). Elsevier, New York.

Therriault, J.-C. and T. Platt. 1980. Environmental control of phytoplankton patchiness. J. Fish. Res. Board Can. 38: 638-641.

Turner, J. S. 1973. Buoyancy effects in fluids. Cambridge Univ. Press, New York.367 p.

Vandevelde, T., L. Legendre, J.-C. Therriault, S. Demers and A. Bah. 1987. Subsurface chlorophyll maximum and hydrodynamics of the water column. J. Mar. Res. 45: 377-396.

Walting, L., D. Bottom, A. Pembroke and D. Maurer. 1979. Seasonal variations in Delaware Bay phytoplankton community structure. Mar. Biol. 52: 207-215.

Winter, D. F., K. Banse and G. C. Anderson. 1975. The dynamics of phytoplankton blooms in Puget Sound, a fjord in the northwestern United States. Mar. Biol. 29: 139-176.

Wyatt, T. and S. Z. Quasim. 1973. Application of a model to an estuarine ecosystem. Limnol. Oceanogr. 18: 301-306.

Yancey, P.H., M.E. Clarck, S.C. Hand, R. D. Bowlus and G. N. Somero. 1982. Living with water stress: evolution of osmolyte system. Science 217: 1214-1222.

Chapter 13

Zooplankton of the St. Lawrence Estuary: The Imprint of Physical Processes on its Composition and Distribution

J.A. Runge and Y. Simard

Institut Maurice-Lamontagne, Ministère des Pêches et Océans, C.P. 1000, Mont-Joli, Québec, Canada G5H 3Z4

ABSTRACT

Studies on the composition and distribution of zooplankton in the St.Lawrence estuary are reviewed and discussed with special attention to the dominant role of physical oceanographic processes. The 350 km-long estuary is divided in two parts by a 50 m-deep sill, which isolates the 350 m-deep Lower estuary from the shallow (<100m) Upper estuary. At the upstream end of the Upper estuary, planktonic (*Eurytemora affinis*) and epibenthic (*Ectinosoma curticorne*) copepods and mysids (*Neomysis americana*) maintain endemic populations at the front of the salt wedge (S = 1-15°/oo). Further downstream, the zooplankton abundance appears to pass through a minimum before increasing in a 100 m-deep basin adjacent to the sill (S = 25-30 °/oo). There the copepod composition is dominated by *Acartia longiremis*, *Eurytemora herdmani* and *Calanus finmarchicus*. In the Lower estuary, the dominant copepods belong to the genus *Calanus*, mainly *C. finmarchicus* and some *C. hyperboreus* in their later developmental stages (CIV-CVI). A few smaller copepod species are frequent, but the genus *Pseudocalanus* is surprisingly rare. Other important taxonomic groups in the Lower estuary are the the euphausiids *Thysanoessa raschi* and *Meganyctiphanes norvegica*, which concentrate in dense sound scattering layers at the head and along the deep channel.

The composition in the Upper estuary is explained by the maintenance of populations of estuary adapted species. In contrast, the Lower estuary contains zooplankton whose local recruitment is hindered by the combination of a strong, surface-flushing circulation and cold temperature structure; their population dynamics seem to depend largely on the estuarine-driven import/export processes. Many physical processes inherent to this high-energy environment such as internal tides, tidal advection and upwelling, sill dynamics, high-frequency internal waves, eddies and fronts are responsible for the generation of high variability in the small- and mesoscale distributions.

Key Words: Copepods, mysids, abundance, recruitment, physical processes, space-time variability

RESUME

Les travaux récents sur le zooplancton de l'estuaire du Saint-Laurent sont passés en revue et discutés en mettant l'accent sur l'influence dominante des processus physiques sur la composition et la répartition spatiale. Long de 350 km, l'estuaire est divisé en deux moitiés par un seuil de 50 m qui sépare le profond estuaire maritime (350 m) de l'estuaire moyen (<100 m). A l'extrémité amont de l'estuaire moyen, des espèces de copépodes planctoniques

Coastal and Estuarine Studies, Vol. 39
M. I. El-Sabh, N. Silverberg (Eds.)
Oceanography of a Large-Scale Estuarine System
The St. Lawrence
© Springer-Verlag New York, Inc., 1990

(*Eurytemora affinis*) et suprabenthiques (*Ectinosoma curticorne*) et le mysidacé *Neomysis americana* maintiennent des populations endémiques au front salin (S = 1-15 °/oo). Plus en aval, l'abondance devient minimale avant d'augmenter à l'intérieur d'un bassin profond de 100 m adjacent au seuil (S = 25-30 °/oo). Là, le zooplancton est dominé par *Acartia longiremis*, *Eurytemora herdmani* et *Calanus finmarchicus* . Dans l'estuaire maritime, les copépodes dominants appartiennent au genre *Calanus* , principalement *C. finmarchicus* et, de façon secondaire, *C. hyperboreus* , représentés surtout par des stades avancés CIV-CVI. Quelques espèces de copépodes de plus petite taille y sont fréquentes, mais le genre *Pseudocalanus* est étonnamment rare. Les autres taxon importants sont les euphausides *Thysanoessa raschi* et *Meganyctiphanes norvegica* , qui se concentrent en denses couches diffusantes à la tête et le long du profond chenal.

La composition dans l'estuaire moyen est expliquée par le maintien de populations endémiques d'espèces bien adaptées aux estuaires. Par contre, le zooplancton de l'estuaire maritime est composé de populations dont le recrutement est entravé par l'évacuation rapide des eaux de surface combinée à la temperature froide de la colonne d'eau. Sa composition et sa répartition spatiale semblent dépendre en grande part de processus d'importation et d'exportation controlés par la circulation estuarienne. Plusieurs processus physiques inhérents à cet environment à forte énergie, tels que les marées internes, les résurgences d'eau profonde et l'advection dues à la marée, la dynamique des seuils, les ondes internes de haute fréquence, les tourbillons et les fronts, génèrent une forte variabilité dans la répartition spatiale du zooplancton à petite et moyenne échelle.

1. Introduction

Our knowledge of zooplankton in the St. Lawrence estuary (hereafter termed the SLE) is based on relatively few studies, a good number of which are unpublished master's or doctoral theses. Apart from a few rudimentary taxonomic lists published earlier than 1950 (Herdman et al., 1898; Willey, 1932; Préfontaine, 1936; Tremblay, 1942: reviewed in Bousfield et al., 1975), most research on zooplankton in the estuary has been carried out since about 1970. In this chapter, we present the first overview of these studies, making a special attempt to record the significant findings from the unpublished work.

While there is not a broad foundation of zooplankton research in the estuary, there are nevertheless certain distinctive characteristics of its zooplankton communities that emerge from the studies that have been done. A prevailing theme of our review is that these features are related to the estuary's dynamic physical environment. We will proceed first by summarizing the relevant physical oceanography of the SLE, then we will describe the observed characteristics of the zooplankton composition and abundance in each of the two natural divisions (Upper and Lower) of the estuary. To some extent we also describe zooplankton characteristics of the Saguenay fjord, although in many ways this is a separate system, which is treated in more detail in a separate chapter (Schafer et al., this volume). Following each description there is a discussion that combines previous research with some speculation of our own on how the zooplankton may be interacting with the physical environment. For the Upper estuary we focus on the relation of zooplankton population modes to the salinity gradient and the residual downstream transport. For the Lower

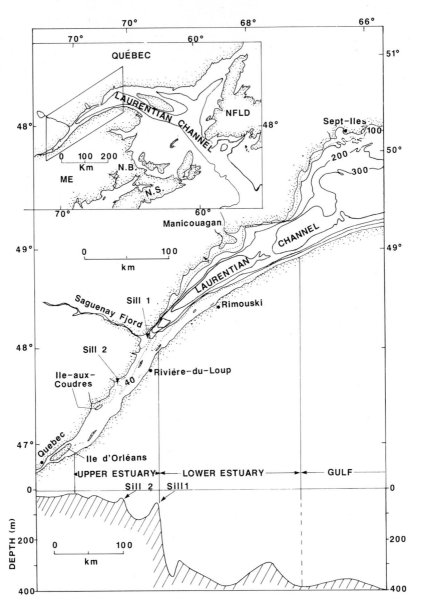

Figure 1. Map of the St. Lawrence Estuary and longitudinal section of main channel showing the two major sills.

estuary, we develop a hypothesis on the role of temperature and the two-layer estuarine circulation in determining zooplankton population dynamics and consequently the distinctive pattern in species composition. Finally, we review recent research on the influence of tidal currents, internal waves, and fronts on local zooplankton distribution in

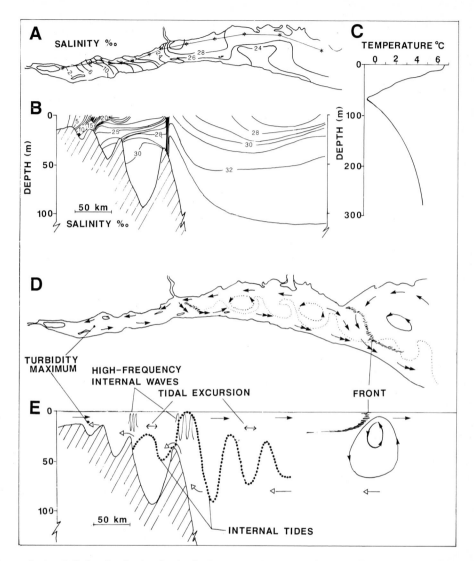

Figure 2. A: Salinity distribution in the St. Lawrence Estuary at the end of June at the surface. B: Salinities along a section, shown in A, in the main channel. C: Typical temperature profile in the Laurentian Channel. D and E: Mean horizontal and vertical circulation patterns showing fronts, eddies, and internal waves. A-C redrawn from Kranck (1979) and Greisman and Ingram (1977). D-E summarized from Bousfield et al. 1975, Deguise (1977), El-Sabh (1976; 1979), El-Sabh et al. (1982), Forrester (1974), Gratton et al. 1988, Ingram (1975; 1976; 1978), Koutitonsky (1979), Koutitonsky and El-Sabh (1985), Lavoie et al. (1984; 1985), Lie and El-Sabh (1983), Mertz et al. (1988), Muir (1979), Tang (1980; 1983).

the SLE, particularly in the region of transition between the Upper and Lower estuary. The variability induced by these physical phenomena must be considered when planning sampling programs and interpreting their results.

1.1. Relevant physical oceanography

Located in a temperate-subarctic climate between 47°and 49° N latitude, the SLE links the St. Lawrence River to the Gulf of St. Lawrence, which separates (or buffers) the estuary from the North Atlantic proper by a distance of about 1000 km. The estuary itself stretches 350 km, expanding from a depth of 20 m and width of 25 km at Ile d'Orléans to a depth of 350 m and width of 50 km at Manicouagan (Fig. 1). The deep Laurentian Channel extends unobstructed from the Gulf of St. Lawrence, tapering along the northern side of the estuary to the first sill, 50 m deep. This partial barrier effectively separates the estuary into upper and lower halves (Fig. 1). At this point, a long, narrow, and deep (350 m) fjord connects the Saguenay River to the estuary. A second sill, 50 km upstream and 30 m deep, isolates a 100 m deep basin. Mixing of the enormous freshwater discharge ($1-2 \times 10^4$ m^{-3} s^{-1}) from the St. Lawrence River starts just east of Québec City and salinities in the partially mixed Upper estuary range from 0°/oo near Ile d'Orléans to ~30°/oo at depth at the first sill (Fig. 2 a,b). A prominent feature in the region of mixing is a zone of maximum turbidity, located between Ile d'Orléans and Ile-aux-Coudres and caused by flocculation and resuspension of sediment (Krank, 1979; Gobeil et al.,1981; Lucotte and d'Anglejan, 1986). In the more stratified Lower estuary, which has in summer a three layer temperature structure with a minimum temperature as low as -1.5 °C at ~ 75-100 m (Fig. 2c), the salinity at depth increases up to 34°/oo (Fig. 2 b). The freshwater input from the St. Lawrence discharges principally along the southern side of the estuary. It is joined in the Lower estuary by the outflow of the Saguenay and these currents feed the Gaspé Current, a strong, narrow, but often unstable coastal jet (mean flow 50-75 cm s^{-1}) running along the Gaspe peninsula (Fig. 2d: El-Sabh, 1979; Therriault and Levasseur, 1985; Mertz et al., 1988). To counteract this outward flow, salt water moves in mainly at depth along the northern side in the Laurentian Channel. In addition to the residual circulation, tidal currents up to 300 cm s^{-1} occur as the tidal range increases from ~2 to ~4 m from the Gulf to Quebec City in response to the funnel-like shape of the estuary (Godin, 1979; El-Sabh, 1979).

2. Results and Discussion

2.1. Zooplankton in the Upper estuary

Data on abundance and composition of zooplankton in the Upper estuary can be found in Bousfield et al. (1975: reviewed in Miller,1983), Gagnon and Lacroix (1981; 1982; 1983), Courtois et al. (1982), Maranda and Lacroix (1983), and Dodson et al. (1989). The most extensive study along the entire stretch of the Upper estuary was carried out by Bousfield et al. (1975). They collected nearly 200 samples with a 2.5 cm diameter pump (strained through a 158 μm net) at approximately 40 stations on unspecified dates between May and September 1971. About 25 species from various phyla and classes were identified. From data in Table 1 in Bousfield et al. (1975), calanoid copepods constituted

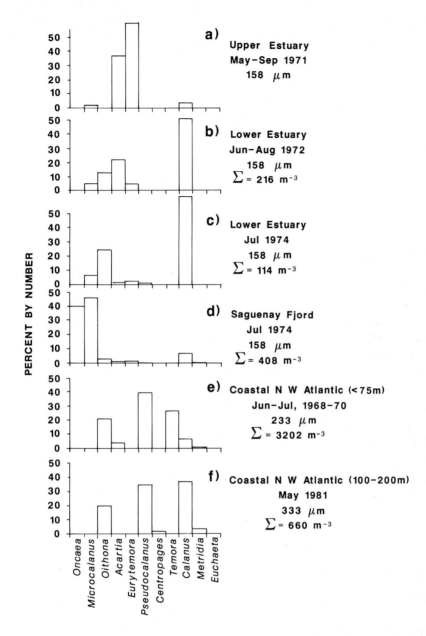

Figure 3. Average copepod species composition (percent by number of copepodite and adult stages) in various regions of the St. Lawrence Estuary and neighboring northwest Atlantic waters. Date of sampling period, mesh size of plankton net, and total average number of copepods during sampling period shown. Copepod genera arranged according to approximate adult body size, increasing from left to right. Data from a) Bousfield et al.(1975: does not include harpacticoid species), b) Côté (1972), c) Rainville (1979: Station 90), d) Rainville (1979: Station 290), e) Paranjape and Conover (1973), f) Gardner and Howell (1983).

roughly 60%, the harpacticoid copepod *Ectinosoma curticorne* about 20%, and the cladoceran *Bosmina longirostris* and larvae of the barnacle *Balanus crenatus* about 15% of the average maximum densities observed. The calanoid copepod community included *Eurytemora affinis*, *Acartia longiremis*, and *E. herdmani* accounting for 43%, 35%, and 17%, respectively, of all the calanoids in the plankton catches (Fig. 3a). Because of its large size, the oppossum shrimp, *Neomysis americana*, although constituting a relatively small proportion of the total number of plankton, figured prominently in terms of zooplankton biomass.

The distribution of these species was not uniform throughout the Upper estuary; rather, there were gradients in the mean distribution of the predominant crustacean taxa related to the progression from warm, brackish water at the head of the estuary to cold, saline water east of Sill 2 (Fig. 2). The relationship between species distribution and the horizontal salinity gradients in the Upper estuary has been discussed already in Miller's (1983) review of the zooplankton of estuaries, so we present here only a summary of the major results from Bousfield et al. (1975) in the form of Fig. 4. In general, freshwater species, especially *Bosmina*, *Ectinosoma*, and *Neomysis* were most abundant in the upstream portion just east of Ile d'Orléans. The mode of the *E. affinis* population was located in the brackish water between Ile d'Orléans and Ile-aux-Coudres, concentrated particularly at the front of the salt wedge at stations with average salinities of 1-5°/oo. *Acartia longiremis* and *Eurytemora herdmani* were most abundant in the higher salinity waters, from 20-28°/oo between Ile-aux-Coudres and east of the second sill. *Calanus finmarchicus* was most abundant in the cold, saline water > 25°/oo downstream of the sill. Barnacle larvae, not shown in Fig. 4, were most abundant downstream but were found as far upstream as the north channel east of Ile-aux-Coudres in waters as low as 6°/oo. Data covering smaller sections of the Upper estuary from the other studies mentioned earlier are consistent with this general pattern.

Because the same method was always used, Bousfield et al. could make valid comparisons of relative zooplankton abundance along the Upper estuary. They found that areas of highest concentration occur in the maximum turbidity zone east of Ile d'Orléans and further downstream in the deeper channels. Dodson et al. (1989) also concluded that the maximum turbidity zone was an important site of zooplankton biomass, especially for epibenthic macrozooplankton like *Neomysis* and to a lesser extent *Mysis stenolepsis* and the sand shrimp, *Crangon septemspinosus*. The area of lowest concentration occurred in the shallower middle portion of the Upper estuary upstream of sill 2 south and west of Ile-aux-Coudres, although it should be noted that this region was not well sampled. Bousfield et al. (1975) note that zooplankton concentrations were low relative to the Miramichi Estuary in the southern Gulf of St. Lawrence; however, the accuracy of the absolute abundance estimates (in terms of number per 450 L) is somewhat uncertain, as they found an 8-fold difference in abundance when comparing pump samples with the catch from a standard Clarke-Bumpus plankton net. More quantitative estimates of zooplankton abundance (no.m^{-3}) in summer downstream of Ile-aux-Coudres may be found in other

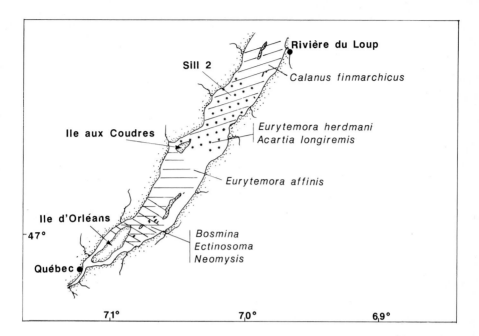

Figure 4. Approximate average distribution of predominant zooplankton taxa, summer, 1971. Zones delimit mode of maximum concentrations of each particular species or genus, using data in Bousfield et al. (1975: Table 1).

papers mentioned above, but overall the data are insufficient to confirm whether zooplankton biomass and productivity are indeed depressed relative to other estuaries.

2.2. Species gradients and retention mechanisms in the Upper estuary

The general distribution patterns shown in Figure 4 are based only on data averaged over 4 months during one summer, so there is likely to be considerable variability at any station and sampling time, depending on the tide (e.g. Gagnon and Lacroix, 1981: see section on variability below), season, and the particular hydrodynamic, climatic, and ecological characteristics of any given year. Nevertheless, it is clear that several species, including *Eurytemora affinis* , *Ectinosoma curticorne* , and *N. americana* are endemic to the Upper estuary, despite the net transport of water downstream. As well, *E. herdmani* and *A. longiremis* were actively reproducing in the Upper estuary, as evidenced by the abundance of juvenile stages, and it appears that *E. herdmani* and perhaps *Acartia* were maintaining a population mode upstream of the first sill (Bousfield et al., 1975; Miller, 1983; Gagnon and Lacroix, 1983).

Eurytemora , *Acartia* , and *Neomysis* are common estuarine taxa, and species from one or more of these groups maintain populations in most, if not all, major north temperate estuaries, including for example Narragansett Bay, Delaware Bay, Yaquina Bay, the

Sacremento-San Joaquin Estuary, and the Severn Estuary (Collins and Williams, 1982; Miller, 1983; Orsi, 1986). With regard to dominant zooplankton taxa, then, the Upper St. Lawrence estuary is not substantially different from other estuaries at similar latitudes. The same unresolved question -- how these species maintain their position in the face of the residual downstream transport (Miller,1983) -- is also fundamental to understanding zooplankton dynamics in the Upper estuary.

Mechanisms by which zooplankton populations may be retained in estuaries (along with representative references) involve the relationship of net flushing rate to population birth rate (Ketchum, 1954, reviewed in Miller, 1983), passive and behavioral interaction with tidal currents and net two-layer circulation patterns (review in Miller, 1983; Cronin and Forward, 1979; Forward and Cronin, 1980; Gagnon and Lacroix, 1983; Orsi, 1986; Soltonpour-Gargari and Wellershaus, 1987), and an epibenthic or partially epibenthic lifestyle, including egg stages (Grice and Marcus, 1981; Marcus, 1984; Miller, 1983). Only a few studies have explicitly addressed mechanisms of retention in the Upper St. Lawrence estuary. Dodson et al.(1989) point out that the same hydrodynamic processes responsable for maintaining the maximum turbidity zone, that is, the repeating cycle of seaward transport of sediment in the surface layer, settlement to the upstream-moving bottom layer, and rapid resuspension during flood tide at the head of the estuary, are probably also used by pelagic animals to maintain their position. They found that the center of distribution of *N. americana* coincided with the estimated position of the null zone, but could not evaluate whether this was a result of passive accumulation or active positioning. Based on results of an intensive sampling program at a station near Sill 2, Gagnon and Lacroix (1983) argued that high concentrations of *E.herdmani* upstream of the sill could be explained by an accumulation of individuals in the sill area resulting from the nature of the interaction between the vertical distribution of *Eurytemora* with turbulent tidal exchange across the sill. This hypothesis works less effectively for *A. longiremis* , which did not have an exchange coefficient consistent with the proposed mechanism of differential transport across the sill (at least for adult and copepodid stages).

While the behavioral components to retention mechanisms have not yet been emphasized in studies of Upper estuary zooplankton populations, they are undoubtedly important. Zooplankton have the capacity to control their vertical position in response to light gradients and probably, for species adapted to estuarine environments, in response to tidal rhythms (e.g. Cronin and Forward 1979; Forward and Cronin 1980; Orsi 1986), thereby also interacting with the two-layer circulation in more complicated and precise ways than nonliving particles. In addition, *Neomysis* and *Ectinosoma* are epibenthic species and presumably can seek refuge from advection in the bottom boundary layer. To some extent this is probably also true for species of *Eurytemora* (Miller 1983). For example, *E. affinis* is the dominant calanoid copepod found in tidal salt marsh pools along the southern shore downstream of the Upper estuary at Ile Verte (Castonguay, 1988). This is a localized distribution pattern that would not have been sampled by Bousfield et al. (1975). *Eurytemora* was found in high densities among the epibenthos in these shallow pools and may persist there despite periodic flooding during spring tides because of an ability to stay in the bottom epilayer. Nauplii of this species may also have a similar behavior enabling them to avoid export downstream. Finally, eggs released by at least some estuarine species (e.g. *Acartia* : Landry, 1978) can hatch or remain dormant on the bottom, which depending again on the behavior of the nauplii, could be effective in influencing and maintaining position. These behavioral aspects need to be examined in order to arrive at a complete understanding of the distribution and dynamics of zooplankton populations in the Upper estuary.

While estuarine species like Eurytemora may be adapted to cope with salinity changes and circulation patterns of the Upper estuary, freshwater-adapted species from upstream (e.g. Bosmina) and marine-adapted species from downstream (e.g. *Calanus*), are likely to be physiologically stressed if advected into the zone of mixing in the Upper section of the estuary. Bousfield et al. (1975) suggested that this zone is a "graveyard" of bodies of marine and freshwater animals that were trapped in water whose temperature and salinity characteristics were beyond their physiological capacity to survive. Studies of salinity and temperature tolerances of zooplankton (reviewed by Miller 1983) are not sufficient to indicate the effect of thermal and osmoregulatory shock on growth, survival, and susceptibility to predation of the zooplankton that are found in the Upper estuary. Maranda and Lacroix (1983) measured ATP content and dry weight in an attempt to assess tidal advection and mixing on variability of living zooplankton biomass, but the extent of *in situ* mortality related to rapid mixing of advected zooplankton in the Upper estuary is still unknown.

2.3. Zooplankton in the Lower estuary

Studies of the zooplankton of the Lower St. Lawrence estuary are limited to a few unpublished master's theses (Côté, 1972; Ouellet-Larose, 1973; Rainville, 1979), and two doctoral dissertations (Simard, 1985; Rainville, 1990.), parts of which have been published (Simard et al., 1985; Simard et al., 1986 a;b; Rainville and Marcotte, 1985). Aspects of the distribution and abundance of zooplankton in the Saguenay fjord were studied by Rainville (1979), De Ladurantaye and Lacroix (1980) and De Ladurantaye et al. (1984). A few other studies in the Gulf of St Lawrence and Lower estuary also include measurements of zooplankton displacement volume or biomass (Steven, 1974; Jacquaz et al., 1977; de Lafontaine et al., 1984) or euphausid abundance (Sameoto, 1976; Berkes, 1976).

2.3.1. The copepod community

About 30 calanoid, cyclopoid, and harpacticoid species of copepods have been reported in the Lower estuary; together they constituted 79-90% of the total zooplankton catch taken between May and October (Rainville, 1979). Our analysis of the copepod species composition in the Lower estuary is based on studies by Côté (1972) and Rainville (1979), which contain the most extensive data sets presently available. Both investigators made horizontal tows during the daytime with 158 μm mesh plankton nets at stations located between Rivière-du-Loup and Rimouski. Côté sampled at 8 nominal depths (0, 10, 20, 30 m and, depending on station depth, 50, 75, 100, and 150 m) during 4 cruises between 17 June and 22 August. Rainville towed at two nominal depths (25 and 100 m) during 6 cruises between 20 May and 9 October.

In both studies, eight species accounted for about 99% of the total number of copepods captured in summer (Fig. 3 b-c). A distinctive characteristic of the summer composition of the copepod community is the overwhelming predominance of species of *Calanus* . Over 50% of the copepods in the total catch were *Calanus* species, principally *Calanus finmarchicus* stages CV and CVI but also a substantial number of *C. hyperboreus* stages CIV to adult and probably some *C. glacialis* . *Pseudocalanus* , an important genus in temperate world oceans (Corkett and McLaren, 1978) made up only about 1% of the total catch, approximately the same level of abundance as the much larger *Metridia longa* and *Euchaeta norvegica* . Two calanoid species abundant in the lower portion of the Upper estuary, *Acartia longiremis* and *Eurytemora herdmani* , as well as the small *Microcalanus pygmaeus* and the cyclopoid, *Oithona similis* , were also common in the Lower estuary. *Scolecithricella minor*, another small calanoid, was frequently present but at low concentrations. Although *Microcalanus* and *Eurytemora* were relatively more abundant in September, the same copepod community was observed in early fall (Rainville, 1979). It should be noted that these results apply particularly to the deep (> 40 m) part of the Lower estuary. Species composition at the shallow margins close to shore, which has not been well documented, may comprise relatively more of the smaller taxa.

An entirely different composition was observed in July in the interior basin of the Saguenay fjord. Using the same mesh size and sampling design, Rainville (1979) found that three species, *Oncaea borealis* , *Oncaea similis* (two cyclopoid copepods which were not abundant in the Lower estuary) and *Microcalanus pygmaeus* , made up close to 90% of the copepods in the catch. *Pseudocalanus* was present in higher concentrations than in the Lower estuary, especially in late summer and early fall, but never exceeded 3% of the total number of copepods. In the first basin of the fjord, at its mouth, zooplankton composition is closer to that of the estuary, where *Calanus* , *Acartia* , *Eurytemora* , and *Oithona* predominate (Rainville, 1979; DeLadurantaye et al., 1984) DeLadurantaye et al. (1984) concluded that regular exchange of zooplankton occurred across the exterior sill, with a net export of *Acartia longiremis* from surface waters of the Saguenay and a net import of mesozooplankton from the Lower estuary. This exchange affects mainly the plankton composition in the outer two basins of the Saguenay fjord, however. The interior basin is more isolated from these exchange processes and contains water with arctic characteristics similar to the intermediate layer of the Lower estuary, supporting growth of endemic populations of *Microcalanus* and *Oncaea* (Rainville, 1979).

Because these results indicate important differences with respect to the copepod communities in neighboring waters (as discussed below), it is important to consider the potential sources of error that may have led to an inaccurate representation of the actual species composition. The mesh size of the plankton net is a factor: the larger the mesh size, the lower the relative abundance of smaller copepod species, including *Pseudocalanus* . However, both Côté and Rainville used a 158 μm mesh, which, while possibly undersampling juveniles and adults of smaller copepod species, is relatively fine compared to other studies on which our understanding of species composition in the North Atlantic is based (Fig. 3: see below). The sampling design is another factor: both investigations are based on horizontal net tows at depths that bias sampling toward surface water, which probably undersampled species whose daytime distribution was deeper than 100 m. It is reasonable to expect, however, that the bias would favor the smaller species, rather than the much larger *Calanus* species that typically occupy depths greater than 80 m during daylight hours. For these reasons, and because observations in other studies (Ouellet-Larose, 1973; Rainville and Marcotte, 1985; J.-C. Therriault, Institut Maurice Lamontagne, Mont-Joli, Québec, unpubl.) corroborate the basic findings of Côté and Rainville, we conclude that these results are an accurate representation of copepod species composition in these areas.

The distinctness of these summer copepod communities in the Lower estuary and Saguenay fjord can best be illustrated by comparison with species compositions from neighboring waters of the northwest Atlantic (Fig.3 e,f). Two studies, one in shelf water less than 75 m deep just outside St. Margaret's Bay, Nova Scotia (Paranjape and Conover, 1973) and one in shelf break water greater than 100 m on the southeast shoal of the Newfoundland Grand Banks (Gardner and Howell, 1983) illustrate the typical pattern. These data show a characteristic, shared predominance of *Oithona similis* , *Pseudocalanus* species, and either *Calanus finmarchicus* in deeper water or *Temora longicornis* or possibly *Centropages hamatus* and *Calanus* in the shallower coastal waters in the northwest Atlantic in summer. Despite the larger mesh sizes used in these studies (233-333-μm mesh), which would undersample the smaller species, this composition is not observed in the Lower estuary, where *Calanus finmarchicus* and *C. hyperboreus* are relatively more abundant than all smaller species combined and where *Pseudocalanus* is only a minor component of the community. While not conclusive, the data on total copepod concentration provided in Figure 3 do suggest that this pattern is due to a low abundance of the small species rather than an unusually high density of *Calanus*. Comparative data for the Gulf of St. Lawrence plankton in the deep Laurentian Channel do not exist. However, using a 366 μm mesh net, Lacroix and Filteau (1970) found that *Pseudocalanus* species were third in mean relative abundance after *Temora* and *Calanus* in summer in the Baie-des-Chaleurs in the southern Gulf, supporting the argument that the species composition in the Lower estuary is unusual. Summer copepod communities somewhat similar to the Lower estuary may occur in some regions in the high Canadian Arctic (Grainger, 1965; Mohammed and Grainger, 1974; Longhurst et al., 1984) and in the high latitude waters of northwestern Norway (Wiborg, 1954).

2.3.2. Other zooplankton

Zooplankton other than copepods in the Lower estuary includes protozoans, ostracods, gelatinous zooplankton, euphausiids, and ichthyoplankton. This last group is covered separately (see deLafontaine, Chapter 14) and will not be discussed here. Tintinnids in the Estuary and Gulf of St. Lawrence were studied by Cardinal et al. (1977), who identified 12 species, measured lorica dimensions, and discussed distribution characteristics, especially the absence of tintinnids with hyaline lorica in the Upper estuary. Simard (1985) correlated small scale changes in abundance of prominant taxa of ciliated protozoans with environmental variables along 3 transects in the northern part of the Lower estuary. Apart from these studies, microzooplankton has not been investigated. According to Rainville and Marcotte (1985), the ostracod, *Conchoecia elegans* is an important component of the zooplankton community in the Lower estuary, ranking third behind *Calanus finmarchicus* and *C. hyperboreus* in relative abundance. Therefore, it can be as abundant as smaller copepods in summer, but as yet we don't know enough about *Conchoecia* to evaluate its significance.

The prominence of euphausiids is a distinguishing characteristic of the non-copepod zooplankton community in the Lower estuary relative to other estuarine systems. Three species, *Meganyctiphanes norvegica* , *Thysanoessa raschi* , and *T. inermis* , are present (Berkes, 1976; Sameoto, 1976; Simard et al., 1986a;b). Using a high frequency echosounder, Simard et al. (1986a) found that, in July 1982, euphausiids were aggregated in a patch 1-7 km wide and at least 100 km long along the northern side of the Laurentian Channel (Figs. 5, 6c,d). The biomass within the patch was greater than 1 g dry wt..m^{-2} and concentrations up to 57 individuals.m^{-3} (840 mg dry wt.m^{-3}) were observed in net samples (Simard et al., 1986a;b). Euphausiid densities shoreward of the patch and in midchannel were typically less than 0.25 g dry wt.m^{-2}, but increased somewhat along the southern channel edge. The euphausiids caught were adults, predominantly *T. raschi* , which was found in highest daytime concentrations between 50 and 100 m. The daytime distribution of the larger *M. norvegica* was deeper, between about 100 and 175 m. Both species underwent a normal diel vertical migration (Simard et al., 1986b). While usually low in terms of relative numerical abundance, euphausiids typically make up more than 90% of the zooplankton biomass in areas where they are abundant (Simard et al., 1986a;b).

The apparent low abundance and infrequent occurrence of chaetognaths and gelatinous zooplankton (jellyfish, ctenophores, and larvaceans) in the Lower estuary is also notable. Côté (1972) reported finding chaetognaths, larvaceans and coelenterates in less than 10% of his samples. Rainville (1979) reports very low densities of *Oikopleura* and *Fritillaria* in the Lower estuary (in contrast, these larvaceans reached densities of up to 15.m^{-3} in August and September at stations in the Saguenay fjord). Steven (1974) did not quantitatively analyze zooplankton composition, but does report finding *Aglantha* and *Sagitta* in samples taken from the Lower estuary. Based on samples taken at a station off Rimouski, Rainville and Marcotte (1985) report that ctenophores and jellyfish are almost exclusively confined to the deep layer (greater than 150 m), which may explain the

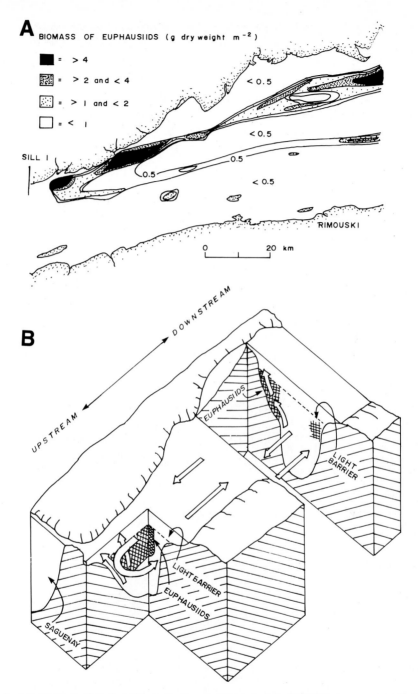

Figure 5. A: Aggregations of euphausiids in the Lower estuary in July 1982 based on an acoustic (104 kHz) survey. B: Conceptual model of the role of advection, upwelling, and light avoidance behavior in determining observed euphausiid distributions in the Lower estuary. (Redrawn from Simard et al. 1986b).

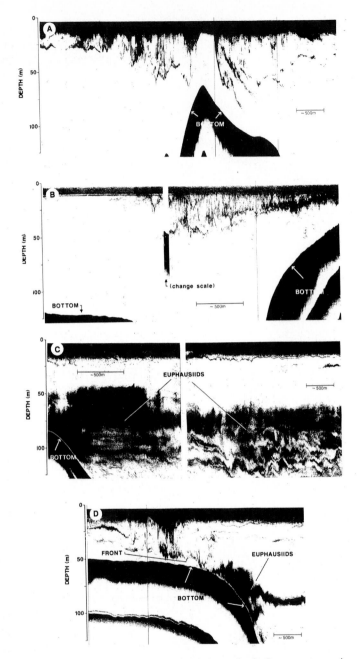

Figure 6. Echogram (104 kHz) of zooplankton scattering layers in the Lower estuary. A and B: Intensive vertical mixing of zooplankton scattering layers down to ~100 m in the vicinity of the first sill at the head of the Laurentian Channel in July (A) and May (B) 1982. C: Dense euphausiid scattering layer along the northern side of the channel at Les Escoumins. D: Same euphausiid layer along the channel further downstream with a vertical mixing front on the shelf.

uncommon occurrence of these taxa and chaetognaths in the studies of Côté (1972) and Rainville (1979). In any case, the record indicates a distinct scarcity of gelatinous zooplankton and chaetognaths in the upper layers (> ~ 150 m) of the Lower estuary.

2.4. Factors controlling zooplankton populations in the Lower estuary

In our discussion of zooplankton in the Upper estuary, we focused on mechanisms of retention of populations of zooplankton species adapted to estuarine conditions. In contrast, we will argue here that, despite the much larger volume of the Lower estuary, its particular physical characteristics restrain the development of endemic zooplankton populations and are responsible for the unusual mix of zooplankton found there.

The precursor for this hypothesis has been elaborated for species of euphausiids (Simard et al., 1986a) in the Lower estuary. They proposed that the observed spatial distribution of euphausiid aggregations results from the upstream advection in the Laurentian Channel of deep water containing adults of the three species (Fig. 5b). As with other vertical migrators, these animals avoid the surface 50 m during daylight, presumably an evolved behavior to escape from visual predation. As a consequence, they accumulate in dense aggregations, as observed (Fig. 5a, 6c,d), at the head of the Laurentian Channel in the vicinity of the first sill and along the northern side of the channel due to cross-channel transport. However, young stages produced by these individuals undergo much weaker and shallower vertical migrations (Lacroix, 1961), and consequently would be dispersed by cross channel currents toward the rapidly moving downstream surface layer along the southern side of the estuary. Instability in the surface flow and the frequent establishment of large estuary-wide gyres would complicate the pattern of advection of organisms in the surface layer. Nevertheless, net transport in the upper ~50 m in the Lower estuary is outward (El-Sabh, 1988). In summer, the retention time of the surface layer may be on the order of 1 month (El-Sabh, 1979) and sometimes as little as 1 week, according to drogue experiments (Fortier and Leggett, 1985), much shorter than larval development times. A similar outward drift of surface-dwelling capelin larvae has been discussed by Fortier and Leggett (1982; 1983; 1985). Therefore, it appears that larval stages produced from adult euphausiids in the Lower estuary are transported to the Gulf of St. Lawrence population, perhaps in the vicinity of the Magdalen plateau (Berkes, 1976). Conceivably, a portion of these individuals eventually return to the Lower estuary after developing to deeper-dwelling stages, during their second year.

This basic mechanism may also account for many of the other previously mentioned features of the zooplankton community. Species found deep in the water column in the Gulf, like older stages of *Calanus* (which overwinter at depth and typically undergo deep, diel vertical migrations at least during spring to fall: Simard et al., 1985) and possibly the ostracod *Conchoeica* (Rainville and Marcotte, 1985), would be advected into the Lower estuary along the Laurentian Channel. Smaller species like *Pseudocalanus* and *Oithona*, which are expected to be found in the Gulf but are more abundant in the upper water column, are not imported into the Lower estuary in the same relative proportion as *Calanus*. Other smaller species, like *Eurytemora* and *Acartia* , coming from the Upper estuary or the Saguenay and also existing mostly in the upper water column, would be subject to export from the Lower estuary (this implies that mechanisms allowing retention of species in the Upper estuary do not work well in the much deeper Lower estuary). The net result on the copepod community would be a tendency for a composition biased toward the larger, deeper-dwelling species like *Calanus* .

Other physical factors operating in the Lower estuary must also be considered. Summer temperatures in the surface waters of Lower estuary (3^0 to 11 ^{o}C: Fig. 2c and Mertz et al.,1988; Simard et al., 1986a) are considerably colder than in the Upper estuary or the Gulf of St. Lawrence. Consequently, eggs and early life history stages of copepods will take longer to develop and will in the meantime be subject to export from the Lower estuary because they normally tend to reside in the upper water column. An added complication is the very low temperature (to -1 ^{o}C: Fig. 2c) of the underlying water, which would result in even longer development times and possibly arrested development and increased mortality (Tande, 1988). In addition, the spring bloom occurs late, in June or July, in most parts of the Lower estuary (Therriault and Levasseur, 1985; Therriault et al., Chapter 12 this volume), so that the period of sufficient food availability required for high production of eggs and growth of young stages in many species (Mullin, 1988; Runge, 1988) is relatively short. We propose that these conditions hinder recruitment (i.e. annual input of young copepodite stages) of copepod species in the Lower estuary. While this would apply as well to *Calanus finmarchicus* , which may explain the observed rarity of early *Calanus* developmental stages, it may have the overall effect of reinforcing the differential transport mechanism discussed above.

Less is known about the vertical distribution and reproduction of gelatinous zooplankton in the region and therefore statements on the role of the circulation pattern in determining the low abundance of gelatinous species would be premature. Most coelenterates have a benthic phase (polyps) which normally settle in shallow water (< 50 m). The scarcity of such habitats in the Lower estuary may be a factor limiting their presence there (Y. deLafontaine, Institut Maurice-Lamontagne, Mont-Joli, Québec, pers. comm.).

In summary, then, we hypothesize that immigration and emigration rates dominate over birth (and mortality) rates of many zooplankton species in the Lower estuary, resulting in an aggregation of older stages of euphausiids at head and along the northern edge of the Laurentian Channel and a copepod community dominated by *Calanus* . This is in contrast to the Upper estuary, where several species maintain local populations despite the net downstream flow. However, the Upper estuary is a more typical situation, containing estuarine species adapted to this kind of environment, whereas we propose that the Lower estuary is a more special situation, containing oceanic and estuarine species not particularly

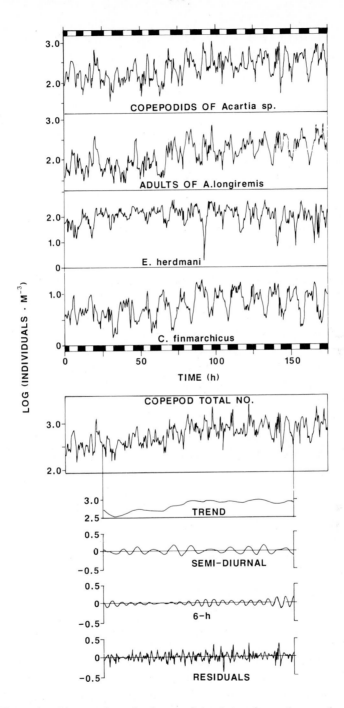

Figure 7. A: Time series of log transformed estimates of abundance of several copepod species at a station (water depth, 45 m) on the eastern edge of sill 2 during part of a neap-to-spring tide cycle, July, 1975. B: Filtered series showing the main components of the signal variability. (Redrawn from Gagnon and Lacroix, 1981).

adapted to its combination of physical conditions (circulation, temperature, depth) in a way that allows high local recruitment.

2.5. Zooplankton distribution in relation to tidal currents, internal waves, and fronts

In the high-energy environment of this large estuary, several physical processes affect the spatio-temporal variability of the distribution of zooplankton along the estuarine gradient, which generate large variance in samples. Among these processes are the vertical excursions of the water masses under the forcing of internal tides, the back and forth tidal horizontal displacements, the physical discontinuities due to the frontal dynamics, the flow instabilities and eddies, and the high-frequency internal waves and the localized turbulent mixing (e.g. Fig. 2D,E).

In the Lower estuary, an important component of the tidal variability is related to the strong internal tide (up to 120 m in amplitude) generated at the first sill and propagating seaward (Fig. 2e, Forrester, 1974; Therriault and Lacroix, 1976). Côté (1972) has found important tidal fluctuations of the biomass of zooplankton in his multi-strata sampling plan at the head of the Lower estuary. In this area the vertical distribution of the animals is strongly affected by this energic physical phenomenon, which seems to not only move the animals up and down but also mixes them over the water column as suggested by the high-frequency acoustic records in this area (Fig. 6a,b). The combination of this tide-induced upwelling of cold intermediate waters (Ingram, 1975; Levasseur and Therriault, 1987) with the local circulation, marked by the cross-channel flow driven by the Saguenay discharge (Ingram, 1983; Lebel et al., 1983), appears to result in the transport of a significant fraction of the upwelled biomass to the southern shelf (Ouellet-Larose, 1973) in the flushing jet of the estuary (Fig. 5b, see Simard et al., 1986a; Fortier and Leggett, 1985, Gratton et al., 1988; Mertz et al., 1988). This area at the head of the Lower estuary is obviously strongly dominated by complex physical phenomena generating a high variability in zooplankton samples, which limits our ability to interpret the data at scales smaller than ~50 km, corresponding to the wavelength of the internal tide.

The vertical displacement of the water masses by the internal tide plays a significant role in the exchange processes with the Saguenay fjord. The upstream progression of the deep zooplankton community is limited by the obstacle of the first sill at the head of the Laurentian Channel. However, due to the interaction of the mean and tidal circulation, the internal tide and the diel vertical migration behaviour of the animals, significant biomass passes over the sill to the upstream basin and the Saguenay fjord (Therriault et al., 1984; De Ladurantaye et al., 1984; Rainville, 1979). Exchange also occurs over the second sill; Gagnon and Lacroix (1983) have shown that the process at this location was controlled more by the turbulent flow (which included tidal and higher frequencies) than by the mean flow. As discussed by Gagnon and Lacroix (1983), the exchange would depend on the vertical distribution of the various species.

Tidal fluctuations of zooplankton biomass and species composition have been investigated in the Upper estuary in an extensive study of estuarine variability at the second sill and the adjacent basin (Gagnon and Lacroix, 1981; 1982; 1983; Maranda and Lacroix, 1983).Long-time series (175 h) with a high sampling frequency (once every 30 min.) at anchor stations have shown that the high level of zooplankton variability could be broken down into various components of 6, 12 and 24-h cycles superimposed on a low frequency trend (Fig. 7: Gagnon and Lacroix, 1981). The 12-h cycle reflects the semi-diurnal advection of long horizontal gradients (longer than the tidal excursion ~ 10 km) at the station. The 6-h cycle has been attributed to vertical motion of the water masses in response to an internal tide (Fig. 2e, Muir, 1979). The 24-h cycle was related to the diel vertical migration patterns of the species involved. These studies showed that the variance of zooplankton observations, as function of their temporal or spatial separation in the estuary, increases up to a plateau which corresponds to the scale of the main advective process. The statistical dispersion of zooplankton at a given sampling scale is not permanent but varies in time and space in response to advection and mixing (Gagnon and Lacroix, 1982).

In addition to vertical migration and the internal tide (12 h cycle), well known phenomena affecting the vertical distribution of zooplankton in the vicinity of sills and banks are high-frequency (< 1 h⁻¹) internal waves (Haury et al., 1979; 1983; Orr, 1981; Farmer and Smith, 1980). At both sills high-frequency internal waves are generated in response to the forcing of tidal currents over the obstacle (Fig. 2e, Deguise, 1977; Ingram, 1978). This obviously contributes to the dispersion of zooplankton at the sill and probably affects the rate of exchange between the basins as well as the sampling variability (Gagnon and Lacroix, 1981; 1982; 1983). An example of this dispersion is given in Fig. 6a,b which shows echograms of the channel cross-section at the first sill.

In the high-energy environment of the estuary, several physical processes act to generate fronts, creating discontinuities of water masses in many areas (Ingram,1976; 1978; 1983; 1985; Deguise, 1977; Gagnon and El-Sabh, 1980; Lavoie et al., 1984; 1985; Tang, 1980; Gratton et al., 1988; Mertz et al., 1988 - see also Ingram and El-Sabh, this volume). Most of them result from the turbulent breakdown of tidal energy and are consequently short-lived, with periods less than or equal to that of the tide. Significant exceptions are the transient (< 1 month) mesoscales eddies, with a diameter about half the width of the Lower estuary (Fig. 2D, Gratton et al., 1988), and the density front at the mouth of the estuary (Tang, 1980; 1983), which is a large-scale persistent feature. Convergence, mixing, and local upwelling phenomena are associated with these discontinuities (e.g. Tang, 1980; 1983; Ingram, 1976; 1978; 1985). The interaction of these turbulent processes with zooplankton behavior (e.g. buoyancy, swimming, phototaxis, temperature and salinity preference etc.) is known to generate plankton patchiness (Owen, 1981; Zeldis and Jillett, 1982; Yamamoto and Nishizawa, 1986). Small-scale zooplankton structures associated with these discontinuities in the estuary have been studied by Simard (1985), examples of which are shown in Fig. 6. Though they are transient structures, it is postulated that because of their high frequency of recurrence (mainly tidal M₂) they might play an important role in the local trophic network, because of their ability to generate aggregation of preys and predators (Simard, 1985). The more permanent density front at the mouth of the estuary most likely has a more important and persistant, but as yet unstudied, influence on zooplankton aggregations that occur there. The significance of the eddies in the retention of surface zooplankton is also unknown.

Acknowledgements

We thank Y. deLafontaine, G. Lacroix and L. Rainville for their comments on earlier versions of the manuscript.

REFERENCES

Berkes, F. 1976. Ecology of euphausids in the Gulf of St. Lawrence. J. Fish. Res. Bd. Can. 33, 1894-1905.

Bousfield, E.I., G. Filteau, M. O'Neill, and P. Gentes. 1975. Population dynamics of zooplankton in the middle St. Lawrence estuary, in Estuarine research, Vol I., L.E. Cronin ed., Academic Press, New York, 325-351.

Cardinal, A. P.-E. Lafleur, and E. Bonneau. 1977. Les tintinnides (Ciliata: Tintinnida) des eaux marines et saumâtres du Québec. I. Formes hyalines. Acta Protozoologica. 16, 15-22.

Castonguay, M. 1988. Le rôle de la prédation dans la structuration de la population d'*Eurytemora affinis* (Copepoda) d'un marais salé. M.Sc. thesis. Univ. Laval, Québec, 45 pp.

Collins, N. R. and R. Williams. 1982. Zooplankton communities in the Bristol Channel and Severn Estuary. Mar. Ecol. Prog. Ser. 9, 1-11.

Corkett, C.J. and I.A. McLaren. 1978. The biology of *Pseudocalanus* . Adv. Mar. Biol. 15, 1-231.

Côté, R. 1972. Influence d'un mélange intensif de différents types d'eau sur la distribution spatiale et temporelle du zooplancton de l'estuaire du Saint-Laurent. M.Sc. thesis. Univ. Laval, Québec. 266 pp.

Courtois, R., M. Simoneau, and J.J. Dodson. 1982. Interactions multispecifiques: répartition spatio-temporelle des larves de capelan (*Mallotus villosus*), d'éperlan (*Osmerus mordax*) et de hareng de l'Atlantique (*Clupea harengus harengus*) au sein de la communauté planctonique de l'estuaire moyen du Saint-Laurent. Can. J. Fish. Aquat. Sci. 39, 1164-1174.

Cronin, T.W. and R.B. Forward. 1979. Tidal vertical migration: an endogenous rhythm in estuarine crab larvae. Science. 205, 1020-1022.

Deguise, J.-C. 1977. High-frequency internal waves in the St.Lawrence estuary. M.Sc. thesis, McGill University, Montréal, Québec. 93 p.

de Lafontaine, Y.M., M. Sinclair, M.I. El-Sabh, C. Lassus and R. Fournier. 1984. Temporal occurrence of ichthyoplankton in relation to hydrographic and biological variables at a fixed station in the St. Lawrence estuary. Estuar. Coast. Shelf Sci. 18, 177-190.

de Ladurantaye, R. et G. Lacroix. 1980. Répartition spatiale, cycle saisonnier et croissance de *Mysis litoralis* (Banner, 1948) (Mysidacea), dans un fjord subarctique. Can J. Zool. 58, 693-697.

de Ladurantaye, R., J.-C. Therriault, G. Lacroix, and R. Côté. 1984. Processus advectifs et répartition du zooplankton dans un fjord. Mar. Biol. 82, 21-29.

Dodson, J.J., J.-C. Dauvin, R. G. Ingram, and B. d'Anglejan. 1989. Abundance of larval rainbow smelt (*Osmerus mordax*) in relation to the maximum turbidity zone and associated macroplanktonic fauna of the middle St. Lawrence estuary. Estuaries 12, 66-81.

El-Sabh, M.I. 1976. Surface circulation pattern in the Gulf of St. Lawrence. J. Fish. Res. Bd. Can. 33: 124-138.

El-Sabh, M.I. 1979. The Lower St.Lawrence estuary as a physical oceanographic system. Naturaliste can. 106, 55-73.

El-Sabh, M.I. 1988. Physical Oceanography of the St. Lawrence Estuary. in Hydrodynamics of Estuaries, Vol. II,, B. Kjerfve, ed., CRC Press, Boca Raton, 61-78.

El-Sabh, M.I., H.-J. Lie, and V.G. Koutitonsky. 1982. Variability of the near-surface residual current in the Lower St.Lawrence estuary. J. geophys. Res. 87, 9589-9600.

Farmer, D.M., and J.D. Smith. 1980. Tidal interaction of stratified flow with a sill in Knight Inlet. Deep-Sea Res. 27, 239-254.

Fortier, L., and W.C. Leggett. 1982. Fickian transport and the dispersal of fish larvae in estuaries. Can. J. Fish. Aquat. Sci. 39, 1150-1163.

Fortier, L., and W.C. Leggett. 1983. Vertical migrations and transport of larval fish in a partially mixed estuary. Can. J. Aquat. Sci. 40, 1543-1555.

Fortier, L., and W.C. Leggett. 1985. A drift study of larval fish survival. Mar. Ecol. Prog. Ser. 25, 245-257.

Forrester, W.D. 1974. Internal tides in the St.Lawrence estuary. J. Mar. Res. 32, 55-66.

Forward, R.B. and T.W. Cronin. 1980. Tidal rhythms of activity and phototaxis of an estuarine crab larva. Biol. Bull. 158, 295-303.

Gagnon, M. and M.I. El-Sabh. 1980. Effets de la marée interne et des oscillations de basse fréquence sur la circulation côtière dans l'estuaire du Saint-Laurent. Naturaliste Can. 107, 159-174.

Gagnon, M. and G. Lacroix. 1981. Zooplankton sample variability in a tidal estuary: an interpretative model. Limnol. Oceanogr. 26: 401-413.

Gagnon, M. and G. Lacroix. 1982. The effects of tidal advection and mixing on the statistical dispersion of zooplankton. J. Exp. Mar. Biol. Ecol. 56: 9-22.

Gagnon, M. and G. Lacroix. 1983. The transport and retention of zooplankton in relation to a sill in the Upper St. Lawrence estuary. J. Plankt. Res. 5: 289-303.

Gardner, G.A. and E.T. Howell. 1983. Zooplankton distribution across the shelf break on the southeast shoal of the Newfoundland Grand Banks in May 1981. Can. Ms. Rep. Fish. Aquat. Sci. no. 1724. 60 pp.

Gobeil, C., B. Sundby and N. Silverberg. 1981. Factors influencing particulate matter geochemistry in the St. Lawrence turbidity maximum. Mar. Chem. 10, 123-140.

Godin, G. 1979. La marée dans le golfe et l'estuaire du Saint-Laurent. Naturaliste can. 106, 105-121.

Grainger, E.H. 1965. Zooplankton from the Arctic Ocean and adjacent Canadian waters. J. Fish. Res. Bd. Can. 22, 543-564.

Gratton, Y., G. Mertz and J. Gagné. 1988. Satellite observations of tidal upwelling and mixing in the St. Lawrence estuary. J. geophys. Res. 93, 6947-6954.

Greisman, P. and R.G. Ingram. 1977. Nutrient distribution in the St. Lawrence Estuary. J. Fish. Res. Board Can. 34, 2117-2123.

Grice, G.B. and N. H. Marcus. 1981. Dormant eggs of marine copepods. Oceanogr. Mar. Biol. Ann. Rev. 19, 125-140.

Haury, L.R., M.G. Briscoe, and M.H. Orr. 1979. Tidally generated internal wave packets in Massachusetts Bay. Nature 278, 312-317.

Haury, L.R., P.H. Wiebe, M.H. Orr and M.G. Briscoe. 1983. Tidally generated high-frequency internal wave packets and their effects on plankton in Massachusetts Bay. J. Mar. Res. 41, 65-112.

Herdman, W.A., I.C. Thompson, and A. Scott. 1898. On the plankton collected continously during two traverses of the north Atlantic in the summer of 1989. Proc. Trans. Liverpool Biol. Soc. 12, 33-90.

Ingram. R.G. 1975. Influence of tidal-induced vertical mixing on primary productivity in the St.Lawrence estuary. Mem. Soc. R. Liège 6, 59-74.

Ingram, R.G. 1976. Characteristics of a tide induced estuarine front. J. geophys. Res. 81, 1951-1959.

Ingram, R.G. 1978. Internal wave observation off Ile Verte. J. mar. Res. 36, 715-724.

Ingram, R.G. 1983. Vertical mixing at the head of the Laurentian Channel. Estuar. coast. Shelf. Sci. 16, 333-338.

Ingram, R.G. 1985. Frontal characteristics at the head of the Laurentian Channel. Naturaliste can. 112, 31-38.

Jacquaz, B., K.W. Able, and W.C. Leggett. 1977. Seasonal distribution, abundance, and growth of larval capelin (*Mallotus villosus*) in the St. Lawrence estuary and northwestern Gulf of St. Lawrence. J. Fish. Res. Bd. Can. 34, 2015-2029.

Ketchum, B.H. 1954. Relation between circulation and planktonic populations in estuaries. Ecology 35, 191-200.

Koutitonsky, V. 1979. Transport de masses d'eau à l'embouchure de l'estuaire du Saint-Laurent. Naturaliste can. 106: 75-88.

Koutitonsky, V. and M.I. El-Sabh. 1985. Estuarine mean flow estimation revisited: application to the St.Lawrence estuary. J. Mar. Res. 43: 1-12.

Kranck, K. 1979. Dynamics and distribution of suspended particulate matter in the St.Lawrence Estuary. Naturaliste can. 106, 163-173.

Lacroix, G. 1961. Les migrations verticales journalières des euphausides à l'entréé de la Baie des Chaleurs. Naturaliste. can. 88, 257-316.

Lacroix, G. and G. Filteau. 1970. Les fluctuations quantitatives du zooplancton de la Baie des Chaleurs. II. Composition des copépodes du genre *Calanus*. Naturaliste can. 97, 711-748.

Landry, M.R. 1978. Population dynamics and production of a planktonic marine copepod, *Acartia clausi*, in a small temperate lagoon on San Juan Island, Washington. Int. Revue ges. Hydrobiol. 63, 77-119.

Lavoie, A., F. Bonn, J.M.M. Dubois et M.I. El-Sabh. 1984. Etude thermique des eaux de l'estuaire du Saint-Laurent à l'aide du satellite HCMM: résultats préliminaires. Proc. 8e Symposium Canadien de Télédétection, Montréal, mai 1983, A.Q.T./C.R.S.S. pp. 321-330.

Lavoie, A., F. Bonn, J.M.M. Dubois et M.I. El-Sabh. 1985. Structure thermique et variabilité du courant de surface de l'estuaire maritime du Saint-Laurent à l'aide d'images du satellite HCMM. Can. J. Remote Sensing 11, 70-84.

Lebel, J., E. Pelletier, M. Bergeron, N. Belzile, and G. Marquis. 1983. Le panache du Saguenay. Can. J. Fish.Aquat. Sci. 40, 52-60.

Levasseur, M., and J.-C. Therriault. 1987. Phytoplankton biomass and nutrient dynamics in a tidally induced upwelling: the role of the $NO_3:SiO_4$ ratio. Mar. Ecol. Prog. Ser. 39, 87-97.

Lie, H., and M.I. El-Sabh. 1983. Formation of eddies and transverse currents in a two-layer channel of variable bottom with application to the Lower St. lawrence Estuary. J. Phys. Oceanogr. 13: 1063-1075.

Longhurst, A., D. Sameoto, and A. Herman. 1984. Vertical distribution of Arctic zooplankton in summer: eastern Canadian archicpelago. Jour. Plankt. Res. 6, 137-168.

Lucotte, M. and B. D'Anglejan. 1986. Seasonal control of the Saint-Lawrence maximum turbidity zone by tidal-flat sedimentation. Estuaries. 9, 84-94.

Maranda, Y., and G. Lacroix. 1983. Temporal variability of zooplankton biomass (ATP content and dry weight) in the St.Lawrence estuary: advective phenomena during neap tide. Mar. Biol. 73, 247-255.

Marcus, N. 1984. Recruitment of copepod nauplii into the plankton: importance of diapause eggs and benthic processes. Mar. Ecol.-Prog. Ser. 15, 47-54.

Marshall, S.M. and A.P. Orr. 1955. The biology of a marine copepod. Oliver and Boyd. Edinburgh. 195 p.

Mertz, G., M.I. El-Sabh, D. Proulx and A.R. Condal. 1988. Instability of a buoyancy-driven coastal jet: The Gaspé Current and its St. Lawrence precursor. J. geophys. Res. 93: 6885-6893.

Miller, C.B. 1983. The zooplankton of estuaries, in Estuaries and enclosed seas, B.K. Ketchum, ed., Elsevier, Amsterdam, 103-149.

Mohammed, A.A. and E.H. Grainger. 1974. Zooplankton data from the Canadian Archipelago, 1962. Fish. Res. Bd. Can. Tech. Rep. No. 460.

Muir, L.R. 1979. Internal tides in the middle estuary of the St.Lawrence. Naturaliste can. 106, 45-54.

Mullin, M.M. 1988. Production and distribution of nauplii and recruitment variability -- putting the pieces together, in Toward a theory on biological-physical interactions in the world ocean, B.J. Rothschild, ed., Kluwer Academic, 297-320.

Orr, M.H. 1981. Remote acoustic detection of zooplankton response to fluid processes, oceanographic instrumentation, and predators. Can. J. Fish. Aquat. Sci. 38, 1096-1105.

Orsi, James J. 1986. Interaction between diel vertical migration of a mysidacean shrimp and two-layered estuarine flow. Hydrobiologia 137. 70-87.

Ouellet-Larose, D. 1973. Influence des marées sur les fluctuations à court terme des biomasses planctoniques dans l'estuaire du St.-Laurent. M.Sc. Thesis. Univ. Laval. 266 p.

Owen, R.W. 1981. Front and eddies in the sea: mechanisms, interactions and biological effects, in Analysis of marine ecosystems, A.R. Longhurst ed., Academic Press, London, 197-233.

Paranjape, M.A. and R.J. Conover. 1973. Zooplankton of St. Margaret's Bay, 1968-1971. Fish. Res. Bd. Can. Tech. Rep. no. 401. 83 pp.

Préfontaine, G. 1936. Additions à la liste d'espèce animales de l'estuaire du St. Laurent dans la région de Trois Pistoles. Trans. Roy. Soc. Canada 3, 205-209.

Rainville, L. 1990. First study of the zooplankton community of the laurentian Trough (Lower St. Lawrence Estuary): composition, vertical structure and ecology. Ph.D. thesis, McGill University, 175 pp.

Rainville, L. 1979. Etude comparative de la distribution verticale et de la composition des popuations de zooplancton du fjord du Saguenay et de l'estuaire maritime du Saint-Laurent. MSc. thesis. Univ. Laval, Québec. 175 pp.

Rainville.L.A. and B.M. Marcotte. 1985. Abundance, energy, and diversity of zooplankton in the three water layers over slope depths in the Lower St. Lawrence Estuary. Naturaliste can. 112, 97-103.

Runge, J.A. 1988. Should we expect a relationship between primary production and fisheries? The role of copepod dynamics as a filter of trophic variability. Hydrobiologia 167/168: 61-71.

Sameoto, D.D. 1976. Distribution of sound scattering layers caused by euphausiids and their relationship to chlorophyll a concentrations in the Gulf of St.Lawrence Estuary. J. Fish. Res. Board Can. 33, 681-687.

Simard, Y. 1985. Exploitation de l'acoustique sous-marine pour l'étude de la répartition spatiale et des rythmes de broutage du plancton dans un environnement turbulent. Ph.D. thesis. Univ. Laval, Québec.

Simard, Y. G. Lacroix, and L. Legendre. 1985. In situ twilight grazing rhythm during diel vertical migrations of a scattering layer of *Calanus finmarchicus* . Limnol. Oceanogr. 30, 598-606.

Simard, Y., R. de Ladurantaye, and J.-C. Therriault. 1986a. Aggregration of euphausiids along a coastal shelf in an upwelling environment. Mar. Ecol.-Prog. Ser. 32, 203-215.

Simard, Y., G. Lacroix, and L. Legendre. 1986b. Diel vertical migrations and nocturnal feeding of a dense coastal krill scattering layer (*Thysanoessa raschi* and *Meganyctiphanes norvegica* in stratified surface waters. Mar. Biol. 91, 93-105.

Soltonpour-Gargari, A., and S. Wellershaus. 1987. Very low salinity stretches in estuaries- the main habitat of *Eurytemora affinis* , a planktonic copepod. Meeresforsch. 31, 199-208.

Steven, D.M. 1974. Primary and secondary production in the Gulf of St. Lawrence. McGill Univ. Marine Sciences Center, Ms. Report No.26, 116 pp.

Tande, K. 1988. Aspects of developmental and mortality rates in *Calanus finmarchicus* related to equiproportional development. Mar. Ecol. Prog. Ser. 44, 51-58.

Tang, C.L. 1980. Mixing and circulation in the northwestern Gulf of St.Lawrence: a study of buoyancy-driven current system. J. geophys. Res. 85, 2787-2796.

Tang, C.L. 1983. Cross-front mixing and frontal upwelling in a controlled quasi-permanent density front in the Gulf of St.Lawrence. J. Phys. Oceanogr. 13, 1468-1481.

Therriault, J.-C., and G. Lacroix. 1976. Nutrients, chlorophyll and internal tides in the St.Lawrence estuary. J. Fish. Res. Bd. Can. 33, 2747-2757.

Therriault, J.-C., and M. Levasseur. 1985. Control of phytoplankton production in the Lower St.Lawrence Estuary: distribution and controlling factors. Naturaliste can. 112, 77-96.

Therriault, J.C., R. de Ladurantaye, and R.G. Ingram. 1984. Particulate matter exchange across a fjord sill. Est. Coast. Shelf Sci. 18, 51-64.

Tremblay, J.-L. 1942. Plancton. Rapp. Gén. Sta. Biol. Saint-Laurent, 1936-1942. 11 p.

Wiborg, K.F. 1954. Investigations on zooplankton in coastal and offshore waters of western and northwestern Norway. Fiskeridir. Skrifter. 11, 5-246.

Willey, A. 1932. Preliminary report on copepod plankton collection by the Station de Biologique du St.-Laurent à Trois Pistoles in July, 1931. Rapp. Sta. Biol. Saint-Laurent. 1, 82-84.

Yamamoto, T., and S. Nishizawa. 1986. Small-scale zooplankton aggregations at the front of a Kuroshio warm-core ring. Deep-Sea Res. 33, 1729-1740.

Zeldis, J.R., and J.B. Jillett. 1982. Aggregation of pelagic *Muninda gregaria* (Fabricius) (Decapoda, Anomura) by coastal fronts and internal waves. J. Plankt. Res. 4, 839-857.

Chapter 14

Ichthyoplankton Communities in the St. Lawrence Estuary: Composition and Dynamics

Yves de Lafontaine

Pêches et Océans Canada, Division de l'Océanographie Biologique Institut Maurice Lamontagne 850 Route de la Mer, Mont-Joli, Qué., Canada, G5H 3Z4

ABSTRACT

A review of the literature on the species composition and dynamics of ichthyoplankton in the St. Lawrence estuary is presented. In comparison with other estuarine systems, the ichthyoplankton fauna of the St. Lawrence estuary, which includes 35 species, is characterized by one of the lowest proportion (<0.05%) of larvae with pelagic eggs and the lowest diversity index (H'<1.0). Community structure analysis indicates that the St. Lawrence estuary is one of the most "estuarine" systems with regards to ichthyoplankton. The spatial distribution of the different species permits to divide the estuary into three distinct zones which correspond to well-established physical structures (frontal systems). The upstream Zone I is considered to be a nursery area for freshwater spawning species, principally smelt (*Osmerus mordax*) and tomcod (*Microgadus tomcod*). Both capelin and herring spawn within Zone II, but only herring are effectively retained in this area, which is considered to be a nursery ground for this species. Capelin larvae, due to their near-surface distribution, drift seaward into Zone III and eventually reach the adjacent waters of the Gulf of St. Lawrence. An analysis of ichthyoplankton communities and the evidence of a cross-channel gradient in the abundance and stages of various species suggest that the drift of ichthyoplankton is linked to the residual circulation along the south shore of the estuary and the Gaspe current in the Gulf. Temporal variability at any station is high at both low (seasonal) and high (tides, winds) frequencies and results in horizontal displacement of larval populations. Overall abundance peaks after the onset of seasonal stratification, but abundance maxima of different species are, however, distinct in time. In conclusion, the ichthyoplankton distribution in the St. Lawrence estuary is largely controlled by estuarine hydrodynamics, but factors regulating recruitment in the area remain unknown and the hypothesis that larval survival may be controlled by physical processes still needs to be tested.

Key Words: Fish larvae, fish eggs, diversity, transport, distribution, abundance, space-time variability

1. Introduction

In general, estuaries are considered to be 1) transitional zones between freshwater and marine ecosystems, 2) zones of pronounced environmental stress and 3) zones of high

Coastal and Estuarine Studies, Vol. 39
M. I. El-Sabh, N. Silverberg (Eds.)
Oceanography of a Large-Scale Estuarine System
The St. Lawrence
© Springer-Verlag New York, Inc., 1990

productivity. These three characteristics greatly affect the nature of the fish fauna found in estuaries (Haedrich, 1983). While the high productivity is a favourable condition, the environmental stress introduces physiological constraints that reduce the number of niches. The diversity of fish species in estuaries is therefore generally low and the degree of dominance by a few species is high (Oviatt and Nixon, 1973; Haedrich, 1983).

The preponderance of juvenile fishes in estuaries suggests that estuaries play an important role as nursery grounds for many fish species (Rogers, 1940; McHugh, 1967; Saila, 1973; Weinstein, 1979). Many adult fish species use estuaries primarily as feeding grounds and migrate to more "stable" areas for spawning, whereas others reproduce and complete their life cycle within estuaries. The sources of ichthyoplankton in estuaries can be two fold: either from local spawning linked to retention mechanisms or from later stages returning to the estuary from outside.

The life cycle and reproductive strategies of estuarine fishes appear adapted to ensure population continuity and integrity (Strathmann, 1981; Hedgecock, 1981). The gradients in estuarine physical properties delineate the zonation of fish fauna and in spawning sites, and consequently influence the initial distribution of fish eggs and larvae. Subsequently, the movement and the fate of ichthyoplankton in estuaries will be strongly related to rates of advection and diffusion. Since favourable distribution both in time and space is of utmost importance for survival, the relationship between ichthyoplankton and estuarine circulation may determine the year-class strength of estuarine-dependent fish stocks (Stevens, 1977; Crecco et al., 1986). It is therefore not surprising that much effort has been dedicated to elucidate the mechanisms permitting the retention of pelagic fish eggs and larvae within estuaries (Rogers, 1940; Dovel, 1971; Graham, 1972; Weinstein et al., 1980; Fortier and Leggett, 1982, 1983; several papers in Kennedy, 1981). It is now generally accepted that the movement of ichthyoplankton in estuaries is controlled by both active and passive behavior (Saila, 1973; Weinstein et al., 1980; Fortier and Leggett, 1982,1983).

Traditionally, the St. Lawrence estuary has not been a major fishing ground. This is probably why the Canadian Fisheries Expedition in 1914-15, which was the first ichthyoplankton survey in Canadian waters, sampled primarily in the Gulf of St. Lawrence and did not move up into the St. Lawrence estuary (Dannevig, 1919). Ichthyoplankton studies in the St. Lawrence estuary were initiated by Tremblay (1942), who collected herring larvae near a main spawning site at Isle-Verte and Trois-Pistoles along the south shore of the estuary. Leggett and colleagues from McGill University accomplished the first intensive ichthyoplankton sampling program in the estuary in 1973-75 (Jacquaz et al., 1977; Bailey et al., 1977; Able, 1978). Since then, numerous studies have contributed to our present knowledge and understanding of the ichthyoplankton communities in the region. The scope of this paper is to synthesize these studies of ichthyoplankton and their dynamics in relation to the physical and biological characteristics of the St. Lawrence estuary.

323

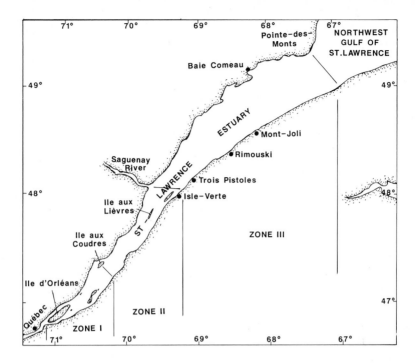

Figure 1. The St. Lawrence estuary divided into three zones based on ichthyoplankton information (see text).

2. Physical zonation of the estuary

Historically, the St. Lawrence estuary has been divided into two subareas (Upper and Lower), related to topographic differences (Brunel, 1970; El-Sabh, 1979). However, for the purpose of this paper I define three distinct zones within the estuary, separated by two well described frontal systems as previously suggested by Neu (1970) (Fig. 1). Zones I and II can be delineated by a salinity gradient observed near Ile-aux-Coudres (d'Anglejan and Smith, 1973; Kranck, 1979). A second frontal system near the mouth of the Saguenay river (Ingram, 1976,1985; Pingree and Griffiths, 1980) marks the boundary between zones II and III. Zone I, extending from Québec City to Ile-aux-Coudres, is characterized by low salinities (0-10 ^{o}/oo), a relatively homogeneous water column and high turbidity (d'Anglejan and Smith, 1973). Zone II, located between Ile-aux-Coudres and the mouth of the Saguenay Fjord, is partially mixed, with salinity values ranging from 10 to 25 ^{o}/oo (El-Sabh, 1979,1988). Zone III, typically referred to as the Lower estuary, represents the maritime portion of the estuary and is characterized by high salinities (>25 ^{o}/oo) and pronounced stratification during the summer months (El-Sabh, 1988). The three zones as defined here represent natural divisions in ichthyoplankton spatial distribution (Able, 1978; Powles et al., 1984) and are also evident in phytoplankton (Therriault et al., this volume) and zooplankton (Runge and Simard, this volume) distribution patterns.

3. Species composition

The ice cover precludes net sampling operations in the St. Lawrence estuary from late-November to the end of March. Because all fish egg and larval surveys have been conducted during the spring and summer months, the temporal description of ichthyoplankton fauna remains incomplete. In total, 35 species belonging to 20 families have been reported (Table 1). A few freshwater species found only in zone I are represented. The majority (27) of species found are typical inshore marine species of the north-western Atlantic (Leim and Scott, 1966). This is partially due to the sampling effort, which mainly concentrated in zones II and III, whereas zone I has been poorly investigated in the past. The high number of marine species in zone III includes many species which are more abundant and widely distributed in the adjacent waters of the Gulf of St. Lawrence (Kohler et al., 1977). Some of these marine species were probably found in zone II, but due to their low numerical abundance, the exact collection sites have not been reported in other studies (Able, 1978; Powles et al., 1984). Vladykov and McAllister (1961) indicated that 108 adult fish species (teleosts only) have been reported in Québec marine waters (seaward of Québec City). Twenty-nine species were of freshwater origin and 43 out of the 79 marine species were considered as common or frequent in the St. Lawrence estuary. The fact that only 27 larval species have been found so far (Table 1) suggests that other common marine species may not spawn in the estuary, or may spawn during the winter months.

TABLE 1.
Species composition of fish eggs (E) and larvae (L) for the St.Lawrence estuary. Types of egg deposition are pelagic (P), benthic (B) or ovoviviparous (O). An asterisk (*) indicates that the species was not collected in a particular zone. Reference sources are: (1) Able 1978; (2) de Lafontaine et al. 1984a,b; (3) Powles et al. 1984; (4) Johnson and Tremblay 1978; (5) de Lafontaine 1980; (6) Able and Irion 1984.

Family/Species	Egg Deposition	I	II	III	Source
Ammodytidae					
Ammodytes sp.	B	*	L	L	1,2,3
Anarhichadidae					
Anarhicas sp.	B			L	1
Anguillidae					
Anguilla rostrata	-			L	4
Aspidophoroidae					
Aspidophoroides monopterygius	B			L	1,2,3
Catostomidae					
Catostomus sp.	B	L	*	*	3
Clupeidae					
Clupea harengus	B		L	L	1,2,3
Alosa sapidissima	B	L	*	*	3
Cottidae					
Cottus sp.	B	L	*	*	3
Myoxocephalus aeneus	B			L	1,2,3
Myoxocephalus scorpius	B			L	1,2
Gymnocanthus tricuspis	B			L	1,2,3

TABLE 1. continued...

Family/Species	Egg Deposition	Zones I	II	III	Source
Cyclopteridae					
Cyclopterus lumpus	B			L	1,2,3
Gadidae					
Enchelyopus cimbrius	P			E,L	1,2,3
Gadus morhua	P		E	E,L	1,2,3
Microgadus tomcod	B	L	L	*	1,3
Urophycis tenuis	P			E,L	1,2
Gasterosteidae					
Gasterosteus aculeatus	B	L			3
Liparidae					
Liparis atlanticus	B		L	L	6,2,3
Liparis gibbus	B		L	L	6,2,3
Lumpenidae					
Lumpenus maculatus	B			L	1,2
Lumpenus lampretaeformis	B			L	1,2
Osmeridae					
Mallotus villosus	B	L	L	L	1,2,3
Osmerus mordax	B	L	L	*	1,3
Percidae					
Perca flavescens	B	L	*	*	3
Percichthyidae					
Morone americana	B	L	*	*	3
Pholidae					
Pholis gunnellus	B			L	1,2,3
Pleuronectidae					
Hippoglossoides platessoides	P	*	*	E,L	1,2
Limanda ferruginea	P			E,L	1,2
Liopsetta putnami	B	*	L	L	2,3
Pseudopleuronectes americanus	B	*	L	L	2,3
Reinhardtius hippoglossoides	P	*	*	L	5
Scombridae					
Scomber scombrus	P	*	*	E,L	1,2
Scorpaenidae					
Sebastes sp.	O	*	*	L	1,2,3
Stichaeidae					
Stichaeus punctatus	B			L	2,3
Ulvaria subbifurcata	B			L	1,2,3

The majority of fish species found in the St. Lawrence estuary are demersal spawners (Table 1). In terms of numerical abundance, the proportion of fish larvae with pelagic eggs in the St. Lawrence estuary is extremely low (<0.05%) relative to other estuarine systems (Table 2). The dominance of species with demersal eggs in estuaries is generally interpreted as an adaptation to decrease loss of eggs by drifting during incubation or to enhance retention of newly-hatched larvae via the deep landward circulation (Pearcy and Richards, 1962; Able, 1978). Table 2 reveals that the numerical proportion of fish larvae with pelagic

TABLE 2.
Number of families (F), species (S), S/F ratio values, Shannon-Weiner diversity index values (H') and percentage in abundance of fish larvae issued from pelagic eggs for year-round surveys of ichthyoplankton from different estuaries and other marine coastal and offshore regions. Annual range of temperature (Temp. ^{O}C) and salinity (Sal. ^{O}/oo) are given for each location when available.

Areas	Latitude	Z	F	S	S/F	%Pelagic	H'	Sal.	Temp	Source
ESTUARIES										
St. Lawrence	48-49N	300	20	36	1.75					This review
"	"		13	21	1.61	<0.01-0.2	.06-.34	20-31	2-14	deLafontaine et al.1984a
"	"		15	21	1.40	<0.01	.50-1.05	0-33	0-17	Powles et al. 1984
"	"		14	21	1.50	<0.05-2.0				Able 1978
Eastmain River	52N	7	9	11	1.22	<0.01		0-20	-1- 18	Ochman & Dodson1982
Yaquina River	44N		18	45	2.50	1.3	1.16			Pearcy & Myers 1974
Maine estuaries	43N	60	18	25	1.39	0.53	1.88	24-32	1-13	Chenoweth 1973
Mystic River	41N	6	4	33	1.46	2.9	1.01	3-30	0-25	Pearcy & Richards1962
Sandy Hook		9	17	21	1.41	8.2	1.74	8-28	-1-26	Croker 1965
Upper Chesapeake	38-39N	30	21	44	2.09	19.0	1.44	0-14	3-27	Dovel 1971
Lower Chesapeake	37-38N		22	34	1.54	90.5	1.10	16-28	3-26	Olney 1983
Chesapeake Bay (all)			30	58	1.93					
Cape Fear (a)	33N	12	27	45	1.67	93.3	1.86	0-35	1-28	Weinstein 1979
Cape Fear (b)	33N	12	26	46	1.77	70.5	1.77	"	"	Weinstein 1979
" (both a & b)	33N	12	31	59	1.90	84.3	2.00	"	"	Weinstein 1979
North Inlet Creek	33N	2	12	17	1.42	98.3	1.10	29-34	0-25	Bozeman & Dean 1980
Whangateau est.	36S	9	23	1	1.35				13-22	Roper 1986
Swartkops est.	33S	3	15	19	1.27		1.13	0-34	15-28	Melville-Baird & Smith 1980
"	33S	2	26	43	1.65			30-35	5-25	Beckley 1983
COASTAL ZONES & EMBAYMENTS										
Gulf St. Lawrence	46-49N	100	23	40	1.74			25-33	2-20	Kohler et al 1977
Gulf of Maine	41-44N	350	27	40	1.48	65.1	2.42	30-35	2-15	Marak et al. 1962
Salem harbor,Mass	42N		25	23	38	1.65	33.7	2.37	-1-22	Elliott et al. 1977
Cape Cod Bay	42N		30	21	34	1.62	61.3	2.35		Elliott et al. 1977
Narragansett Bay	41N		30	26	39	1.50	44.8	2.60	22-32	-1-24	Herman 1963
Block Island Snd.	41N		14	20	1.43	85.2	2.17			Merriman & Sclar 1952
Long Island Snd.	41N	20	26	40	1.54	84.3	2.76	26-35	15-26	Perlmutter 1939
"	41N		17	26	1.53	56.6	1.70			Richards 1959
Biscayne Bay	26N	3	29	>40	>1.38		2.22	26-36	14-32	Houde & Lovdal 1984
Coastal Louisiana	29N	12	48	107	2.23			20-35	16-31	Ditty 1986
Oregon coast	44N	60	21	73	3.47	29.4	2.00			Richardson & Pearcy 1977
Oregon offshore	44N	1000	21	52	2.47	61.5	1.76			"
Otsuchi Bay	39N	70	51	130	2.55		2.46			Yamashita & Aoyama 1984
North Pacific Gyre	28N	4000	54	183	3.39		3.87			Loeb 1979
"	28N	4000	36	113	3.13		3.22			Loeb 1979
Hawaii	21N	40	64	299	4.67					Miller 1974

eggs increases with decreasing latitude, exceeding 75% in low latitude estuaries (i.e. Cape Fear and North Inlet). Given the latitudinal gradient in water temperature, these results suggest that fish spawning habits in estuaries may be adapted to thermal conditions under which egg development proceeds. At low latitudes, warm waters will induce rapid development (maximum 2 days) of fish eggs which can hatch within the estuarine limits. In contrast, in cold-water, high latitude estuaries, such as the St. Lawrence, the incubation time for pelagic eggs would be too long relative to the flushing time of the estuary, thus favouring species with benthic eggs. At a given latitude, the numerical proportion of larval fish with pelagic eggs in estuarine systems is, however, slightly higher than for adjacent coastal marine regions, suggesting that estuarine circulation acts as an additional selective force.

Haedrich (1983) observed that the ratio of the number of species to families (S/F ratio) averaged 1.8 (range 1.5-1.9) for adult fish assemblages from different estuaries varying in location, size, type and phylogenetic composition. He suggested that, because of the high ratios (2.8-5.5) calculated for coral reefs and open ocean fish communities, only a moderate degree of diversification was possible in estuaries. The S/F ratio for the St. Lawrence ichthyoplankton community is 1.75 (data pooled from all sources -Table 1). The ratios for the 3 main surveys in this area are slightly lower (1.33-1.67) than the global estimate but are within the range of those calculated for other estuaries (Table 2). This is similar to Haedrich's (1983) observations for adult fish assemblages. However, the S/F ratios for estuarine ichthyoplankton are not very distinct from those calculated for marine coastal environments and embayments (Table 2). In this regard, the St. Lawrence is not less diversified than the adjacent waters of southern Gulf of St. Lawrence. Similarly, the S/F ratios for the Mystic river (Connecticut, USA) estuary and the adjacent waters of Long Island Sound or Block Island Sound are not different. Thus, the low S/F values do not specifically characterize estuarine systems and the S/F ratio appears too crude a measurement to distinguish between estuaries and coastal ichthyoplankton communities.

Species diversity indices and the pattern of the relative abundance of species are generally used to characterize and compare biological communities (Whittaker, 1970; May, 1976; Washington, 1984). Species diversity indices (using the Shannon-Weiner index H'), calculated from various ichthyoplankton surveys (fish larvae only), indicate that estuaries are significantly less diverse (H'<2.0) than marine coastal zones and bays (H'>2.0, Mann-Whitney U = 7.0, p<0.001) and open ocean regions (H'>3.0) (Table 2). Other diversity indices, such as the Simpson's D or McIntosh's M, (Washington, 1984) yield identical results and the ranking of the ichthyoplankton communities was found similar for all indices (Kendall tau > 0.95, p<0.0001). Moreover, among estuaries, the St. Lawrence has the lowest H' value (Table 2). The ichthyoplankton species diversity in the St. Lawrence estuary varies little during spring and summer months and is always less than that observed in the adjacent southern Gulf of St. Lawrence (de Lafontaine et al., 1984b).

From the analysis of the pattern of relative abundance of species, two major points arise. First, the relative abundance of ichthyoplankton (fish larvae only) as a function of species rank for the St. Lawrence estuary (Fig. 2a) approaches the geometric series pattern that theoretically characterizes communities structured by a high degree of dominance

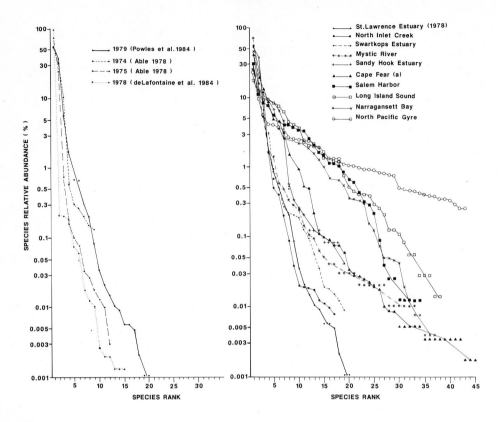

Figure 2. Log-percent relative abundance of larval fish against species rank for (A) four different surveys in the St. Lawrence estuary and for (B) other estuaries and marine coastal and offshore environments.

(Whittaker, 1970; May, 1976; Hubbell, 1979). Secondly, the curves for the St. Lawrence estuary are similar to those observed for other estuaries (i.e. Mystic river, North Inlet, Swartkops) (Fig. 2b). In fact, all estuarine ichthyoplankton communities show a more "geometric" pattern than the log-normal pattern depicted by marine coastal (i.e. Salem Harbor, Long Island Sound, Narragansett Bay) or offshore open communities (North Pacific gyre) (Fig. 2b). The transition from the geometric series to log-normal pattern parallels the increase in the species diversity index (H') from estuarine to offshore marine ichthyoplankton communities (Table 2). It can also be shown that, in estuaries, 5 to 15% of the overall number of species account for 90% of the total number of larvae present while between 25 and 45% of all species is required for marine coastal assemblages. The strong dominance in the St. Lawrence estuary is due to three species, capelin (*Mallotus villosus*) smelt (*Osmerus mordax*) and herring (*Clupea harengus*), which contribute to more than 95% of the total larval abundance (Able, 1978; de Lafontaine et al., 1984a; Powles et al., 1984).

Community structure analysis is normally done on adult individuals or sessile organisms and is rarely applied for transient stages such as ichthyoplankton (Loeb, 1979). The geometric pattern theoretically describes communities governed by a few species pre-empting part of the available niche space through strong competitive dominance (the niche pre-emption hypothesis - Whittaker, 1970; May, 1976). However, Hubbell (1979) hypothesized that such dominance-diversity patterns arise in ecosystems subjected to a high degree of large-scale, repetitive physical disturbance. For example, observed latitudinal changes in dominance patterns of forest communities are predicted by Hubbell's (1979) model. Although the exact interpretation of these patterns is still not clear (May, 1976; Hubbell, 1979) and is beyond the scope of this paper, the analysis of ichthyoplankton data indicates that 1) all estuarine ichthyoplankton communities are characterized by low H' values and geometric series patterns and 2) the St. Lawrence estuary appears to be the most "estuarine" system in regards to ichthyoplankton fauna. Due to the large energy input from freshwater run-off, tides and wind, estuaries appear more physically dynamic than adjacent marine regions (Sinclair et al., 1981; Legendre and Demers, 1984,1985). Considering the repetitive and large-scale effect of these energy sources in estuaries, it is suggested, as predicted by Hubbell (1979), that the structure of estuarine communities would be more strongly determined by physical forces than by biotic competitive influences. This reasoning possibly explains why the St. Lawrence estuary, which is characterized by strong semi-diurnal and neap-spring tidal cycles and by one of the continents largest drainage basins (run-off), appears to be one of the most "estuarine" systems as defined by relative abundance of ichthyoplankton species.

4. Spatial distribution

The above analyses consider the St. Lawrence estuary as a whole, but it is evident that each zone is characterized by a unique community structure, dominated by different species (Table 1).

The large-scale spatial distribution of ichthyoplankton in estuaries is determined by two factors: spawning location, and passive and active movements of ichthyoplankton in relation to circulation patterns. The vertical distribution in response to photoperiod and tidal flows will dictate the spatial distribution of ichthyoplankton in estuaries (Rogers, 1940; Wilkens and Lewis, 1971; Graham, 1972; Weinstein et al., 1980; Shenker and Dean, 1979; Shenker et al., 1983). Fortier and Leggett (1982,1983) demonstrated that newly-hatched larvae act as passive particles in the water. This results in a variable vertical distribution of fish larvae related to the variability in stratification and tidal mixing (Fortier and Leggett 1982). Large scale spatial dispersion for these very early larval stages is principally controlled by circulation processes that regulate salt transport. As swimming capabilities increase during larval growth, active vertical migration gradually develops and influences the horizontal gradients in ichthyoplankton abundance (Fortier and Leggett 1983).

The most landward zone (I) constitutes the main nursery ground for smelt (*Osmerus mordax*) and Atlantic tomcod (*Microgadus tomcod*) which are the two most abundant species in this zone (Able, 1978; Ouellet and Dodson, 1985; Powles et al., 1984). Both species are freshwater spawners. Smelt spawn in small tributaries along the south shore of zone I (Fréchet et al., 1983). The main spawning location for tomcod is at Ste-Anne-de-la-Pérade, about 75 km upstream of Québec (Roy et al., 1975). After hatching, larvae of both species drift rapidly toward the estuary where they are effectively retained, as suggested by their high numerical abundance (peak densities 1/m^3) and by the concentration of the largest individuals within zone I (Roy et al., 1975; Ouellet and Dodson, 1985). The vertical position of smelt larvae is dependent on light conditions (migration toward the surface at night) and tidal currents (concentration near the bottom during ebb tides and higher surface abundance during flood regimes). For most of zone I, residual flow is seaward over the entire water column but decreases from surface to bottom (Ouellet and Trump, 1979). This contributes to the downstream drift of larvae of freshwater origin (Ouellet and Dodson, 1985). Near Ile-aux-Coudres, the estuary becomes partially mixed and a net upstream flow of salt water exists near the bottom (Ouellet and Trump, 1979). The vertical migratory behavior in combination with the two-layer circulation allows smelt larvae to remain within zone I. The seaward limit of both smelt and tomcod in the estuary coincides with the strong salinity gradient near Ile-aux-Coudres (Able, 1978; Powles et al., 1984). Information on other species in this area is totally lacking.

Zone II is dominated by capelin (*Mallotus villosus*) and herring (*Clupea harengus*) larvae with peak densities of 10 and 1 per m^3 respectively. Beaches and tidal flats on both shores of zone II constitute the main spawning areas for capelin (Parent and Brunel, 1977) while herring spawn at a few sites between Ile-aux-Lièvres and Trois-Pistoles along the south shore (Coté et al., 1980; Henri et al., 1985). Hatching capelin larvae are rapidly exported toward the Lower estuary (zone III), eventually reaching the northwestern Gulf of St. Lawrence (Jacquaz et al., 1977; de Lafontaine et al., 1981, 1984b; Fortier and Leggett, 1982,1983,1985). In contrast, the high concentration of large herring larvae at the end of summer and during the fall suggests that these larvae are effectively retained within zone II, which is considered to be the main nursery area for the St. Lawrence estuary herring population (Able, 1978; Fortier and Leggett, 1982,1983,1984; Powles et al., 1984). There is, however, some evidence indicating that some herring larvae are transported seaward and eventually reach the Gulf waters (see below - Fig. 5; also Jacques Gagné, Fisheries and Oceans Canada, pers. comm.). The proportion of larvae being advected out relative to those retained in Zone II has not been quantified. The difference in horizontal dispersion of capelin and herring larvae is linked to their vertical distribution. Capelin larvae exhibit little vertical migration and their near-surface (0-10m) concentration above the depth of null longitudinal velocity (20-25m -Fortier and Leggett, 1982,1983) accelerates their seaward drift. On the other hand, herring larvae are more uniformly distributed over the water column and their vertical position fluctuates on a semi-diurnal basis. However, the vertical position of herring larvae, averaged over tidal cycles, is at the depth of null longitudinal velocity, which favours the maintenance within zone II (Fortier and Leggett, 1983). Fortier and Leggett (1984) argued, based on correlative approaches, that the semi-diurnal migrations of herring larvae are a behavioral response to fluctuations in the vertical distribution of larval food items and do not represent an active behavior related to the tidal flow. An alternative interpretation would be that both zooplankton and fish larvae exhibit a similar response to some environmental cues which promote retention. It is known that the zooplankton community in Zone II is unique and maintains its spatial integrity over time (Runge and Simard, this volume).

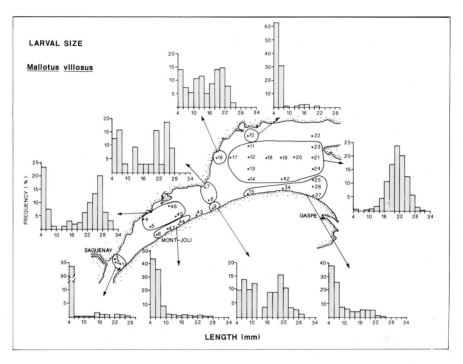

Figure 3. Length frequency distribution of capelin (*Mallotus villosus*) larvae from the Lower St. Lawrence estuary and the northwestern Gulf of St. Lawrence during mid-summer (July 25- August 3 1978).

Larvae of smooth flounder (*Liopsetta putnami*), winter flounder (*Pseudopleuronectes americanus*) and sand lance (*Ammodytes* sp.) are also frequently found in zone II, especially in nearshore habitats. The abundance of these species is relatively low, but increases seaward, with maximum concentration near Ile-aux-Lièvres (Powles et al., 1984). Spawning activity and the transport or retention of these larvae within the region are at present poorly described. The sporadic presence of pelagic fish eggs and fish larvae hatched from pelagic eggs indicates little spawning activity by these species in zone II (Able, 1978).

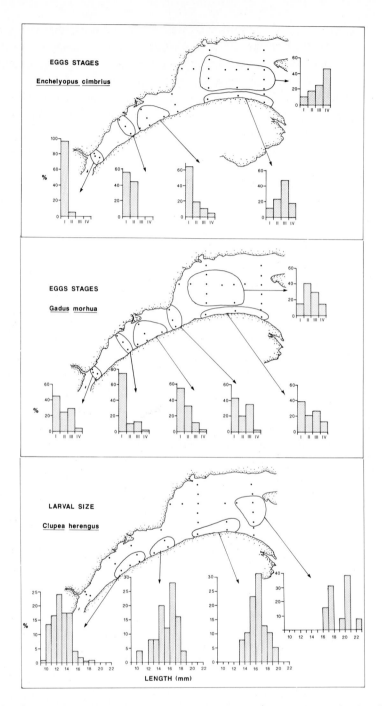

Figure 4. Frequency distribution of developmental stages (I to IV) of cod (*Gadus morhua*) and fourbeard rockling (*Enchelyopus cimbrius*) pelagic eggs and larval size of herring (*Clupea harengus harengus*) in the lower St. Lawrence estuary and the northwestern Gulf of St. Lawrence during mid-summer in 1978.

Although the number of fish egg and larval species is high in the Lower estuary (Zone III), this region is not a major retention area or nursery ground for larval fish. Sand lance dominate during the spring (>85% of all larvae); capelin, herring, redfish (*Sebastes* sp.), cod (*Gadus morhua*), and fourbeard rockling (*Enchelyopus cimbrius*) are the most common species during the summer. Capelin are most abundant, due to the seaward drift of larvae from upstream and to local spawning along the south shore (Parent and Brunel, 1977; de Lafontaine et al., 1984a). This downstream drift is clearly illustrated by the seaward increase in the modal length of capelin larvae (Fig. 3) and by the concentration of larger larvae in the offshore cyclonic gyre of the northwestern Gulf, where they spend their first year of life (Bailey et al., 1977). Large capelin larvae also occurred along the north shore in Zone III, but their abundance was one order of magnitude less than that observed in the northwestern Gulf (de Lafontaine et al., 1981). The seaward transport of spawning products in zone III is also evident from the spatial distribution of various stages of cod and rockling eggs and the seaward increase in size of herring larvae (Fig. 4). The presence of herring larvae may result from larval transport from upstream, although the existence of local spawning cannot be ruled out. The existence of a definite north-south gradient in the abundance and size of capelin larvae (Fig. 3) as well as in concentrations of other fish eggs and larvae (Fig. 5; see also Able, 1978; de Lafontaine et al., 1981,1984b) supports the hypothesis that larval drift is linked to the residual seaward circulation along the south shore within the Lower estuary (de Lafontaine et al., 1984b; Fortier and Leggett, 1985). This is also consistent with the observation that the Lower St. Lawrence estuary is one of the most laterally stratified estuaries in the world (Larouche et al., 1987), resulting in a strong southward cross-channel current at the surface and a seaward outflow along the south shore (El-Sabh et al., 1982). The low densities of ichthyoplankton along the north shore in zone III is also attributed to the lack of shallow coastal habitats suitable for spawning. These data support the hypothesis that the Lower St. Lawrence estuary (Zone III) does not represent a major retention area for ichthyoplankton and that the larval populations in Zone III are not discrete but are members of larger populations distributed in the adjacent Gulf of St. Lawrence.

A strong density front at the seaward limit of zone III separates the offshore Gulf waters and the estuarine waters which are diverted toward the Gaspe coastal current (Tang, 1980; de Lafontaine et al. 1981,1984b; El-Sabh and Benoit, 1984). An analysis of the ichthyoplankton community based on percent similarity indices

$$PSI = \Sigma min[a_i, b_i]$$

where a_i and b_i are the percent of the total individuals that belong to the ith species in samples A and B respectively (Loeb, 1979; McGowan and Walker, 1985), revealed a strong degree of similarity between stations of the Lower estuary (Zone III) and the Gaspe current (Fig. 6). Stations from the offshore northwestern Gulf formed an homogeneous and distinctly different group. Results obtained from spring and summer surveys suggest that seasonal changes in this large-scale pattern are minimal. The Gaspe current clearly represents the main pathway for the drift of larvae out of the estuary (Figs. 5 & 6).

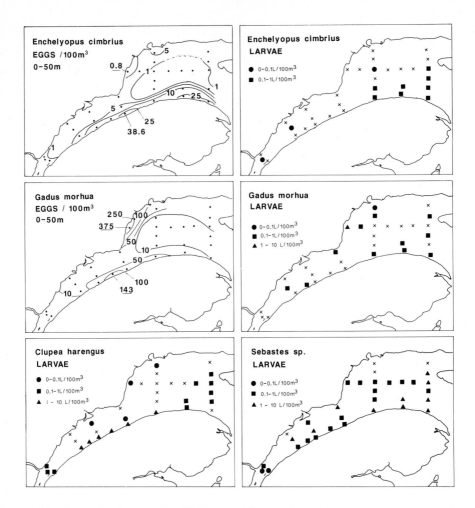

Figure 5. Abundance of various species of fish eggs and larvae in Lower St. Lawrence estuary and adjacent northern Gulf of St. Lawrence during mid-summer (July 25 - August 3, 1978).

While the spatial gradient in species composition and abundance of estuarine ichthyoplankton is not unique to the St. Lawrence estuary, the existence of three distinct zones separated by well defined frontal systems is not frequently observed in other estuaries. Moreover, the St. Lawrence estuary features two main retention areas (zones I and II) utilized by different species. In many estuaries along the western Atlantic the nursery niche is occupied by larval species spawned outside the estuary, which, through active migration, subsequently invade the estuary (Dovel, 1971; Wilkens and Lewis, 1971; Graham, 1972; Weinstein, 1979; Shenker and Dean, 1979; Shenker et al., 1983). Such a situation does not prevail in the St. Lawrence estuary, which serves as the nursery ground for freshwater spawners (Zone I - smelt and tomcod) and some local spawners (Zone II - herring). Presumably these nursery zones permit the existence of locally endemic fish

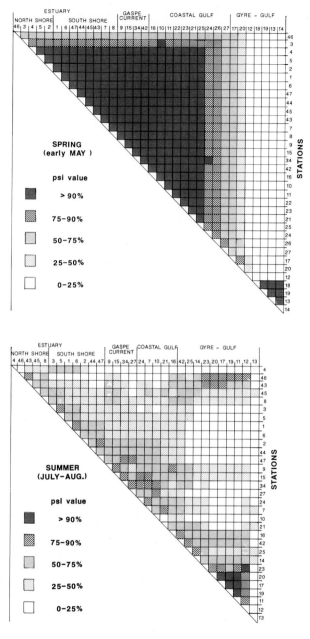

Figure 6. Matrices of percentage similarity indices (psi) for ichthyoplankton communities among stations in the Lower St. Lawrence estuary and the northwestern Gulf of St. Lawrence during spring and midsummer. Stations numbers correspond to those in Figure 3.

populations within the St. Lawrence estuary, although their genetic relationship to other populations in adjacent waters is far from clear (Rivière et al., 1985).

The exact mechanisms delineating the observed spatial transition of ichthyoplankton at the two frontal systems between the three zones are not well understood. Research in progress (Louis Fortier, Univ. Laval, pers. comm.) on the role of the tidally induced front separating zones II and III suggests that the seaward limit of larval herring dispersion coincides with the predicted location of the front, although significant numbers of larvae may be transported past the front along the south shore in zone III (see Fig. 5 for example). Similarly, the drift of capelin larvae across two frontal systems and the transport mechanisms prevailing at the convergence zone near the seaward limit of zone III deserve more detailed study.

TABLE 3.
Signs of correlations between abundance of various species of ichthyoplankton and physical water characteristics in the St. Lawrence estuary. Parentheses indicate that the correlation was not significant at p=0.05. Temp. = Temperature; Sal. = Salinity; Current = Longshore current (U).

Species	Zone	Temp.	Sal.	Current (U)	Secchi Depth	References
Herring	I	+	+	+		Fortier & Leggett 1982,1983
	II	+	+		+	Powles et al. 1984
	III	+	(-)			de Lafontaine et al. 1984a
Capelin	II	+	+	+		Fortier & Leggett 1982,1983
	II	(-)	+		+	Powles et al. 1984
	III	+	-			de Lafontaine et al. 1984a
Smelt	II	+	-			Fortier & Leggett 1982
	I & II	+	-	-		Powles et al. 1984
	I			+		Ouellet & Dodson 1985
Sand lance	III	-	(+)			de Lafontaine et al. 1984a
Cod eggs	III	-	+			de Lafontaine et al. 1984a
Hake eggs	III	(-)	+			de Lafontaine et al. 1984a

5. Temporal variability

High temporal variability is normally expected in estuarine systems. In the St. Lawrence estuary, freshwater runoff contributes primarily to the low frequency or seasonal variability, while winds and tides, principally the neap-spring (14.8 days) and the semi-diurnal (12.4 hours) cycles, are responsible for short-term or high frequency variations in ichthyoplankton abundance at a given station (de Lafontaine et al., 1984a). At present, the seasonal occurrence of ichthyoplankton is incompletely described for the upstream zone I,

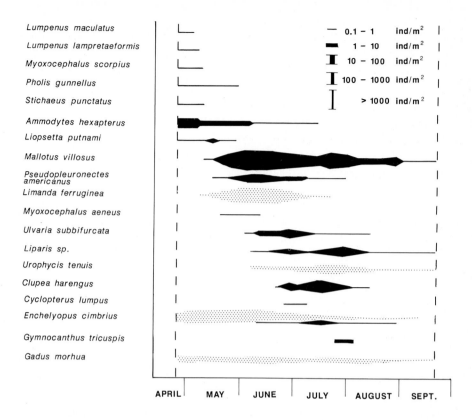

Figure 7. Seasonal occurrence of fish eggs and larvae off Rimouski in the St. Lawrence estuary. Black and cross-shaded lines represent larvae and fish eggs respectively.

but the ichthyoplankton calendar (Fig. 7), established at a fixed station in front of Rimouski, appears valid for zones II and III. Although total larval abundance peaks in June when newly hatched capelin larvae are present, the abundance maxima of different species are very distinct in time and exhibit little temporal overlap (de Lafontaine et al., 1984a). It was concluded that the breeding strategy of fish, through fixed spawning times and resulting hatching times, ensures a temporal succession in the appearance of larvae. When temporal overlap exists, the different spatial distributions of co-occurring species contribute to minimize interspecific competition during the larval drift (Courtois et al., 1982; de Lafontaine et al., 1984a). The seasonal peak in the abundance of most larval species in zone III occurs after the onset of stratification but before the seasonal peak in primary production. This pattern differs from that reported in adjacent southern Gulf of St. Lawrence waters (de Lafontaine et al., 1984b) and other northern temperate marine waters (Cushing, 1975). The timing of ichthyoplankton occurrence in the St. Lawrence estuary does not appear directly linked to the production cycle. Presumably fish larvae in zone III would take advantage, through larval drift, of highly productive zones located further downstream (de Lafontaine et al., 1981,1984a,b). The question as to whether the temporal succession confers any selective advantage to ichthyoplankton remains open.

As expected from the tidal hydrodynamics of the St. Lawrence estuary, short-term variability in ichthyoplankton abundance is high at neap-spring and semi-diurnal tidal frequencies (Fortier and Leggett, 1982,1983,1984; Courtois et al., 1982; de Lafontaine et al., 1984a; Ouellet and Dodson, 1985; Henri et al., 1985). Using time-series analyses, Fortier and Leggett (1982,1983,1984) demonstrated that short-term fluctuations in ichthyoplankton abundance correspond to small-scale tidal displacements of water masses occupied by larvae. Significant correlations between larval abundance and the physical properties of water masses have been reported for most common species, as expected by postulating a passive dispersion (Table 3). Although the signs of the correlations for a given species are similar within a zone, they vary, however, between sampling zones. This probably reflects the large-scale horizontal gradients that exist in the St. Lawrence estuary as a result of localized spawning sites and particular larval drift patterns.

5. Conclusions

Considerable progress has been made in our understanding of the distribution of ichthyoplankton in the St. Lawrence estuary during the past 15 years. Much remains to be done on the population dynamics of these species, principally on the trophic links between larval fish, their prey and predators and their effects on growth, survival and recruitment. The diets of the most common species (smelt, herring and capelin) have been described (Vésin et al., 1981; Courtois and Dodson, 1986) but additional work on the importance of food abundance and availability on larval survival is required. Larval predators are still unknown for the St. Lawrence estuary. Large macrozooplankton (i.e. Coelenterates and Ctenophores), which are recognized as voracious predators in other environments (Moller, 1984; Purcell, 1985; de Lafontaine and Leggett, 1988), are virtually absent in the St. Lawrence estuary; so adult fish and juveniles are presumably the major predators of ichthyoplankton in this area.

The highly dynamic nature of estuaries makes even more difficult the verification of some general hypotheses on larval fish ecology. The work achieved by Fortier and Leggett (1984), who showed that the coherence between herring larvae and their microzooplanktonic prey varies with the scale of observations, clearly indicates that our understanding of estuarine ichthyoplankton dynamics and survival requires that appropriate sampling designs be a function of the circulation and hydrodynamic properties. Studies on the role of the nursery zones and frontal systems on the feeding strategies and survival of fish larvae have been recently initiated (Louis Fortier, Univ. Laval and Jacques Gagné, Fisheries and Oceans Canada) and will hopefully bring additional information on the meso-scale ecological relationships considered important during larval development. It is however clear that the St. Lawrence estuary is not as important a breeding ground for marine fish as other estuaries along the east coast of United States. The nursery role played by the St. Lawrence is principally restricted to zone I which is used by freshwater spawning species like smelt and tomcod.

Factors regulating the recruitment of fish populations in the St. Lawrence estuary (as well as in many other areas) are still largely unknown. In contrast with other estuarine systems where large-scale correlative approaches have suggested that recruitment variability is related to hydrodynamic and climatic forces (Crecco et al., 1986), the lack of suitable data on adult fish populations (due to low exploitation rates) has hindered such analyses for the St. Lawrence estuary. This lack of appropriate data is also foreseen as a major drawback for future testing of possible relationships suggested by ichthyoplankton research in this area. Information on interannual variations in larval fish abundance and growth is lacking. Although it is clear that the ichthyoplankton distribution in the St. Lawrence estuary is largely controlled by estuarine hydrodynamics, the hypothesis that larval survival is also controlled by similar processes remains to be tested.

REFERENCES

Able, K.W. 1978. Ichthyoplankton of the St. Lawrence estuary: Composition, distribution and abundance. J. Fish. Res. Bd Can., 35, 1517-1531.

Able, K.W. and W. Irion. 1984. Distribution and reproductive seasonality of snailfishes and lumpfishes in the St. Lawrence estuary and the Gulf of St. Lawrence. Can. J. Zool., 63, 1622-1628.

Bailey, R.F.J., K.W. Able and W.C. Leggett. 1977. Seasonal and vertical distribution and growth of juvenile and adult capelin (*Mallotus villosus*) in the St. Lawrence estuary and western Gulf of St. Lawrence. J. Fish. Res. Bd Can., 34, 2030-2040.

Beckley, L.E. 1983. The ichthyoplankton fauna associated with *Zostera capensis* Setchell in the Swartkops estuary, South Africa. S. Afr. J. Zool., 18, 15-24.

Bozeman, E.L. and J.M. Dean. 1980. The abundance of estuarine larval and juvenile fish in a South Carolina intertidal creek. Estuaries, 3, 89-97.

Brunel, P. 1970. Bibliographie choisie sur l'océanographie de l'estuaire du Saint-Laurent. Revue Géographie Montréal, 24, 277-282.

Chenoweth, S.B. 1973. Fish larvae of the estuaries and coast of central Maine. U.S. Fish. Bull., 71, 105-113.

Côté, G., P. Lamoureux, J. Boulva and G. Lacroix. 1980. Séparation des populations de hareng de l'Atlantique (*Clupea harengus harengus*) de l'estuaire du Saint-Laurent et de la péninsule gaspésienne. Can. J. Fish. Aquat. Sci., 37, 66-71.

Courtois, R. and J.J. Dodson. 1986. Régime alimentaire et principaux facteurs influençant l'alimentation des larves de capelan (*Mallotus villosus*), d'éperlan (*Osmerus mordax*) et de hareng (*Clupea harengus harengus*) dans un estuaire partiellement mélangé. Can. J. Fish. Aquat. Sci., 43, 968-979.

Courtois, R., M. Simoneau and J.J. Dodson. 1982. Interactions multispécifiques: répartition spatio-temporelle des larves de capelan (*Mallotus villosus*), d'éperlan (*Osmerus mordax*) et de hareng de l'Atlantique (*Clupea harengus harengus*) au sein de la communauté planctonique de l'estuaire moyen du Saint-Laurent. Can. J. Fish. Aquat. Sci., 39, 1164-1174.

Crecco, V., T. Savoy and W. Whitworth. 1986. Effects of density-dependent and climatic factors on American shad, *Alosa sapidissima* , recruitment: a predictive approach. Can. J. Fish. Aquat. Sci., 43, 457-463.

Croker, R.A. 1965. Planktonic fish eggs and larvae of Sandy Hook estuaries. Chesap. Sci., 6, 92-95.

Cushing, D.H. 1975. Marine Ecology and Fisheries. Cambridge Univ. Press, Cambridge, 278p.

d'Anglejan, B.F. and E.C. Smith. 1973. Distribution, transport and composition of suspended matter in the St. Lawrence estuary. Can. J. Earth Sci., 10, 1380-1396.

Dannevig, A. 1919. Biology of Atlantic waters of Canada. Canadian Fish Eggs and Larvae. in: J. Hjort, Canadian Fisheries Expedition, 1914-15, in the Gulf of St. Lawrence and Atlantic waters of Canada. Dept. Nav. Serv. Can., Ottawa, Ont.

de Lafontaine, Y. 1980. First record of Greenland halibut larvae (*Reinhardtius hippoglossoides*) in the lower St. Lawrence estuary. Nat. Can., 107, 285-287.

de Lafontaine, Y. and W.C. Leggett. 1988. Predation by jellyfish on larval fish: an experimental evaluation employing *in-situ* enclosures. Can. J. Fish. Aquat. Sci., 45, 1173-1190.

de Lafontaine, Y., M. Sinclair, S.N. Messieh, M.I. El-Sabh and C. Lassus. 1981. Ichthyoplankton distributions in the northwestern Gulf of St. Lawrence. Rapp. P-v. Réun Cons. Int. Explor. Mer, 178, 185-187.

de Lafontaine, Y., M. Sinclair, M.I. El-Sabh, C. Lassus and R. Fournier. 1984a. Temporal occurrence of ichthyoplankton in relation to hydrographic and biological variables at a fixed station in the St. Lawrence estuary. Est. Coast. Shelf Sci., 18, 177-190.

de Lafontaine, Y., M.I. El-Sabh, M. Sinclair, S.N. Messieh and J-D. Lambert. 1984b. Structure océanographique et distribution spatio-temporelle d'oeufs et larves de poissons dans l'estuaire et la partie nord-ouest du Golfe Saint-Laurent. Sci. Tech. Eau, 17, 43-50.

Ditty, J.G. 1986. Ichthyoplankton in neritic waters of the northern Gulf of Mexico off Louisiana: composition, relative abundance and seasonality. U.S. Fish. Bull., 84, 935-946.

Dovel, W.L. 1971. Fish eggs and larvae of the upper Chesapeake bay. NRI Spec. rept 4, Nat. Res. Inst. Univ. Maryland, Contrib 460, 71p.

Elliott, E.M., D. Jimerez, C.O. Anderson Jr and D.J. Brown. 1979. Ichthyoplankton abundance and distribution in Beverly-Salem harbor, March 1975 through February 1977. Div. Mar. Fish., Dept. Fish Wildl. Recr. Vehicles Comm., Mass. 174 p.

El-Sabh, M.I. 1979. The St. Lawrence estuary as a physical oceanographic system. Nat. Can., 106, 55-73.

El-Sabh, M.I. 1988. Physical oceanography of the St. Lawrence estuary, in Hydrodynamics of Estuaries, Volume II, Chapter 15, B. Kjerfve, ed., CRC Press Inc, Boca Raton, Florida, USA, 61-78.

El-Sabh, M.I. and J. Benoit. 1984. Variabilité spatio-temporelle du courant de Gaspé. Sci. Tech. Eau., 17, 55-61.

El-Sabh, M.I., H.J. Lie and V.G. Koutitonsky. 1982. Variability of the near-surface residual current in the lower St. Lawrence estuary. J. Geophys. Res., 87, 9589-9600.

Fortier, L. and W.C. Leggett. 1982. Fickian transport and the dispersal of fish larvae in estuaries. Can. J. Fish. Aquat. Sci., 39, 1150-1163.

Fortier, L. and W.C. Leggett. 1983. Vertical migrations and transport of larval fish in a partially mixed estuary. Can. F. Fish. Aquat. Sci., 40, 1543-1555.

Fortier, L. and W.C. Leggett. 1984. Small-scale covariability in the abundance of fish larvae and their prey. Can. J. Fish. Aquat. Sci., 41, 502-512.

Fortier, L. and W.C. Leggett. 1985. A drift study of larval fish survival. Mar. Ecol. Prog. Ser., 25, 245-257.

Fréchet, A., J.J. Dodson and H. Powles. 1983. Use of variation in biological characters for the classification of anadromous rainbow smelt (*Osmerus mordax*) groups. Can. J. Fish. Aquat. Sci., 40, 718-727.

Graham, J.J. 1972. Retention of larval herring within the Sheepscot estuary of Maine. U.S. Fish. Bull., 70, 299-305.

Haedrich, R.L. 1983. Estuarine Fishes. in: B.H. Ketchum (ed.), Estuaries and Enclosed Seas. Elsevier Scientific Publ., New-York, 183-208.

Hedgecock, D. 1981. Genetic consequences of larval retention: theoretical and methodological aspects. in: V.S. Kennedy (ed.), Estuarine comparisons, Academic Press, New-York, 553-570.

Henri, M., J.J. Dodson and H. Powles. 1985. Spatial configurations of young herring (*Clupea harengus harengus*) larvae in the St. Lawrence estuary: importance of biological and physical factors. Can. J. Fish. Aquat. Sci., 42, 91-104.

Herman, S.S. 1963. Planktonic fish eggs and larvae of Narragansett bay. Limnol. Oceanogr., 8, 103-109.

Houde, E.D. and J.A. Lovdal, 1984. Seasonality of occurrence, foods and food preferences of ichthyoplankton in Biscayne bay, Florida. Est. Coast. Shelf. Sci., 18, 403-419.

Hubbell, S.P. 1979. Tree dispersion, abundance and diversity in a tropical dry forest. Science, 203, 1299-1309.

Ingram, R.G. 1976. Characteristics of a tide-induced estuarine front. J. Geophys. Res., 81, 1951-1959.

Ingram, R.G. 1985. Frontal characteristics at the head of the Laurentian channel. Nat. Can., 112, 31-38.

Jacquaz, B., K.W. Able and W.C. Leggett. 1977. Seasonal distribution, abundance and growth of larval capelin (*Mallotus villosus*) in the St. Lawrence estuary and northwestern Gulf of St. Lawrence. J. Fish. Res. Bd Can., 34, 2015-2029.

Johnson, G. and C. Tremblay. 1978. Première capture de civelles d'anguille, *Anguilla rostrata*, au large des côtes dans l'estuaire maritime du Saint-Laurent. Nat. Can., 105, 485-486.

Kennedy, V.S. 1981. Estuarine comparisons. Academic Press, New-York.

Kohler, A.C., D.J. Faber and N.J. McFarlane. 1977. Eggs, larvae and juveniles of fishes from plankton collections in the Gulf of St. Lawrence during 1972 to 1975. Fish. Mar. Serv. Tech Rept 747: 180p.

Kranck, K. 1979. Dynamics and distribution of suspended matter in the St. Lawrence estuary. Nat. Can., 106, 163-173.

Larouche, P., V.G. Koutitonsky, J-P. Chanut and M.I. El-Sabh. 1987. Lateral stratification and dynamic balance at the Matane transect in the Lower St. Lawrence estuary. Est. Coast. Shelf Sci., 24, 859-871.

Legendre, L. and S. Demers. 1984. Toward dynamic biological oceanography and limnology. Can. J. Fish. Aquat. Sci., 41, 2-19.

Legendre, L. and S. Demers. 1985. Auxiliary energy, ergoclines and aquatic biological production. Nat. Can., 112, 5-14.

Leim, A.H. and W.B. Scott. 1966. Fishes of the Atlantic coast of Canada. Bull. Fish. Res. Bd. Can., 155, 485p.

Loeb, V.I. 1979. Larval fishes in the zooplankton community of the North Pacific central gyre. Mar. Biol., 53, 173-191.

Marak, R.R., J.B. Colton Jr., D.B. Foster and D. Miller. 1962. Distribution of fish eggs and larvae, temperature and salinity in the Georges bank - Gulf of Maine area 1956. U.S. Fish. Wildl. Serv. Spec. Sci. Rept. Fish 412, 95p.

May, R.M. 1976. Patterns of multi-species communities. in: R.M. May (ed.), Theoretical ecology, principles and applications. W.B. Saunders Cie, Philadelphia, 142-162.

McGowan, J.A. and P.W. Walker. 1985. Dominance and diversity maintenance in an oceanic ecosystem. Ecol. Monogr., 55, 103-118.

McHugh, J.L. 1967. Estuarine nekton. in: G.H. Lauff (ed.), Estuaries. p.581-620, AAAS Spec. Publ. No 83, Washington, D.C.

Melville-Smith, R. and D. Baird. 1980. Abundance, distribution and species composition of fish larvae in the Swartkops estuary. S. Afr. J. Zool., 15, 72-78.

Merriman, D. and R.C. Sclar. 1952. The pelagic fish eggs and larvae of Block Island Sound. Bull. Bingham, Oceanogr. Coll., 13, 165-219.

Miller, J.M. 1974. Nearshore distribution of Hawaiian marine fish larvae: effects of water quality, turbidity and currents. in: J.H.S. Blaxter (ed.),The Early Life History of Fish, Springer-Verlag, Berlin, 217-231.

Möller, H. 1984. Reduction of a larval herring population by jellyfish predator. Science, 224, 621-622.

Neu, H.J.A. 1970. A study on mixing and circulation in the St. Lawrence estuary up to 1964. Atl. Oceanogr. Lab., Bedford Inst., Rep. Ser. 1970-9: 31p.

Ochman, S. and J. J. Dodson. 1982. Composition and structure of the larval and juvenile fish community of the Eastmain river and estuary, James Bay. Nat. Can., 109, 803-813.

Olney, J.E. 1983. Eggs and early larvae of the bay anchovy, *Anchoa mitchilli* and the weakfish *Cynoscion regalis* in lower Chesapeake bay with notes on associated ichthyoplankton. Estuaries, 6, 20-35.

Ouellet, P. and J. J. Dodson. 1985. Dispersion and retention of anadromous rainbow smelt (*Osmerus mordax*) larvae in the middle estuary of the St. Lawrence river. Can. J. Fish. Aquat. Sci., 42, 332-341.

Ouellet, Y. and C. Trump. 1979. La circulation hydrodynamique dans la zone de mélange estuarienne du Saint-Laurent. Nat. Can., 106, 13-26.

Oviatt, C.A. and S.W. Nixon. 1973. The demersal fish of Narragansett Bay: an analysis of community structure, distribution and abundance. Est. Coastal Shelf Sci., 1, 361-378.

Parent, S. and P. Brunel. 1977. Aires et périodes de fraye du capelan (*Mallotus villosus*) in the St. Lawrence estuary and northwestern Gulf of St. Lawrence. Trav. Pêche Québec., 45, 1-46.

Pearcy, W.G. and S.W. Richards. 1962. Distribution and ecology of fishes of the Mystic river estuary, Connecticut. Ecology, 43, 248-259.

Pearcy, W.G. and S.S. Myers. 1974. Larval fishes of Yaquina Bay, Oregon: a nursery ground for marine fishes? U.S. Fish. Bull., 72, 201-213.

Perlmutter, A. 1939. A biological survey of the salt waters of Long Island Sound, 1938. Section I. An ecological study of young fish and eggs identified from tow net collections. Suppl. 28, Ann. Rept. New-York Conserv. Dept. Part II: 11-71.

Pingree, R.D. and D.K. Griffiths. 1980. A numerical model of the M2 tide in the Gulf of St. Lawrence. Oceanol. Acta., 3, 221-226.

Powles, H., F. Auger and G.J. Fitzgerald. 1984. Nearshore ichthyoplankton of a north temperate estuary. Can. J. Fish. Aquat. Sci., 41, 1653-1663.

Purcell, J.E. 1985. Predation on fish eggs and larvae by pelagic cnidarians and ctenophores. Bull. Mar. Sci., 37, 739-755.

Richardson, S.L. and W.G. Pearcy. 1977. Coastal and oceanic fish larvae in the area of upwelling off Yaquina bay, Oregon. U.S. Fish. Bull., 75, 125-145.

Rivière, D., D. Roby, A.C. Horth, M. Arnac and M.F. Khalil. 1985. Structure génétique de quatre populations de hareng de l'estuaire du Saint-Laurent et de la Baie des Chaleurs. Nat. Can., 112, 105-112.

Rogers, H.M. 1940. Occurrence and retention of plankton within the estuary. J. Fish. Res. Bd. Can., 5, 164-171.

Roper, D.S. 1986. Occurrence and recruitment of fish larvae in a northern New Zealand estuary. Est. Coast. Shelf Sci., 22, 705-717.

Roy, J-M, G. Beaulieu and G. Labrecque. 1975. Observations sur le poulamon, *Microgadus tomcod*, de l'Estuaire du Saint-Laurent et la Baie des Chaleurs. Cahiers d'Information No 70, Min. Ind. Comm., Dir. Peches Mar., Quebec.

Saila, S.B. 1973. Some aspects of fish production and cropping in estuarine systems. in: L.E. Cronin (ed.), Estuarine Research. Vol. 1, Academic Press, New-York, 473-493.

Shenker, J.M. and J.M. Dean. 1979. The utilization of an intertidal salt marsh creek by larval and juvenile fishes: abundance, diversity and temporal variation. Estuaries, 2, 154-163.

Shenker, J.M., D.J. Hepner, P.E. Frere, L.E. Currence and W.W. Wakefield. 1983. Upriver migration and abundance of naked goby (*Gobiosoma bosci*) larvae in the Patuxent river estuary, Maryland. Estuaries, 6, 36-42.

Sinclair, M., D.V. Subba Rao and R. Couture. 1981. Phytoplankton temporal distributions in estuaries. Oceanol. Acta, 4, 239-246.

Stevens, D.E. 1977. Striped bass (*Morone saxatilis*) year class strength in relation to river flow in the Sacramento-San Joaquin estuary, California. Trans. Am. Fish. Soc., 106, 34-42.

Strathmann, R.R. 1981. Selection for retention or export of larvae in estuaries. in: V.S. Kennedy (ed.), Estuarine Comparisons, Academic Press, New-York, 521-536.

Tang, C.L. 1980. Mixing and circulation in the northwestern Gulf of St. Lawrence: a study of buoyancy driven current system. J. Geophys. Res., 95, 2787-2796.

Tremblay, L. 1942. Rapport de la Station Biologique du Saint-Laurent, 1936-42. Université Laval, Québec.

Vésin, J-V., W.C. Leggett and K.W. Able. 1981. Feeding ecology of capelin (*Mallotus villosus*) in the estuary and western Gulf of St. Lawrence. Can. J. Fish. Aquat. Sci., 38, 257-267.

Vladykov V.D. and D.E. McAllister. 1961. Preliminary list of marine fishes of Québec. Nat. Can., 88, 53-78.

Washington, H.G. 1984. Diversity, biotic and similarity indices. A review with special relevance to aquatic ecosystems. Water Res., 18, 653-694.

Weinstein, M.P. 1979. Shallow marsh habitats as primary nurseries for fishes and shellfish, Cape Fear river, North Carolina. U.S. Fish. Bull., 77, 339-357.

Weinstein, M.P., S.L. Weiss, R.G. Hodson and L.R. Gerry. 1980. Retention of three taxa of postlarval fishes in an intensively flushed tidal estuary, Cape Fear River, North Carolina. U.S. Fish. Bull., 78, 419-435.

Whittaker, R.H. 1970. Communities and ecosystems. The Macmillan Co., London. 152p.

Wilkens, E.P.H. and R.M. Lewis. 1971. Abundance and distribution of young Atlantic menhaden (*Brevoortia tyrannus*) in the White Oak river estuary, North Carolina. U.S. Fish. Bull., 69, 783-789.

Yamashita, Y. and T. Aoyama. 1984. Ichthyoplankton in Otsuchi bay on northern Honshu with reference to the time-space segregation of their habitats. Bull. Jap. Soc. Sci. Fish., 50,: 189-198.

Chapter 15

The Macrobenthic Fauna of the St. Lawrence Estuary

Bruno Vincent

Groupe de Recherche en Océanographie Côtière, Département d'Océanographie, Université du Québec à Rimouski, Rimouski, Québec, Canada   G5L 3A1

ABSTRACT

Most of the the studies on the macrobenthos of the St. Lawrence Estuary have been carried out since 1970, and these have been almost exclusively in the Lower Estuary. The intertidal zone fauna is species poor but densities are often elevated. The most common species occuring in the tidal flats are the bivalves *Macoma balthica* and *Mya arenaria*, the gasteropod *Hydrobia totteni* and the polychaete *Nereis virens* . The principal species of the rocky substrates of the intertidal zone are the barnacle *Semibalanus balanoides* at mid-tide levels and *Mytilus edulis* at lower levels, both species mainly colonizing crevices. The green sea urchin, *Strongylocentrotus droebachiensis*, is the dominant invertebrate in the rocky infralittoral zone and this species is important in determining the structure of the infralittoral community. The seastar *Leptasterias polaris* is also abundant in the infralittoral and the presence of this predator explains the rarity of mussels and barnacles at these levels. The circalittoral and bathyal zones display evident spatial variation in the structure of the benthic communities. The abundance and diversity of the communities , as well as their spatial heterogeneity, are greater at the head of the Laurentian Trough. Polychaetes (64.9%) and bivalves (15.8%) dominate the fauna of the Trough. The longitudinal differences in the community structure may be related to variations in the sand content of the sediments and to differences in the supply of nutrients falling from the upper layers of the water column.

Keywords: benthos, intertidal zone, infralittoral zone, circalittoral zone, bathyal zone, Laurentian Trough, reproduction, ice

1. Introduction

Nearly all studies of the macrobenthos of the St. Lawrence have been carried out in the Lower Estuary. Some local studies on the biology and the ecology of amphipods were conducted on the north shore of the Upper Estuary (Sainte-Marie 1986a, 1986b, 1987) but there are few data on macrobenthic species distribution. In a study of longitudinal variations of the macrobenthos in the Fluvial Estuary, Vincent (1979) sampled at the

Coastal and Estuarine Studies, Vol. 39
M. I. El-Sabh, N. Silverberg (Eds.)
Oceanography of a Large-Scale Estuarine System
The St. Lawrence
© Springer-Verlag New York, Inc., 1990

landward limit of the Upper Estuary and found mainly chironomid larvae and oligochaetes. Fradette and Bourget (1980) showed that numbers of organisms and biomass of invertebrates on buoys decreased markedly on passing from the Lower Estuary to the Upper Estuary and a similar decrease was observed by Himmelman et al. (1983) in the abundance of the green sea urchin, *Strongylocentrotus droebachiensis*, in the infralittoral zone. Our knowledge of the macrobenthic fauna of the Saguenay Fjord is limited to a list of species sampled between 1958 and 1970 (Drainville et al., 1978).

For a long time, our knowledge of benthic communities in the Lower St. Lawrence Estuary was also based on lists of species (Bousfield, 1955; Préfontaine, 1931; Préfontaine and Brunel, 1962; Whiteaves, 1901) which are still very useful as references (Robert, 1979). Most of the ecological studies of the macrobenthos of the Lower Estuary were carried out after 1970, and they are much more numerous in the intertidal and the infralittoral zones than in the circalittoral and bathyal zones.

2. The intertidal zone

2.1. Tidal flats

The Lower St. Lawrence Estuary is a macrotidal environment where the semi-diurnal tide has a maximal amplitude of 3 to 5 m. On the north shore, the tidal flats are sandy and very homogenous but most studies of the macrobenthos of the intertidal zone have been conducted on the very heterogenous tidal flats of the south shore, where major salt marshes are also located. These flats are characterized by numerous tidal pools of various dimensions, boulders, and muddy-sandy-gravel sediments containing around 2.5 % organic matter (Vincent et al., 1987). Boulders remain numerous in the central part of the flats, where the sediment is an heterogenous muddy-gravelly-sand with about 2 % organic content. In the lower part of the flats the sediment is an homogenous muddy sand with low organic matter content.

Surface-sediment temperatures decrease from August to December and remain around -1°C from November to March (Harvey & Vincent, 1989). Thick shore-ice (> 1 m) covers the intertidal flats from mid-December to March and protects surface sediments from the very cold air temperatures. For this reason, Bourget (1977) hypothesized that the minimum temperature of the surface sediments would be reached in autumn, before ice formation, or in spring, after ice breakup. In July, Joly (1987) observed maximum temperatures of about 25 °C and daily variations of 15 °C.

On the tidal flats of the south shore, the benthic community is dominated by *Macoma balthica* , with the main accompanying species being the polychaete *Nereis virens* and the bivalve *Mya arenaria* (Desrosiers & Brêthes, 1984). High densities of the gastropod

Hydrobia totteni are found on fine sediments with high organic matter content. The polychaete *Nephtys caeca* and the amphipod *Gammarus oceanicus* are associated with medium salinity waters, whereas the polychaete *Marionina* sp. and the amphipod *Gammarus lawrencianus* occur in low salinity waters. Because of the abundance of boulders and pebbles in the upper two-thirds of the flats, the littorinids *Littorina saxatilis* and *Littorina obtusata* can be locally abundant. The same species are found on the sand flats of the north shore where there are also some dense populations of the bivalve *Mesodesma arctatum* (Giguère & Lamoureux, 1978). Ward & Fitzerald (1983) noted 21 species in tidal pools of the Ile-Verte salt marsh and they detected few or no significant effects of pool size upon the physical-chemical environment or the macrobenthic fauna.

Macoma balthica is abundant in all tidal flats of the Lower St. Lawrence Estuary. Joly (1987) studied spatial variations in density and shell growth in a population where the mean density was 1302 m^{-2}. Densities were higher and more variable near mean water level, with a maximum density of 3680 m^{-2}. Similar results were found by Vincent et al. (1987) in a neighbouring bay. The shell growth rate and mean shell length of each generation increased with immersion time and there was a strong density effect on shell growth, which increased with tidal level (Vincent et al., 1987, 1989). Joly (1987) was able to explain 40 to 70% of the spatial variation of mean shell length of each generation by multiple stepwise regression, in which density and tidal level were the first two independant variables introduced . A field experiment confirmed this effect of density on shell growth rate and showed a stronger density effect on tissue dry weight and on gonad size (Vincent et al., 1989). This indicated that, despite the severe environmental conditions affecting the upper two-thirds of the flat, the population was mainly food limited.

In the first study of *Macoma balthica* populations of the St. Lawrence Estuary, Lavoie (1970a) and Lavoie et al. (1968) noticed an absence of gonads in many adults sampled in March near the mean water level. This led them to conclude that spawning occured in March-April. Harvey & Vincent (1989) studied reproduction and energy allocation at two tidal levels in a population located near Rimouski, 100 km seaward of the population studied by Lavoie (1970a). They observed that individuals began spawning in May and at the same moment at the two tidal levels. Spawning ends in mid-June and larvae are mainly found in June and July (Belzile et al., 1984), during the chlorophyll maximum in the water column (Ferreyra, 1987). Shell length at sexual maturity was the same at both tidal levels. However, a high proportion of older individuals at the upper level did not contain sexual products and fecundity of younger adults of similar shell length is greater at the lower tidal level. For this reason, gamete production for the whole population is largely assured by individuals from the lower level, which is a more stable environment. Large inter-annual variations of fecundity showed that individuals use an opportunistic strategy which allows investment of any surplus energy in gamete production (Harvey & Vincent, 1989; Vincent et al., 1989).

Dense populations of the soft-shell clam, *Mya arenaria*, are found throughout the Lower Estuary and its potential economical interest justified several studies of its distribution (Giguère & Lamoureux, 1978; Lamoureux, 1977; Lavoie, 1967, 1969, 1970b) and of potential catch estimations (Brêthes & Desrosiers, 1981). Presently, there is no appreciable commercial exploitation of this species but clam digging is important in some areas, especially in spring and in autumn when risks of toxicity due to *Gonyaulax tamarensis* are low. In most tidal flats of the Lower Estuary, soft-shell clams are more abundant near the mid-water level (Roseberry, 1988). Jacques et al. (1984) observed that shell growth rate of *Mya arenaria* is greater on sandy bottoms at the lower tidal level than on the muddy bottom of the higher tidal levels, but it is difficult to distinguish between the effects of sediment type and of tidal level. There is also a density effect on shell growth at the mean water level, which is particularly visible on older clams (Roseberry, 1988). The latter suggest the possibility of a horizontal depletion of food at small scales (< 1 m). The importance of advection and small scale process for this species is also demonstrated by the preferential orientation of individuals in relation to tidal current direction (Vincent et al., 1988). In the Lower St. Lawrence Estuary, *Mya arenaria* spawn in May-June (Roseberry, 1988) and larvae are found in littoral waters from June to August (Belzile et al., 1984). However, reproductive success can be very low for long periods, as shown by Roseberry (1988) and Jacques (1987), who found few individuals of the 1978 to 1985 generations in two neighbouring populations.

The polychaete *Nereis virens* spawns in May. Post-larvae were found in sediments at the beginning of June by Olivier (1989) who studied the growth of juvenile *Nereis virens* and *Eteone longa* at three sites on a tidal flat. Spatial variations in density, individual body-weight and sexual maturity of *N. virens* are correlated with tidal level: density increases in the onshore direction and older and sexually mature individuals are found at the lower tidal level (Miron, 1988; Miron & Desrosiers, in press). On the other hand, these authors noted an inverse relationships between the density of *Nephtys caeca* and the tidal level, although the larger individuals were also found in the lower part of the flat.

2.2 Intertidal rocky shores

Algal and faunal diversity are low on the intertidal rocky shores of the Lower St. Lawrence Estuary. Major animal species are mussels (*Mytilus edulis*), barnacles (*Semibalanus balanoides*) and littorinids (*Littorina obtusata, Littorina saxatilis*). Complete descriptions of intertidal communities along the south shore of the Estuary can be found in the studies of Archambault & Bourget (1983), Bergeron & Bourget (1984) and Bourget et al. (1985).

Barnacles are found at mid-tide levels where competition is not an important factor in structuring the community (Archambault & Bourget, 1983). The upper limit of their distribution is determined by physical factors, while the lower limit is determined by competition for space with mussels, which are abundant in the lower part of the zone

(Bergeron, 1985). The succession of barnacles and mussels along the tidal gradient can also be explained by the fact that barnacles are more opportunistic species than mussels. *Semibalanus balanoides* dominates where there are severe and regular disturbances, while the dominance of *Mytilus edulis* requires a long period without disturbance (Archambault & Bourget, 1983). Barnacles and mussels occur mainly in crevices, where they occupy distinct zones along the vertical gradient of the walls: *Mytilus edulis* colonize the bottom of the crevices, while *Semibalanus balanoides* colonize the zone immediately above the mussels. The other main species are littorinid grazers, which have a negligible influence on the community (Archambault & Bourget, 1983).

Because of the absence of predators, ice and topography are the main factors explaining spatio-temporal variations of structure in this community, which is biologically accomodated in summer and physically controlled in autumn and winter (Bergeron, 1985; Bourget et al., 1985). The direct scouring of ice during ice breakup explains the significant reductions in the abundance of many species (e.g. *Semibalanus balanoides* and *Mytilus edulis*) on intertidal rocky shores (Bourget et al., 1985). Bergeron & Bourget (1984) suggested that the direct ice effect can also be important during the automnal sequences of the ice cover formation. Bourget et al. (1985) considered the effect of cold temperatures on intertidal communities of rocky shores to be less important than earlier supposed (Bourget, 1983) because of 1) thick ice cover during the periods of very low temperature, 2) low lethal temperature of the species and 3) short emersion times. Ice may also have an indirect effect associated with the selection of cryptic individuals (Bourget et al., 1985).

The settlement period of *Semibalanus balanoides* is very short and concentrated in July and August (Bourget & Lacroix, 1973). The direct effects of ice could explain the presence of extensive areas of bare rock but the absence of barnacles outside crevices is due to the lack of settlement (Bergeron & Bourget, 1986; Chabot & Bourget, 1988). Le Tourneux and Bourget (1988) showed that there was a shift from biological cues (algae) at large scales (1 m), to physical and biological cues at medium scales (1 mm) and to physical cues (microheterogeneity) at the smallest scale (< 300 µm). Bourget (1988) summarized the main events occuring during the settlement of *Semibalanus balanoides* and compared it with those of the subtidal *Balanus crenatus* (Hudon et al., 1983).

Growth rates of *Mytilus edulis* in the Lower St. Lawrence Estuary are comparable to those reported for other subarctic regions, with the greatest rates found on the north shore (Cossa & Bourget, 1985). These authors compared this geographical variation to that observed for benthic epifauna of buoys (Fradette & Bourget, 1980; 1981) and explained it by spatial variations of water temperature and productivity. Fréchette & Bourget (1985a, 1985b) and Fréchette et al. (1989) showed that food is depleted above the very dense mussel beds often found in the lower intertidal zone, and that water movement and small-scale heterogeneity can determine food availability and growth for mussels (Fréchette & Bourget, 1987). Dense mussel populations could also account for the low phytoplancton biomass and its high productivity in the littoral zone (Demers et al., 1989).

Due to its wide distribution in the Lower Estuary and in the Gulf of St. Lawrence, and its widespread acceptance as an indicator of pollution, the trace-metal content of *Mytilus edulis* has been intensively studied (Bourget & Cossa, 1976; Cossa, 1987; Cossa & Bourget, 1980, 1985; Cossa & Rondeau, 1985; Cossa et al., 1979, 1980). For additional details see Chapter 11, this volume.

3. The Infralittoral zone

Our knowledge of the macrobenthos of the infralittoral zone of the Lower St. Lawrence Estuary is limited to hard substrates. Detailed descriptions of communities of the rocky infralittoral zone are given by Lavergne & Himmelman (1984). A general model for their organization was proposed by Himmelman & Lavergne (1985), who described four infralittoral communities and discussed their relative importance in the estuary. -1) The fringe algal community is found in the upper part of the infralittoral zone where the more abundant invertebrates are mussels (*Mytilus edulis*), littorinids (*Littorina obtusata* and *L. saxatilis*) and barnacles (*Balanus crenatus* and *Semibalanus balanoides*). Mussels can be abundant in the upper part of the fringe but they disappear when depth increases. Barnacles are much less abundant than in the intertidal zone. The green sea urchin is scarce in this fringe but is a dominant species in the three other communities. -2) The grazing-resistant algae community begins just below the fringe algal community and the abundance of sea urchins is intermediate. -3) The barren substrate community is lacking in species and unproductive, with rocky surfaces covered by corraline algae and with a few limpets, chitons and urchins. -4) In the deeper part of the infralittoral zone, they find a filter feeder community, where the density of sea urchins is reduced and where the characteristic species are anemones, alcyonarians, sea cucumbers, tunicates and sponges.

The green sea urchin, *Strongylocentrotus droebachiensis*, is the most abundant invertebrate in the rocky infralittoral zone. Its growth rate is inversely related to densities, which are very high near the seaward extremity (Himmelman et al., 1983a). Population structures show longitudinal changes which can be related to salinity. In the seaward portion of the Lower Estuary, small sea urchins are abundant in shallow water and there is no indication of periodic mortalities, whereas there are only a few small and large individuals and no medium-sized individuals in the upper portion of the Lower Estuary. The salinity gradient in the estuary and periodic decreases of salinity in shallow water could account for the marked changes of population structure, through greater mortality of the smaller sea urchins in low salinity conditions (Himmelman et al., 1984). The importance of this species in determining the structure of the infralittoral community is clearly demonstrated by the removal experiment performed by Himmelman et al. (1983b). Moreover, there are good spatial relationships between algal densities, sea urchin densities and sea urchin growth rates (Himmelman & Lavergne, 1985).

Echinoderm larvae are found in the Estuary in June (Belzile et al., 1984). After observing the June spawning of *S. droebachiensis* and of the mottled red chitón *Tonicella marmorea*, Jalbert et al. (1982) hypothesized that phytoplankton may be their spawning stimulus. The delayed spring phytoplankton bloom in the Estuary thus would account for the relatively late spawning of these two species.

The subarctic seastar *Leptasterias polaris* is abundant in the infralittoral zone of the Lower Estuary. It spawns over several months in the autumn (Boivin et al., 1986) and the female remains on her brood for 4 to 5 months (Himmelman et al., 1982). Himmelman & Lavergne (1985) explain the rarity of mussels and barnacles in the deeper part of the infralittoral zone, despite the intense colonization observed on artificial substrates (Bourget & Lacroix, 1972, 1973; Brault & Bourget, 1985), through intense predation by *Leptasterias polaris*.

4. The circalittoral and bathyal zones

Most of our early knowledge of the macrobenthic fauna from the circalittoral and bathyal zones derives from samples collected throughout the Lower Estuary during the summers of 1970, 1971 and 1972. An anchor-type dredge was used to study the distribution of polychaetes (Massad, 1975; Massad & Brunel, 1979) and mollusks (Robert, 1974, 1979) and a suprabenthic sampler (Brunel et al., 1978) was used to study the distribution and the ecology of the amphipod *Arrhis phyllonyx* (Sainte-Marie & Brunel, 1983). Our knowledge of the macrobenthic fauna of the Lower St. Lawrence Estuary can be further completed from the studies of Ledoyer (1975a, 1975b) and Bellan (1977, 1978) on the circalittoral and bathyal macrobenthos of the Gaspé Strait.

Robert (1979), using the data from 73 stations, distinguished six major mollusk communities. There is a complete distinction between stations situated above or below the 75 m isobath, with the shallow water (< 75 m) community having the greater diversity. In deeeper water, diversity is low with little variation but lower values for the whole estuary have been found in the mid-depth stations of the north shore. The five deep water communities show high regional clumping: the mid-depth community from the south shore differs from the community of the north shore and there is a longitudinal succession of three different communities along the Trough.

Massad & Brunel (1979) studied polychaete distributions using 36 of the 73 stations of Robert (1979). They identified six polychaete communities and found good correspondance between three of these communities and the mollusk communities of the shallow waters, the seaward Trough floor and the head of the Trough, as described by Robert (1979). The other three polychaete communities occurred only very locally, landward of the Trough, on the sand-gravel bottom at its head, on deep sand of the slopes, and on deep sandy muds. A general comparison of polychaete and mollusk distributions and a detailed analysis of longitudinal variations of the polychaete fauna in the whole Lower Estuary is difficult because of the scarcity of stations sampled by Massad & Brunel (1979) upstream of Rimouski. However, species diversity decreases as depth and mud content increase and there is a general decrease of abundance and species richness in the seaward direction. The high diversity at the head of the Trough could be explained by sand enrichment at all depths and by an increase in the spatial heterogeneity of sediments. Moreover, proportions of filter-feeding and selective deposit-feeders are higher on sandy bottoms than on muddy bottoms, where the importance of Errantia and non-selective deposit-feeders is greater. Massad & Brunel (1979) and Robert (1979) explained spatial variations of mollusk and polychaete communities in terms of spatio-temporal variations of primary productivity related to circulation patterns and upwelling.

The longitudinal variation of the macrobenthic fauna in the Laurentian Trough was confirmed by Ouellet (1982) who sampled macrofauna with a Van Veen grab and box-corer at twenty four stations concentrated in sectors located along the landward half of the Lower Estuary. This is the first and only study of the entire macrobenthos assemblage of the Trough. It provides a good general understanding of the relative importance of the main taxonomic groups: - polychaetes (64.9 %), bivalves (15,8 %), amphipods (8.1 %), sipunculids (4.3 %) and ophiurids (3.5 %). There is a regular variation of the community structure in the seaward direction along the Trough and the most striking feature of this longitudinal gradient is the decrease in the total abundance (from about 750 to 290 ind.m^{-2}) and of the species richness of deposit-feeders. These data also confirm, for the whole macrobenthic fauna, the higher species richness and abundance previously observed for mollusks (Robert, 1979) and polychaetes (Massad & Brunel, 1979) at the head of the Trough. However, sharp decreases in species richness and total abundances then occur when proceeding seaward along the Trough. Species richness then remains quite low in the seaward half of the Lower Estuary, but there is a strong local increase in total abundance of macrobenthic fauna just landward of Rimouski. Ouellet (1982) also observed a shift in the general feeding behaviour of the (statistically-derived) characteristic species as one progressed seaward along the Trough: diverse modes of nutrition at the head of the Trough, then vertically-mobile deposit feeders, followed by horizontally-mobile deposit feeders and carnivores in the seaward portion of the Lower Estuary. Bouchard (1983) used these benthic faunal trends together with those obtained from sediment chemical characteristics to suggest that benthic biogeochemistry followed geographic patterns similar to that of the circulation in the surface waters of the Lower Estuary.

The above studies of infaunal macrobenthos of the Lower St. Lawrence Estuary have been complemented by direct submersible observations at six different stations (Syvitski et al., 1983). The authors defined five depth-dependent benthic zones on the floor and slopes of the Laurentian Trough. The floor and some portions of the adjoining slope are visually dominated by scattered sea pens and anemones but there also are many polychaete tubes and ophiuroids. A succession of four other zones can be observed on the side wall slopes: 1) the Infaunal zone near the base of the slope is dominated by polychaetes 2) the Ophiura zone is dominated by ophiurids but urchins (*Strongylocentrotus droebachiensis*), polychaetes and anemones are also abundant, 3) the Ice Rafting zone is dominated by basket stars (*Gorgonocephalus* sp.) and 4) the sandy-bottomed Wave Base zone has a scarce epibenthic fauna.

Sainte-Marie and Brunel (1983) studied the distribution of the suprabenthic amphipod *Arrhis phyllonyx* at 97 stations covering the whole Lower Estuary. This species occurs between 20 and 300 m but is more abundant at stations on the shelf and at the head of the Trough (20-170 m). All bottom types are inhabited and frequencies are similar on the southern and northern shelves. These authors compared a population from the Lower Estuary with a population from the Gulf of St. Lawrence (Chaleur Bay) and concluded that the lower density, smaller size and delayed breeding (November-December) of the Estuary population are probable adaptations to the later-occuring and reduced level of primary production. On the other hand, the detritivorous and necrophagous amphipod *Hippomedon propinquus* grows faster in the Lower Estuary than in the Gulf of the St. Lawrence (Chaleur Bay), and it seems better adapted to the poorer and less predictable primary productivity in the St. Lawrence Estuary than the more strictly detrivorous amphipods (Lamarche & Brunel, 1987).

5. Concluding remarks

The study of the benthos of the St. Lawrence Estuary is relatively young, and one can forsee the need for more detailed examination of the ecology of the various populations and benthic communities and a better understanding of their role in the estuarian ecosystem as a whole.

All of the current studies reveal significant spatial variations in benthic communities on the scale of the estuary, and in several demographic parameters, such as the age structure, or the size structure of the populations and growth rates of individuals. Several results suggest that there may be regional patterns of variation applicable to many different species and to different zones of the Lower Estuary.

Some general characteristics of the Lower Estuary benthos appear to be discernable, mainly from the many studies in the littoral zone. These include slow annual growth, late reproduction, and irregular recruitement. The presence of ice has a definite impact in the intertidal zone, but we still know very little about the functioning of the benthos during the 4-5 months of ice cover elsewhere in the estuary. Research on the functioning of benthic communities is both long and difficult because of i) the slowness and irregularity of benthic processes and ii) the great spatio-tempopral variability in the physical, biological and geological processes that have a direct influence on the benthos (e.g. circulation, pimary production, sediment distribution). A proper understanding of the benthos will require a better knowledge of these other factors

REFERENCES

Archambault D. & E. Bourget 1983. Importance du régime de dénudation sur la structure et la succession des communautés intertidales de substrat rocheux en milieu subarctique. Can. J. Fish. Aquat. Sci., 40: 1278-1292

Bellan G. 1977. Contribution à l'étude des Annélides Polychètes de la Province du Québec (Canada). I - Les facteurs du milieu et leur influence. Thétys, 7: 365-373

Bellan G. 1978. Contribution à l'étude des Annélides Polychètes de la province du Québec (Canada). II - Etude synécologique. Thétys, 8: 231-240

Belzile S., J.-C. Brêthes & G. Desrosiers 1984. Evolution saisonnière de l'abondance des larves planctotrophiques dans un système côtier. Sci. Tech. Eau, 17: 87-90

Bergeron P. 1985. Hétérogénéité du substrat et structure d'une communauté épibenthique intertidale. Thèse de maîtrise, Université Laval, Québec, 117 p.

Bergeron P. & E. Bourget 1984. Effet du froid et des glaces sur les peuplements intertidaux des régions nordiques, particulièrement dans l'estuaire du Saint-Laurent. Océanis, 10: 279-304

Bergeron P. & E. Bourget 1986. Shore topography and spatial partitioning of crevice refuges by sessile epibenthos in an ice disturbed environment. Mar. Ecol. Prog. Ser., 28: 129-145

Boivin Y., D. Larrivée D. & J.H. Himmelman 1986. Reproductive cycle of the subarctic brooding asteroïd *Leptasterias polaris*. Mar. Biol., 92: 329-338

Bouchard, G., 1983. Variations des paramètres biogéochimiques dans les sédiments du chenal Laurentien. M. Sc. thesis, Université du Québec à Rimouski, 161 p.

Bourget E. 1977. Observations of the effects of frost on beach sediments at Rimouski, Lower St. Lawrence Estuary, Québec. Can. J. Earth Sci., 14: 1732-1739

Bourget E. 1983. Seasonal variation of cold tolerance in intertidal mollusk and relation to environmental condition in the St. Lawrence estuary, Canada Can. J. Zoo., 61: 1193-1201

Bourget E. 1988. Barnacles larval settlement: the perception of cues at different spatial scales. In: Behavioral Adaptation to Intertidal Life. Ed by G. Chelazzi & M. Vannini, Plenum Publishing Corporation: 153-172

Bourget E. & G. Lacroix 1972. Colonisation et inhibition de la colonisation des Cirripèdes dans l'estuaire du Saint-Laurent. Naturaliste can., 99: 279-285

Bourget E. & G. Lacroix 1973. Aspects saisonniers de la fixation de l'épifaune benthique de l'étage infralittoral de l'estuaire du Saint-Laurent. J. Fish. Res. Bd Can., 30: 867-880

Bourget E. & D. Cossa 1976. Mercury Content of Mussels from the St. Lawrence Estuary and Northwestern Gulf of St. Lawrence, Canada. Mar. Poll. Bull., 7: 237-239

Bourget E., D. Archambault & P. Bergeron 1985. Effet des propriétés hivernales sur les peuplements benthiques intertidaux dans un milieu subarctique, l'estuaire du Saint-Laurent. Naturaliste can., 112: 131-142

Bousfield E. 1955. Studies on the shore fauna of the St. Lawrence Estuary and Gaspé Coast. Bull. Natl. Mus. Can., No 136: 95-101

Brault S. & E. Bourget 1985. Structural changes in an estuarine subtidal epibenthic community: biotic and physical causes. Mar. Ecol. Prog. Ser., 21: 63-74

Brêthes J.-C. & G. Desrosiers 1981. Estimation of potential catches of an unexploited stock of soft-shell clam (*Mya arenaria*) from length composition data. Can. J. Fish. Aquat. Sci., 38: 371-374

Brunel P., M. Besnerr, D. Messier, L. Poirier, D. Granger & M. Weinstein 1978. Le traîneau suprabenthique Macer-GIROQ: appareil amélioré pour l'échantillonnage quantitatif étagé de la petite faune nageuse au voisinage du fond. Int. Revue ges. Hydrobiol., 63: 815-829

Chabot R. & E. Bourget 1988. Influence of substratum heterogeneity and settled barnacle density on the settlement of cypris larvae. Mar. Biol. 97: 45-56

Cossa D. 1987. Cadmium et mercure dans les eaux côtières: biogéochimie et utilisation de Mytilus edulis comme indicateur quantitatif. Thèse de doctorat, Université Paris, Paris, 383 p.

Cossa D. & E. Bourget 1980. Trace element in Mytilus edulis from the estuary and gulf of St. Lawrence (Canada): lead and cadmium concentrations. Environ. Pollut. Ser. A. Ecol. Biol., 23: 1-8

Cossa D. & E. Bourget 1985. Croissance et morphologie de la coquille de Mytilus edulis L. dans l'estuaire et le golfe du Saint-Laurent. Naturaliste can., 112: 417-423

Cossa D. & J.-G. Rondeau 1985. Seasonal, geographical and size induced variability in mercury content of Mytilus edulis in an estuarine environment: a re-assessment of mercury pollution level in the Estuary and Gulf St. Lawrence. Mar. Biol., 88: 43-39

Cossa D., E. Bourget & J. Piuze 1979. Sexual maturation as a source of variation in the relationship between cadmium concentration and body weight of Mytilus edulis. Mar. Poll. Bull., 10: 174-176

Cossa D., E. Bourget, D. Pouliot, J. Piuze & J.P. Chanut 1980. Geographical and seasonal variation in the relationship between trace metal content and body weight in Mytilus edulis .Mar. Biol., 58: 7-14

Demers S., J.-C. Therriault, E. Bourget & H. Desilets 1989. Small-scale gradients of phytoplancton productivity in the littoral fringe. Mar. Biol., 100: 393-399

Desrosiers G. & J.-C. Brêthes 1984. Etude bionomique de la communauté à Macoma balthica de la batture de Rimouski. Sci. Tech. Eau, 17: 25-31

Drainville G., L.-M. Lalancette & L. Brassard 1978. Liste préliminaire d'Invertébrés marins du fjord du Saguenay recueillis de 1958 à 1970 par le Camps des Jeunes Explorateurs. Min. Ind. Com. Dir. Gén. pêch. mar., Cah. Inf. No 83

Ferreyra G. 1987. Etude spatio-temporelle des transports particulaires vers une communauté intertidale de substrat meuble (Estuaire maritime du Saint-Laurent). Thèse de maîtrise, Université du Québec à Rimouski, Rimouski, 98 p.

Fradette P. & E. Bourget 1981. Groupement et ordination appliqués à l'étude de la répartition de l'épifaune benthique de l'estuaire maritime et du golfe du Saint-Laurent. J. Exp. Mar. Biol. Ecol., 50: 133-152

Fredette P. & E. Bourget 1980. Ecology of benthic epifauna of the estuary and gulf of the St. Lawrence, Canada: factors influencing their distribution and abundance on buoys. Can. J. fish. Aquat. Sci., 37: 979-999

Fréchette M. & E. Bourget 1985a. Food limited growth of Mytilus edulis in relation to the benthic boundary layer. Can. J. Fish. Aquat. Sci., 42: 1166-1170

Fréchette M. & E. Bourget 1985b. Energy flow between the pelagic and benthic zone: factors controlling particulate organic matter available to an intertidal mussel Mytilus edulis bed. Can. J. Fish. Aquat. Sci., 42: 1158-1165

Fréchette M. & E. Bourget 1987. Significance of small-scale spatio-temporal heterogeneity in phytoplankton abundance for energy flow in Mytilus edulis. Mar. Biol., 94: 231-240

Fréchette M., C.A. Butman & W.R. Geyer 1989. The importance of boundary layer flows in supplying phytoplancton to the benthic suspension feeder, Mytilus edulis L. Limnol. Oceanogr., 34: 19-36

Giguère M. & P. Lamoureux 1978. Présence et abondance de certains mollusques, plus particulièrement Mytilus edulis, Macoma balthica et Mesodesma arctatum, sur les bancs de myes au Québec. Min. Ind. Com. Dir. Gén. pêch. mar., Cah. Inf. No 85

Harvey M. & B. Vincent 1989. Spatial and temporal variations of the reproduction cycle and energy allocation of the bivalve Macoma balthica (L.) on a tidal flat. J. Exp. Mar. Biol. Ecol., 129: 199-217

Himmelman J.H. & Y. Lavergne 1985. Organization of rocky subtidal communities in the St. Lawrence estuary. Naturaliste can., 112: 143-154

Himmelman J.H., Y. Lavergne, A. Cardinal, G. Martel & P. Jalbert 1982. Brooding behavior of the northern sea-star *Leptasterias polaris*. Mar. Biol., 68: 235-240

Himmelman J.H., Y. Lavergne, F. Axelsen, A. Cardinal & E. Bourget 1983a. Sea-urchins in the St-Lawrence estuary, Canada: their abundance size structure and suitability for commercial exploitation. Can. J. Fish. Aquat. Sci., 40: 474-486

Himmelman J.H., A. Cardinal & E. Bourget 1983b. Community developement following removal of sea-urchins *Strongylocentrotus droebachiensis* from the rocky zone of the St-Lawrence estuary, eastern Canada. Oecologia, 59: 27-39

Himmelman J.H., H. Guderley, G. Vignault, G. Drouin & P.G. Wells 1984. Response of the sea-urchin *Strogylocentrotus droebachiensis* to reduced salinities: importance of size acclimatation and interpopulation difference. Can. J. Zoo., 62: 1015-1021

Hudon C., E. Bourget & P. Legendre 1983. An integrated study of the factors influencing the choice of the settling site of *Balanus crenatus* cyprid larvae. Can. J. Fish. Aquat. Sci., 40: 1186-1194

Jacques A. 1987. Etude de la dynamique de la population des myes (*Mya arenaria)* de la Batture de Rimouski. Thèse de maîtrise, Université du Québec à Rimouski, Rimouski, 181 p.

Jacques A., J.-C. Brêthes & G. Desrosiers 1984. La croissance de *Mya arenaria* en relation avec les caractéristiques du sédiment et de la durée d'immersion, sur la batture de Rimouski. Sci. Tech. Eau, 17: 95-99

Jalbert P., D. Larrivée & J.H. Himmelman 1982. Reproductive cycle of the mottled red chiton *Tonicella marmorea* in the St-Lawrence estuary, Québec, Canada. Naturaliste can., 109: 33-38

Joly D. 1987. Variations spatiales du profil démographique à l'intérieur d'une population de *Macoma balthica* dans la zone intertidale du Saint-Laurent. Thèse de maîtrise, Université du Québec à Rimouski, Rimouski, 133 p.

Lamarche G. & P.Brunel 1987. Cycle de développement, écologie et succès d'*Hippomedon propinquus* (Amphipoda, Gammaridae) dans deux écosystèmes du golfe du Saint-Laurent. Can. J. Zoo., 65: 3116-3132

Lamoureux P. 1977. Estimation des stocks commerciaux de mye (*Mya arenaria*) au Québec. Min. Ind. Com. Dir. Gén. pêch. mar., Cah. Inf. No 83

Lavergne Y. & J.H. Himmelman 1984. Localisation of sea-urchin stocks of the St-Lawrence estuary and their situation in the benthic community Qué. Min. Ind. Com. Dir. Gén. pêch. mar., Cah. Inf. No 108: 1-40

Lavoie R. 1967. Inventaire des populations de myes communes (*Mya arenaria* L.) de l'estuaire du Saint-Laurent. Sta. Biol. mar., Grande-Rivière, Rapp. ann. 1966: 107-113

Lavoie R. 1969. Inventaire des populations de myes communes (*Mya arenaria*) de Grandes-Bergeronnes à Portneuf-sur-mer, été 1968. Sta. Biol. mar., Grande-Rivière, Rapp. ann. 1968: 103-118

Lavoie R. 1970a. Contribution à la biologie et à l'écologie de *Macoma balthica* L. de l'estuaire du Saint-Laurent. Thèse de doctorat, Université Laval, Québec., 156 p.

Lavoie R. 1970b. Inventaire des populations de coques (*Mya arenaria*) de Forestville à Papinachois, été 1969. Sta. Biol. mar., Grande-Rivière, Rapp. ann. 1969: 107-125

Lavoie R., J.L. Tremblay & G. Filteau 1968. Age et croissance de *Macoma balthica* L. à Cacouna-est dans l'estuaire du Saint-Laurent. Naturaliste can., 95: 887-895

Le Tourneux F. & E. Bourget 1988. Importance of physical and biological settlement cues used at different spatial scales by the larvae of *Semibalanus balanoides*. Mar. Biol., 97: 57-66

Ledoyer M. 1975a. Les peuplements benthiques circalittoraux de la Baie des Chaleurs (Golfe du Saint-Laurent). Trav. Pêch. Qué., 42: 1-141

Ledoyer M. 1975b. Aperçu sur les peuplements benthiques des vases profondes du détroit de Gaspé (Golfe du Saint-Laurent). Trav. Pêch. Qué., 44: 1-27

Massad R. 1975. Distribution et diversité endobenthiques des Polychètes dans l'estuaire maritime du Saint-Laurent. Thèse de maîtrise, Université de Montréal, Montréal, 101 p.

Massad R. & P. Brunel 1979. Associations par stations, densités et diversité des Polychètes du benthos circalittoral et bathyal de l'estuaire maritime du Saint-Laurent. Naturaliste can., 106: 229-253

Miron G. 1988. Distribution et variation intra-population de trois espèces de polychètes de la baie de l'Anse à l'Orignal, Parc du Bic. Thèse de maîtrise, Université du Québec à Rimouski, Rimouski, 155 p.

Miron G.Y. & G.L. Desrosiers . Distributions and population structures of two intertidal estuarine polychaetes in the Lower St. Lawrence Estuary, with special reference to environmental factors. Mar. Biol. (in press),

Olivier 1989. Aperçu du recrutement larvaire benthique et de la première saison de croissance de quatre espèces de la communauté à *Macoma balthica* de l'Anse à l'Orignal. Thèse de maîtrise, Université du Québec à Rimouski, Rimouski, 125 p.

Ouellet G. 1982. Etude de l'interaction des animaux benthiques avec les sédiments du Chenal Laurentien. Thèse de maîtrise, Université du Québec à Rimouski, Rimouski, 148 p.

Préfontaine G. 1931. Notes préliminaires sur la faune de l'estuaire du Saint-Laurent dans la région de Trois-Pistoles. Naturaliste can., 59: 213-227

Préfontaine G. & P. Brunel 1962. Liste d'invertébres marins recueillis dans l'estuaire du Saint-Laurent de 1929 à 1934. Naturaliste can., 89: 237-263

Robert G. 1974. The sublittoral Mollusca of the St. Lawrence Estuary, east coast of Canada. Ph.D. thesis, Dalhousie University, Halifax, 174 p.

Robert G. 1979. Benthic molluscan fauna of the St. Lawrence estuary and its ecology as assessed by numerical methods. Naturaliste can., 106: 211-227

Roseberry L. 1988. Etude de la croissance et de la reproduction de *Mya arenaria* (Bivalvia: Mollusca) dans la zone intertidale de l'estuaire du Saint-Laurent. Thèse de maîtrise, Université du Québec à Rimouski, Rimouski, 112 p.

Sainte-Marie B. 1986a. Feeding and swimming of lysianassid amphipods in a shallow cold-water bay. Mar. Biol., 91: 219-229

Sainte-Marie B. 1986b. Effect of bait size and sampling time on the attraction of the Lysianassid amphipods *Anonyx sarsi* Steele and Brunel and *Orchomenella pinguis* (Boeck). J. Exp. Mar. Biol. Ecol., 99: 63-77

Sainte-Marie B. 1987. Meal size and feeding rate of the shallow water Lysianassid *Anonyx sarsi* (Crustacea, Amphipoda). Mar. Ecol., 40: 209-219

Sainte-Marie B. & P. Brunel 1983. Differences in life history and success between suprabenthic shelf populations of *Arrhis phyllonyx* (Amphipoda Gammaridae) in two ecosytems. J. Crustacean Biol., 3: 45-69

Syvitski J.P.M., N. Silverberg, G. Ouellet & K.W.Asprey 1983. First observations of benthos and seston from a submersible in the lower St. Lawrence estuary. Géographie Physique et Quaternaire, XXXVII: 227-240

Vincent B. 1979. Etude du benthos d'eau douce dans le haut-estuaire du Saint-Laurent (Québec). Can. J. Zool., 57: 2171-2182

Vincent B., C. Brassard & M. Harvey 1987. Variations de la croissance de la coquille, et de la structure d'âge du bivalve *Macoma balthica* (L.) dans une population intertidale de l'estuaire du Saint-Laurent (Québec). Can. J. Zool., 65: 1906-1916

Vincent B., G. Desrosiers & Y. Gratton 1988. Orientation of the infaunal bivalve *Mya arenaria* L. in relation to local current direction on a tidal flat. J. Exp. Mar. Biol. Ecol., 124: 205-214

Vincent B., D. Joly & C. Brassard 1989. Effets de la densité sur la croissance du bivalve *Macoma balthica* (L.) en zone intertidale. J. Exp. Mar. Biol. Ecol., 126: 145-162

Ward G. & J. Fitzgerald 1983. Macrobenthic abundance and distribution in tidal pools of a Québec salt marsh. Can. J. Zool., 61: 1071-1085

Whiteaves J.F. 1901. Catalogue of the marine invertebrates of eastern Canada. Geol. Surv. Canada, Pub., No 772: 271 p.

Chapter 16

Marine Fisheries Resources and Oceanography of the St. Lawrence Estuary

Jacques A. Gagné[1] and Michael Sinclair[2]

[1] Institut Maurice Lamontagne, Ministère des Pêches et des Océans, 850 Route de la Mer, Mont-Joli, Québec, Canada, G5H 3Z4

[2] Biological Sciences Branch, Fisheries and Oceans Canada, Bedford Institute of Oceanography, Dartmouth, Nova Scotia, B2Y 4A2

ABSTRACT

Analyses of the limited landing statistics available on commercially exploited species indicate that less than 1% of the marine fish and invertebrate catches of Atlantic Canada come from the St. Lawrence Estuary. Physical processes appear to be more influential than trophodynamic interactions in determining the presence and distribution of most marine species exploited in the Estuary. The effect of hydrodynamic processes operating within the Estuary is not confined to the estuarine boundaries but is felt far beyond by marine organisms from the Gulf of St. Lawrence to possibly the Gulf of Maine.

Key Words: exploitation, fish, fisheries, invertebrates, resources, runoffs.

1. Introduction

The high productivity characterizing most estuaries generally supports large biomasses of fish and invertebrates (Headrich, 1983). Because these resources are easily accessible to coastal communities, some of the oldest and most developed fisheries of the world are based on the harvesting of estuarine species.

In many ways the Estuary of the St. Lawrence system is atypical (El-Sabh, 1979; Therriault et al., this book) and its primary productivity is more similar to that of coastal embayments (Therriault and Levasseur, 1985). While its shores were intimately associated with the colonization of central North-America, its fisheries have only recently evolved into intensive and modern operations.

Coastal and Estuarine Studies, Vol. 39
M. I. El-Sabh, N. Silverberg (Eds.)
Oceanography of a Large-Scale Estuarine System
The St. Lawrence
© Springer-Verlag New York, Inc., 1990

The original objective of this paper was to evaluate the commercial resources of the St. Lawrence Estuary and relate them to its oceanographic characteristics. It soon became obvious to us that this was an impossible task as the fisheries data available for the area proved to be desperately limited both in scope and accuracy. Landing statistics constitute the bulk of that information and, as pointed out by Powles (1980), are generally so deficient that they can only be used as rough indicators of relative species importance. Until recently, most harvesting operations were conducted merely to satisfy local requirements (Anderson and Gagnon, 1980) and catches were often not reported. Because exploitation rates were low and without broad economic significance fisheries research efforts remained minimal. As a result, little information is available on accumulated and/or exploitable biomasses.

In this chapter we briefly describe general trends in the various fisheries of marine organisms prosecuted in the Lower St. Lawrence Estuary since 1968, as they appear in the landing data collected by the Quebec Statistics Bureau and, since 1983, by the federal Department of Fisheries and Oceans. We also summarize the information available from the primary literature and various government reports on the most important commercial species and, where possible, relate their presence and distribution to characteristics of the Estuary. Only species with a completely marine life cycle are reviewed. We end with a discussion of ways whereby oceanographic processes occurring in the Estuary may influence fisheries production downstream from the Gulf of St. Lawrence to the Gulf of Maine.

2. The Commercial Fisheries

Catches of marine fish and invertebrates in the Lower St. Lawrence Estuary have always represented less than 10% of the total commercial fisheries landings in Quebec. In 1986 they constituted less than 4% of that total; this amounts to about 0.3% of the total for Atlantic Canada in that year.

Several factors are responsible for such low figures. For most species the available biomass is low and often distributed over areas hardly accessible to fishing gear (Anderson and Gagnon, 1980). Several fish species move into the Estuary only for a few months in summer and are therefore available for only short periods of time. In several areas fishing effort is still controlled by local market demand and often, only a fraction of the available biomass of relatively abundant species such as capelin (*Mallotus villosus*) and herring (*Clupea harengus*) is harvested (Maranda et al., 1981). Still recently we witnessed on several occasions herring fishermen neglecting to purse their weirs for lack of demand for their catch.

The shores of the St. Lawrence River system and those of the Great Lakes which flow into it are heavily populated and industrialized. Biological and chemical pollutants contaminate

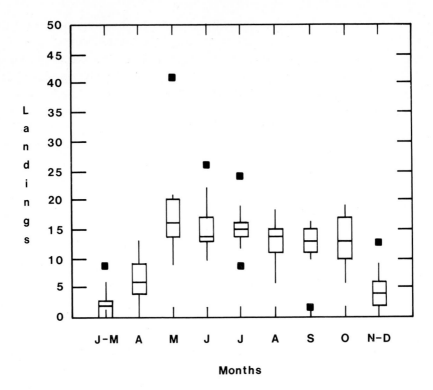

Figure 1.   Box-and-whisker plot showing the monthly distribution of landings in metric tons for all the species combined (see Table 1) over the period 1968 to 1986. Dark squares indicate extreme observations.

the estuarine food web as dramatically demonstrated by the severe health condition and heavy mortality rates of one of its top predators, the beluga whale *Delphinapterus leucas* (Massé et al., 1986; Martineau et al., 1988). These contaminants are responsible for the closing of large areas to shellfish harvesting and the associated significant losses in yield (Cejka, 1976; Soucy et al., 1978; Cossa, this book).

The exploitation of shellfish beds in the Estuary is also restricted by the presence of the toxic dinoflagellate *Protogonyaulax tamarensis* (Therriault et al., 1985; Cembella et al., 1988; Cembella and Therriault, in press). This microflagellate produces a paralytic toxin which is concentrated by filter-feeders rendering them unsuitable for human consumption.

Finally, large areas of the St. Lawrence Estuary are covered by ice for several months during winter (El-Sabh, 1979) forcing fishing activities to a complete halt.

These combined factors restrict the significance of commercial fisheries in the Estuary. This does not mean however that the Estuary cannot play a key role in the dynamics of several marine populations.

2.1. Commercial Landings

The serious limitations associated with the landing statistics available cause us to limit our analysis to the seasonal distribution of catches and to the relative abundance of the dominant species. Table 1 lists the species included in the statistics.

TABLE 1.
List of commercial species included in the landing statistics considered here.

Demersal fish species

Haddock	*Melanogrammus aeglefinus*
Atlantic halibut	*Hippoglossus hippoglossus*
Greenland halibut	*Reinhardtius hippoglossoides*
Pollock	*Pollachius virens*
Hakes	*Urophysis* sp
Cod	*Gadus morhua*
Flounders	Family *Pleuronectidae*
Atlantic tomcod	*Microgadus tomcod*
Redfish	*Sebastes* sp.

Pelagic and diadromous fish species

American shad	*Alosa sapidissima*
American eel	*Anguilla rostrata*
Capelin	*Mallotus villosus*
Rainbow smelt	*Osmerus mordax*
Sturgeons	*Acipenser* sp.
Atlantic herring	*Clupea harengus*
Atlantic mackerel	*Scomber scombrus*
Salmonids	Family *Salmonidae*

Invertebrate species

Whelks	*Buccinum* sp.
Snow crab	*Chionoecetes opilio*
Shrimp	*Pandalus borealis*
Mussels	*Mytilus edulis*
Soft-shelf clams	*Mya arenaria*
Scallops	*Placopecten magellandicus* and *Chlamys islantica*

Most of the landings occur between the months of May to October and few catches are reported for the winter period (Fig. 1). The annual trends in the relative abundance of the three species groups as they evolved from 1968 to 1986 are presented in Figure 2. Pelagic and diadromous species constituted more than 50% of the catch in the late 60's. American eel (*Anguilla rostrata*) and Atlantic herring (*Clupea harengus*) together made up 80% of that fraction (Fig. 3b). The relative importance of the pelagic and diadromous species decreased between 1970 and 1975 and has levelled at slightly less than 20% from 1977 to the present. Eels and herring still dominate these catches, but capelin (*Mallotus villosus*) now represents about 20% of the total.

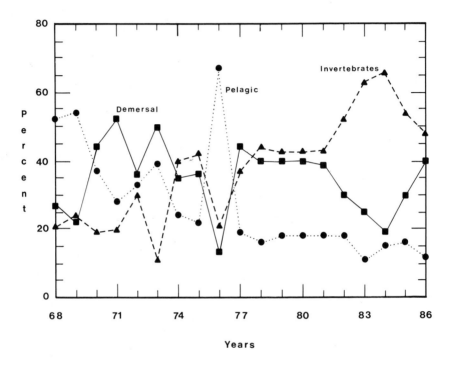

Figure 2. Annual variations in the percent distribution of commercial landings for each of the three major species groups for the period 1968 to 1986.

Landings of demersal fishes increased from about 20% in 1968-1969 to between 40 and 50% from 1970 onward. They fluctuated around that level until 1981 when a large increase in the catches of invertebrate species reduced their relative importance. The reported landings of demersal fishes were dominated by Atlantic cod (*Gadus morhua*) and redfish (*Sebastes* spp.) between 1968 and 1976 (Fig. 3a). Greenland halibut (*Reinhardtius hippoglossoides*) began to be exploited in the Estuary in 1977 and has represented between 20 and 70% of the total catch of demersal fishes recently.

Invertebrates made up between 20 and 40% of the total reported landings between 1968 and 1981 with soft-shell clams (*Mya arenaria*) and shrimp (*Pandalus borealis*) as the major contributors (Fig. 3c). That proportion rose to 65% in the following three years as landings of snow crab (*Chionoecetes opilio*) rapidly increased. In 1986 the proportion of invertebrates was again about equal to that of demersal fishes at around 40%.

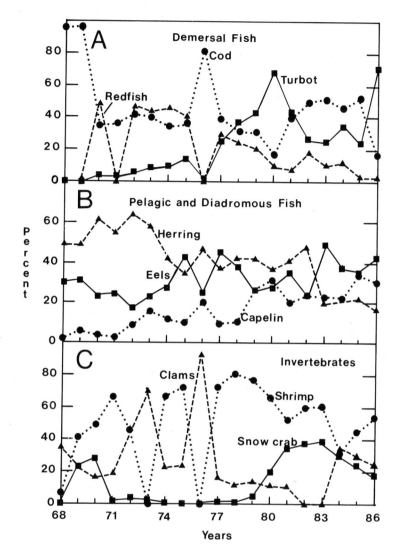

Figure 3. Annual variations in the percent contribution of major species to the total landings of each species group. A: demersal fish. B: pelagic and diadromous fish. C: invertebrate species.

This brief survey of the fishery since the late 60's has identified the major species of commercial interest landed in the Estuary. We will now summarize the information currently available on each of these estuarine populations.

2.2. Ecology of the major species exploited in the Estuary

The dependence of species on estuarine conditions varies in many ways. Only a few fish species rely exclusively on estuaries for reproduction (Headrich, 1983). Estuaries are however very important as nurseries for the young of many species. Because they are generally very productive, estuaries are also used as summer feeding areas by several migratory species.

Of all the major commercial species in the St. Lawrence Estuary herring (*Clupea harengus*) and capelin (*Mallotus villosus*) have generated the most scientific interest. They have also constituted a major component of the landings throughout the time series considered. Because they are mostly harvested for private and local consumption they are likely to be the most under-reported species in the official statistics.

Both species migrate into the Estuary to spawn and their early life history has been and still remains the object of several studies (see review by de Lafontaine in this book). The presence of adult capelin in the Estuary in November-December was confirmed by their abundance in the diet of harp seals (*Pagophilus groenlandicus*) over-wintering near the mouth of the Saguenay River (Sergeant, 1973). This was interpreted by Bailey et al., (1977) as possibly the beginning of the spawning migration. Bailey et al., (1977) also reported large biomasses of maturing capelin in April and May all along the Quebec north shore from Iles aux Coudres in the Upper Estuary down to Anticosti Island. Adult capelin captured prior to spawning by Vésin et al., (1981) were feeding heavily which suggests that the waters over the Laurentian Channel play an important role in providing the energy required for gonad development. Mature capelin spawn primarily on beaches along the north shore, but also along the southern coast of the Estuary (Parent and Brunel, 1976) as well as in deeper waters (Able et al., 1976). Spawning begins near Iles aux Coudres around mid-April and progresses downstream until the end of June (Able et al., 1976). Recently hatched larvae emerge in the surface waters and are transported downstream by the residual currents to the northwestern Gulf of St. Lawrence were they concentrate in a major nursery area west of Anticosti Island (Bailey et al., 1977; Jacquaz et al., 1977). Large concentrations of 1 and 2 year-old juveniles were also observed by Bailey et al. (1977) near the mouth of the Saguenay River suggesting the existence in this area of a secondary nursery. Mortality is very high among spent capelin. Those that survive apparently migrate back to summer-feeding areas in the western Gulf in late June and July (Bailey et al., 1977).

Atlantic herring (*Clupea harengus*) migrate into the St. Lawrence Estuary from the Baie des Chaleurs area in April-May en route towards the spring spawning grounds located in the Iles Verte-Iles aux Lièvres area (Côté, 1979). Spawning occurs in May (Tremblay, 1942; Côté et al., 1980) and most spent fish appear to return to the Baie des Chaleur where they mix with other populations of the southern Gulf on summer feeding grounds (Côté, 1979). Herring also spawn in autumn in the Estuary as demonstrated by the results of population genetic studies (Rivière et al., 1985) and the presence of yolk-sac larvae in great numbers in September (Fortier and Gagné, in press). This fall spawning population is poorly known but is under investigation.

The spawning grounds of both populations appear to be located in the same region, upstream from a tidal front system. Larvae are retained in this area for several months (Courtois et al., 1982; Henri et al., 1985; Fortier and Gagné, in press) in contrast with capelin larvae which emerge in great numbers in the same waters but are swept out of the Estuary by the surface currents (Jacquaz et al., 1977; Courtois et al., 1982; Fortier and Leggett, 1983). Frontal dynamics and the two layer circulation coupled with the vertical migration behaviour of herring larvae are involved in this retention (Fortier and Leggett, 1983; Henri et al., 1985; Fortier and Gagné, in press). Such interactions were proposed by Iles and Sinclair (1982) as a key factor in determining the number and integrity of herring populations.

Primary production within the retention area is very low by estuarine standards (Therriault and Levasseur, 1985). This may explain the very slow growth characteristics of herring in this region (Able, 1978; Sinclair and Tremblay, 1984). Jean (1967) showed that spring spawners in the Estuary were smaller at age than individuals from neighbouring populations. Côté et al. (1980) confirmed that these herring indeed constituted a population of "pygmies" and suggested that slow larval growth was the cause. Little is known about herring juveniles other than that they are caught in great numbers by local weirs and do not appear to leave the Estuary before the age of three (Côté, 1979). The population size of the spring spawners was very roughly estimated at between 1000 (Auger and Powles, 1980) and 10000 metric tons (Côté and Powles, 1978).

The American eel *Anguilla rostrata* is the third most important species of the pelagic-diadromous group. Eels residence time in estuaries is brief and is related to physiological adjustments to salinity changes experienced during migrations to or from the spawning grounds. As other estuarine processes do not appear to influence their abundance and survival in important ways (Dutil et al., 1987) we will not discuss them any further in the present context.

The ecological status of the three species that make up most of the groundfish catches landed in the St. Lawrence Estuary is poorly known. Of the five species of flatfish present in the Lower Estuary only Greenland halibut (*Reinhardtius hippoglossoides*) and possibly winter flounder (*Pseudopleuronectes americanus*) have sufficient biomasses to warrant commercial exploitation. Halibut has been intensively exploited since 1975, first as a bycatch in the shrimp fishery, then as a target species since 1977 (Tremblay and Axelsen, 1980). Signs of overexploitation were reported by Tremblay (1982). Winter flounder remains an unwanted bycatch in other fisheries (Vaillancourt et al., 1985).

Adult Greenland halibut move into the Lower Estuary in the spring and are distributed rather uniformly below 200 m in the deep oceanic water layer of the Laurentian Channel (Steele, 1958). They appear to move upstream as the summer progresses and are more abundant off Rimouski in July and August. They apparently leave the Estuary in the fall, as indicated by the movements of the fishing fleet (Tremblay et al., 1983). While in the

Estuary they feed on other fish species, primarily Atlantic soft pout (*Melanostigma atlanticum*) but also on capelin and redfish (Tremblay et al., 1983). The stock affinity of these seasonal migrants is still unclear. Bowering (1980; 1981) suggested that Greenland halibut caught in the northwestern Gulf of St. Lawrence are immature animals emigrating from spawning grounds along the Labrador - Eastern Newfoundland shelf. If true, the Estuary would constitute the southwestern limit of that ontogenic migration (Fréchet, 1987), a limit occupied by most of the Gulf individuals in summer (Tremblay et al., 1983). This hypothesis is supported by the finding that mostly juvenile fish were present in the Estuary at the time of the investigation (Tremblay et al., 1983). However mature fish are also caught in the Estuary and a few larvae were captured there by de Lafontaine (1980). The possibility that the Greenland halibut exploited in the Estuary come from a mixture of populations cannot therefore be rejected.

Very little is known about cod (*Gadus morhua*) and redfish (*Sebastes mentella*) in the Estuary. These species constitute a small fraction of the large populations inhabiting the southern and northwestern Gulf of St. Lawrence (Jean, 1963; Tremblay et al., 1983). The Estuary is the southwestern limit of their summer migration. Redfish are distributed at depths between 200 and 300 m in the Laurentian Channel while cod appear to be restricted to shallower waters above the 100 m isobath (Steele, 1958; Tremblay et al., 1983). These species colonize both sides of the Estuary. Redfish are sparse above Rimouski (Tremblay et al., 1983) but cod migrate as far up as Trois Pistoles (Bell, 1859). Cod caught by Tremblay et al., (1983) were mostly adult fish while a large proportion of the redfish were still juveniles.

Large quantities of shrimp, principally *Pandalus borealis*, are landed each year in the St. Lawrence Estuary but these catches are small relative to those taken in the northwestern Gulf (Savard et al., 1988). Although shrimp can be found in very dense concentrations along the northern slope of the Laurentian Channel (Couture, 1967; Tremblay et al., 1983; Y. Simard, Maurice Lamontagne Institute, unpubl. data), their overall biomass is too low to support a viable fishery (Couture, 1965, 1966; Couture and Trudel, 1968; Tremblay et al., 1983). *P. borealis* is mostly distributed at depths between 100 and 300 m (Couture, 1967; Tremblay et al., 1983).

Catches of the snow crab *Chionoecetes opilio* in the St. Lawrence Estuary increased rapidly in the early 1980's following a series of successful resource inventories and stimulated by a sudden improvement in market demand (Lafleur et al., 1984). Snow crabs are quite uniformly distributed on both sides of the Estuary mainly between the 40 and 200 m isobaths (Greendale and Bailey, 1982; Bailey and Coutu, 1987). However an important beaching event on the north shore of the Estuary in March 1988 lead Sainte-Marie et al. (1989) to suggest that this bathymetric distribution may not hold for the winter months and that snow crabs could migrate into shallow waters during that period. Snow crabs seem to prefer mud or sand-mud bottoms and temperatures ranging from -0.5 to 5.0 °C (Dufour, 1988).

As for snow crab most of the research conducted in the St. Lawrence Estuary on the soft-shell clam *Mya arenaria* has been devoted to obtaining biomass estimates for management purposes. Little has been carried out on specific ecological questions. Soft-shell clams are widely distributed in dense colonies in the intertidal zone of sheltered embayments along both shores of the Estuary (Lavoie, 1967, 1968, 1969, 1970; Lamoureux, 1974, 1977). Immersion period and percentage of mud in the sediments are important factors influencing the distribution and growth of these benthic organisms (Lamoureux, 1977; Jacques et al., 1984). Pollution from anthropogenic sources and toxicity from dinoflagellates are greater impediments to the full exploitation of the available biomasses than market demand and resource availability (Lamoureux, 1977; Mercier et al., 1978). In spite of these problems, soft-shell clams have generally represented an important fraction of the landings of invertebrate species in the Estuary (Fig. 3c). The large variations seen over the time series probably resulted more from technical problems in harvesting and toxicity outbreaks than from actual biomass fluctuations (Lamoureux, 1977).

2.3. Species distribution and oceanographic features

There appears to be no relationship between the distribution, and perhaps the abundance, of commercially exploited species and the spatio-temporal patterns of primary production in the St. Lawrence Estuary (as described by Therriault and Levasseur, 1985). Physical characteristics of the environment seem to be the determining factors.

Capelin move in the Estuary during winter along the north shore and probably feed on the energy-rich zooplankters overwintering in the area (Rainville and Marcotte, 1985). However they also reproduce in the Estuary and the location of their spawning grounds is strongly dependent on bottom type and surface currents. Capelin larvae emerge along the shores in each one of the regions defined by Therriault and Levasseur (1985), from the poorest to the richest.

Herring spring spawners (Côté et al., 1980) and probably fall spawners as well (J.A. Gagné, unpubl. data), come into the Estuary to spawn and move out soon afterwards. The spawning grounds are located in the poorest primary production area (Powles et al., 1984; Henri et al., 1985) near a tidal front which is hypothesized to be essential to the maintenance of the populations identity (Iles and Sinclair, 1982; Henri et al., 1985; Fortier and Gagné, in press). Juveniles appear to remain in the vicinity along the south shore, still in the same poor region.

Demersal fish migrate into the Estuary via the Laurentian Trough. Cod, Greenland halibut and to a lesser extent redfish distribute themselves throughout the Lower Estuary again independently of the boundaries of the primary production regions. Bottom type and temperature preferences have been suggested as possible determining factors (Steele, 1958;

Tremblay et al., 1983). This is in good agreement with Oviatt and Nixon (1973) who showed that depth, sediment organic content and average wind speed were the three most important physical factors influencing the distribution of demersal fish in Narraganset Bay. Demersal fish of the St. Lawrence Estuary may also follow copepods and euphausids and/or their predators as they move into the Estuary. Runge and Simard (this book) hypothesize that there is no endemic zooplankton population in the Lower St. Lawrence Estuary and that these too are migrants, mainly from adjacent Gulf waters.

The same physical parameters, water temperature and bottom type, are suspected to determine the distribution of snow crab in the Estuary (Dufour, 1988). Jacques et al., (1984) suggested that sediment type and exposure time are key determinants of the distribution of soft-shell clams. These factors combine with tidal amplitude to determine the amount of space available to a species and they are likely to play an important role in the distribution of all species colonizing the littoral zone.

Primary production therefore appears to play a distant secondary role to physical processes in determining the presence and distribution of commercially important species in the St. Lawrence Estuary. This could be specific to this Estuary which is atypical in many respects (see other reviews in this book) and where physical processes contribute to maintain a low level of primary production relative to other estuarine systems (Therriault and Levasseur, 1985).

The influence of hydrodynamic processes operating within the Estuary is not confined to the estuarine boundaries but is felt much further downstream as demonstrated in the next section of this review.

3. Importance of the St. Lawrence Estuary on Fisheries Fluctuations Downstream

The importance of the St. Lawrence drainage system on fisheries production and its variability downstream of the Estuary itself has been extensively reviewed in recent years by Bugden et al. (1982), Dickie and Trites (1983), Sinclair et al. (1986) and Drinkwater (1987). This summary extracts material from the above reviews. We will consider the impact of St. Lawrence runoff on downstream fisheries production for the Gulf of St. Lawrence and the Scotian Shelf and Gulf of Maine areas.

3.1. Gulf of St. Lawrence

In order to evaluate the impact of St. Lawrence runoff on fisheries production in the Gulf it is of interest to consider the degree to which the Gulf of St. Lawrence is an identifiable ecosystem within which variability in fisheries production is internally generated. There is good evidence that a number of important fish populations complete their life cycle within the Gulf and the Shelf areas immediately adjacent to Cabot Strait. In addition, there are marked differences in relative year-class strengths, and in some cases timing of spawning, between the Gulf populations and populations of the same species in adjacent downstream waters. Biomass trends, as well as fish species composition and their rank order, are also markedly different on the Scotian Shelf and within the Southern Gulf of St. Lawrence.

However, the average trawlable biomass in the two areas are essentially identical (Sinclair et al., 1984). This, as well as comparative estimates of zooplankton and benthos biomasses (Bugden et al., 1982, Table 2), suggest that at trophic levels above the primary, the Gulf is not more productive than contiguous Shelf areas. Analyses of lobster landings since the turn of the century also suggest that the Gulf of St. Lawrence and Cape Breton areas have different temporal patterns in abundance than those observed along the Scotian Shelf and in the Gulf of Maine (Campbell and Mohn, 1983; Harding et al., 1983).

There are, however, complex patterns in the landing trends within the Gulf itself, suggesting that the overall population complex is not responding to the environment and fishing in a coherent fashion. In summary, even though biological productivity at higher trophic levels may be similar to that of adjacent Shelf areas, there is considerable evidence that the Gulf of St. Lawrence can be viewed as a distinct ecosystem. Thus one might expect interannual differences in fisheries production to be a function of internal events rather than a reflection of large scale processes along the Continental Shelf.

Several attempts have been made to evaluate the effect of variability in the St. Lawrence runoff on fisheries production in the Gulf. This work was summarized by Drinkwater (1987) who states (p. 41):

> "Significant positive relationships were found between Quebec landings of halibut, haddock, lobster and soft-shell clams and St. Lawrence River discharge (Sutcliffe, 1972). Maximum correlation coefficients occurred when the species landings lagged the river discharge by an amount approximately equal to the age at maturity (i.e. the age at recruitment). This suggested that runoff was correlated with survival during the species' first year of life and is consistent with the concept that reproductive success is established during early life stages. Even higher correlation coefficients resulted from the analysis of monthly St. Lawrence River discharge and landings of the various species with the spring discharge correlating most closely with the lagged fish abundance (Sutcliffe, 1973)."

Budgen et al., (1982) considered the time series of year-class strengths data for the southern Gulf of St. Lawrence cod and RIVSUM (defined by Sutcliffe et al. (1976) as discharge of the St. Lawrence, Ottawa and Saguenay Rivers). Two cod recruitment time series (1947 to 1976 year classes, as well as 1957 to 1976) were analyzed in relation to RIVSUM and the Magdalen Shallows summer surface layer (0-10 m) temperature (May to September) and salinity (June). The Magdalen Shallows hydrographic data were used in the statistical analysis since this is a major distribution area for the early life history stages of the southern Gulf cod population. Bugden et al., (1982) had shown that St. Lawrence runoff influences Magdalen Shallows surface layer temperature and salinity.

Cod recruitment was not statistically correlated with any of the temperature or salinity time series but a significant correlation was found with runoff the summer prior to spawning. It was recognized, however, that the usual statistical confidence levels are not applicable when high correlations are being searched for. Nevertheless, the "suggestion" that St. Lawrence runoff events the year prior to spawning may be important is consistent with: 1) the finding that the number of oocytes are defined the year before spawning, and 2) the conclusion by Doubleday and Beacham (1982) that adult cod productivity in the Southern Gulf is partially a function of RIVSUM. Bugden et al. (1982) concluded, however, that for the best finfish data set in the Gulf of St. Lawrence there was not a strong case (in a statistical sense) to be made about a major impact of freshwater variability on year-class strength.

Drinkwater (1987) updated the analysis of Sutcliffe (1972) to determine how the relationships found by Sutcliffe and his coworkers between environment and fish had fared in recent years. Linear regression analyses of fish landings and environmental indices were performed. Only data contained in the original analyses were considered. The regressions obtained were then used to predict fish landings during the intervening years.

Sutcliffe (1972) had considered Quebec landings of haddock, halibut, soft-shell clam and lobster in his correlation analyses with St. Lawrence runoff. The haddock landings dropped to zero shortly afterwards. It is believed that the haddock caught in the Gulf are from the Scotian Shelf populations. When the abundance on the Scotian Shelf declined markedly during the mid 1970's, the geographic area of the summer feeding migration shrunk. Thus Drinkwater did not consider haddock in his predictive analysis. Results showed that the trend and amplitudes of both lobster and soft-shell clams were relatively well predicted from the linear regressions but not those of halibut.

Predictions of Quebec lobster landings from RIVSUM had been published previously by Sheldon et al. (1982). Drinkwater's updated analysis shows the high similarity between the predictions from their equation and the actual landings for the past seven years. The relationship was remarkably good for 30 years. The most recent years, however, are not being well predicted with the Sheldon et al. (1982) regression.

In summary, the accumulated descriptive information and statistical analyses suggest that St. Lawrence discharge has an influence on interannual variability in fisheries production within the Gulf. However, since the marked increase in regulation of freshwater discharge in the 1970's (Budgen et al., 1982) there has not been a dramatic impact on fisheries yield; for example cod year classes have been strong and lobster landings have not decreased dramatically during the late 1970's and early 1980's. Regulation of freshwater discharge does not seem to have impaired the long-term productivity of the populations.

The runoff - lobster landing regression continues to imply that St. Lawrence discharge is critical to recruitment variability of at least one population of a commercially-important species. The underlying processes remain to be identified and possible mechanisms are suggested in Sinclair et al. (1988).

3.2. Scotian Shelf and Gulf of Maine

Sutcliffe and colleagues have further argued that St. Lawrence discharge has an influence on interannual variability in fisheries production on the Scotian Shelf and on the Gulf of Maine. Drinkwater (1987) has again summarized this work and he states (p. 42):

> "The relationship of fish catches in the Gulf of Maine to both local sea surface temperatures (SST) and RIVSUM were explored by Sutcliffe et al. (1977). Examining 17 commercial marine species of fish and shellfish, they found that catches of 10 of these species were significantly correlated (p < 0.05) with SST's at St. Andrews, New Brunswick, or Boothbay Harbour, Maine. Of these, six were also correlated with RIVSUM, as were two other species. On average, the environmental indices explained 65% of the variability in the fish records. Like for the Gulf of St. Lawrence, maximum correlations occurred if the fish were lagged by a time equivalent to their age at commercial size. Greater than half of the significant correlations were positive and only four were negative. The species exhibiting negative correlations (cod, redfish, yellowtail flounder and soft-shell clams) are considered to be 'cold' water species and near the southern limit of their geographical distribution. Thus, in high discharge ('warm') years the 'cool-water' species do poorly. In contrast, many of the positively correlated species are generally near the northern limits of their range (e.g. butterfish, menhaden) and do well during 'warm' years. Inclusion of fishing effort improved correlations for those species (cod, haddock, yellowtail flounder and menhaden) where effort data were available (Sutcliffe et al., 1977)."

In an approach similar to that followed for the Gulf of St. Lawrence relationships, Drinkwater (1987) evaluated the explanatory power of the statistical relationships between RIVSUM/SST and fisheries landings for the Gulf of Maine area. Regressions derived from the data available to Sutcliffe et al. (1977) were used to predict subsequent landings. In general the regressions did not fare as well in predicting landings as for lobster and soft-shell clams in the Gulf of St. Lawrence.

The conclusion of Sutcliffe et al. (1976, 1977) that St. Lawrence discharge influences the hydrography and fisheries variability in the Gulf of Maine is obviously not as strong as that for the Gulf of St. Lawrence. First, the effect of RIVSUM on the interannual variability of the circulation and water mass characteristics of the Gulf of Maine is not well established. The proportional effects of, respectively, the Labrador Current and RIVSUM have not been evaluated. If Labrador Current and RIVSUM vary in a similar manner the lagged RIVSUM correlations with SST's in the Gulf of Maine may not necessarily reflect a causal relationship. Koslow (1984) and Koslow et al. (1987) have argued that large scale oceanographic processes influence variability in gadoid fisheries from the Labrador Shelf to the Gulf of Maine. It is to be expected that the fisheries variability in inshore waters of the Scotian Shelf (those influenced by the Nova Scotia Current driven by Cabot Strait outflow) are more influenced by St. Lawrence discharge than are the more distant Gulf of Maine stocks and the offshore fisheries on the Scotian Shelf. Pringle (1986) has hypothesized that St. Lawrence discharge variability is involved in the major benthic community fluctuations (including lobster landings) along the eastern and southern shores of Nova Scotia.

In summary, there is considerable evidence that freshwater discharges from the St. Lawrence, as modulated by processes in the Estuary, influence fisheries variability in the Gulf of St. Lawrence. There is further evidence of an expanded influence along the coastal waters of the Scotian Shelf and the Gulf of Maine. The strength of these relationships become understandably weaker as distance from the putative source increases. The mechanisms underlying the statistical relationships between RIVSUM and fisheries landings from particular populations of fish and invertebrates have not been identified.

In conclusion, even though the St. Lawrence Estuary does not itself contribute significantly to the overall fisheries yield of Atlantic Canada, hydrodynamic processes within its boundaries appear to exert strong influences on the abundance of commercial and other species beyond the limits of the estuary itself.

Acknowledgements

We gratefully acknowledge the assistance of the "Bureau de la statistique du Québec" in providing us with the commercial statistics used in this paper. Additional information gathered in data bases developed by the Fish Habitat Division of DFO's Quebec Region was also very useful. Jean-Guy Rondeau collected and compiled most of these data and provided assistance throughout the whole exercise; we thank him for his invaluable help.

REFERENCES

Able, K.W. 1978. Ichthyoplankton of the St. Lawrence Estuary: composition, distribution, and abundance. J. Fish. Res. Board Can., 35, 1518-1531.

Able, K.W., R.F.J. Bailey, B. Jacquaz et J.P.Vesin.1976. Biologie du capelan (*Mallotus villosus*) de l'estuaire et du golfe du Saint-Laurent. M.I.C., D.P.M., Dir. Rech., Cah. Inform. No. 75, 24 P.

Andersen, A. and M. Gagnon. 1980. Les ressources halieutiques de l'estuaire du Saint-Laurent. Rapp. Can. Ind. Sci. Halieut. Aquat., 119, iv + 56p.

Auger, F. and H. Powles. 1980. Estimation of the herring spawning biomass near Isle Verte in the St. Lawrence Estuary from an intensive larval survey in 1979. CAFSAC Res. Doc., 80/59, 29 p.

Bailey, R.F.J., K.W. Able and W.C. Leggett. 1977. Seasonal and vertical distribution and growth of juvenile and adult capelin (*Mallotus villosus*) in the St. Lawrence Estuary and western Gulf of St. Lawrence. J. Fish. Res. Board Can., 34, 2030-2040.

Bailey, R.F.J. and J.-M. Coutu. 1987. Crabe des neiges (*Chionoecetes opilio*) de l'estuaire et du nord du golfe Saint-Laurent: évaluation de 1986. CAFSAC Res. Doc., 87/70, 42 p.

Bell, R. jr. 1859. Catalogue of animals and plants collected and observed on the south-east side of the St. Lawrence, from Québec to Gaspé and in the counties of Rimouski, Gaspé and Bonaventure. Geol. Surv. Can., Rept. Progr., 1858, 243-263.

Bowering, W.R. 1980. The Greenland halibut fishery in the Gulf of St. Lawrence. CAFSAC Res. Doc. 80/24.

Bowering, W.R. 1981. Greenland halibut in the Gulf of St. Lawrence - from immigrants to emigrants. CAFSAC Res. Doc., 81/55, 11 p.

Bugden, G.L., B.T. Hargrave, M.M. Sinclair, C.L. Tang, J.-C. Therriault and P.A. Yeats. 1982. Freshwater runoff effects in the marine environment: the Gulf of St. Lawrence example. Can. Tech. Rep. Fish. Aquat. Sci., 1078, 89 p.

Campbell, A. and R.K. Mohn. 1983. Definition of American lobster stocks for the Canadian Maritimes by analysis of fishery-landing trends. Trans. Amer. Fish. Soc., 112, 744-759.

Cejka, P. 1976. Relevé bactériologique des zones coquillières, districts de pêche maritime; no-1: Comté Bonaventure; no-3: Comtés Rivière-du-Loup et Rimouski, Québec, 1976. Pour le service de protection de l'environnement par le Centre écologique de Port-au-Saumon, 103 p.

Cembella, A.D. and J.C. Therriault. In press. Population dynamics and toxin of *Protogonyaulax tamarensis* from the St. Lawrence Estuary. In Red tides: biology, environmental science and toxicology, T. Okaichi, D.M. Anderson, T. Nemoto, eds., Elsevier, New york, 73-82.

Cembella, A.D., J.C. Therriault and P. Béland. 1988. Toxicity of cultured isolates and natural populations of *Protogonyaulax tamarensis* (Lebour) Taylor from the St. Lawrence Estuary. J. Shellfish Res., 7, 611-621.

Côté, G. 1979. Etude de la population de hareng (*Clupea harengus harengus* L.) de l'estuaire et du golfe du Saint-Laurent. M. Sc. Thesis, Biol. Depart., Laval Univ., Ste-Foy, Québec. 43 p.

Côte, G., P. Lamoureux, J. Boulva and G. Lacroix. 1980. Séparation des populations de hareng de l'Atlantique (*Clupea harengus harengus*) de l'estuaire du Saint-Laurent et de la péninsule gaspésienne. Can. J. Fish. Aquat. Sci., 37, 66-71.

Côté, G. and H. Powles. 1978. The herring population of the St. Lawrence Estuary. CAFSAC Res. Doc., 78/14, 13p.

Courtois, R., M. Simoneau and J.J. Dodson. 1982. Interactions multispécifiques: répartition spatio-temporelle des larves de capelan (*Mallotus villosus*), d'éperlan (*Osmerus mordax*) et de hareng de l'Atlantique (*Clupea harengus harengus*) au sein de la communauté planctonique de l'estuaire moyen du Saint-Laurent. Can. J. Fish. Aquat. Sci., 39, 1164-1174.

Couture, R. 1965. Pêche expérimentale aux crevettes. Rapp. Ann. 1964., Stat. Biol. Mar. Grande Rivière, Québec, 93-95.

Couture, R. 1966. Pêche expérimentale aux crevettes, juillet-août 1965. Rapp. Ann. 1965., Stat. Biol. Mar. Grande Rivière, Québec, 117-128.

Couture, R. 1967. Pêche expérimentale aux crevettes, été-automne 1966. Rapp. Ann. 1966., Stat. Biol. Mar. Grande Rivière, Québec, 75-89.

Couture, R. and P. Trudel. 1968. Les crevettes des eaux côtières du Québec. Nat. Can., 95, 857-885.

De Lafontaine, Y. 1980. First record of Greenland halibut larvae (*Reinhardtius hippoglossoides*) in the lower St. Lawrence Estuary. Nat. Can., 107, 285-287.

Dickie, L.M. and R.W. Trites. 1983. The Gulf of St. Lawrence. In Estuaries and Enclosed Seas, B.H. Ketchum, ed., Elsevier, Amsterdam, 403-425.

Doubleday, W.G. and T. Beacham. 1982. Southern Gulf of St. Lawrence cod: a review of multi-species models and management advice. In Multispecies Approaches to Fisheries Management Advice, M.C. Mercer, ed., Spec. Publ. Can. J. Fish. Aquat. Sci., 59, 133-140.

Drinkwater, K.F. 1987. "Sutcliffe revisited": previously published correlations between fish stocks and environmental indices and their recent performance. In Environmental Effects on Recruitment to Canadian Atlantic Fish Stocks, R.I. Perry and K.T. Frank, eds., Can. Tech. Rep. Fish. Aquat. Sci., 1556, 44-61.

Dufour, R. 1988. Overview of the distribution and movement of snow crab (*Chionoecetes opilio*) in Atlantic Canada. In Proceedings of the International Workshop on Snow Crab Biology, December 8-10, 1987, Montréal, Québec, G.S. Jamieson and W.D. McKone, eds., Can. MS Rep. Fish. Aquat. Sci., 2005, 75-81.

Dutil, J.D., M. Besner and S.D. McCormick. 1987. Osmoregulatory and ionoregulatory changes and associated mortalities during the transition of maturing American eels to a marine environment. Amer. Fish. Soc. Symp., 1, 175-190.

El-Sabh, M.I. 1979. The lower St. Lawrence Estuary as a physical oceanographic system. Nat. Can., 106, 55-73.

Fortier, L. and J.A. Gagné. In Press. Larval herring dispersion, growth and survival in the St. Lawrence Estuary: match/mismatch or membership/vagrancy? Can. J. Fish. Aquat. Sci.

Fortier, L. and W.C. Leggett. 1983. Vertical migrations and transport of larval fish in a partially-mixed estuary. Can. J. Fish. Aquat. Sci., 40, 1543-1555.

Fréchet, A. 1987. Exploitation du flétan du Groenland (*Reinhardtius hippoglossoides*) du golfe du Saint-Laurent en 1986. CAFSAC Res. Doc., 87/56, 23 p.

Greendale, R. and R.F.J. Bailey. 1982. Résultats d'inventaire du crabe des neiges (*Chionoecetes opilio*) dans l'estuaire et le golfe du Saint-Laurent. Can. Tech. Rep. Fish. Aquat. Sci., 1099, 41 p.

Harding, G.C., K.F. Drinkwater and W.P. Vass. 1983. Factors influencing the size of American lobster (*Homarus americanus*) stocks along the Atlantic coast of Nova Scotia, Gulf of St. Lawrence, and the Gulf of Maine: a new synthesis. Can. J. Fish. Aquat. Sci., 40, 168-184.

Headrich, R.L. 1983. Estuarine fishes. In Estuaries and enclosed seas, B.H. Ketchum, ed., Elsevier Sci. Publ., New York, 183-208.

Henri, M., J.J. Dodson and H. Powles. 1985. Spatial configurations of young herring (*Clupea harengus harengus*) larvae in the St. Lawrence Estuary: importance of biological and physical factors. Can. J. Fish. Aquat. Sci. 42 (Suppl. 1), 91-104.

Iles, T.D. and M. Sinclair. 1982. Atlantic herring: stock discreteness and abundance. Science:215, 627-633.

Jacquaz, B., K.W. Able and W.C. Leggett. 1977. Seasonal distribution, abundance, and growth of larval capelin (*Mallotus villosus*) in the St. Lawrence Estuary and northwestern Gulf of St. Lawrence. J. Fish. Res. Board Can., 34, 2015-2029.

Jacques, A., J.C.F. Brêthes and G. Desrosiers. 1984. La croissance de *Mya arenaria* en relation avec les ca-ractéristiques du sédiment et la durée d'immersion, sur la batture de Rimouski. Sci. Tech. Eau, 17, 95-99.

Jean, Y. 1963. Where do Seven-Island cod come from? Fish. Res. Board Can., St-Andrews Biol. Stat., N.B., Gen. Ser. Circ., 39, 2 p.

Jean, Y. 1967. A comparative study of herring (*Clupea harengus*) from the estuary and gulf of St. Lawrence. Nat. Can., 94, 7-27.

Koslow, J.A. 1984. Recruitment patterns in northwest Atlantic fish stocks. Can. J. Fish. Aquat. Sci., 41, 1722-1729.

Koslow, J.A., K.R. Thompson and W. Silvert. 1987. Recruitment to northwest Atlantic cod (*Gadus morhua*) and haddock (*Melanogrammus aeglefinus*) stocks: influence of stock size and climate. Can. J. Fish. Aquat. Sci., 44, 26-39.

Lafleur, P.E., R.F.J. Bailey, J.C. Brêthes and P. Lamoureux. 1984. Le crabe des neiges (*Chionoecetes opilio O. Fabricius*) de la côte nord de l'estuaire et du golfe du Saint-Laurent: état des stocks et perspectives d'exploitation. Trav. Pêch. Qué., 50, 1-53.

Lamoureux, P. 1974. Inventaire des stocks commerciaux de myes (*Mya arenaria*) au Québec: 1971-1973. Québec, Min. Ind. Comm., Cah. Inf., 62, 24 p.

Lamoureux, P. 1977. Estimation des stocks commerciaux de myes (*Mya arenaria*) au Québec: biologie et aménagement des pêcheries. Québec, Min. Ind. Comm., Cah. Inf., 78, 109 p.

Lavoie, R. 1967. Inventaire de myes communes (*Mya arenaria* l.) de l'estuaire du Saint-Laurent. Rapp. Ann. 1966., Stat. Biol. Mar. Grande Rivière, Québec.

Lavoie, R. 1969. Les coques de l'estuaire du Saint-Laurent. Actualités Marines, vol. 13, No 2, 2-11.

Lavoie, R. 1969. Inventaire des mollusques de la région de Tadoussac. Québec, Min. Ind. Comm., Cah. Inf., 49, 23 p.

Lavoie, R. 1969. Inventaire des populations de Mye commune (*Mya arenaria*) de Grandes- Bergeronnes à Portneuf-sur-Mer, été 1968. Rapp. Ann. 1968, Stat. Biol. Mar. Grande Rivière, Québec, 103-118.

Lavoie, R. 1970. Inventaire des populations de coques (*Mya arenaria*) de Forestville à Papinachois, été 1969. Rapp. Ann. 1969, Stat. Biol. Mar. Grande Rivière, Québec, 107-125.

Maranda, Y., S.S.M. Labonté and H. Powles. 1981. Exploitation potentielle du capelan au Québec: débarquements (1950-1979) et caractéristiques biologiques (1979-1980). Can. MS. Rep. Fish. Aquat. Sci., 1604, 43p.

Martineau, D., P. Béland, C. Desjardins and A. Lagacé. 1988. Levels of organochlorine chemicals in tissues of beluga whales (*Delphinapterus leucas*) from the St. Lawrence estuary, Québec, Canada. Arch. Envir. Contam. Toxicol., 16, 137-147.

Massé, R., D. Martineau, L. Tremblay and P. Béland. 1986. Levels and chromatographic profiling of DDT metabolites and PCB residues in stranded beluga whales (*Delphinapterus leucas*) from the St. Lawrence middle estuary - Canada. Arch. Environ. Contam. Toxicol., 15, 567-579.

Mercier, Y., P. Lamoureux and J. Dubé. 1978. Nouvelle estimation des stocks commerciaux de myes (*Mya arenaria*) de la région de Rivière Portneuf sur la côte du Saint-Laurent en 1977. Québec, Min. Ind. Comm., Cah. Inf., 87, 23 p.

Oviatt, C.A. and S.W. Nixon. 1973. The demersal fish of Narragansett Bay: an analysis of community structure, distribution and abundance. Est. Coastal Shelf Sci., 1, 361-378.

Parent, S. et P. Brunel. 1976. Aires et périodes de frai du capelan (*Mallotus villosus*) dans l'estuaire et le golfe du Saint-Laurent. M.I.C., D.G.P.M., Serv. Biol., Trav. Pêch. Québec No 45, 46 p.

Powles, H. 1980. Preface. In A. Andersen and M. Gagnon. Les ressources halieutiques de l'estuaire du Saint-Laurent. Rapp. Can. Ind. Sci. Halieut. Aquat. No. 119:V.

Powles, H., F. Auger and G.J. Fitzgerald. 1984. Nearshore ichthyoplankton of a north temperate estuary. Can. J. Fish. Aquat. Sci., 41, 1653-1663.

Pringle, J.D. 1986. A review of urchin/macro-algal associations with a new synthesis for nearshore, eastern Canadian waters. Monogr. Biol., 4, 191-218.

Rainville, L.A. and B.M. Marcotte. 1985. Abundance, energy, and diversity of zooplankton in the three water layers over slope depths in the lower St. Lawrence Estuary. Nat. Can., 112, 97-103.

Rivière, D., D. Roby, A.C. Horth, M. Arnac and M.F. Khalil. 1985. Structure génétique de quatre populations de hareng de l'estuaire du Saint-Laurent et de la baie des Chaleurs. Nat. Can., 112, 105-112.

Sainte-Marie, B., R. Dufour and C. Desjardins. In press. Beaching of snow crabs (*Chionoecetes opilio*) on the north shore of the Gulf of St. Lawrence. Nat. Can.

Savard, L., Y. Lavergne and J. Lambert. 1988. Evaluation des stocks de crevette (*Pandalus borealis*) du golfe du Saint-Laurent. CAFSAC Res. Doc. 88/60, 49 p.

Sergeant, D.E. 1973. Feeding, growth and productivity of Northwest Atlantic harp seals. J. Fish. Res. Board Can., 30, 17-29.

Sheldon, R.W., W.H. Sutcliffe and K. Drinkwater. 1982. Fish production in multispecies fisheries. In Multispecies Approaches to Fisheries Management Advice, M.C. Mercer, ed., Can. Spec. Public. Fish. Aquat. Sci., 59, 28-34.

Sinclair, M., J.T. Anderson, M. Chadwick, J.A. Gagné, W.D. McKone, J.C. Rice and D. Ware. 1988. Report from the national workshop on recruitment. Can. Tech. Rep. Fish. Aquat. Sci., 1626, 261 p.

Sinclair, M., G.L. Bugden, C.L. Tang, J.-C. Therriault and P.A. Yeats. 1986. Assessment of effects of freshwater runoff variability on fisheries production in coastal waters. In The Role of Freshwater Outflow in Coastal Marine Ecosystems, S. Skreslet, ed., NATO ASI Series, vol. 67, Springer-Verlag, Berlin, 139-160.

Sinclair, M., J.-J. Maguire, P. Koeller and J.S. Scott. 1984. Trophic dynamic models in light of current resource inventory data and stock assessment results. Rapp. Proc. Verb. Réun., Cons. Intern. Explor. Mer, 183, 269-284.

Sinclair, M. and M.J. Tremblay. 1984. Timing of spawning of Atlantic herring (*Clupea harengus harengus*) populations and the match-mismatch theory. Can. J. Fish. Aquat. Sci., 41, 1055-1065.

Soucy, R., P. Lamoureux, C. Blaise and Y. Dussault, 1978. Les coquillages marins comestibles du Québec.Secteurs contaminés et ouverts.Gouv. Québec, MIC, DGPM, 86 p.

Steele, D.H. 1958. Fishes taken in the Laurentian Channel, Gulf of St. Lawrence, between Birds Rocks and the Saguenay River, 1953 and 1954. Fish. Res. Board Can. MS Rep. Ser. No 651, 32p.

Sutcliffe, W.H. Jr. 1972. Some relations of land drainage, nutrients, particulate material, and fish catch in two eastern Canadian bays. J. Fish. Res. Board Can., 29, 357-362.

Sutcliffe, W.H. Jr. 1973. Correlations between seasonal river discharge and local landings of American lobster (*Homarus americanus*) and Atlantic halibut (*Hippoglossus hippoglossus*) in the Gulf of St. Lawrence. J. Fish. Res. Board Can., 30, 856-859.

Sutcliffe, W.H. Jr., K. Drinkwater and B.S. Muir. 1977. Correlations of fish catch and environmental factors in the Gulf of Maine. J. Fish. Res. Board Can., 34, 29-30.

Sutcliffe, W.H. Jr., R.H. Loucks and K. Drinkwater. 1976. Coastal circulation and physical oceanography of the Scotian Shelf and the Gulf of Maine. J. Fish. Res. Board Can., 33, 98-115.

Therriault, J.C. and M. Levasseur. 1985. Control of phytoplankton production in the lower St. Lawrence Estuary: light and freshwater runoff. Nat. Can., 112, 77-96.

Therriault, J.C., J. Painchaud and M. Levasseur. 1985. Factors controlling the occurrence of *Protogonyaulax tamarensis* and shellfish toxicity in the St. Lawrence Estuary: freshwater runoff and the stability of the water column. In Toxic Dinoflagellates, D.M. Anderson, A.W. White and B.C. Baden eds., Elsevier, New York, 141-146.

Tremblay, C. 1982. Le flétan du Groendland du golfe du Saint-Laurent (4RST): conséquences de son exploitation et évaluation de son recrutement. CAFSAC Res. Doc., 82/18, 32 p.

Tremblay, C. and F. Axelsen. 1980. Données sur la pêche, la biologie et l'abondance du flétan du Groendland (*Reinhardtius hippoglossoides*) dans le golfe du St-Laurent. CAFSAC Res. Doc , 80/34, 27 p.

Tremblay, C., B. Portelance and J. Fréchette. 1983. Inventaire au chalut de fond des espèces de poissons et crustacés dans l'estuaire maritime du Saint-Laurent. Québec, M.A.P.A.Q., Dir. Rech. Sci. Tech., Cah. Inf., 103, 96 p.

Tremblay, J.L. 1942. Quatrième rapport annuel de la station biologique du Saint-Laurent, 1936-1942. Univ. Laval, Ste-Foy, Québec. 100 p.

Vaillancourt, R., J.C. Brêthes and G. Desrosiers. 1985. Croissance de la plie rouge (*Pseudopleuronectes americanus*) de l'estuaire maritime du Saint-Laurent. Can. J. Zool., 63, 1610-1616.

Vésin, J.V., W.C. Leggett and K.W. Able. 1981. Feeding ecology of capelin (*Mallotus villosus*) in the estuary and western gulf of St. Lawrence. Can. J. Fish. Aquat. Sci., 38, 257-267.

Chapter 17

The Saguenay Fiord: A Major Tributary to the St. Lawrence Estuary

C.T. Schafer [1], J.N. Smith [2] and R. Côté [3]

[1] Geological Survey of Canada, Bedford Institute of Oceanography, Dartmouth, N.S., B2Y 4A2
[2] Atlantic Oceanographic Laboratory, Bedford Institute of Oceanography, Dartmouth, N.S., B2Y 4A2
[3] Sciences Fondamentales, Université du Québec à Chicoutimi, Chicoutimi, Québec, G7H 2B1

ABSTRACT

The temporal and spatial pattern of water circulation processes in the Saguenay Fiord, and the fall and winter incursion of water from the St. Lawrence Estuary, help to explain the arctic affinity of its indigenous marine assemblages. Industrialization of the Saguenay Region during the 20th century has caused the development of benthic anoxic conditions near the head of the fiord as a result of the disposal of large volumes of organic waste from pulp and paper mills. Disposal of Hg-contaminated waste in the fiord prior to 1971 appears to be an ongoing problem with respect to concentrations of this metal observed in shrimp tissue as recently as 1985. Industrial waste discharges appear to be an important factor in controlling the toxicity of copper with respect to the freshwater phytoplankton population of the Saguenay River and the brackish/marine phytoplankton of the north arm.

Landslides and river discharge events are preserved as distinctive sediment layers in unbioturbated prodelta deposits that have accumulated in a basin near the head of the fiord. These deposits also contain a high resolution (annual) record of pollution inputs that reflect the scale of 20th century anthropogenic activity in this area of Quebec. Percent sand variations in cores of these prodelta deposits show a direct relationship to spring freshet magnitude that may be useful for reconstructing the 19th century discharge history of the Saguenay River.

Key Words: Circulation, sedimentary processes, historical, marine mammals, zooplankton, phytoplankton, pollution, biological considerations

1. Introduction

1.1. General Relationship of the Saguenay to Some Other Canadian Fiords

The Saguenay is a classic fiord in the geomorphic sense, being a long (170 km) and narrow (1-6 km) feature with a typical U-shaped cross-section and having a shallow 20 m deep sill at its mouth which intersects the St. Lawrence Estuary near Tadoussac (Figs. 1 & 2). An 80 m deep sill located 18 km upstream from Tadoussac subdivides the fiord into two distinct basins (Fig. 1C). The "outer" or eastern basin is relatively small but nevertheless

Coastal and Estuarine Studies, Vol. 39
M.I. El-Sabh, N. Silverberg (Eds.)
Oceanography of a Large-Scale Estuarine System
The St. Lawrence

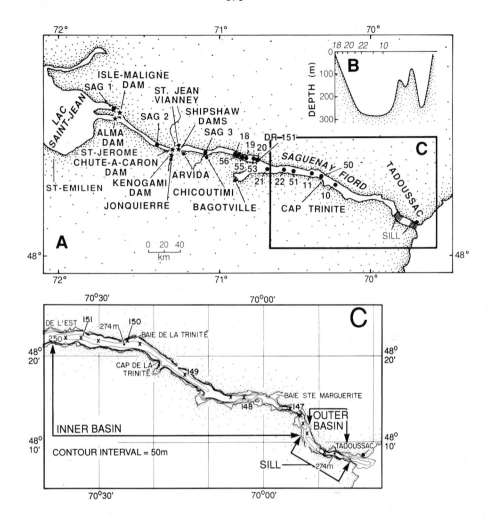

Figure 1. (A) Location of cities, towns and power dams along the Saguenay Fiord. Stations 10 and 11, and 18 to 22, refer to core locations reported in Smith and Walton (1980). Stations SG-1 to SAG-3, and DR-151 relate to phytoplankton productivity investigations discussed in this chapter. (B) Location of Smith and Walton's (1980) coring stations with respect to water depth and the location of sills near the mouth of the fiord. (C) Bathymetry of the main channel of the fiord showing relative size of the inner and outer basins. Station numbers refer to cruise 74-006.

reaches a maximum depth of about 250 m. The larger western or "inner" basin extends from the sill westward to Baie des Ha! Ha! reaching a maximum depth of 275 m. The main fiord valley branches into two comparatively shallow arms near its head at Bagotville and about 15 km east of Chicoutimi. These arms were once hanging valleys from which glaciers merged to continue their excavation of the deep inner basin of the fiord (Fig. 2). The river systems that discharge into the fiord at the western ends of these arms drain a basin that covers more than 78,000 km^2. About 90% of this runoff is derived from the Saguenay River system which discharges into the north arm about 8 km east of Chicoutimi.

This river contributes annually more than 5.0×10^9 m^3 of fresh water to the St. Lawrence system (Schafer et al., 1983). Most of this volume reaches the fiord during the annual spring freshet between April and June.

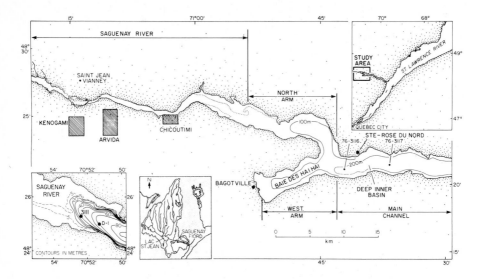

Figure 2. Upper reaches of the Saguenay Fiord showing north and west arms in relation to the inner basin (main channel). Insets show the location of the fiord in relation to the St. Lawrence Estuary, the drainage basin of the Saguenay River, and detailed bathymetry of the Saguenay river delta foreslope showing the location of core 76-3111 (Schafer et al., 1980) and D-1 (Smith and Schafer, 1987).

The size of the Saguenay Fiord valley is comparable to many of the fiords that lie along the east coast of Baffin Island (e.g., Syvitski and Schafer, 1985). However, the more southerly location of the Saguenay is reflected by the greater thicknesses of soil that covers its emerged flanks, and also by its mature flora. These features contrast profoundly with the virtual absence of soil and the meager occurrences of dwarfed trees and sparse ground cover that mark the impact of arctic climate and the comparatively recent deglaciation of the less mature Baffin Island fiords (Syvitski et al., 1986). Compared to the Saguenay, the mainland fiords of British Columbia are distinctive in terms of their generally greater relief and, like their Baffin Island counterparts, by the annual occurrence of two prominent freshet events derived from rainfall and snow melt in the late spring and from ice melt during the summer. Because of their reduced levels of vegetation, the detrital sediment input rates for the mainland B.C. fiords are on a par with some of the Baffin Island fiords and can be about an order of magnitude higher than observed in the Saguenay (Syvitski et al., 1986, Figs. 3.3 and 8.16). The hydrological characteristics of the 60 fiords found along the Labrador coast (Nutt, 1963) tend to fall between the Baffin Island types and the drowned "fiards" that characterize the Nova Scotia coast (e.g. Piper et al., 1983) and are, in general, comparable to those observed for the Saguenay.

Figure 3. (A) Steeply sloping walls of the inner basin section of the fiord near Cap Trinite (for location, see Fig. 1). (B) Satellite image of the Saguenay Fiord taken on August 29, 1973. At the head of the fiord, the image shows the city of Chicoutimi and the surrounding industrial complexes and towns that are situated along the Saguenay River channel about 10 km west of the river mouth. The suspended sediment plume appears to be more concentrated on the south side of the north arm and extends to the main channel.

In this chapter, we review some of the oceanographic and geologic characteristics of the Saguenay Fiord that have been elucidated by researchers over the past several decades and focus on some of the unique sedimentological, biological and geochemical features of this marine system.

1.2. Geological Perspective

The valley of the Saguenay Fiord has been carved into a series of Precambrian anorthosites and acidic igneous rocks that underlie sedimentary rocks of Paleozoic age. The location of the valley may mark the position of old Precambrian lines of weakness in the earth's crust (Kumarapeli and Saul, 1966) that offered a path of least resistance to advancing continental glaciers. LacSaint-Jean (Fig. 1A) occupies a depression near the head of the fiord whose flanks are bounded by 500 m escarpments that reflect a Precambrian graben structure (Ouellet and Jones, 1987). These escarpments run E-W and converge toward the head of the fiord. During early Paleozoic time, the area of the present day fiord must have had a basin-like character in the Lac St. Jean region in which the marine sediment equivalents of Ordovician age fossiliferous shales and limestones were deposited. These sedimentary rocks appear to dip toward the centre of the depression that forms the present day lake (LaSalle, 1968). This structural relationship supports the idea that this area was the site of a depression over a long interval of geologic time.

Presumably, geologic agents such as glacier ice have eroded these older sedimentary deposits from the main channel of the fiord (see: LaSalle and Tremblay, 1978). During the Quaternary, the fiord basin was overdeepened and its walls steepened and polished as a consequence of the repeated advance of glaciers along its axis (Fig. 3A). The last of these advances appears to have occurred prior to 9500 years BP (LaSalle, 1968).

At the time of the retreat of late Wisconsinan ice, the fiord was flooded by marine water which provided a medium for the deposition of the Mer de Laflamme or Leda Clay sediments (Gadd, 1960). In parts of the Lac St. Jean basin, these fossiliferous clayey silt deposits are found at elevations of as much as 180 m above sea level (LaSalle, 1968; Legget, 1945). During the early interval of postglacial rebound, the clays were mantled by sandy prograding beach and deltaic deposits. Between 12,000 and 8,000 years BP, many of these unconsolidated sandy sediments were reworked by winds to form well sorted aeolian sands. In some localities these sands have given rise to well-defined dune forms (LaSalle, 1968). In this regard, the Saguenay environment of late glacial time is almost an exact analog to modern environments found near the heads of several Baffin Island fiords (e.g., Itirbilung and McBeth: in Syvitski et al., 1986, p. 73). Since the end of early Holocene time, the Saguenay Fiord has taken on its north temperate character. Bog deposits have formed in areas that are poorly drained and rich stands of spruce and hardwood tree species carpet the slopes lying adjacent to its steep and polished walls.

1.3. Historical Perspective

A brief historical overview of the Saguenay region has been compiled by LaSalle (1968). It seems that all the territory of the Saguenay River and Lac St. Jean districts was known to the Indians well before the arrival of Europeans. Jacques Cartier was one of the first explorers to show an interest in the Saguenay River during his visit to Canada in 1535. The arrival of European traders during the 17th century resulted in the establishment of a number of trading posts near the city of Chicoutimi, and on the southern shore of Lac St. Jean near St. Jerome (Fig. 1A). By 1837, a settlement was developed on Baie des Ha! Ha! (Blanchard, 1935) which provided labour and logs for some of the saw mills that had been established along the Saguenay River by William Price. By 1852, Chicoutimi had become the business centre of the entire Lac St. Jean - Saguenay Fiord region. Colonization of the Lac St. Jean lowlands started in the late 1840's and many of the larger towns in the Saguenay - Lac St. Jean area were first settled in the middle and late 1800's. This interval of history is also marked by the great fire of 1870 which burned a large area around Chicoutimi.

The early 20th century history of the Saguenay region is dominated by industrial expansion related to the exploitation of forest resources. The first paper mill was built in the town of Jonquiere by the Price Brothers Company and began operations in 1909. It was soon followed by the construction of a large paper mill at Kenogami in 1913. Aspects of the environmental impact of the discharge of organic waste by this mill, as well as that of other industries located on the Saguenay River, are covered in a later section of this chapter. The 1920 to 1950 interval is marked by the construction of several power dams along the main channel of the Saguenay River. The first of these was built in 1926 by Mr. Duke Price at Isle Maligne near the mouth of Lac St. Jean (Fig. 1A). This structure raised the level of the lake by about 9 m. Much of its hydroelectrical output was used for the smelting of aluminum ore; the metal was apparently first produced in the Saguenay Region by ALCOA at its Arvida facility. The town of Arvida, named after ALCOA's first president (Arthur Vining Davis), is located between Chicoutimi and Jonquiere. It enjoyed significant expansion during the World War II years because of the aviation industry's large demand for aluminum. Power for the Arvida industrial complex was augmented in 1931 by completion of the Chute-a-Caron dam and power station (224 Mw), and by the Shipshaw dams (896 Mw) in 1945. The installation of these dams has significantly altered the hydrograph of the Saguenay river (see below). Over the years, the Saguenay - Lac St. Jean area has continued to develop as an important industrial centre for the production of aluminum and a variety of pulp and paper products.

2. Watermass Sources and Circulation

Many of the salient features of Saguenay Fiord circulation are treated in unpublished reports by Taylor (1975) and Loucks and Smith-Sinclair (1975). In general, the spring tides carry St. Lawrence Estuary water into the outer basin at a rate of about 1.5 to 2.0 m s⁻¹. Flood tide currents are dissipated in the lower reaches of the fiord such that they are virtually imperceptible about 23 km west of Tadoussac. The rate of ebb flow near Tadoussac varies from 1.5 to 2.5 m s⁻¹ and is primarily a function of fiord breadth. Ebb flow is strongest at the sill near the fiord mouth where spring tides can reach flow rates of 3.0 to 3.6 m s⁻¹. Data in Jordan and de la Ronde (1974) indicate that relatively high salinity (>29 ⁰/oo) and low temperature (<1°C) water is transported over the 20 m deep sill into the outer basin during spring flood tides. Stratification is reduced or eliminated at this location by tidal mixing during most of the year.

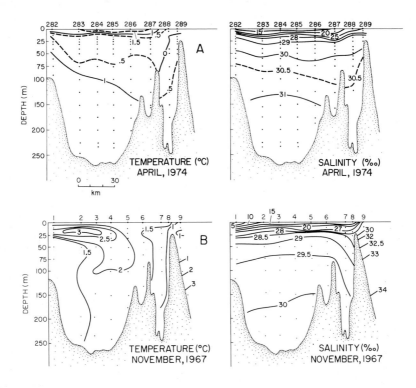

Figure 4. Temperature and salinity profiles from (A) April, 1974 and (B) November, 1967 showing the effect of deep water renewal processes over the summer season (adapted from Taylor, 1975, cruises 74-006 and 67-008).

Surface currents in the north arm just east of Chicoutimi are mostly related to the river discharge rate and are strongest during the spring freshet. A Coriolis effect forces the low density river water toward the south side of the arm from whence it flows into the main channel where it can often be delineated on satellite images (Fig. 3B) because of its characteristic load of suspended particulate matter (SPM). In the deeper parts of both fiord basins, the base of the halocline occurs at about 26 to 29 º/oo during the spring and fall (Fig. 4A and B). Its slope tends to be more pronounced near the western end of the inner basin reflecting the source direction of Saguenay River water which dominates the shallow part of the water column in the spring and early summer. River discharge promotes considerable stratification that can be detected throughout most of the fiord until early autumn (Fig. 5). In the autumn season, the temperature distribution of inner basin water sometimes reflects the influence of summer heating. For example, the +3°C watermass observed in the upper 25 m interval during November, 1967 is interpreted by Loucks and Smith-Sinclair (1975) as a remnant of the past summer's heating (Fig. 4B). This watermass must eventually lose its heat through mixing with the surrounding waters and, in the November, 1967 profile, appears to be undergoing upward displacement by newly advected cold water.

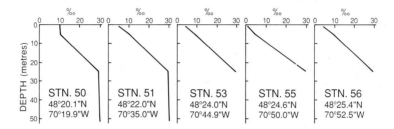

Figure 5. Salinity profile traverse collected on October 1, 1979 between Cap Trinite (station 50) and the head of the north arm (station 56). The low surface salinity at station 55 probably reflects the confinement of the Saguenay River buoyant plume to the south side of the north arm (CSS Dawson Cruise 79-024).

In the outer basin section of the fiord, the occurrence of gravel deposits, as opposed to sand and muddy sediments in the inner basin, is suggestive of relatively high near-bottom turbulence. Vigorous mixing in the outer basin is also suggested by the relatively vertical orientation of salinity and temperature contours (e.g., Fig. 4B) and by comparatively uniform SPM concentration distributions (Sundby and Loring, 1978). Temperature patterns of bottom water in the outer basin show seasonal variations that are consistent with surface water variations. In contrast, at the western end of the inner basin, this response seems to be completely attenuated and cold bottom water is present at all times (Loucks and Smith-Sinclair, 1975). The presence of distinct temperature and salinity structure in vertical profiles from the inner basin reflects the reduced level of mixing in this part of the fiord.

Drainville (1968) speculated that the surface water characteristics of St. Lawrence Estuary water near the mouth of the fiord during the winter season, and at high tide, are virtually identical to those of deep inner basin water thereby suggesting a winter replenishment

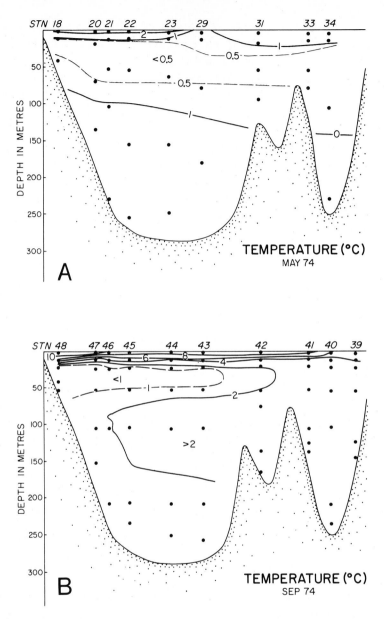

Figure 6. (A) Temperature and salinity distributions in May (cruise 74-006) and (B) September (cruise 74-032) of 1974 showing evidence of the development of a shallow cold water core in the western part of the inner basin (adapted from Sundby and Loring, 1981).

process (Fig. 6A). Using Mn-rich particulate matter as a watermass tracer, Sundby and Loring (1978) showed that this relatively cold bottom water is displaced upward and westward as the result of the incursion of relatively warm water at 75-150 m depth during the summer months (Fig. 6B). The intruding water seems to originate somewhat higher in the Estuary water column and is sufficiently dense to be carried towards the bottom of the basin. Stratification in the inner basin is enhanced during this season as a consequence of the formation of a warm low salinity surface water layer. Loucks and Smith-Sinclair (1975) concluded that the rate of deep water replenishment in the fiord appears to be highest in the winter and near the mouth of the fiord. They observed further that the range of water densities presented at the second sill that separates the inner and outer basins 18 km west of Tadoussac is reduced by mixing in the outer basin so that the inner basin is also vulnerable to frequent intrusion of Estuary water. Spring temperature profiles suggest that these intrusions are relatively large during the winter period because of the involvement of relatively dense (cold) water. While the summer intrusions reach only halfway up the inner basin, the winter influx appears to effect the entire basin (e.g., Figs. 4B and 6B versus 4A and 6A). Like many other Canadian fiords (e.g., Syvitski et al., 1986), the fresh water inflows of the spring freshet occur over a much shorter time interval compared to the bottom water incursion processes.

SPM distribution in the western part of the inner basin has a three layer configuration with relatively high concentrations in the top and bottom layers that are consistent with observed water temperature and current velocity distributions. Surface layer SPM concentrations can be easily related to a river water (upstream) source. Conversely, the high SPM values observed in the bottom layer have been attributed to bioturbation, and to several physical mechanisms including internal waves generated at the sills and density currents resulting from water intrusions over the sills (Sundby and Loring, 1978).

Seaward of the Saguenay River mouth, a suspended sediment (SPM) "wash load" is carried into the north arm and beyond by the river plume. The mixing of this sediment-laden river plume with fiord basin water follows a linear trend with distance from the river mouth when delineated by conservative parameters such as water temperature and salinity. However, a map of surface SPM concentration will usually show a characteristic exponential relationship with respect to distance because of particle settling that occurs while the river water is mixing with fiord water. This exponential distribution may be enhanced by flocculation, agglomeration and pelletization processes (Syvitski et al., 1986). These depositional mechanisms tend to be more pronounced in two layer estuarine settings such as those that prevail in the north arm. An example of the combined effects of settling, and of other sediment "scavenging" processes, is provided by the distribution of sediment thicknesses in the north arm arising from a single well documented SPM depositional event that has been related to the 1971 St. Jean Vianney landslide (Fig. 7).

During periods of extremely low discharge (late fall and winter), when only fine clay size mineral particles and organic matter are carried into the north arm of the fiord, the concentration field of SPM may be expected to follow that of temperature and salinity, i.e. it behaves as a conservative water property that reflects the expected slower response between the surface layer and the underlying watermass (e.g., Syvitski et al., 1985).

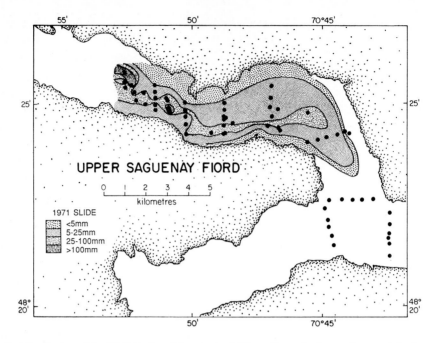

Figure 7. Distribution of the 1971 St. Jean Vianney landslide sediment in the north arm of the Saguenay Fiord. The relatively thick part of the deposit tends to follow the thalweg of the north arm basin (adapted from Schafer and Smith, 1987).

Total SPM concentrations near the head of the Saguenay Fiord range from 10 to 20 mg L^{-1} and decrease exponentially to less than 0.3 mg L^{-1} in the deep parts of the inner basin (Sundby and Loring, 1978). Sediments reaching the deep inner basin as part of the suspended load are composed of quartz (up to 70%), feldspar (20-30%) and a range of other rock forming minerals (Loring, 1975). In addition, particulate organic matter of both natural and anthropogenic origins is carried into the fiord as part of the SPM load during all seasons of the year. Studies by Pocklington and Leonard (1979) have shown that organic matter reaching the deep inner basin has a C/N ratio of about 20. This value is indicative of organic material associated with pulp and paper mill discharges. Tan and Strain (1979) also concluded that this organic matter is primarily terrestrial in origin and that the autochthonous carbon component of the sediments in this part of the fiord is negligible. Deep basin environments near the mouth of the fiord appear to be depocentres for particulate organic carbon derived from the St. Lawrence Estuary through "tidal pumping" mechanisms (Therriault et al., 1980; 1984; Drainville, 1968). In general, the distribution of SPM in the fiord is a function of dispersal induced by the dynamics of the Saguenay River, and of the resuspension of bottom sediments by biological (infauna and epifauna) and physical (waves and tidal currents) agents. During the spring freshet, the spatial distribution of the SPM field is highly correlated to the sedimentation rate field (e.g., compare the sediment plume in Fig. 3B to the thickness distribution of May, 1971 landslide sediments in Fig. 7).

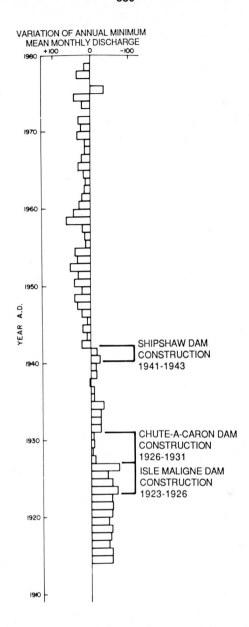

Figure 8. Percent variation of annual minimum mean monthly discharge as the result of the release of stored water from a series of reservoirs in the Saguenay River drainage basin during the summer season. (Data from station 062901, Service de l'Hydrométrie, Ministère des Richesses Naturelles, Quebec.)

3. Impact of Power Dams

The natural hydrograph of the Saguenay River was altered significantly during the 20th century as the result of the construction of hydroelectric dams at several locations between

Lac St. Jean and Chicoutimi (Fig. 1). Prior to 1926, the unregulated river discharge cycle typically exhibited a sharp maximum (freshet) in May followed by a continuously decreasing mean monthly discharge rate (MMD) through March of the following year. The minimum of this discharge cycle occurred at values of approximately 400 m^3 s^{-1} during the winter months when the river was ice covered and the major component of precipitation was stored as snow. The completion of the Isle Maligne dam and power station in 1926 was marked by a doubling of the minimum MMD from 300-400 m^3 s^{-1} to 600-800 m^3 s^{-1} (Fig. 8). Construction of the Chute-a-Caron facility in 1931 near Kenogami does not appear to have increased the minimum MMD significantly. However, the completion of the Shipshaw station near Kenogami in 1943 raised the minimum MMD to 1000 m^3 s^{-1}. The increasing demand for power during the post World War II period was marked by a gradual increase in minimum MMD rates which attained maximum values of 1200 m^3 s^{-1} after 1960. In contrast, the magnitude of maximum MMD's during the spring freshets of 1973, 1974 and 1976 (Smith and Schafer, 1987) suggest that the power stations have had little effect on this aspect of the Saguenay River hydrograph.

4. Sedimentary Processes

During late glacial time, the basins of the Saguenay Fiord were initially depocentres for clayey marine silts, and later for riverine and periglacial sediments. During the Holocene sea level rise, slumping of these deposits from the submerged walls of the fiord has occurred. This process has contributed significantly to the buildup of sediments on the floor of the fiord basin which, at some locations, may exceed 1.5 km in thickness (J.P.M. Syvitski, pers. comm.) (Fig. 9). Modern sedimentary processes evident in the geologic record of the past several centuries can be classified into fluvial, hypopycnal (i.e. the transport and deposition of riverine suspended sediment into a basin containing stratified water) and mass transport categories. Hypopycnal processes are treated in general terms in the previous section on watermass sources. Mass transport events may be broken down further into those of terrestrial (landslides) or submarine (slope failure) origin.

4.1. Fluvio-deltaic depositional processes

210Pb dating of a series of Lehigh gravity cores collected over the entire length of the fiord (Smith and Walton, 1980) has shown that the combined effect of fluvio-deltaic and hypopycnal depositional processes is marked by an exponential decrease in sedimentation rates with distance from the mouth of the Saguenay River which is taken arbitrarily as core station 3111 (Fig. 2). Values near the base of the foreslope of the river delta may exceed 10 cm per year. In contrast, the rates in the inner basin near station 76-3117 are almost totally a function of hypnopycnal processes and are of the order of 0.1 cm per year (Fig. 10A). Leclerc et al. (1986) have shown that the porosity of fiord surficial sediments correlates

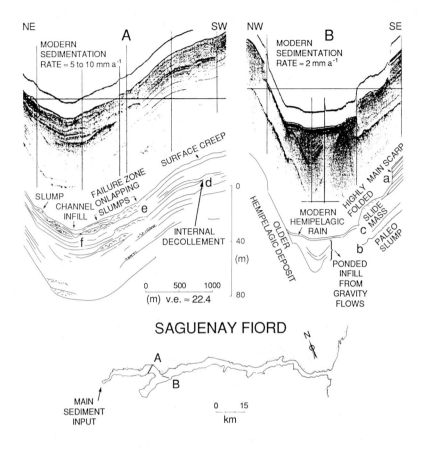

Figure 9. Reflection seismic records and interpretations from two transects (A and B) in the upper reaches of the fiord. The records show distinctive evidence of slumping from the fiord walls (adapted from Syvitski and Farrow, 1988).

directly with this sedimentation rate pattern, and that sedimentation rates determined in their 0.5 m long gravity cores vary inversely with compaction parameters derived from porosity values.

In the part of the fiord basin adjacent to the mouth of the Saguenay River, relatively high sedimentation rates, in conjunction with high organic matter fluxes from both natural and man-made sources, have created a benthic environment that has, on occasion, been virtually devoid of bioturbating organisms. Here coarse sand and organic matter is deposited during the annual spring freshet. Finer particles, that reach the floor of the basin during the subsequent low runoff and ice-covered fall and winter periods, stand in contrast to sandy spring freshet deposits that give rise to distinctive rhythmites which are clearly defined in the X-radiographs of sediment cores (Fig. 11 and 12A). These deposits represent a "proxy" record of the variations of the Saguenay River hydrograph and, in particular, of the relative intensity of spring freshet discharges that can often be resolved on

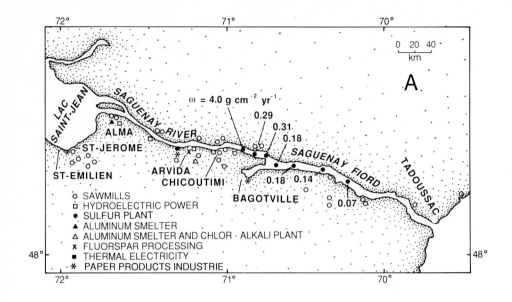

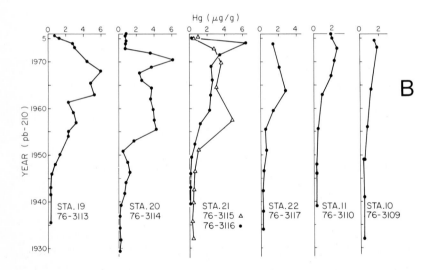

Figure 10. (A) Outline map of the Saguenay Fiord showing the location and type of various industrial installations. Values of the sediment accumulation rate (w) for stations 10, 11 and 18 through 22 (see Fig. 1B) are also noted (adapted from Schafer et al., 1980). (B) Temporal distribution of Hg in sediment cores collected in the north arm and inner basin (see Fig. 1 for station locations) of the Saguenay Fiord (adapted from Smith and Loring, 1981).

a time scale of years (Schafer et al., 1980; Smith and Schafer, 1987). The energy gradient of sediment transport processes between the head of the north arm and the mouth of the fiord controls the distribution of basin floor sediment characteristics (Loring and Nota,

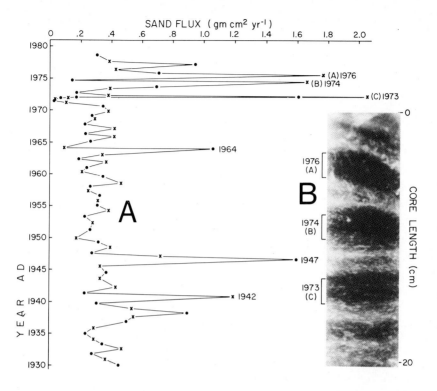

Figure 11. Relationship between sand flux (A) and spring freshet layers observed in x-radiographs (B) of the 1973, 1974 and 1976 sandy rhythmites of core D-1. The comparatively high spring freshet discharges of 1942, 1947 and 1964 are characterized by a sand flux > 1.0 gm cm^2 y^{-1} (adapted from Smith and Schafer, 1987 and Schafer et al., 1983).

1973). Dark organic-rich sandy deposits are indicative of the proximity of fluvial sources. Near the mouth of the fiord, clean sands occur in shallow sill environments that are scoured by fast flowing tidal currents. Deep inner basin settings are somewhat removed from both of these influences. Consequently, deposits observed there tend to be comparatively muddy and are well bioturbated by a community of indigenous infaunal species. Sediment color ranges from black in areas that are proximal to a sources of organic matter, to grey in the deeper parts of the basin east of Baie des Ha! Ha!. This distribution contrasts with that observed in Howe Sound and Bute Inlet, B.C., where mass flows of nearshore sandy deltaic sediments have traversed deep and typically mud covered basin floors for distances in excess of 50 km (Syvitski and MacDonald, 1982; Schafer et al., 1988).

During exceptionally intense spring freshet events, the modal size of the sand fraction of spring freshet sediment deposited near the head of the north arm may decrease by several tenths of a phi unit (Fig. 12A and B). This coarsening effect is a consequence of the higher river velocities generated during high discharge intervals. During the freshet, the competence of the river increases thereby enabling it to carry a relatively coarse SPM load containing greater numbers of higher settling velocity particles (i.e. fine to medium sand).

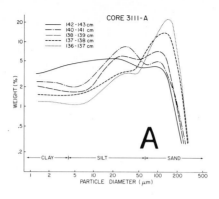

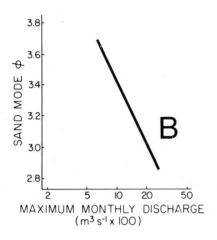

Figure 12. (A) Variation of particle size distribution during the 1964 spring freshet showing the increase in the proportion of sand size particles as river flow approaches maximum velocity (adapted from Smith and Ellis, 1982). (B) Relationship between modal diameter of sand and river velocity based on textural data from core D-1 (adapted from Schafer et al., 1983)

At this time, medium to coarse sand may be carried to the edge of the river delta as bedload, or perhaps as near-bottom intermittant suspension load. From that point it is deposited rapidly on the delta foreslope where it forms a convex pattern of sand waves that have their steep side facing down slope (Fig. 13).

Given a relatively unrestricted sediment source, the load of SPM in the river water increases directly with current velocity so that it becomes possible to use parameters such as sedimentation rate and sand flux (e.g., Fig. 11A) as proxy indices of spring freshet intensity. In several west coast fiords (e.g., Howe Sound as discussed by Syvitski, 1980 and Syvitski and Murray, 1981), the vertical flux of SPM has been shown to be about two orders of magnitude higher during the late summer freshet (ice melt) maximum than in the

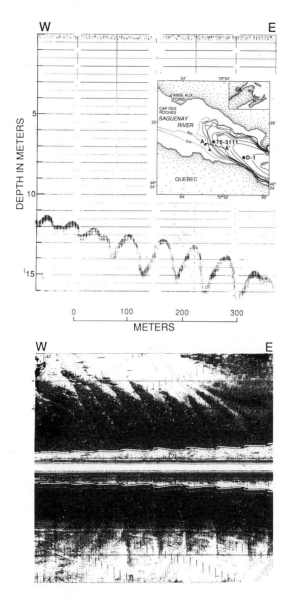

Figure 13. Echogram (A) and sidescan sonar record (B) of current ripples developed on the foreslope of the Saguenay River delta. The records are located along the eastern half of transect A-A' in 10 m to 15 m water depth (adapted from Schafer et al., 1983).

low runoff (winter) interval. When the textural characteristics of spring freshet sediment deposited in the basin near the head of the north arm of the Saguenay Fiord are considered together with sedimentation rate data derived from Pb-210 activity profiles (Smith and Ellis, 1982), they show good potential for reconstructing a proxy series of minimum annual mean flow and spring freshet intensity for the period prior to the systematic

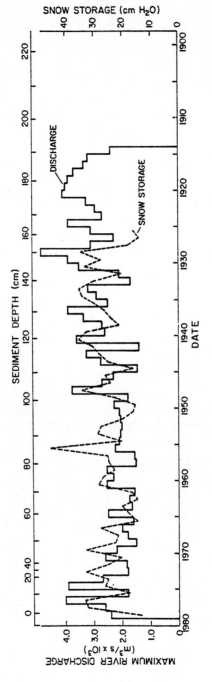

Figure 14. Relationship between annual maximum mean monthly discharge and snow storage for the Saguenay River drainage basin between 1926 and 1979. The snow storage parameter shows a distinctive increase after 1965, i.e. after culmination of the early 20th century Northern Hemisphere warming trend (adapted from Smith and Schafer, 1987).

measurement (1914) of the interannual variation of these parameters (Smith and Schafer, 1987). 20th century Saguenay prodelta facies contain sediment deposited primarily through hemipelagic fallout from the surface plume with no indication of deposition from small forset bed failures. In this sense, the depositional setting seems to be comparable generally to that of Bute Inlet (e.g., Schafer et al., 1989).

The unique annual sediment signal of spring freshet discharge intensity observed in the north arm prodelta deposits does not appear to be resolvable in the Baffin Island fiords for several reasons. Firstly, the nature of river discharge in the absence of significant vegetation is highly variable in the arctic so that the rhythmic character of annual deposits is not as regular as is observed in cores from the head of the Saguenay. Short rainfall events and the occurrence of both ice and snowmelt freshets further complicates the proxy signal of annual runoff recorded in arctic fiord prodelta deposits. Secondly, the flux of sediment to the basins at the head of many arctic fiords is relatively large. Consequently, submarine mass flows originating on the river delta foreslope can occur several times each year thereby mixing and redepositing a large proportion of the total annual sediment volume (Syvitski et al., 1986). Finally, in the absence of significant vegetation and snowfall, the transport of windblown sand size particles into the Baffin fiord basins is of much greater importance compared to the Saguenay (Gilbert, 1983).

Discharge rates have been monitored at several stations along the course of the Saguenay River. The longest record is from the station at Isle Maligne (#062901). Monitoring was initiated there in 1914 perhaps in response to design requirements for the dam that was constructed near this location in the early 1920's. Between 1960 and 1975, the mean annual discharge of the river at Isle Maligne was 1581 $m^3 s^{-1}$. Annual maximum MMD's in excess of the 1914-1978 grand average of 3132 $m^3 s^{-1}$ are evident before 1950 and after 1970. Values 40% higher than the grand average occurred in 1928, 1929, 1937, 1942, 1947, and 1976 (e.g., Fig. 14).

Lehigh gravity core D-1 (Fig. 2) contains a record of spring freshet sediment deposition that extends back to about 1901 (Smith and Schafer, 1987). The precise dating of this core was accomplished using the Pb-210 isotopic method (Smith and Walton, 1980). Pb-210 enters the water column from the atmosphere where it is scavenged rapidly by fine suspended particles and organic matter, accumulating as an "unsupported excess" in the sediments. The excess Pb-210 activity level of a particular sediment interval can be related to the time of deposition using a constant flux model relationship that is described in detail in Smith and Schafer (1987). The Pb-210 chronology of core D-1 was verified through comparison with profiles of other isotopes that have been deposited from the atmosphere as the result of nuclear weapons testing (e.g., Cs-137) and, as well, through the measurement of some anthropogenic geochemical indicators that were also observed in D-1 subsamples (Smith and Loring, 1981; Smith and Ellis, 1982). In conjunction with X-radiograph data from which annual rhythmites can be identified, this integrated dating technique has provided a resolution of one year.

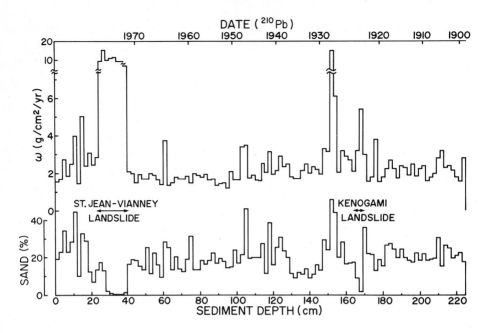

Figure 15. Relationship between sediment accumulation rate (w) and percent sand in core D-1 (Smith and Schafer, 1987) which was collected on the Saguenay river delta foreslope (see: Fig. 2). The highest percent sand and sediment accumulation rate values are associated with the 1928 spring freshet. Sediment accumulation rates are also > 10 g cm2 y-1 during the deposition of the 1971 St. Jean Vianney landslide sediment layer.

The relationship of annual freshet discharges to sediment texture was first noted by Smith and Walton (1980) and by Schafer et al. (1980) during the study of core 76-3111. Particle size analysis of a sediment interval deposited during the relatively intense spring freshet of 1974 indicated that this material had been enriched in fine sand (0.125-0.250 mm). This aspect of the discharge-sediment texture relationship was confirmed for the 1964 spring freshet rhythmite by Smith and Ellis (1982) (Fig. 12A). In their initial evaluation of core D-1, Schafer et al. (1983) found that most of the core subsamples containing more than 35% sand corresponded to deposition during comparatively high mean annual river discharge years (i.e. 17×10^3 m^3 s^{-1}). These high discharge years were always associated with above average maximum MMD's. In a later detailed evaluation of core D-1, Smith and Schafer (1987) focused on a % sand parameter (-1 to +4 phi) as a potential proxy indicator of spring freshet intensity (Fig. 15). When the landslide sediment intervals are omitted from the data base, a direct correlation of r=0.78 (P=0.001; N=102) is obtained between % sand and the sediment accumulation rate (based on Pb-210 activity results). This relationship reflects the spring freshet interval of relatively high sediment accumulation rates adjacent to the river delta foreslope that involves the deposition of an increased proportion of fine sand. Sand deposition is superimposed on an ambient and relatively constant depositional flux of finer sized particles and organic matter. The actual sand flux (F-sand in g cm^2 yr^{-1}) for 2 cm thick increments of core D-1 is the product of the % sand

and the sediment accumulation rate. F-sand provides the link between river discharge and sediment texture because it reflects temporal changes in fluvial particle transport intensity (Fig. 11A).

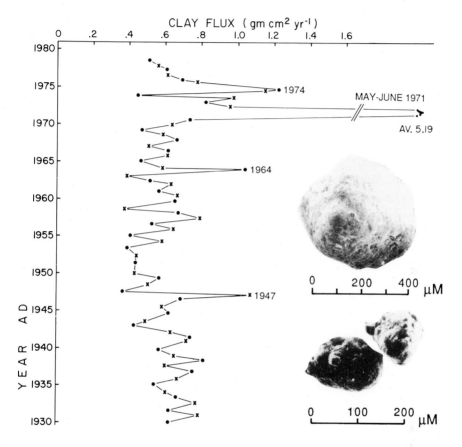

Figure 16. (A) Variation in clay-size particle flux observed from core D-1 textural data. High clay flux values in 1973 and 1974 represent resuspension and transport of residual 1971 landslide sediment that was initially deposited in the Saguenay River channel. (B) Examples of clay "pellets" derived from landslide sediments (see: Schafer et al, 1980; 1983).

The magnitude of the Saguenay River spring freshet decreased through the first half of the 20th century reaching a low during the 1960's. One important climate-related element controlling the magnitude of the freshet is the amount of precipitation stored as snow during the previous winter. Both snow storage and maximum MMD show a comparable trend during the 20th century (Fig. 14). The correspondence of these parameters suggests that the general decrease in MMD between 1950 and 1969 was governed largely by local climatic factors that may reflect broader Northern Hemisphere trends in atmospheric circulation (e.g., Hayden, 1981; Jones et al., 1982).

4.2. Landslides and Submarine Mass Flows

Over the past several centuries, there have been a number of major landslides in the Saguenay valley that have had a significant impact on sediment deposition in the north arm and inner basin of the fiord. All of these slides have involved sensitive Champlain or "Leda" marine clayey silt deposits that accumulated in inland seas which flooded the fiord valley during the waning of late Wisconsinan ice between 8,000 and 11,000 years BP (Gadd, 1960). These clayey sediments have been associated with numerous landslides throughout southern Quebec (e.g. Chagnon, 1968; LaRochelle et al., 1970). The most well known 20th century slide occurred on May 4th, 1971 at the village of St. Jean Vianney which is located on the north side of the Saguenay valley across from Arvida (Fig. 2). The slide followed a period of heavy rainfall and displaced an estimated 6.9×10^6 m^3 of sediment. It caused the death of 31 people and the destruction of 40 homes. About 5.4×10^6 m^3 of sediment was channeled into the Saguenay River through the Rivière Petit Bras (Tavenas et al., 1971). Its force was sufficient to carry a section of a concrete bridge that forms part of the coastal road system in this area into the middle of the Saguenay River channel. Sediment derived from the 1971 slide was deposited as an exponentially thinning tongue of clayey silt that can be traced almost to the mouth of the north arm (Fig. 7). The thickness of this deposit is greatest on the south side of the north arm basin because of (i) the response of the river plume to the Coriolis force and (ii) the location of the deepest (dredged) section of the river channel on the south side of the river delta (Fig. 2). The slide sediment appears as a distinctive grey colored layer of clayey silt in the 20-40 cm section of core D-1. An additional feature of this section of the core is its high concentration of oval-shaped clay pellets (Schafer et al., 1980; 1983) that are thought to represent the breakdown of massive, overconsolidated landslide-displaced clay boulders (Fig. 16B). Additional indicators of 1971 landslide sediment in north arm basin cores include an allochthonous microfossil assemblage of calcareous benthonic foraminifera (Schafer et al., 1980) and an unusual pollen subzone containing relatively high concentrations of *Sphagnum* (Fig. 17).

The second largest 20th century landslide in the Saguenay region occurred during the fall of 1924 at Kenogami. In contrast to the 1971 slide, which was related to climatological factors (LaRochelle, 1973), the Kenogami slide was caused by the release of liquid industrial waste from a paper mill into a canyon that had been excavated by previous waste water discharges. The eventual failure of the canyon walls resulted in a mud flow that descended down the canyon with a velocity of 15 km hr^{-1} eventually reaching the Saguenay through the Rivière Aux Sables (Terzaghi, 1950). The Kenogami slide displaced approximately 1.9×10^6 m^3 of material or about 28% of the volume displaced by the St. Jean Vianney slide. Evidence of this event occurs in the 165-169 cm section of core D-1. The sediment in this interval is comparatively clayey (<2% sand) and shows the highest clay pellet flux in sediment deposited prior to 1971. Its most distinctive feature is an assemblage of calcarous foraminifera many of which are unique to the 1924 core interval (Smith and Schafer, 1987). The identification of the Kenogami landslide horizon provided an additional important chronostratigraphic marker for dating the early 20th century interval of Saguenay River prodelta deposits.

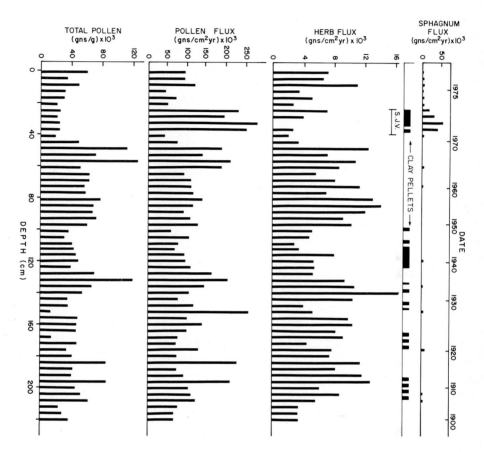

Figure 17. Pollen data for core D-1 showing a general inverse relationship between clay pellet presence and the flux of herb pollen suggestive of a dilution of the typical suspended load of the river by older Champlain clay sediments. The flux of Sphagnum increases significantly in the 1971 St. Jean Vianney landslide sediment apparently reflecting the mixing and transport of bog vegetation along with Champlain clays during the landslide.

An ancient landslide which appears to have taken place in 1663 (Schafer and Smith, 1987) formed the basin in which the 1971 St. Jean Vianney slide occurred. Evidence that the sediment displaced during the ancient slide (210×10^6 m^3) actually reached the main channel of the Saguenay River was first discovered during the excavation of the foundation pits for the Shipshaw power station during the early 1940's (Legget, 1945).

Lehigh gravity cores collected near the head of the inner basin just east of the confluence of the north and west (Baie des Ha! Ha!) arms appear to contain a record of the ancient

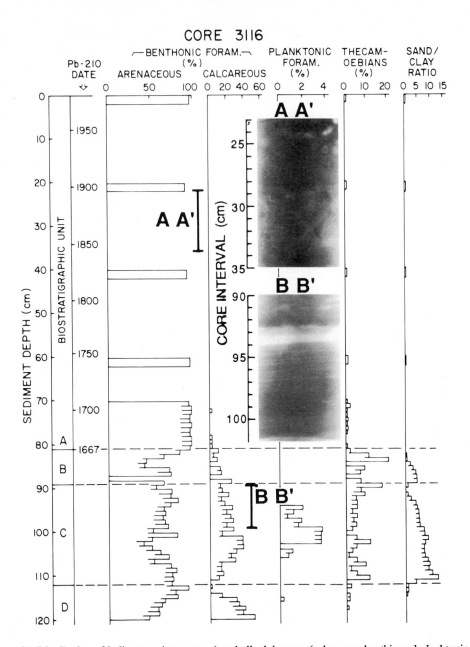

Figure 18. Distribution of indigenous (arenaceous) and allochthonous (calcareous benthic and planktonic foraminifera) in core 3116 (see Fig. 2 for location) and their relationship to submarine mass flow events associated with a 17th century landslide. X-radiograph BB shows a section of the core that appears to represent suspended sediment associated with the ancient landslide that was deposited under the influence of tidal currents. Note also the bioturbated character of the late 19th century hemipelagic sediment (AA).

landslide. The top of the slide deposit was found in core 3116 at a depth of 84 cm (Figs. 2 and 18). A similar discontinuity was observed in an adjacent core (3117) at a depth of 64 cm. In terms of Pb-210 rates, these depths corresponded respectively to dates of deposition of 1667 and 1686. Schafer and Smith (1987) hypothesized that this slide could have been triggered by a large earthquake that is known to have occurred in this part of Quebec in 1663 (Smith, 1962). Support for an earthquake-triggering mechanism rests on contentions such as (a) the timing of the 1663 earthquake lies well within an uncertainty of approximately ±10% estimated for the 300 year extrapolation of the 20th century Pb-210 dating chronology determined for the two cores, (b) there are no records of an earthquake larger than Richter 3 between 1673 and 1731 and (c) the large volume of sediment displaced by the slide (30 times greater than the 1971 slide) could have easily reached the head of the inner basin. Schafer and Smith (1987) were able to divide the cores into four correlatable units based on their contained foraminifera and thecamoebian assemblages (Fig. 18). The estuarine character of the modern inner basin environment is reflected in the uppermost unit (A) by an indigenous foraminifera assemblage that consists of more than 90% arenaceous species. Calcareous foraminifera characteristic of Champlain clays (e.g. Corliss et al., 1982) become prominent in older units (B, C and D). In units B and C, the increased proportion of calcareous benthonic foraminifera, thecamoebians, and planktonic foraminifera, in conjunction with their graded textures, the absence of bioturbation traces, and Pb-210 extrapolated ages, constitutes strong evidence for an upstream 17th century source for these apparently rapidly deposited sediments. Planktonic foraminifera specimens, representing speices which do not live in the fiord today, seem to have been preserved in the redeposited landslide sediments as the result of rapid transport and burial during this mass flow event. Furthermore, their co-occurrence with increased numbers of thecamoebian specimens suggest that the slide material could have included both shallow nearshore estuarine and deep water offshore marine facies of Champlain clay (e.g. Hillaire-Marcel, 1981).

The sequence of events during the 1663 earthquake probably involved: (1) collapse of local subaqueous deposits of Champlain clay and Holocene age estuarine sediments (unit D of core 3116) and their transport to the floor of the inner basin as a cohesive debris flow; (2) transport of fluvially-reworked landslide sediments (unit C of core 3116) from the delta foreslope near the head of the north arm to the head of the inner basin as a cohesionless mass flow; (3) resuspension and deposition of the fine fraction of residual displaced landslide sediments by river and tidal currents for at least six weeks after the first cohesionless mass flow was deposited; (4) deposition of a second smaller cohesionless mass flow of reworked landslide sediment (unit B). This final deposit could reflect slope failure resulting from the transport of landslide sediment to the delta foreslope during the spring freshet of 1663, or of the breaching of a slide sediment dam near the mouth of the Shipshaw River at about this time (Legget and LaSalle, 1978). No comparable mass flow events are evident during the interval represented by unit A sediments which suggests that there were probably no landslides of comparable scale in the lower reaches of the Saguenay River valley since the 17th century. However, given the inverse relationship between sedimentation rate and sediment compaction proposed by Leclerc et al. (1986), it would not be surprising if future surveys in the Saguenay show evidence of major submarine slumping as a result of the magnitude 6.0 earthquake that occurred 36 km south of

Chicoutimi on November 25, 1988. This event was described by Munro and North (1989) as the most significant earthquake in eastern North America in over 50 years. Its motion was felt as far south as Washington, D.C.

The interval of very fine laminae that occurs below the interface of units B and C (Fig. 18; BB) is thought to reflect depositional processes associated with the 60 cycle per month diurnal tidal regime of the fiord (Canadian Tide and Current Tables, 1987). The total number of laminae could represent about six weeks of hemipelagic sedimentation following the 17th century earthquake/landslide event. This element of the hypothesis is consistent with reports of high siltation levels occurring in streams for up to several months following the 1663 earthquake (see Doig, 1986).

5. Pollution History and Impacts

Industrialization of the upper reaches of the Saguenay valley during the 19th and 20th centuries has resulted in a significant pollution impact on the benthic environment of the fiord (Fig. 10A). Interest in this aspect of the fiord was stimulated in large part by the discovery of high levels of Hg in the sediments of this region (Loring, 1975). Elevated levels of Hg had also been detected in fish collected in the fiord (Tam and Armstrong, 1972), and it appeared that the fish and the sediment contamination had developed from the same source. Loring's (1975) results indicated that an industrial source of Hg was located somewhere upstream from the head of the fiord, probably in the Saguenay River. A chlor-alkali plant, several pulp and paper mills, and a range of other industries are located in this heavily populated region which made the precise identification of the Hg source difficult. Pocklington and Leonard (1979) found that the organic-rich sediment deposited near the head of the fiord contained elevated C/N ratios and high levels of lignin. Pocklington (1976) had previously identified this region as a depositional site for pulp mill wastes. Loring's (1975) measurements of Hg in sediments from this part of the fiord indicated that woody fibres were the primary "carrier phase" for the industrially-derived Hg. In subsequent studies, Loring (1976a and 1976b) observed that elevated levels of Pb and Zn (derived from industrial sources) detected in fiord sediments were associated with the same carrier phase as the Hg. Although Hg discharges from chlor-alkali plants located near the Saguenay River had been restricted in by new government regulations in 1971, the results of a study by Loring and Bewers (1978) showed that high levels of Hg were still present in the water column of the fiord as late as 1973. Concern about possible Hg contamination of the large commercial fishery in the Gulf of St. Lawrence was highlighted by the findings of Bourget and Cossa (1976). They demonstrated that the Hg content of mussels decreased in a seaward direction through the St. Lawrence Estuary thereby implying that this concentration gradient could be related to Hg discharges from the Saguenay Fiord.

Despite all of the studies summarized above, the source function of Hg (i.e. the rate of input or flux of Hg to the fiord) remained conjectural until the development of methods capable of establishing the precise depositional history of sediments carried into the fiord by the Saguenay River. Application of the Pb-210 dating technique to this problem (Smith and Walton, 1980), in conjunction with the measurement of other time-stratigraphic horizons associated with fallout radionuclides such as Cs-137 from weapons tests (e.g. Smith and Ellis, 1982), contributed substantially to the reconstruction of a chronostratigraphic record of Hg deposition. Measurement of Hg in the Pb-210 dated cores showed that the anthropogenic threshold of this metal corresponded to a depositional date of approximately 1950 (Smith and Loring, 1981). The largest Hg fluxes were measured in cores from the north arm. However, the most significant finding was that the onset of anthropogenic Hg deposition had occurred in 1947 ±3 years, thereby linking this metal to the chlor-alkali plant at Arvida (Fig. 10B). Subsequent research (Barbeau et al., 1981a; 1981b), in which timing for the onset of Hg input to the fiord was calibrated to a Cs-137 chronology, confirmed that the contamination of the fiord by industrial inputs of Pb and Zn had occurred during the post-World War II period. According to Smith and Loring (1981), both the timing and magnitude of Hg fluxes observed near the 1971 horizons of their gravity cores indicated compliance by the chlor-alkali plant with government regulations that imposed restrictions on the industrial release of Hg in the plant's liquid effluent (see also: Gobeil and Cossa, 1984).

The burial of the relatively highly contaminated Hg sediment in the north arm by clayey sediment from the 1971 St. Jean Vianney landslide was viewed as a positive aspect of this otherwise tragic event. However, the 1971 landslide sediment was not carried beyond the north arm in appreciable quantities and recent studies by Pelletier et al (1988) have shown that since 1976, the concentration of Hg in the surficial sediments of the inner basin near Sainte-Rose-du-Nord (Fig. 2) has ranged between 0.75 to 1.20 $\mu g\ g^{-1}$ with a mean value of 0.93 $\mu g\ g^{-1}$. This range of values overlaps those observed in pre-1971 sediments (Smith and Loring, 1981). Pelletier et al. (1988) suggested two possibilities to explain their observations. One is that bioturbation has contributed to the remobilization of Hg from deeper (pre-1971) sediment layers and its transportation to the sediment-water interface. This hypothesis is consistent with observations of bioturbation features (e.g. worm burrows) evident in the X-radiographs of cores from this part of the basin (e.g. Fig. 18; AA). Their alternative proposal was that the anthropogenic up-stream discharge of Hg was not really stopped in 1971, or that one or many unidentified sources are still active along the Saguenay River. The results of Smith and Loring (1981) and the X-radiograph evidence of bioturbation support the first hypothesis. Pelletier et al. (1988) concluded that the average Hg level observed in all surface samples that they collected throughout the fiord is well above the pre-industrial level and seems to indicate a steady state of Hg concentrations in fiord surface sediments over the last ten years. Regardless of which hypothesis is invoked, Pelletier et al. (1988) note that mercury concentrations measured in the edible parts of shrimp collected in November of 1985 were ranging from 0.13 to 0.58 $\mu g\ g^{-1}$ (wet weight) with an average value of 0.36 ±0.11 $\mu g\ g^{-1}$. This mean value was not significantly different from one reported in 1982. Adult shrimp fed on a diet of contaminated mussel tissues showed an uptake rate in their edible parts of 0.09 $\mu g\ g^{-1}$ per day during the first 14 days of a bioassay. Pelletier et al. (1988) point out that these findings clearly indicate the fragility of the equilibrium existing between the biota and the physical environment. Given a typical bioturbation depth of 20 to 30 cm, and an annual maximum deposition rate from hypopycnal sources in this part of the inner basin of about 0.2 cm per year (Smith and Walton, 1980), it would appear that the shrimp contamination by Hg derived from older remobilized inner basin sediments (and the effect on their related food chain) could continue well into the 22nd century.

In addition to heavy metals such as Hg, many other types of contaminants have been identified in Saguenay Fiord sediments. A number of researchers have drawn attention to the large quantities of terrestrial organic matter that have been deposited near the head of the fiord (Loring and Nota, 1968; Loring, 1975; Pocklington and MacGregor, 1973; Tan and Strain, 1979). Ironically, the high flux of organic matter to the north arm basin during the 20th century is probably responsible for limiting colonization of the western part of this area by bioturbating species, thereby reducing post-depositional sediment mixing and the remobilization of toxic metal contaminants.

The historical record of the impact of the pulp and paper industry on the benthic environment of the upper part of the fiord is evidenced in the organic matter profiles of Pb-210 dated cores (Schafer et al., 1980; Smith and Schafer, 1987). An increase in organic matter concentration above background levels is first evident in sediments deposited around 1912 (Fig. 19A). This increase occurs within one year of the introduction of the first large pulp machines to the paper mill at Kenogami. Expansion of the mills at Alma and Kenogami during the 1920's is reflected by enhanced sediment organic matter concentrations for this period. In contrast, the bankruptcy of pulp mills in the Saguenay region during the 1930's is marked by a sudden decline in the organic matter concentration profile. The flux of organic matter to the sediments increases again during the post-World War II period of pulp mill expansion and then declines for a second time during the 1960's, perhaps in response to the introduction of pollution control devices in local pulp mills at this time. The historical impact of organic matter deposition on the north arm benthonic foraminifera community is one of reduction in both population density and species diversity (Collins and Schafer, 1987). Similar relationships have been observed in some organically-polluted Norwegian fiords (Nagy and Alve, 1987).

Recently, other contaminants found in Saguenay Fiord sediments have become major sources of environmental concern. Poly aromatic hydrocarbons (PAH) are extremely carcinogenic substances that have been detected at high concentrations in Saguenay Fiord sediments (Martel et al., 1986). Identification of the source function for these contaminants within the Saguenay Region is particularly important because Chicoutimi, the largest city in the area, has one of the highest incidences of cancer in Canada (Mortality Atlas of Canada, 1984). The inferred source of the PAH releases is the large aluminum refining facility at Arvida and Alma (Fig. 10A). At this facility, large quantities of PAH's are released as a bi-

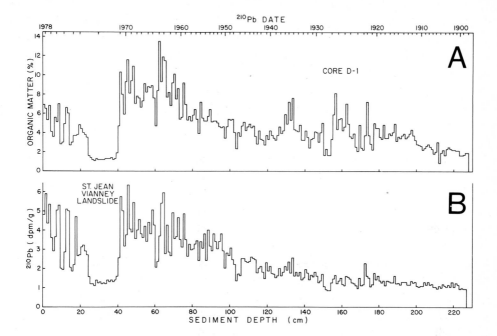

Figure 19. (A) Organic matter concentrations in core D-1. The increase in concentration to levels >3% marks the initiation of significant pulp and paper mill discharges into the Saguenay River channel. (B) Pb-210 activity in core D-1. Pb-210 activity shows a direct relationship to organic matter percentages pointing to this component of the sediment as the "carrier phase" for the Pb isotope (see Smith and Schafer, 1987).

product of the aluminum smelting process. Smith and Levy (1988) determined the source function for PAH fluxes to the sediments of the north arm through detailed analysis of a core that had been dated to a resolution level of better than one year using the Pb-210 method. Their results show that the PAH inventory is closely associated with the same terrestrial organic matter carrier phase which is responsible for the temporal Hg distribution described earlier. The major PAH flux to the fiord sediments is related to the direct discharge into the Saguenay River of wastes recovered from atmospheric filters or "scrubbers". A smaller component of the sediment PAH flux is attributed to atmospheric releases of PAH's that are deposited locally within the fiord drainage basin. Reduced inputs of PAH's into the Saguenay River since the late 1970's appear to reflect the use of more efficient cleaning and filtering procedures that were initiated by the refinery in 1976 (Smith and Levy, 1988).

6. Biological Considerations

Many aspects of the biological characteristics of the Saguenay Fiord marine fauna are a reflection of watermass conditions and circulation. Between May and October, the water of the Saguenay Fiord is well stratified (e.g., Figs. 5 and 6B). The thin surface layer (5-10 m) is warm and brackish whereas the deep layer (down to 275 m) is characterized by cold, saline and well oxygenated water. In the surface layer, the highest levels of incident solar radiation coincide with the period of highest river discharge. This influx of fresh water modifies the thickness of the photic layer, the geometry of the mixing layer, and the concentrations of nutrients in the surface water (Côté and Lacroix, 1978a). As noted earlier, the physical chemical and biological characteristics of the deep layer of the fiord are influenced by the adjacent surface waters (10-30 m) of the St. Lawrence Estuary (e.g., Therriault et Lacroix, 1976) which penetrate over the shallow sill during the rising tide (e.g., Therriault and Lacroix, 1975; Loucks and Smith-Sinclair, 1975).

6.1. Phytoplankton dynamics

Abnormally high chlorophyll values occur in the deep layer of the fiord (down to 100 m in the downstream region of the inner basin) during the entire summer season. This "deep phytoplankton" population is potentially productive because, given an appropriate light level, its P/B ratios (production/biomass) are as high as those of the surface phytoplankton (Côté and Lacroix, 1978b). The deep aphotic layer phytoplankton takes on a greater importance in the fiord because the phytoplankton of the photic layer is characterized by a relatively low level of productivity. The photic layer of the fiord (maximum depth of 10 m) shows a strong seasonal variation in its physical, chemical and biological properties (Côté and Lacroix, 1978a). The maximum rate of primary production in this layer is as low as 165 mg $C.m^{-2}.d^{-1}$ (Côté and Lacroix, 1979). In contrast, a maximum value of 6460 mg $C.m^{-2}.d^{-1}$ is observed for the adjacent surficial waters of the St. Lawrence Estuary (Steven, 1974), 2,960 mg $C.m^{-2}.d^{-1}$ for Baie des Chaleurs (Legendre, 1971), and 3,700 mg $C.m^{-2}.d^{-1}$ for some of the British Columbia fiords (Gilmartin, 1964).

The seasonal dynamics of the fiord biota are characterized by four elements: (1) the late start of production processes; (2) the low level of primary production; (3) the persistently poor phytoplankton biomass; and (4) the absence of definite spring (or summer) blooms. In the upstream and central region of the fiord (near Cap Trinité), the initiation of the annual production process is clearly related to the slow establishment of a favorable thickness ratio between the photic layer and the mixing layer. In the downstream region, the intrusion of nutrient-rich saline water from the St. Lawrence Estuary is considered an important factor. Both of these conditions follow the spring freshet (late spring to early summer) and are directly associated with decreasing summer freshwater runoff (Côté and Lacroix, 1979). In the surface water of the Saguenay River channel (Station SAG2 and SAG3, Fig. 1A), the mean seasonal biomass of phytoplankton is twice as high, and the rates of primary productivity are only half as high, as in the brackish water of the lower north arm (station DR151). Values of photosynthetic capacity (P/B ratios) are about 8 times higher in the fiord than in the river channel (Table 1).

TABLE 1

Comparison of seasonal variability of chlorophyll values and of rates of primary production between the surface waters of the Saguenay River (Stations SAG1, SAG2 and SAG3) and the brackish waters of the upstream part of the fiord (Station DR151) from May to October 1983.

Station	N	Total Biomass mg m^{-2}	Total Primary Production mg C m^{-2}h^{-1}	P/B ratios mg C.mg Chl^{-1} h^{-1}
SAG 1	8	0.43	0.38	1.29
SAG 2	8	0.47	0.46	1.59
SAG 3	8	0.33	0.42	1.62
Mean	23	0.38 ±0.15	0.42 ±0.20	1.50 ±1.01
DR 151	7	0.23 ±0.15	0.85 ±0.60	12.14 ±10.21

It has already been mentioned that parts of the Saguenay Fiord are seriously affected by high levels of industrial pollution. To study the toxicity of elements such as copper in its ionic, complexed (to humic acid) and adsorbed (on cellulose fiber waste from a local paper mills) forms on the photosynthesis of fiord phytoplankton, experiments in chemostats with natural aggregations (either composed mainly of green algae or diatoms) have been carried out. Although copper is an essential ion for the metabolism of phytoplankton cells (Manahan and Smith, 1973), it becomes toxic at very low concentrations (Saifullah, 1978). There is a significant relationship between cupric ion activity and the toxicity of copper for the phytoplankton of the Saguenay. Results indicate that diatoms are more sensitive to Cu than green algae and that their rates of primary production are more strongly inhibited by copper than are chlorophyll concentrations (Canterfordand Canterford, 1980; Côté, 1983; Thompson and Côté, 1985). It appears, therefore, that phytoplankton species of the Saguenay have different tolerance mechanisms to ionic pollution. For example, with *Scenedesmus quadricauda*, a dominant species of the phytoplankton of Lac St. Jean (Constant and Duthie, 1978) and of the surface waters of the Saguenay River, Bastien and Côté (1988a, 1988b) have shown that the concentration of copper is significantly correlated ($\alpha = 0.05$) to the frequency of appearance, and to the mean number per cell, of intravacuolar inclusions. These inclusions are used by the cell to store copper ions in its organelles, thereby reducing the toxicity effect. Other results indicate that tolerance mechanisms are related to the synthesis of metal-binding molecules of the metallothionein type. Several studies on the variability of the chemical composition of the cellular matter (carbohydrates, proteins) (Foy and Smith, 1980; Raven and Beardall, 1981) suggest that tolerance is an expression of metabolic activation processes. The relative importance of these tolerance mechanisms is the object of current investigations. The Saguenay Fiord receives a mixture of industrial wastes, dissolved organic matter and suspended cellulose fibers discharged by local pulp and paper mills. This material seems to be an important factor in controlling the toxicity of copper on the freshwater phytoplankton of the Saguenay River, and on the brackish and marine phytoplankton of the north arm (Thompson and Côté, 1985).

In general, the surface layer primary production in the fiord is very low. The structure of the phytoplankton populations and their low productivity are attributed to very high river discharge and runoff during the spring freshet season, but they are also modulated by summer river flow levels which tend to remain too high to allow a significant building-up of plankton biomass. From previous studies (Côté, 1977; Cloutier and Côté, 1985), there is a suggestion that the increase in minimum mean monthly discharge by power dam water regulation (Schafer et al., 1983) could influence phytoplankton biomass in the fiord particularly in its upstream region. Many freshwater and even some euryhaline species are eliminated along the salinity gradient that is established in the upper reaches (100 km) of the fiord, or are carried too rapidly into water of unsuitable salinity to allow high productivity. Primary productivity in the fiord is also limited by industrial pollution which influences the density, biomass and the metabolism of cells. It is difficult to evaluate the effects of pollution on phytoplankton in different regions of the fiord because different species of this group may exhibit different tolerance levels. Yet, from the downstream end of the inner basin to the mouth of the fiord, the upper 100 m of the water column is rich in potentially productive and longliving cells. These are replenished twice daily in the fiord by water flowing over the sill from the lower relatively productive part of the St. Lawrence Estuary upwelling region.

6.2. Zooplankton dynamics

The structure and distribution of the mesoplankton populations in the deep waters of the inner fiord basin are influenced by the intrusion of St. Lawrence Estuary surface waters (10-30 m) into the deep layer of the fiord. This exchange process between the estuary and the fiord has a tidal periodicity (Seibert et al., 1979). Such an enrichment mechanism has a potential ecological importance for the fiord. It is responsible for the very high standing stock of zooplankton observed in the deep waters of the fiord basin, as compared to that of the adjacent St. Lawrence Estuary.

Rainville (1979) and DeLadurantaye et al. (1984) have shown that the deep layer of the fiord is characterized by an abundant and diverse zooplanktonic community. They observed 83 different species distributed among 12 phyla. Crustacea account for about 80% (66 species) of them and many species have an arctic affinity. In their study, DeLadurantaye et al. (1984) demonstrated the influence of advective processes on the structure of the zooplankton community of the fiord. In the deep inner basin, and in the smaller outer basin of the fiord, the zooplankton community shows a strong similarity to that of the adjacent ecosystem of the St. Lawrence Estuary. In both ecosystems, the community is dominated by copepods *Calanus hyperboreus*, *C. finmarchicus*, *Metridia longa* and *Oithona similis* (Rainville, 1979). In the upstream region of the inner basin, advective processes are less intense and here endogeneous mesoplankton populations thrive under local temperature conditions; smaller species such as *Microcalanus pygmaeus* and *Oncaea borealis* are present in large numbers. The distribution of these microzooplankton species reflects the fact that the deep waters of the western part of the inner basin are relatively stable physically and chemically during the summer season.

The cold watermass enclave (<0.5 °C) that occurs in the inner basin impinges on its slopes during spring and summer (Côté, 1977). It occupies a depth interval of 20-100 m and is carried upstream in the autumn because of the reduced influx of fresh water (Taylor, 1975; Therriault et al., 1984). The permanent presence of this cold watermass in the upstream region of the fiord supports the growth of microzooplankton (Rainville, 1979) and the coexistence of exogeneous and endogeneous zooplankton populations in this part of the fiord results in a diversity and population density which are significantly higher than observed in the adjacent St. Lawrence Estuary (Rainville, 1979).

Mesoplankton populations observed in the fiord (either exogeneous or endogeneous), are dependent on the availability of phytoplankton as a food source. Because of the sharp physico-chemical stratification of the water column during the summer season, herbivorous mesoplankton do not have easy access to the low productive surface water near the head of the fiord. Only "deep phytoplankton" are available as a food source for the deep mesoplankton populations. Yet, the biomass of the deep phytoplankton group is only relatively abundant in the downstream region of the fiord. In the upstream and central regions, this deep phytoplankton is limited to the surface 15-20 m (Côté et Lacroix, 1978a). Because herbivorous mesoplankton are relatively abundant in the fiord, the penetration of the surface phytoplankton of the St. Lawrence Estuary into fiord basin waters plays a prominent part in the trophic chain even though it is a comparatively short term phenomenon that only occurs during the flood stage of the tide. The deep water mesoplankton community of the fiord shows many dietary modes other than herbivorous; such as omnivorous, carnivorous and ectoparasitic (Rainville, 1979).

In the plankton population of the Saguenay Fiord, two groups (Gammaridean amphipods and Mysidacea) show some particular biological characteristics. Brunel et al. (1980) have reported that the fiord plankton community includes 34 species of gammarideans. Their frequency and density are significantly higher in the fiord than in three adjacent ecosystems. Gammaridean amphipods show a high spatial disparity (Brunel et al., 1980). In the downstream rocky region of the fiord, species are more abundant than in the upstream muddy region. According to Brunel et al. (1980), species observed in upstream muddy slope environments are predominantly predators and may be attracted to these settings by the abundant and diverse assemblage of small copepods which occurs there (Rainville, 1979). However, this muddy environment may be less attractive for other Gammaridean species because of the anoxic condition of the sediments that has developed there as a consequence of organic pulp mill waste and petroleum pollution (Pocklington and Leonard, 1979). In the downstream region, Gammaridean species are usually pelagic or are associated with algae or hydroïds that are typical of hard offshore substrates (Brunel et al., 1980).

Among the nine species of Mysidacea found in the deep waters of the inner fiord basin, two are characterized as arctic-subarctic species: *Mysis litoralis* and *Boreomysis nobilis*. These species are endemic to the fiord and show a high salinity and temperature tolerance (De Ladurantaye and Lacroix, 1980). *M. litoralis* is found in the waters of the lower fiord while *B. nobilis* inhabits the upstream region. The adaptive physiology of *M. litoralis* is higher in the fiord than in the arctic (DeLadurantaye and Lacroix, 1980). Its reproductive cycle occurs once a year (October-November) and lasts about 15 days; the release of juveniles from the marsupium of the female takes place in April or May. In the Saguenay, the period of incubation of this species is shorter than in the arctic. Physical conditions and the planktonic food source available in the downstream region of the fiord may be contributing factors for stimulating the growth of juvenile *Mysidacea* (Tattersall and Tattersall, 1951).

TABLE 2
Marine invertebrate taxa observed in the Saguenay Fiord (From Drainville et al., 1978).

Taxa	Number of species	
Crustacea	112	66.8 %
Molluscs	47	
Annelids Polychaeta	34	
Echinodermata	22	
Cnidarians	12	

Other taxa
  Protozoa
  Porifera
  Chaetognatha
  Brachiopoda
  Ectopracta
  Appendicularia
  Urochordata
	11	33.2%

	238	

6.3. Benthic invertebrates

The invertebrate community of the fiord is highly diversified. More than 238 species have been classified (Drainville et al., 1978) with Crustacea (112 species) and Molluscs (47 species) accounting for about 66% of them. Polychaeta (34 species), Echinodermata (22 species), Cnidarians (12 species), and a mixed group of 11 taxa account for the other 34% (Table 2). The few species found in the surface layer are boreal, eurythermal, freshwater and euryhaline whereas the majority of species observed in the deep layer are arctic, stenothermal and stenohaline. The great majority of benthic invertebrates are typically arctic and many of these species are absent in the St. Lawrence Estuary and in the Gulf of St. Lawrence. The abundance of these endemic species in the deep layer of the fiord lends further support to Drainville's *biogeographical arctic enclave* hypothesis.

6.4. Fish ecology

The ichthyological fauna of the Saguenay is very diverse. It includes more than 50 fish species (Drainville, 1970) that may be classified into four zoogeographical categories based on temperature and salinity criteria:

1. 10 species (20% are typically arctic);
2. 7 species (14% are typically subarctic);
3. 24 species (48% are typically boreal-arctic); and
4. 9 species (18% are typically boreal).

More than 80% of these species have an arctic affinity. Boreal species occur in the surface (0-5 m) layer while typically arctic and subarctic forms are restricted to a deep (>15 m) layer. Typically, boreal-arctic species clearly dominate in the surface layer but are also observed in the deep layer and in the thermohalocline layer (5-15 m). Many of the arctic species found in the deep layer are endemic to the fiord. To account for the biological characteristics of the fiord's deep layer, Drainville (1970) suggested that the Saguenay appears to be a *biogeographical arctic enclave* in a boreal area.

6.5. Marine mammals

The St. Lawrence Estuary, near the mouth of the Saguenay Fiord, constitutes the main habitat of a resident population of Beluga whales. This population is isolated from those reported from other parts of the world. The most important factors influencing the distribution of these mammals in the St. Lawrence are climatic conditions and an intense water mixing setting that promotes upwelling. Near the mouth of the Saguenay, these factors interact to produce an environment having arctic oceanographic affinities where planktonic productivity is relatively high. At this location, the 15-100 m interval of the water column has higher oxygen concentrations than those observed in the 0-15 m layer (Côté and Lacroix, 1978a). This explains, in part, the biological richness and diversity observed in the deeper parts of the water column which presumably represent a major food source for the resident Beluga population. The presence of Belugas in the fiord can be explained by the fact that these marine mammals appear to prefer marine environments in which the topography of the bottom is irregular rather than flat (Montigny, 1988). Presently, the population of Beluga whales varies between 350 and 500 individuals compared with about 5000 specimens at the end of the 19th century (Reeves and Mitchell: see Béland and Martineau, 1986). This decrease in population appears to be related to an industrial pollution problem. Béland and Martineau (1986) show that Beluga whales inhabiting the Estuary are highly contaminated by a variety of anthropogenic substances such as heavy metals, chlorinated compounds (PBC, DDT, Mirex), and by polycyclic aromatic hydrocarbons (e.g., benzo-a-pyrene). The concentration of chlorinated compounds varies with sex and age of the animals. In adult specimens, the concentration of pollutants observed in adipose tissues shows an exponential increase with age; this increase occurs more rapidly in females than in males. If the rate at which the Beluga population has decreased in the Estuary over the last century continues, the disappearance of this marine mammal from this area is conceivable in the near future.

7. Summary

The water column of the Saguenay Fiord is characterized by a sharp thermohalocline which divides it in two distinct layers: a thin surficial layer and a relatively thick deep layer. The circulation pattern of the fiord is linked to the intrusion of adjacent surface waters from the St. Lawrence Estuary into the deep fiord basins. Intrusions are comparatively large during the winter months and effect both the inner and the outer basin. Studies of the dynamics of Mn-rich particles suggest that relatively large winter intrusions can effect the water column to a depth of about 150 m. This relatively cold intermediate depth winter watermass is displaced upward and westward in the inner basin by water that intrudes during the summer season.

SPM distributions in the western part of the inner basin show a three layer configuration with relatively high concentrations in the surface and bottom layers. Surface layer SPM is derived from freshwater inputs while bottom water SPM appears to be related to either bioturbation, internal waves or density currents.

The watermass exchange process influences the nature of trophic relationships in the fiord. Surface layer primary production tends to be low especially near the head of the fiord where it may be limited by industrial pollutants that can influence the density, biomass and metabolism of cells. The enrichment of deep fiord waters by incursions of cold oxygenated water from the St. Lawrence Estuary is responsible for the very high standing stock of zooplankton observed in the fiord. The coexistence of exogeneous and endogeneous zooplankton populations especially near the head of the fiord results in a diversity and population density which is higher than that observed in the St. Lawrence Estuary. In fiord surface waters, species of invertebrates and fishes are boreal, eurythermal, freshwater and euryhaline whereas the majority of species observed in the deep layer are arctic, stenothermal and stenohaline. Many of the arctic species found in the deep layer of the fiord are endemic. Because of these distributions, the Saguenay Fiord is considered by many biologists as a *biogeographical arctic enclave* in a boreal area.

The upper reaches of the fiord represent a unique depositional setting for the study of fluvio-deltaic and hypnopycnal sediment transport processes. Deposits observed in prodelta environments near the head of the fiord contain "proxy" records that reflect both the natural modulation of the spring freshet by climatic elements over the past 80 years and the impact of man's activities on the benthic biota of this system. Anomalous depositional events (i.e. landslides) have provided calibration points for sediment chronologies obtained through the application of isotope dating methods and, in addition, have helped to elucidate relationships on the nature of mass flow sediment transport processes in this environment.

REFERENCES

Barbeau, C., Bougie, R. and Côté, J.-E., 1981a. Temporal and spatial variations of mercury, lead, zinc, and copper in sediments of the Saguenay fiord. Canadian Journal of Earth Sciences 18, p. 1065-1074.

Barbeau, C., Bougie, R. and Côté, J.-E., 1981b. Variations spatiales et temporelles du cesium-137 et du carbon dans des sediments du fiord du Saguenay. Canadian Journal of Earth Sciences 18, p. 1004-1011.

Bastien, C. et Côté, R., 1989a. Effet du cuivre sur l'ultrastructure de *Scenedesmus quadricauda* et *Chlorella vulgaris*. Internationale Revue der gesamten Hydrogiologie 74, p. 51-71.

Bastien, C. et Côté, R., 1989b. Variations temporelles de l'ultrastructure de *Scenedesmus quadricauda* exposée au cuivre lors d'une expérience à long terme. Internationale Revue der gesamten Hydrogiologie74, p. 207-219.

Béland, P. and Martineau, D., 1986. L'intoxication des bélugas dans le Sait-Laurent. Interface 7, p. 20-25.

Blanchard, R., 1935. L'est du Canada francais: Montréal, Beauchemin; Paris, Masson, tome deuxième, p. 7-155.

Bourget, F. and Cossa, D., 1976. Mercury content of mussels from the St. Lawrence Estuary and northwestern Gulf of St. Lawrence. Marine Pollution Bulletin 7, p. 237-239.

Brunel, P., DeLadurantaye, R. et Lacroix, G., 1980. Suprabenthic Gammaridean Amphipoda (Crustacea) in the plankton of the Saguenay Fiord, Quebec. In: Fjord Oceanography. H.J. Freeland, D.M. Farmer and C.D. Levings (eds.), Plenum Press, NewYork, p. 609-613.

Canadian Tide and Current Tables, 1987. St. Lawrence and Saguenay Rivers, Volume 3,__Fisheries and Oceans Canada, 39 p.

Canterford, G.S. and Canterford, D.R., 1980. Toxicity of heavy metals to the marine diatom *Ditylum brightwelli* (West) Grunom: correlation between toxicity and metal speciation. Journal of Marine Biology Association, U.K. 50, p. 227-242.

Chagnon, J.-Y., 1968. Les coulées d'argile dans la province de Québec. Le Naturalist Canadien 95, p. 1327-1343.

Cloutier, S. and Côté, R., 1985. Etudes expérimentales sur la sensibilité du phytoplancton d'eau douce aux variations de salinité dans le fiord du Saguenay, (Estuaire du St-Laurent), Canada. Internationale Revue der gesamten Hydrobiologie, 70, p. 187-201.

Collins, E.S. and Schafer, C.T., 1987. Environmental impacts on the foraminifera and thecamoebians of the upper Saguenay Fiord, Quebec, Canada (abstract). Geological Society of America, 1987 Annual Meeting, Abstracts with Programs, p. 624.

Constant, H. and Duthie, H.C., 1978. The phytoplanton of Lac-St-Jean, Québec. A.R. Gantner Verlag K.-G., FL-9490 Vadus, Liechtenstein.

Corliss, B.H., Hunt, A.S. and Keigwin, I.D., 1982. Benthonic foraminiferal faunal and isotopic data for the postglacial evolution of the Champlain Sea. Quaternary Research 17, p. 325-338.

Côté, R., 1977. Aspects dynamiques de la production primaire dans le Saguenay, fiord subarctique du Québec. Doctoral thesis, Laval University, Quebec, 194 p.

Côté, R., 1983. Aspects toxiques du cuivre sur la biomasse et la productivité du phytoplancton de la rivière du Saguenay, Québec. Hydrobiologia, 98, p. 85-95.

Côté, R. and Lacroix, G., 1978a. Variabilité à court terme des propriétés physiques, chimiques et biologiques du Saguenay, fiord subarctique du Québec (Canada), Internationale Revue der gesamten Hydrobiologie, 63, p. 25-39.

Côté, R. and Lacroix, G., 1978b. Capacité photosynthétique du phytoplancton de la couche aphotique du Saguenay. Internationale Revue der gesamten Hydrobiologie, 63, p. 233-246.

Côté, R. and Lacroix, G., 1979. Influence de débits élevés et variables d'eau douce sur le régime saisonnier de production primaire d'un fiord subarctique. Oceanologica Acta, 2, p. 299-306.

DeLadurantaye, R. and Lacroix, G., 1980. Répartition spatiale, cycle saisonnier et croissance de *Mysis litoralis* (Banner, 1948) (Mysidacea) dans un fiord subarctique. Canadian Journal of Zoology 58, p. 693-697.

DeLadurantaye, R., Therriault, J.-C., Lacroix, G. and Côté, R., 1984. Processus advectifs et répartition du zooplancton dans un fiord. Marine Biology, 82, p. 21-29.

Doig, R., 1986. A method for determining the past frequency of large magnitude earthquakes using lake sediments. Canadian Journal of Earth Sciences, 7, p. 930-937.

Drainville, G., 1968. Le fiord du Saguenay: Contribution a l'océanographie. Le Naturaliste Canadien 95, p. 809-855.

Drainville, G., 1970. Le fiord du Saguenay: II. La faune ichtyologique et les conditions écologiques. Le Naturaliste Canadien, 97, p. 623-666.

Drainville, G., Lalancette, L.-M., and Brassard, S., 1978. Liste préliminaire d'Invertébrés marins du fiord du Saguenay recueillis de 1958 à 1970 par le Camp des Jeunes Explorateurs. Min. Industrie et Commerce, Cahier d'information 83, 27 p.

Foy, R.H. and Smith, R.V., 1980. The role of carbohydrate accumulation in the growth of planktonic *Oscillatoria* species. British Phycological Journal, 15, p. 139-150.

Gadd, N.R., 1960. Surficial geology of the Becancour map area, Quebec. Geological Survey of Canada Paper 59-8.

Gilbert, R., 1983. Sedimentary processes of Canadian arctic fiords. Sedimentary Geology, 36, p. 147-175.

Gilmartin, M., 1964. The primary production of a British Columbia fiord. Journal of the Fisheries Research Board of Canada, 21, p. 507-538.

Gobeil, C. and Cossa, D., 1984. Profils des teneurs en mercure dans les sédiments et les eaux interstitielles du fiord du Saguenay (Quebec): données acquises au cours de la période 1978-83. Technical Report of Canadian Hydrography and Ocean Sciences, 53, 23 p.

Hayden, B.P., 1981. Secular variation in Atlantic coast extratropical cyclones. Monthly Weather Review 109, p. 159-167.

Hillaire-Marcel, C., 1981. Paleo-océanographie isotopique des mers post-glaciares du Québec. Palaeogeography, Palaeoclimatology, Palaeoecology, 37, p. 63-119.

Jones, P.D., Wigley, T.M. and Kelly, P.M., 1982. Variations in surface temperatures: Part 1. Northern Hemisphere, 1881-1980. Monthly Weather Review 110, p. 59-70.

Jordan, F. and de la Ronde, M.S., 1974. T.S. observations in the St. Lawrence system from Goose Cap to the Scotian Shelf, May 10-23, 1973. Bedford Institute of Oceanography Data Series BI-D-4, 65 p.

Kumarapeli, P.S. and Saul, V.A., 1966. The St. Lawrence valley system: a North American equivalent of the East African rift valley system. Canadian Journal of Earth Sciences, 3, p. 639-658.

LaRochelle, P., 1973. Rapport de synthèse des études de la coulée d'argile de Saint-Jean-Vianney. Gouvernment du Québec, Ministère des Richesses Naturelles Paper DPV-516, 74 p.

LaRochelle, P., Chagnon, J.Y. and Lefebvre, G., 1970. Regional geology and landslides in the marine clay deposits of eastern Canada. Canadian Journal of Earth Sciences, 7, p. 145-156.

LaSalle, P., 1968. Field trip of Quaternary geology Saguenay River - Lac St. Jean. Contribution to the ACFAS Congress, Quebec Dept. of Natural Resources. 31 p.

LaSalle, P. and Tremblay, G., 1978. Depôts meubles Saguenay Lac St. Jean. Geologic Report 191, Ministere des Richesses Naturelles du Québec, Québec, P.Q., 61 p.

Leclerc, A., Gagnon, M.J., Côté, R., and Rami, A., 1986. Compaction and movement of interstitial water in bottom sediments of the Saguenay Fiord, Quebec, Canada. Sedimentary Geology 46, p. 213-230.

Legendre, L., 1971. Production primaire dans la Baie-des-Chaleurs. (Golfe du Saint-Laurent). Le Naturaliste Canadien, 98, p. 743-774.

Legget, R.F., 1945. Pleistocene deposits of the Shipshaw area, Quebec. Transactions of the Royal Society of Canada, Section 4, Series 3, v. 39, p. 27-39.

Legget, R.F. and LaSalle, P., 1978. Soil studies at Shipshaw, Quebec: 1941 and 1969. Canadian Geotechnical Journal, 15, p. 556-564.

Loring, D.H., 1975. Mercury in the sediments of the Gulf of St. Lawrence. Canadian Journal of Earth Sciences, 12, p. 1219-1237.

Loring, D.H., 1976a. The distribution and partition of zinc, copper and lead in the sediments of the Saguenay Fiord. Canadian Journal of Earth Sciences, 13, p. 960-971.

Loring, D.H., 1976b. The distribution and portition of cobalt, nickel, chromium, and vanadium in the sediments of the Saguenay Fiord. Canadian Journal of Earth Sciences, 13, p. 1706-1718.

Loring, D.H. and Nota, D.J.G., 1968. Occurrence and significance of iron, manganese and titanium in glacial marine sediments from the estuary of the St. Lawrence River. Journal of the Fisheries Research Board of Canada, 25, p. 2327-2347.

Loring, D.H. and Nota, D.J.G., 1973. Morphology and sediments of the Gulf of St. Lawrence. Bulletin of the Fisheries Research Board of Canada, 182, 147 p.

Loring, D.H. and Bewers, J.M., 1978. Geochemical mass balances for mercury

in a Canadian fiord. Chemical Geology, 22, p. 309-330.

Loucks, R.H. and Smith-Sinclair, R.E., 1975. Report on the physical oceanography of the Saguenay Fiord. Manuscript, Chemical Oceanography Division of Bedford Institute of Oceanography, Dartmouth, N.S., 77 p.

Manahan, S.E. and Smith, M.J., 1973. Copper micronutrient requirement for algae. Environmental Sciences and Technology, 7, p. 829-833.

Martel, L., Gagnon, M.J., Masse, R., Leclerc, A., and Tremblay, L., 1986. Polycyclic aromatic hydrocarbons in sediments from the Saguenay Fiord, Canada. Bulletin of Environmental Contaminants and Toxicology, 37 p.

Montiguy, D., 1988. La complainte du béluga. Contact, le magazine de l'Université Laval, Automne 1988, p. 25-28.

Mortality Atlas of Canada, 1984. v. 3, Urban Mortality, Statistics Canada, D.S.S. No. H49-6-3-1984, 139 p.

Munro, P.S. and North, R.G., 1989. The Saguenay earthquake of November 25, 1988 strong motion data. Geological Survey of Canada Open File Report No. 1976, 17 pp.

Nagy, J. and Alve, E., 1987. Temporal changes in foraminiferal faunas and impact of pollution in Sandebukta, Oslo Fiord. Marine Micropaleontology, 12, p. 109-128.

Nutt, D.C., 1963. Fiords and marine basins of Labrador. Polar Notes, Occasional Publications of the Stefansson Collection 5, p. 9-23, Dartmouth College, Dartmouth.

Ouellet, M. and Jones, H.G., 1987. Some physio-chemical aspects of Lake Saint-Jean, Quebec, Canada. International Association of Theoretical and Applied Limnology, 23, pt. 2, p. 961-967.

Pelletier, E. and Canuel, G., 1988. Trace metals in surface sediments of the Saguenay Fiord. Marine Pollution Bulletin 212 (in press).

Pelletier, E., Rouleau, C. and Canuel, G., 1988. Niveau de contamination par le mercure des sédiments de surface et des crevettes du fiord du Saguenay en 1985-86. Revue Internationale de Science de Eau (in press).

Piper, D.J.W., Letson, J.R.J., DeIure, A.M., and Barrie, C.Q., 1983. Sediment accumulation in low sedimentation, wave dominated, glacial inlets. Sedimentary Geology, 36, p. 195-215.

Pocklington, R., 1976. Terrigenous organic matter in surface sediments from the Gulf of St. Lawrence. Journal of the Fisheries Research Board of Canada, 33, p. 93-97.

Pocklington, R. and MacGregor, C.D., 1973. The determination of lignin in marine sediments and in particulate form in seawater. International Journal of Environmental Analytical Chamistry, 3, p. 881-893.

Pocklington, R. and Leonard, J.D., 1979. Terrigenous organic matter in sediments of the St. Lawrence Estuary and the Saguenay Fiord. Journal of the Fisheries Research Board of Canada, 33, p. 1250-1255.

Rainville, L., 1979. Etude comparative de la distribution verticale et de la composition des populations de zooplancton du fiord du Saguenay et de l'estuaire maritime du St-Laurent. Thèse M. Sc., Université Laval, Québec, 175 p.

Raven, J.A. and Beardall, J., 1981. Respiration and photorespiration. In: Physiological Bases of Phytoplankton Ecology, Platt T. (Ed.). Canadian Bulletin of Fisheries and Aquatic Sciences, 210, p. 55-82.

Saifullah, S.M., 1978. Inhibitory effects of copper on marine dinoflagellates. Marine Biology, 44, p. 299-308.

Schafer, C.T., Cole, F.E. and Syvitski, J.P., 1989. Bio- and lithofacies of modern sediments in Knight and Bute Inlets, British Columbia. Palaios, 4, p.

Schafer, C.T. and Smith, J.N., 1987. Hypothesis for a submarine landslide and cohesionless sediment flows resulting from a 17th century earthquake-triggered landslide in Quebec. Geomarine Letters, 7, p. 31-37.

Schafer, C.T. and Smith, J.N., 1988. Evidence of the occurrence and magnitude of terrestrial landslides in recent Saguenay Fiord sediments. p. 137-145, In: Natural and Man-Made Hazards, M.I. El Sabh and T.S. Murty (eds.), Reidel Publishing Co., Amsterdam.

Schafer, C.T., Smith, J.N. and Seibert, G., 1983. Significance of natural and anthropogenic sediment inputs to the Saguenay Fiord, Quebec. Sedimentology, 36, p. 177-194.

Schafer, C.T., Smith, J.N. and Loring, D.H., 1980. Recent sedimentation events at the head of the Saguenay Fiord. Environmental Geology, 3, p. 139-150.

Seibert, G.H., Trites, R.W. and Reid, S.J., 1979. Deepwater exchange processes in the Saguenay Fiord. Journal of the Fisheries Research Board of Canada, 36, p. 42-53.

Smith, W.E.T., 1962. Earthquakes of eastern Canada and adjacent areas 1534-1927. Publications of the Dominion Observatory, 26, p. 271-289.

Smith J.N. and Walton, A., 1980. Sediment accumulation rates and geochronologies measured in the Saguenay Fiord using the Pb-210 dating method. Geochimica et Cosmochimica Acta, 46, p. 941-954.

Smith, J.N. and Ellis, K.M., 1982. Transport mechanism for Pb-210, Cs-137 and Pu fallout radionuclides through fluvial-marine systems. Goechimica et Cosmochimica Acta, 46, p. 941-954.

Smith, J.N. and Schafer, C.T., 1987. A 20th-century record of climatologically modulated sediment accumulation rates in a Canadian fiord. Quaternary Research, 27, p. 232-247.

Smith, J.N. and Levy, E., 1988. A geochronology for PAH contamination recorded in the sediment of the Saguenay Fiord (in prep.).

Smith, J.N. and Loring, D.H., 1981. Geochronology for mercury pollution in the sediments of the Saguenay Fiord, Quebec. Environmental Science and Technology, 15, p. 944-951.

Steven, D.M., 1974. Primary and secondary production in the Gulf of St. Lawrence. MS Rep. Mar. Sci. Centre McGill Univ., 26, 116 p.

Syvitski, J.P.M., 1980. Flocculation, agglomeration and zooplankton pelletization of suspended sediment in a fiord receiving glacial meltwater. In: H.J. Freeland, D.M. Farmer and C.D. Levings (eds.) Fiord Oceanography, Plenum Press, New York, p. 615-623.

Syvitski, J.P.M. and Murray, J.W., 1981. Particle interaction in fiord suspended seidment. Marine Geology, 39, p. 215-242.

Syvitski, J.P.M. and MacDonald, R., 1982. Sediment character and provenance in a complex fiord: Howe Sound, British Columbia. Canadian Journal of Earth Sciences, 19, p. 1025-1044.

Syvitski, J.P.M. and Schafer, C.T., 1985. Sedimentology of arctic fiords experiment (SAFE): 1. Project introduction. Arctic, 38, p. 264-270.

Syvitski, J.P.M., Burrell, D.C. and Skei, J.M., 1986. Fiords Processes and Products. Springer-Verlag, New York, 379 p.

Syvitski, J.P.M., Asprey, K.W., Clattenburg, D.A., and Hodge, G.D., 1985. The prodelta environment of a fiord: suspended particle dynamics. Sedimentology, 32, p. 83-107.

Syvitski, J.P.M. and Farrow, G.E., 1988. Fiord sedimentation as an analog for small hydrocarbon-bearing submarine fans. In: Whateley, M.K.G. and Pickering, D.T. (eds.), Deltas: Sites and Traps for Fossil Fuels. Special Publication, Geological Society of London (in press).

Sundby, B. and Loring, D.H., 1978. Geochemistry of suspended particulate matter in the Saguenay Fiord. Canadian Journal of Earth Sciences, 15, p. 1002-1011.

Tan, F.C. and Strain, P.M., 1979. Organic carbon isotope ratios in recent sediments in the St. Lawrence Estuary and the Gulf of St. Lawrence. Estuarine and Coastal Marine Science, 8, p. 213-225.

Tam, K.C. and Armstrong, F.A.J., 1972. Mercury contamination in fish from Canadian waters. In: Mercury in the aquatic environment, J.F. Uthe (ed.). Fisheries Research Board of Canada Manuscript No. 1167, p. 4-21.

Tattersall, W.M. and Tattersall, O.S., 1951. The British Mysidacea. The Ray Society, London.

Tavenas, F., Chagnon, J.Y. and LaRochelle, P., 1971. The Saint-Jean Vianney landslide: observations and eyewitnesses accounts. Canadian Geotechnical Journal, 8, p. 463-478.

Taylor, G.B.,1975. Saguenay River sections from fifteen cruises 1961-1974. Unpublished manuscript, Bedford Institute of Oceanography, Data Series, BI-D-75-2, 38 p.

Therriault, J.-C. and Lacroix, G., 1975. Penetration of the deep layer of the Saguenay Fiord by surface water of the St. Lawrence Estuary. Journal of the Fisheries Research Board of Canada, 32, p. 2373-2377.

Therriault, J.-C. and Lacroix, G., 1976. Nutrients, chlorophyll, and internal tides in the St. Lawrence Estuary. Journal of the Fisheries Research Board of Canada, 33, p. 2747-2757.

Therriault, J.C., DeLadurantaye, R. and Ingram, R.G., 1980. Particulate matter exchange processes between the St. Lawrence Estuary and the Saguenay Fiord. In: J.J. Freeland, D.M. Farmer and D.C. Levings (eds.), Fiord Oceanography, Plenum Press, New York, p. 363-373.

Therriault, J.-C., DeLadurantaye, R. and Ingram, R.G., 1984. Particulate matter exchange across the Saguenay Fiord sill. Estuarine and Coastal Shelf Science, 18, p. 51-64.

Thompson, P.A. et Côté, R., 1985. Influence de la spéciation du cuivre sur les populations phytoplanctoniques naturelles de la rivière du Saguenay, Québec, Canada Internationale Revue der gesamten Hydrobiologie, 70, p. 711-731.

Terzaghi, K., 1950. Mechanism of landslides. The Geological Society of America Berkey, p. 83-123.

Chapter 18

The St. Lawrence Estuary: Concluding Remarks

Norman Silverberg[1] and Mohammed I. El-Sabh[2]

[1]Institut Maurice-Lamontagne, Pêches et Océans Canada, C.P. 1000, Mont-Joli (Québec) G5H 3Z4

[2]Département d'Océanographie, Université du Québec à Rimouski, Rimouski (Québec) Canada G5L 3A1

1. For the St. Lawrence Estuary, large-scale means variety

Each of the three principal subdivisions of the St. Lawrence Estuary (the shallow Upper Estuary with strong salinity gradients, the deep, more marine, Lower Estuary, and the Saguenay Fjord) are larger than many of the common coastal plain estuaries, whose lengths may be shorter than the St. Lawrence is wide. The Upper Estuary, which most closely resembles a typical partially stratified estuary, is also deeper than many inundated-river-valley estuaries. A number of estuarine systems rival the St. Lawrence in surface area (e.g.Chesapeake Bay in the United States, Spencer Gulf in south Australia, Gulf of Suez in Egypt), but none include the great depths characteristic of the Lower St. Lawrence Estuary. With a major portion of North America as its drainage basin, the St. Lawrence River delivers a mean freshwater discharge of 11.9×10^3 m^3 s^{-1}, close to that of the Mississippi. The approximately 4 million metric tons per year of suspended matter discharge is, however, two orders of magnitude less than the Mississippi. This is because the Great Lakes trap most of the solids, so that only those introduced below Lake Ontario are available to the Estuary. Because of the low ratio of sediment input to size, the St. lawrence Estuary is very far from being filled in and much of the topography and surficial sediment cover have yet to attain equilibrium with modern conditions.

The size of the St. Lawrence permits the coexistence of numerous smaller sub-environments. These represent different combinations of topography and sediments inherited from late Pleistocene events; modern inputs of freshwater and particulates; tidal, meteorological and inertial forces; and biological and chemical interactions. The practitioners of the various oceanographic disciplins seem to have defined their own subregions, but few have been defined in an interdisciplinary fashion, and the permanence or permeability of their boundaries as physical or biogeochemical frontiers remains to be established.

Coastal and Estuarine Studies, Vol. 39
M. I. El-Sabh, N. Silverberg (Eds.)
Oceanography of a Large-Scale Estuarine System
The St. Lawrence
© Springer-Verlag New York, Inc., 1990

The variety of conditions found in the St. Lawrence is exemplified by the physical regime, for which the principal forcing functions (thermohaline, meteorologic and tidal) have about equal importance. Mertz and Gratton (Chapter 5) describe some of this variety very eloquently - "...*The St. Lawrence Estuary is remarkable in that the full spectrum of oceanic variability, from internal waves to low-frequency wind-driven currents and unstable waves can be found within its confines...*". One has only to refer to the other chapters in this book to confirm the impression of motions at many different time and space scales, and to realize that the richness in environmental variety is not limited to the physical regime. This can perhaps best be demonstrated by a tour through the estuary, summarizing the general conditions and pointing out some of the highlights.

1.1 The Upper Estuary

"...*Owing to the relatively shallow mean depth of the Upper estuary, tidal ranges and currents are very large (up to 10 m and 3 m s^{-1}, respectively)...The Upper estuary, with its rough bottom, and strong tidal flows is an ideal locale for internal wave generation...*" Mertz and Gratton (Chapter 5). Most of the research in the Upper Estuary has been carried out in the navigable North Channel, and on several of the muddy tidal flats. The discontinuous deeper stretches of the South Channel have been much less visited, and the shoals and island margins separating the two are almost undescribed. More is known about the sedimentary deposits and the suspended particulate matter dynamics than about the physics and biology of the Upper Estuary. Although the tides have received much attention, the general water circulation pattern has not really been treated beyond the brief description by Neu (1970),and the residual flow calculations of Meric (1975). These studies indicated that the freshwater moves seaward mostly over the southern portion, while the compensatory landward flow of more saline water is concentrated in the deeper North Channel. Cross-channel flow is also known to occur (d'Anglejan and Ingram, 1976). The general circulation pattern, either for the spring-neap tidal cycle or for the low-high river discharge periods, remains to be described in the literature. A number of studies have referred to the vertical structure within the water column. These indicate that while the head of the estuary exhibits the strongest salinity gradients, the water column is often well-mixed vertically, and a distinctly stratified two-layer structure does not develop until further seaward. Indeed, an argument can be made for the further subdivision of the Upper Estuary into two parts which also reflect differences in biology and geochemistry : a head region and a lower region.

1.1.1. Head Region (Ile d'Orléans to Ile aux Coudres)

This subregion is shallow (<15m), subject to strong currents, and assymetric ebb and flow. Tidal amplitudes are maximal near Ile aux Coudres. The salinitiy varies between 0 and 15 °/oo, and the water column is dominated by a pronounced turbidity maximum, with

SPM concentrations between 10 and 450 mg/L (see d'Anglejan, Chapter 6). Vertical stratification tends to be limited to slack water periods, but there are marked longitudinal and cross-channel salinity gradients, with the fresher, turbid waters extending further seaward along the south shore. The salinity structure is influenced by the Seasonal variations in freshwater discharge, although this seasonality has been reduced to only a factor of two by the series of hydroelectric dams in the drainage basin.

In spite of a well-developed turbidity maximum of fine-grained particles, the bottom sediments are mostly sandy and subject to reworking by the strong currents, leaving some gravel deposits and, in places, exposing Goldthwait Sea clays. Bedload transport may be extensive, as suggested by the presence of sand waves and the need for annual dredging to maintain the shipping channel. Muds occur only rarely in the channels and are probably only temporary deposits. Extensive mud flats occur at Cap Tourment, but these are swept away when the marsh grasses are grazed down by migratory geese. Along the intertidal of the south shore, mud deposits appear to be better developed, particularly at Montmagny, but a balance appears to be maintained between accumulation and erosion, the latter aided by displacements of grounding ice. Overall, the head of the estuary is a region of retardation, but not accumulation, of the approximately 4 million tonnes of suspended matter carried in annually by the St. Lawrence River.

In this zone of sharp gradients in ionic strength, the geochemistry is particularly interesting (see Yeats, Chapter 8) with many phenomena affecting the composition of dissolved and particulate matter (adsorption-desorption, precipitation and flocculation of dissolved and colloidal substances, aggregation and decomposition of organic matter, exchanges with bottom sediments and interstitial waters, size separation of suspended particles). Still, most dissolved substances behave conservatively behaviour, and those disappearing from the water column (e.g. Fe) do not accumulate in the surficial sediments. Increased understanding of the importance of the chemical reactions occuring in this zone of intense water motion and particulate matter bypassing will require obtaining concurrent information from the water column, and the channel, bank and intertidal deposits. Given the apparently strong role of the tidal flats in the particulate matter budget, more information about their chemistry and level of contaminantion will probably prove to be very valuable.

The biology at the head of the estuary is also influenced by the salinity gradient and turbidity maximum. The phytoplankton consist primarily of deteriorating freshwater species, which along with the organic detritus carried in by the river, provide a substrate for enhanced bacterial production, which may provide the principal energy transfer between primary and secondary producers (see Therriault et al., Chapter 12). This may explain why, in spite of the environmental stresses, an important biomass of fresh-brackish water zooplankton (calanoid copepods and epibenthic macrozooplankton) is maintained within the turbidity maximum zone (see Runge and Simard, Chapter 13). Similarly, icthyoplankton studies (see de Lafontaine, Chapter 14) indicate that freshwater spawners (smelt and tomcod) maintain themselves in nursery areas at the head of the estuary.

1.1.2. Partially Stratified Upper Estuary (Ile aux Coudres to the mouth of the Saguenay)

Seaward of Ile aux Coudres the Upper Estuary widens, the very shallow banks and islands separating the north and south channels are much less frequent, and the North Channel in particular deepens in a series of 50-150 m deep basins and broad sills. Tidal energies are still strong and internal waves occur but a distinct two-layer is maintained. Salinities range from 10-25 o/oo in the surface layer and as much as 29 o/oo in the deep waters. Fronts develop at the seaward margin (see Ingram and El-Sabh, Chapter 4), where the Saguenay Fjord outflows and upwelled colder waters from the Laurentian Trough meet the seaward-flowing Upper Estuary surface waters. SPM concentrations drop below 8 mg/L, and again, despite the deep basins, mud deposition apparently does not occur. Instead, lag gravels and coarse sands dominate on the sills and poorly-sorted glacial tdeposits occur in the basin.

The phytoplankton here are dominantly marine but there is sufficient turbidity and deep mixing in the surface layer to cause light limitation, so production is low. Plant cells, unable to respond to rapid changes in vertical stability, adapt their physiological responses to a 24-hr cycle, and productivity responds to fortnightly changes in the tidally-controlled stratification. Zooplankton abundance is low east of Ile aux Coudres and then increases in the deeper basins to seaward. The species are typically estuarine and appear to be endemic to the Upper Estuary. Despite the seaward flow in the surface layers, herring larvae also maintain position in nursery areas near where they are spawned, unlike the capelin larvae, which drift across the boundary into the Lower Estuary. The combinations of biological mechanisms and physical processes required to prevent dispersal of some of these zoo- and ichthyoplankton populations in the different portions of the Upper Estuary remain of great interest.

1.2 Lower (Maritime) Estuary

"... In only 20 km the bottom shoals from a depth of 300 m to a depth of about 50 m. To the upstream propagating barotropic tide, this region appears step-like. This feature couples the baroclinic and barotropic motions to yield strong internal tides, propagating upstream and downstream from the shoaling region... " Mertz and Gratton (Chapter 5). This glacially-overdeepened valley, following an ancient line of weakness between the Canadian Shield and the Appalachians, permits the invasion of the continental shelf and upper continental slope environment 1200 km into the North American landmass. The bulk of the valley is occupied by only slightly altered North Atlantic Water, but the marine conditions are modified by the input of brackish water and terrestrial particulates from the Upper Estuary and the north shore rivers, creating a unique estuarine region.

A two layer structure occurs in the water column during winter when cold (0 to -1.4 °C) waters with salinities of 27-31.5 o/oo, partly advected from the Gulf, persist to depths of 100 m below the ice-covered surface. Spring runoff reduces the salinities to 20-25 o/oo and gradual warming increases the temperature by 12-14°, establishing a third layer at the surface while diminishing the winter layer to a temperature minimum zone about 50 m thick (see Figure 1). The properties of the deep water mass remain almost constant through time with salinities of 33-34.5 °/oo and temperatures increasing to 4.5 °C.

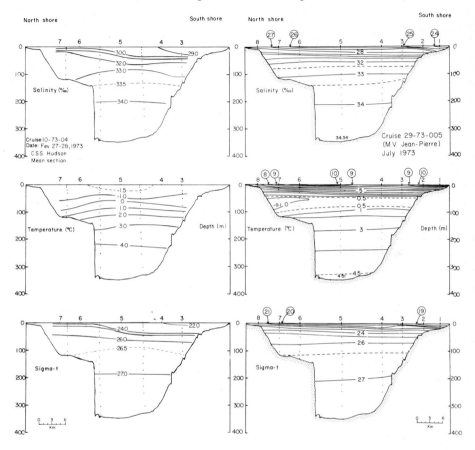

Figure 1   Vertical distribution of temperature, salinity and density for the mean section of Rimouski in (A) Febuary and (B) July, 1973 (after El-Sabh, 1979).

Although the water mass structure in the Lower Estuary s relatively simple, the relative freedom of water motion permitted by its size induces constantly changing flow conditions and it is much more difficult to define subregions with definite boundaries. "...*its width (30 to 50 km) is equal to several internal Rossby radii, the scale characterizing baroclinic motions. This implies that the LSLE can accomodate such features as baroclinic shelf waves and unstable shear waves. The emphasis in going from the Upper to Lower estuaries thus shifts from high-frequency to low-frequency motions, from estuarine to oceanic characteristics...Once one allows for the shorter scales inherent to channel motions, the characteristics (large coherence scales) of the low-frequency motions in the LSLE correspond more closely to open ocean situations than to other estuaries or straits...."* Mertz and Gratton (Chapter 5). Phenomena occurring in three areas: the head of the Laurentian Trough, the southern shelf and the Estuary-Gulf boundary appear to have a strong influence on the internal structure and regional differences in the biological, geological and geochemical characteristics over the Lower Estuary..

Head of the Laurentian Trough: - The steep ramp marking the landward limit of the Laurentian Trough induces the acceleration of tidal currents, enhancement of vertical motion, and the formation of high amplitude internal waves. This leads to periodic transport of intermediate waters towards the surface through upwelling and turbulent mixing . Besides spilling over the sills into the Saguenay Fjord and the deep basins of the Upper Estuary, the colder, nutrient-rich waters can extend eastwards along the Trough as far as Ile du Bic and sometimes along the northern and southern shelves.

Southern Shelf: - The fresher surfaces waters entering from the Upper Estuary and the Saguenay tend to be constrained by Coriolis force along the southern shelf of the Lower Estuary, forming a well-developed coastal jet. .Although subject to instability, this coastal jet persists in some form or another as far as the seaward limit of the Estuary, where it is strongly re-enforced and continues into the Gulf as the Gaspe Current. *"...The combined system consisting of the south shore jet and the Gaspe Current stretches over 400 km and must rank with other, better-known, unstable buoyant jets such as Australia's Leeuwin Current and the Norwegian Coastal Current in terms of interest to the oceanographic community..."* Mertz and Gratton (Chapter 5).

Estuary-Gulf Boundary: - The constriction at Pointe-des-Monts at the mouth of the Estuary is the site of a major density front. Variable but often intense southerly flows separates the waters of the Lower Estuary from the Gulf of St. Lawrence. Meteorological events in both the Gulf and Estuary, spring-neap tidal cycles, and variations in freshwater inputs influence the circulation at the mouth of the Estuary. These cause variations in the strength of the southerly flow and determine whether or not there is inward or outward flow along the north shore. In this manner, the boundary front intensity is intimately connected to the pattern of eddies within the Estuary itself (see Ingram and El-Sabh, Chapter 4, and Mertz and Gratton, Chapter 5).

The physical processes occuring in these regions, combined with the freshwater inflows from the hydroelectrically-controlled rivers in the Manicouagan area, induce variable surface water circulation patterns. Cross-channel flows can develop, with northerly drifts near Ile du Bic and southerly flows east of the Manicouagan. Cyclonic and anticyclonic eddies, separation of north and south waters along the longitudinal axis, and transitional modes of circulation have all been noted from satelite photographs and inferred from direct current measurements.

phytoplankton production in the Lower Estuary is stimulated by the introduction of nutrient-rich intermediate waters at the head of the Trough, but it is also related to the vertical stability of the water column and to turbidity (Therriault et al., Chapter 5) Thus, it is somewhat downstream of the head of the Trough that the plant cells have a chance to develop freely ("Upwelling region"). Continued availability of excess nutrients, and enhanced stability due to the inflow of fresher waters from the Manicouagan delta, produces a zone of maximal productivity (Plume region). The "Near Gulf" region,with its great stability and distance from the head, has lower nutrient contents, lower productivity and is more like the open Gulf of St. Lawrence, with early spring blooms and nutrient limitation in summer. The least productive region is along the southern shelf (Outflow region) where the coastal jet provides high turbidity and short residence times of the shallow waters. Overall, the productivity of the Lower St. Lawrence Estuary is only moderate, although nutrients are rarely limiting. The levels are similar to other estuarine environments (e.g. Hudson River Estuary, Bristol Channel), but much lower than those of productive regions such as Long Island and Puget Sounds. The late onset of the spring bloom is one reason for this, and the understanding of the different factors (e.g. high turbidity during the spring runoff, vertical mixing, cold temperatures) implicated in the low productivity is of ongoing interest.

The physical regime has a strong influence on the variability and the composition of the zoo- and ichthyoplankton communities (see Chapters 13,14). It would appear that the residence times of the surface waters are not long enough, given the relatively cold temperatures in this high-latitude estuary, to permit the complete development of adult forms within the Lower Estuary. This is particularly the case for fish species with pelagic eggs, whose relative numerical abundance is extremely low compared to other estuarine systems. Locally produced offspring, contrary to some species of the Upper Estuary, have developed the mechanisms necessary for retention within the confines of the Lower Estuary. The adult forms in the Lower Estuary are imported from the Gulf of St. Lawrence, and consist of species which spend considerable time in deep water layers and are subject to advection into the Estuary.

The more energetic physical regime near the head region is reflected in a gradient in the grain-size distribution of the muddy bottom sediments, with the sand and silt fractions being more abundant near the head of the Laurentian Trough. The projection of the southern shelf at Ile du Bic can sometimes set up a northerly cross channel flow and when there are westerly flows along the north shore a broad cyclonic circulation can develop between Rimouski and the Saguenay. The increase in the residence time of the surface waters, due to the presence of this eddy, may be responsable for the 4-8 fold increase in sedimentation rate observed in the area off Trois-Pistoles. From Rimouski, the sedimentation rates decrease only gradually towards the Gulf and the grain-sizes also remain stable. The overall sedimentation rate of 2-3 mm y^{-1} confirms the lack of deposition in the Upper Estuary, as the annual accumulation of fine-grained sediments in the Lower

St. Lawrence Estuary is more than enough to account for the suspended matter input from the St. Lawrence River.

The benthic fauna respond to the differences in productivity, sedimentation rate and sediment grain-size distribution along the Laurentian Trough. Species diversity and abundance are high in the head region. These decrease in the area of high sedimentation rate where more vertically-mobile organisms occur, and then stabilize at lower levels seaward of Rimouski with the appearance of more horizontally-mobile organisms and carnivores.

The biogeochemistry of the sediments, similarly to the macrobenthos, responds to regional differences. It is also influenced by a longitudinal gradient of terrestrial versus marine sources of mineralizable organic matter, and the composition of the sediment reflects the diagenetically mobile substances that are swept landward in the compensatory landward flow of the bottom waters. The large dimensions and relatively long flushing times of the Lower Estuary are particularly suited to studies of early diagenesis. It has served well to elucidate the diagenetic behaviour of a number of elements, and to demonstrate that oxidation-reduction sensitive metals are particularly mobile and not completely retained after burial within estuarine sediments. Levels of organic and inorganic toxic substances in the sediments are still quite low, compared to the Great Lakes and the Saguenay Fjord (see Cossa, Chapter 11).

2. Future considerations

Process-oriented research, with an emphasis on mass and energy transfers within the ecosystem, unfortunately continues to be favoured in the competition for research funding over the descriptive studies (e.g. species composition, abundance and behaviour, sediment and suspended particulate matter distribution and composition) needed to provide the background from which to interpret process mechanisms. Much of the Upper Estuary, and the intertidal and shelf regions of the Lower Estuary have therefore yet to be adequately described. Because of the long period of ice cover and the lack of ice-reinforced ships, our knowledge of winter conditions in the Estuary remains very fragmentary.

The difficulties inherent in the study of a large estuary are apparent throughout this volume. A significant problem is obtaining a synoptic view of such large areas. Travel times between sample sites are too great to avoid changing conditions during a survey. Even a simple longitudinal section takes several tidal cycles to complete, and any sampling is certain to be troubled by a fluctuating environment (weather patterns, spring-neap tidal cycles, water column instabilities, nutrient cycles and perhaps even to changes in the plant and animal populations). Satelite imagery, although generally limited to information about a thin surface layer, has begun to reveal broad synoptic patterns and it is becoming apparent that events in one part of the Estuary are not independent of those in a distant part. We do need other observations of the system as a whole. New technological developments (e.g. inexpensive, in-situ measuring devices whose data are easily recoverable) and scientific achievements (e.g. three-dimensional modelling of stratified water bodies) will hopefully relieve somewhat the classic Canadian problem of small numbers of people trying to deal with a vast territory. On the other hand, the considerable spatio-temporal heterogeneity in the environment challenges our ingenuity to develop methodologies to detect and understand fundamental mechanisms operating at small scales. It can be anticipated that recognition of the need for simultaneous observations will produce more interdisciplinary projects at bothlarge and small scales.

This review volume shows the enormous progress in the understanding of the St. Lawrence Estuary that has been achieved over the last twenty years. We are now aware of the great variety of phenomena which occur in this large system. We realize, however, that it shall require another generation of research to attain the state where we may quantitively define the interactions between these phenomena, and to develop truly predictive models.

REFERENCES

d'Anglejan, B. and R.G. Ingram, 1976 Time-depth variations in tidal flux of suspended matter in the St. Lawrence Estuary. Estuarine Coastal Marine Science: 4, 401-416.

El-Sabh, M..I., 1979. The lower St. lawrence Estuary as a physical oceanographic system. Nat. Canadien: 106, 55-73.

Meric, P., 1975 Circulation résiduelle dans l'estuaire moyen du St-Laurent et son influence sur les processus sédimentaire - Univ. Laval CENTREAU, Rap. no. CRE-75/03, 68 p.

Neu, H.J.A., 1970. A study on mixing and circulation in the St. Lawrence estuary up to 1964. Bedford Inst. Oceanogr. AOL Rept. 1970-9. Unpublished manuscript, 31 p.

LIST OF CONTRIBUTORS

d'Anglejan, B. - Department of Geological Sciences, McGill University, Franck Dawson Adams, 3450 University, Montréal, Québec, Canada H3A 2A7

Cossa, D. - Institut Français de Recherche pour l'Exploitation de la Mer, B.P. 1049, F-44037 Nantes, cedex 01, France

Côté, R. - Sciences Fondamentales, Université du Québec à Chicoutimi, Chicoutimi, Québec, Canada G7H 2B1

de Lafontaine, Y. - Pêches et Océans Canada, Division de l'Océanographie Biologique Institut Maurice Lamontagne 850 Route de la Mer, Mont-Joli, Québec,Canada G5H 3Z4

Demers, S. - Institut Maurice-Lamontagne, Ministère des Pêches et des Océans, 850 Route de la Mer, Mont-Joli, Québec, Canada, G5H 3Z4

Drapeau, G. - INRS-Océanologie, Université du Québec, 310 des Ursulines, Rimouski, Québec, Canada G5L 3A1

El-Sabh, M.I. - Département d'Océanographie, Université du Québec à Rimouski, Rimouski, Québec, Canada G5L 3Al

Gagné, J.A. - Institut Maurice Lamontagne, Ministère des Pêches et des Océans, 850 Route de la Mer, Mont-Joli, Québec, Canada, G5H 3Z4

Gearing, J.N. - Fisheries and Oceans Canada, Maurice Lamontagne Institute, C.P. 1000, Mont-Joli, Québec, Canada G5H 3Z4

Gratton, Y. - Institut Maurice-Lamontagne, Pêches et Océans Canada, B.P. 1000, Mont-Joli, Québec, Canada G5H 3Z4

Ingram, R.G. - Department of Meteorology, McGill University, Montreal, Québec, Canada H3A 2K6

Legendre, L. - Département de biologie, Université Laval, Québec, Canada G1K 7P4

Mertz, G. - Northwest Atlantic Fisheries Centre, Fisheries and Oceans Canada, P.O. Box 5667, St. John's, Newfoundland, Canada   A1C 5X1

Murty, T.S. - Institute of Ocean Sciences, Fisheries and Oceans Canada, P.O.B. 6000, Sidney, British Columbia, Canada   V8L 4B2

Pocklington, R. - Fisheries and Oceans Canada, Bedford Institute of Oceanography, P.B. 1006, Dartmouth, Nova Scotia, Canada B2Y 4A2

Runge, J.A. - Institut Maurice-Lamontagne, Ministère des Pêches et Océans, C.P. 1000, Mont-Joli, Québec, Canada G5H 3Z4

Schafer, C.T. - Geological Survey of Canada, Bedford Institute of Oceanography, Dartmouth, Nova Scotia, Canada   B2Y 4A2

Silverberg, N. - Institut Maurice-Lamontagne, Pêches et Océans Canada, C.P. 1000, Mont-Joli, Québec, Canada G5H 3Z4

Simard, Y. - Institut Maurice-Lamontagne, Ministère des Pêches et Océans, C.P. 1000, Mont-Joli, Québec, Canada G5H 3Z4

Sinclair, M. - Biological Sciences Branch, Fisheries and Oceans Canada, Bedford Institute of Oceanography, Dartmouth, Nova Scotia, Canada B2Y 4A2

Smith, J.N. - Atlantic Oceanographic Laboratory, Bedford Institute of Oceanography, Dartmouth, Nova Scotia, Canada B2Y 4A2

Sundby, B. - Institut Maurice-Lamontagne, Pêches et Océans Canada, C.P. 1000, Mont-Joli, Québec, Canada G5H 3Z4

Tee, K-T. - Physical and Chemical Sciences, Department of Fisheries and Oceans, Bedford Institute of Oceanography, P.O. Box 1006, Dartmouth, Nova Scotia, Canada B2Y 4A2

Therriault, J-C. - Institut Maurice-Lamontagne, Ministère des Pêches et des Océans, 850 Route de la Mer, Mont-Joli, Québec, Canada, G5H 3Z4

Vincent, B. - Département d'Océanographie, Université du Québec à Rimouski, Rimouski, Québec, Canada G5L 3A1

Yeats, P.A. - Physical and Chemical Sciences, Department of Fisheries and Oceans, Bedford Institute of Oceanography, P.O. Box 1006, Dartmouth, Nova Scotia, Canada B2Y 4A2

SUBJECT INDEX

Coastal and Estuarine Studies

(formerly Lecture Notes on Coastal and Estuarine Studies)